KU-405-416

0248564-9
ARONOFF, SAMUEL
PHLOEM TRANSPORT
000248564

QK871.A66

WITHDRAWN FROM STOCK
The University of Liverpool

DU

1ε

DUE FO

Phloem Transport

NATO ADVANCED STUDY INSTITUTES SERIES

A series of edited volumes comprising multifaceted studies of contemporary scientific issues by some of the best scientific minds in the world, assembled in cooperation with NATO Scientific Affairs Division.

Series A: Life Sciences

Volume 1 – Vision in Fishes: New Approaches in Research
edited by M. A. Ali

Volume 2 – Nematode Vectors of Plant Viruses
edited by F. Lamberti, C.E. Taylor, and J.W. Seinhorst

Volume 3 – Genetic Manipulations with Plant Material
edited by Lucien Ledoux

Volume 4 – Phloem Transport
edited by S. Aronoff, J. Dainty, P. R. Gorham, L. M. Srivastava, and C. A. Swanson

Volume 5 – Tumor Virus–Host Cell Interaction
edited by Alan Kolber

The series is published by an international board of publishers in conjunction with NATO Scientific Affairs Division

A Life Sciences	Plenum Publishing Corporation
B Physics	New York and London
C Mathematical and Physical Sciences	D. Reidel Publishing Company Dordrecht and Boston
D Behavioral and Social Sciences	Sijthoff International Publishing Company Leiden
E Applied Sciences	Noordhoff International Publishing Leiden

Phloem Transport

Edited by

S. Aronoff
Simon Fraser University
Burnaby, B. C., Canada

J. Dainty
University of Toronto
Toronto, Ontario, Canada

P. R. Gorham
University of Alberta
Edmonton, Alberta, Canada

L. M. Srivastava
Simon Fraser University
Burnaby, B. C., Canada

and

C. A. Swanson
Ohio State University
Columbus, Ohio, U.S.A.

with the cooperation and advice of

P. E. Weatherley
Aberdeen University, Scotland

PLENUM PRESS • **NEW YORK AND LONDON**
Published in cooperation with NATO Scientific Affairs Division

Library of Congress Cataloging in Publication Data

Nato Advanced Study Institute on Phloem Transport, Banff, Alta., 1974.
   Phloem transport.

   (Nato advanced study institutes series: Series A, life sciences; v. 4)
   "Lectures presented at the 1974 NATO Advanced Study Institute on Phloem
Transport, held at the Banff Centre, School of Fine Arts, Banff, Alberta, August
18-29, 1974.
   Includes bibliographies and index.
   1. Plant translocation—Congresses. 2. Phloem—Congresses. I. Aronoff, Samuel.
II. Title. III. Series.
QK871.N37 1974   582'.01'1   75-15501
ISBN 0-306-35604-X

Lectures presented at the 1974 NATO Advanced Study Institute on Phloem
Transport, held at The Banff Centre, School of Fine Arts, Banff, Alberta,
August 18-29, 1974

© 1975 Plenum Press, New York
A Division of Plenum Publishing Corporation
227 West 17th Street, New York, N.Y. 10011

United Kingdom edition published by Plenum Press, London
A Division of Plenum Publishing Company, Ltd.
Davis House (4th Floor), 8 Scrubs Lane, Harlesden, London, NW10 6SE, England

All rights reserved

No part of this book may be reproduced, stored in a retrieval system, or transmitted,
in any form or by any means, electronic, mechanical, photocopying, microfilming,
recording or otherwise, without written permission from the Publisher

Printed in the United States of America

Preface

Ten years ago, at the International Botanical Congress in
Edinburgh, a group of us from various countries discussed the
difficulty of pursuing academic problems in depth at such meetings.
In particular, we were discouraged at the poverty of time for
phloem transport. From long association, we were conscious of the
extraordinary breadth of the problem, from developmental through
anatomical, to biophysical and physiological. Only by a reasonable
understanding of all these components could one hope to come to
some kind of understanding. We decided to establish common plant
material so that data would have a common source. Similarly, we
resolved to exchange information by circulating pre-publication
manuscripts. For awhile, after the meeting was a pleasant memory,
the plan seemed to be working; but, as is so often the case, human
infirmities and foibles played early and, subsequently, predominant
roles. Some became administrators (a punishment for good
behaviour); others concentrated on alternative rings in their
academic circuses.

The next Congress (in Seattle) proved similar to its
predecessor in its neglect and, consequently, succor was sought
elsewhere. A little known, but remarkably understanding group
becoming visible was the Science Committee and the Division of
Scientific Affairs of N.A.T.O. Its sponsorship of Advanced Study
Institutes including phytochemistry and phytophysics, was unusual
both in the generosity of its funding and in the requirements for
academic quality. In time we convinced them of the desirability of
investigating a phenomenon mandatory to all multicellular higher
plants, where the rate of transport of solute was orders of
magnitude faster than diffusion and more rapid than protoplasmic
streaming, and must nevertheless pass through pores of the order of
a hundred nanometers; where energy appears requisite at one end of
the system, but not the other; a phenomenon which involved gross
degeneration of the transport cells, losing nuclei and most
organelles; where the function of companion cells, packed with
inclusion, was completely unknown, as were the protein threads
("P-protein") found in many (but not all) sieve cells.

It was hoped, for a time, that the use of radiotracers would
explain phloem transport; indeed, it was (and still is) a
frequently-used basis for modelling. We were now faced with a

351866

surfeit of theories, and an aim of the A.S.I. involved the study
of these several theories in the light of current knowledge of
anatomists, physiologists, and biophysicists.

This was the major pedagogic device: that these three groups
of investigators should learn to speak each others´ language; to be
aware of artefacts and problems; to devise new approaches.

It was a fruitful meeting to all involved. Supplemented by
the generosity of additional funds from the National Research
Council (Canada) and the National Science Foundation (U.S.A.),
housed and fed by the Banff School of Fine Arts, our surroundings
at Banff, Alberta (Canada) provided a setting of appropriate but
unexcelled beauty.

All sessions were taped (courtesy of Mr. Frank Campbell,
Audio-Visual Centre, Simon Fraser University, Burnaby, B.C.,
Canada) and edited by the Section organizers. Manuscripts were
transcribed onto the S.F.U. computer and subsequently printed
using the WYLBUR system, under the supervision of Ms. Marilyn
Cheveldayoff, assisted by Ms. Lynda Zink. We are grateful to them
for their patience and unusual good sense in coping with this new
facility, and to Ms. Enid Britt for superb assistance in the
organization and operation of the Institute.

The text contains the lectures as given, divided into the
three major Sections: Anatomy, Physiology, Biophysics. Each
lecture was followed by a Discussion session, which, at times,
included briefer, supplementary presentations, termed "Discussion
Papers". These latter are intercollated within their corresponding
Discussion sessions.

We wish also to thank the Plenum Publishers for their
generosity in providing a grant for typing.

S.A.
J.D.
P.G.
L.S.
C.S.
P.W.

Contents

PART I: ANATOMY
Chairman: L.M. Srivastava

Phloem Tissue in Angiosperms and Gymnosperms

Cell Types and Their Spatial Distribution; Longevity of
Sieve Elements; Changes in Old Phloem

Charles H. Lamoureux

Department of Botany, University of Hawaii

Honolulu, Hawaii 96822

I. Introduction

The phloem is that tissue in more highly organized plants
("higher plants") which is characterized by the possession of
certain specialized cells, the sieve elements, and which functions
as the major channel of rapid conduction of sugars over fairly long
distances (more tha a few cells) in the plant body. Phloem is a
complex tissue and, in addition to sieve elements, it always
contains parenchyma cells, usually of more than one type, and
frequently includes sclerenchyma cells.

Since the discovery of the sieve element by Hartig (1837) and
the introduction of the term "phloem" by Nageli (1858), many
botanists have investigated the morphology and physiology of the
phloem tissue and more than a thousand publications have appeared
which deal in some detail with phloem structure and function.   The
voluminous   literature on phloem structure has been reviewed
periodically (Strasburger, 1891; Perrot, 1899; Esau, 1939, 1950),
most recently and extensively by Esau in 1969. In this paper, in
view of limitations of time and space, no attempt will be made to
present a comprehensive review of phloem structure or to trace in
detail the development of concepts and terminology, or to cite all
available references, especially if they have been recently
reviewed (Esau, 1969). Rather, the object of this paper is to
review briefly our current state of knowledge of the structure of
the phloem tissue in angiosperms and gymnosperms, with emphasis on
the kinds of cells found in functioning phloem, and on the changes
which occur in older phloem tissue as it ceases to function in
conduction.

1

In the phloem, perhaps to a greater extent than in other plant tissues, a thorough understanding of the structure of the mature, functioning tissue is possible only after consideration of the developmental stages which have led to the mature state.   However, since aspects of development and differentiation will be covered in detail in other papers in this volume, they will be considered only briefly in this paper.

II. Sieve elements

Sieve elements are typically elongate cells which bear sieve areas on their walls. The sieve areas contain pores through which pass cytoplasmic connections with adjacent cells. Sieve areas, at least when prepared for microscopic examination, usually exhibit deposits of callose. At functional maturity sieve elements have undergone partial autolysis of the protoplast components and typically lack intact nuclei, tonoplasts, and dictyosomes. There are few or no ribosomes but the plasmalemma is intact and somewhat modified endoplasmic reticulum, mitochondria, and plastids are present. In many investigated species various forms of P-protein have been detected.

A. Size and shape

Sieve elements are typically elongated parallel with the axis of the plant organ or the vascular bundle in which they occur. Sieve element length varies in relation to a number of factors (Esau, 1969): the evolutionary level of the taxon; the developmental stage of the tissue, whether primary or secondary, in primary phloem whether protophloem or metaphloem; the position in the plant, whether in root, stem or leaf, whether near the base, middle or top of the stem, etc.; the developmental sequences involved in the formation of the sieve elements from the procambial cell in primary phloem, or from the phloem initial formed by a fusiform cambial initial in the secondary phloem.

While recognizing that the amount of variation is considerable, it is still useful to mention sizes of sieve elements which have been reported. In secondary phloem of conifers sieve element lengths range from 1400 to 4850 μm and widths from 15 to 70 μm, while secondary phloem sieve element lengths in dicotyledons range from 75 to 1290 μm and widths from 20 to 60 μm (Huber, 1939; Holdheide, 1951; Chang, 1954; Zahur, 1959). In the metaphloem of palm stems sieve element lengths range from 500 to 5000 μm, widths from 30 to 45 μm (Parthasarathy and Tomlinson, 1967), while in petioles of palms metaphloem sieve element lengths range from 650 to 2300 μm, widths from 10 to 55 μm (Parthasarathy, 1966), and in roots metaphloem sieve element lengths range from 300 to 4000 m, widths from 25 to 400 μm (Parthasarathy, 1968).

Sieve element shape is most conspicuously related to the
degree of inclination of the end walls. End wall inclination
varies in relation to the same genetic and developmental factors
which affect sieve element length. In most gymnosperms and lower
vascular plants the sieve elements are long tapering cells without
distinct end walls (Fig.1,A). Such cells are called sieve cells.
In most angiosperms sieve elements have distinct end walls which
range in orientation from very oblique to transverse (Fig.1, B-H).
Sieve elements with distinct end walls are typically arranged in
vertical series, the sieve tubes, and each sieve element in a sieve
tube usually possesses certain features discussed below,

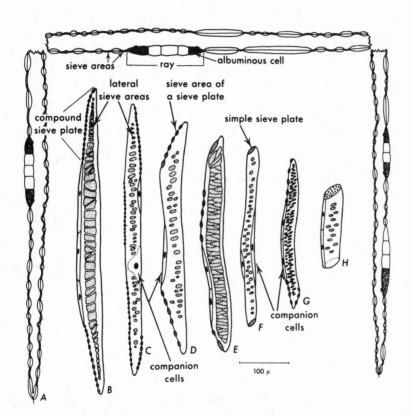

Fig.1. Variations in structure of sieve elements. A. Sieve
cell of Pinus pinea, with associated rays, as seen in tangential
section. Others are sieve tube members with companion cells from
tangential sections of phloem of the following species: B, Juglans
hindsii; C, Pyrus malus; D, Liriodendron tulipifera; E, Acer
pseudoplatanus; F, Cryptocarya rubra; G. Fraxinus americana; H,
Wisteria sp. (After K. Esau, 1960. Anatomy of Seed Plants. John
Wiley and Sons, Inc.)

(specialized sieve areas on end walls, presence of companion cells), which, in combination, permit designation of these cells as sieve tube members.

B. Wall structure
    Walls of very young sieve elements are similar to those of most young parenchyma cells. Rather thin primary walls with a matrix of cellulose microfibrils are joined to walls of contiguous cells by a middle lamella. In many species shortly after sieve element differentiation begins the wall becomes conspicuously thickened. This thickened wall, which in fresh, unstained, section often has a whitish or "pearly" luster, has been termed the nacre or nacreous wall. The nacreous wall develops inside the primary wall, but is absent over the surfaces of sieve areas. It may be relatively thin, or may become so thick as to occlude as much as 80% of the cell lumen (Esau and Cheadle, 1958). Observations with the electron microscope have shown that the microfibrils in the nacreous wall are, at least in earlier stages of development, often arranged parallel with one another and at right angles to the long axis of the cell (Behnke, 1971b; Singh and Srivastava, 1972; Parthasarathy, 1974a). In some plants the nacreous wall may be distinctly layered (Srivastava, 1969); Behnke, 1971b); in other plants such layering is not evident (Singh and Srivastava, 1972).

    Observations with the electron microscope frequently reveal the presence of a nacreous wall layer, but this layer is not always recognizable with the light microscope.

    In the Pinaceae the thick sieve element walls have long been designated as secondary walls (Abbe and Crafts, 1939; Srivastava, 1969), but nacreous walls in other taxa were usually not considered as secondary walls. More recently electron microscope studies have indicated the structural similarity between nacreous walls and secondary walls and various authors have suggested that at least some nacreous walls should be classified as secondary walls (Esau, 1969; Srivastava, 1969).

    Sieve element walls are composed of cellulose, polyuronides and pectins. The relative proportions of these substances and the degree of hydration of the wall vary from species to species and with stage of development. Almost all reports indicate that lignin is absent from sieve element walls, even though other phloem cells may become lignified at some stage of development.

C. Sieve areas and sieve plates
    A sieve area is part of the sieve element wall bearing pores through which the protoplasts of contiguous sieve elements are in communication. Typically, at least in material prepared by conventional methods for observation with both light and electron microscopes, the carbohydrate wall substance callose is found in association with sieve area pores during both developmental and

mature stages. By strict definition a sieve area refers to a structure in the wall of one cell, and one should speak of a sieve area pair when describing the connection between adjacent sieve elements since sieve areas from two cells are involved. However, this terminology can become cumbersome, and, in general, usage of the term sieve area may refer either to a single sieve area or to a sieve area pair.

Pores in sieve areas vary in diameter from a fraction of a micron to as much as 15 μm (Esau and Cheadle, 1959). It is generally considered that larger pores represent a higher degree of evolutionary specialization, and sieve areas bearing such larger pores are said to be more specialized or more highly differentiated. Usually all pores within a single sieve area are of approximately the same diameter.

In sieve cells all sieve areas are of a similar degree of specialization. In sieve tube members some sieve areas are more highly specialized (have much larger pores) than others in the same cell. Such specialized sieve areas usually occur on end walls and the walls bearing the specialized sieve areas are called sieve plates. When specialized sieve areas occur on lateral walls, the wall areas bearing them are referred to as lateral sieve plates. A sieve plate bearing a single sieve area is called a simple sieve plate, while one bearing two or more sieve areas is called a compound sieve plate. In general, the more oblique the orientation of a sieve plate on an end wall, the greater the number of sieve areas which occur on the sieve plate.

Each pore seems to develop at a site where a plasmodesma was present at an earlier stage, and pore differentiation involves callose. When the pore forms it is lined with a cylinder of callose in which is an opening containing the cytoplasmic material connecting adjacent protoplasts. This cytoplasm has frequently been called the connecting strand, or sometimes the pore contents.

Callose is a carbohydrate, a β-1,3 glucan (Feinghold et al, 1958; Clarke and Stone, 1963). The amount of callose associated with a sieve area varies considerably, and available evidence indicates that callose can be both synthesized and degraded within the sieve element. In intact, functioning, phloem tissue the sieve element contents are under postive pressures of as much as 30 atmospheres (Weatherley, 1962). Wounding of the phloem tissue brings about a sudden release of pressure, and an exudation of sieve element contents. Within periods of a few minutes (Engleman, 1965a) massive deposits of callose, which seems to function in plugging or sealing the injured sieve elements, develop over sieve areas, and such deposits are called wound callose. Massive callose deposits also develop in many species in sieve elements reaching the end of their functional life (definitive callose), and also

develop in some species in sieve elements which function for more
than one growing season. These deposits, called dormancy callose,
form at the onset of dormancy and are removed at the start of the
next growing season (Esau, 1948, 1965).

Most workers on phloem believe that dormancy callose and
definitive callose develop in phloem tissue of intact plants in at
least some species, and are thus a "normal" part of the phloem.
Nearly all studies of sieve area development have revealed
platelets or pads of callose involved in pore formation and the
presence of a cylinder of callose lining young pores (Esau et al.
1962; Esau, 1965; Esau and Cheadle, 1965; Engleman, 1965b, Bouck
and Cronshaw, 1965; Wark and Chambers, 1965; Tamulevick and Evert,
1966; Evert et al, 1966; Singh and Srivastava, 1972; Parthasarathy,
1974b, and many others), although Evert et al (1973b) did not find
callose platelets in developing sieve areas in leaf veins of
Welwitschia. However, the question of whether or not each pore in
a mature functioning sieve element in an intact plant is lined with
a hollow callose cylinder has not yet been satisfactorily answered.
While many light and electron microscope studies reveal such
structures (Esau and Cheadle, 1959), a number of other studies
which utilized rapid killing techniques (Zimmermann, 1960; Evert
and Dorr, 1964; Evert and Alfieri, 1965; Ervin and Ever, 1967;
O'Brien and Thimann, 1967; Currier and Shih, 1969; Parthasarathy
1968; Evert et al, 1971) indicate that callose cylinders lining
pores may be absent under such conditions, and suggest that these
structure may be artifacts.

Sieve area pores may be smoothly cylindrical or there may be
an enlargement (the median cavity or median nodule) in the region
of the middle lamella (Kollmann and Schumacher, 1962, 1963;
Northcote and Wooding, 1966).

D. Protoplast
The protoplast of the mature functioning sieve element has
been the subject of considerable investigation but there still
exist many problems in its interpretation. These problems are
related largely to the fragile nature of the protoplast and the
disruption and formation of artifacts which usually occur during
preparation for microscopic investigation.

Immature sieve elements possess the usual complement of
organelles associated with meristematic cells: nucleus and
nucleolus; tonoplast and vacuole; plasmalemma; endoplasmic
reticulum; ribosomes; dictyosomes; plastids or proplastids;
mitochondria; microtubules; etc. During development some of these
structures disappear, others are modified and other structures
appear. Detailed discussions of developmental changes will be
provided later in this volume; here will be given only a brief

description of the structure of the protoplast of the mature sieve
element.

The mature sieve element possesses a thin layer of parietal
cytoplasm, bounded externally by a plasmalemma. The tonoplast
disappears during development and the central part of the sieve
element contains what has been interpreted as a mixture of vacuolar
and cytoplasmic components called mictoplasm (Engleman, 1965b).

The parietal cytoplasm contains endoplasmic reticulum, usually
smooth endoplasmic reticulum without ribosomes, which has been
called sieve tube reticulum (Bouck and Cronshaw, 1965) or sieve
element reticulum (Srivastava and O´Brien, 1966). Various forms of
endoplasmic reticulum have been reported: a network of tubules
(Esau, 1965); Murmanis and Evert, 1966; Esau and Cronshaw, 1968);
stacks of cisternae and single convoluted cisternae (Esau and
Cronshaw, 1968); lattice-like structures (Behnke, 1965, 1968;
Wooding, 1967; Parthasarathy, 1974b, 1974c).

Plastids occur in the parietal cytoplasm of sieve elements.
Behnke (1971a) distinguished two types: S-type plastids which
contain only starch, and P-type plastids which contain protein
inclusions.   Some P-type plastids may also contain starch. The
distribution of different plastid types in angiosperms has recently
been reviewed (Behnke, 1972).   In dicotyledons 40 orders were
examined, of which 32 had only S-type plastids, 1 had only P-type
plastids, and 7 had both the S-type and the P-type. All 19 orders
of monocotyledons studied had only P-type plastids usually with
cuneate protein inclusions. Among monocotyledons starch-containing
P-type plastids occur in some species of Diascoreaceae,
Zingiberales, Arales (Behnke, 1972) and Arecaceae (Parthasarathy,
1974b).

Starch-containing P-type plastids have been reported in Pinus
(Srivastava and O´Brien, 1966; Murmanis and Evert, 1966; Wooding,
1966, 1868), while S-type plastids have been found in Gnetales
(Evert et al, 1973a; Behnke and Paliwal, 1973).

Mitochondria have been reported from mature sieve elements in
nearly all electron microscope studies (Esau, 1969); in those where
no mitochondria were discerned (Ziegler, 1960; Hohl, 1960) it is
possible that the fixation techniques used were inadequate to
preserve them. Sieve element mitochondria are usually described as
more spherical than those in adjacent parenchyma cells, and with
swollen cristae. Mitochondria in mature sieve elements have often
been reported to be partly disrupted or degenerate, but it has been
suggested (Esau, 1969) that sieve element mitochondria may be more
sensitive to injury during fixation than mitochondria in adjacent
nucleate cells, since such disrupted sieve element mitochondria

often are found in the same sections in which normal mitochondria occur in adjacent parenchyma cells.

Microtubules are commonly observed in developing sieve elements, but are rarely observed in mature sieve elements (see Srivastava and O'Brien, 1966). Dictyosomes and ribosomes disappear during development, and have not been reported from mature sieve elements.

Nuclei are reported to disappear during development and the enucleate feature is considered to be characteristic of mature sieve elements, but several exceptions have been reported. Evert, et al (1970) found that 3 species of Taxodiaceae contained nuclei normal in appearance in all sieve elements interpreted as mature and functional. Nuclei in the sieve elements of these species showed visible signs of degeneration only when the sieve elements were undergoing cessation of function. Evert et al (1970) also reported that nuclei were occasionally-to-frequently present in functioning sieve elements of 12 out of 13 angiosperm species studied. In most cases these nuclei in functioning angiosperms sieve elements appeared to be somewhat degenerate, but in 3 of the species some of the nuclei had a normal appearance. They suggested the possibility that the changes in nuclear structure and loss of nuclei typically reported in sieve elements might be induced during manipulation and fixation of the phloem tissue.

During nuclear degeneration in some species the nucleoli are extruded and may remain as components of the sieve element cytoplasm (Esau, 1969), but one recent study (Deshpande and Evert) indicates that further studies of this phenomenon are needed. It has also been suggested that fragments of the nuclear envelope may remain, but these would be difficult, if not impossible, to distinguish from the endoplasmic reticulum.

For more than a century light microscope studies of phloem have revealed a proteinaceous substance called slime (cf. Esau, 1969) to be a characteristic component of sieve element protoplasts in many species. Young sieve elements were observed to possess discrete slime bodies which frequently disappeared during development and were replaced by an amorphous slime. In a few species the slime bodies were described as remaining intact in mature sieve elements.

Within the past decade electron microscope studies have revealed the presence of characteristic proteins in sieve elements. These proteins, called P-proteins (Esau and Cronshaw, 1967; Cronshaw and Esau, 1968a, 1968b) occur in various forms, both tubular and fibrillar in nature. At times P-proteins occur in discrete bodies, the structures formerly called slime bodies and now known as P-protein is dispersed in the cell protoplast, and is considered to represent what used to be called slime.

One or more forms of P-protein has been observed in all dicotyledonous species which have been carefully investigated. P-protein has also been detected in some gynmosperms (Murmanis and Evert, 1966; Parameswaren, 1971) and some monocotyledons (Ervin and Evert, 1965; Currier and Shih, 1968; Behnke, 1969). On the other hand, workers have been unable to detect P-protein in sieve elements of other gymnosperms (Kollmann and Schumacher, 1964; Srivastava and O´Brien, 1966; Evert et al, 1973a) and monocotyledons (Evert et al, 1971; Singh and Srivastava, 1972).

III. Parenchyma cells in phloem

In addition to sieve elements, the phloem tissue of all species contains parenchyma cells. One can recognize ranges of cell types in the parenchyma component of the phloem tissue, based on differing degrees of physiological association with sieve elements, and on differing ontogenetic relationships with sieve elements. At one extreme of the range is the companion cell, at the other the phloem parenchyma cell.

A. Companion cell
A companion cell is usually defined as a parenchyma cell which is derived from the same mother cell as the sieve element, which has extensive protoplasmic continuity with the sieve element, and which functions as long as the sieve element does (Esau, 1969). Companion cells occur in the angiosperms and are associated with sieve tube members. Primary pit fields on companion cell walls have characteristic connections with sieve areas on sieve element walls. Each pore in the sieve element side of the wall is connected with several branched plasmodesmata on the companion cell side of the wall, and there is usually a median cavity in the wall where the connection occurs. Companion cells are connected with each other and with other phloem parenchyma cells by the plasmodesmata.

Companion cells in most species investigated do not form starch grains, although starch has been detected in a few species. Tannins and crystals have not been reported. Nuclei are typically prominent and stain densely. Vacuoles vary in size but companion cell protoplasts are typically described as dense. Ribosomes and mitochondria are abundant, endoplasmic reticulum is conspicuous. Plastids are usually present. Typically they are leucoplasts with few membranes, but in a few cases they may be differentiated as chloroplasts. In some species P-protein bodies occur in younger developmental stages and dispersed forms of P-protein have been recorded in protoplasts of mature cells. Usually the companion cell contents disappear when the associated sieve element ceases to function, and the companion cell collapses along with the sieve element. Companion cell walls are basically cellulosic, although there are a few reports of lignification of companion cells in older phloem.

Most sieve-tube members have one or more companion cells associated with them, but occasionally sieve tube members lack companion cells (Cheadle and Esau, 1958). Companion cells may occur singly, or in longitudinal series.

B. Albuminous cells

Albuminous cells are generally considered to be physiological counterparts of companion cells, as they seem to be closely physiologically related to sieve elements. However, they are usually not derived from the same mother cells as the sieve elements with which they are associated. Albuminous cells are most abundant in gynmosperms, where the sieve elements are sieve cells. Albuminous cells may be derived from both ray initials and fusiform initials in the vascular cambium. In Pinaceae (Srivastava, 1963) they most frequently develop from derivatives of declining fusiform initials and from marginal ray initials.

Albuminous cells at maturity possess nuclei, dictyosomes, vacuoles, much rough endoplasmic reticulum, large numbers of mitochondria with well-developed cristae, and plastids (Srivastava and O'Brien, 1966). Albuminous cells do not usually form starch grains, and are generally recognized by the absence of starch and other ergastic substances, and the presence of protoplasmic connections with sieve elements (Srivastava, 1963). The sieve element-albuminous cell connections are apparently similar to the sieve element-companion cell connections, with a single pore on the sieve element side, a median cavity, and several branched plasmodesmata on the albuminous cell side (Wooding, 1966). An albuminous cell typically loses its contents at the same time that the associated sieve element ceases to function.

It has been suggested that the term "albuminous cell" is somewhat inappropriate (Srivastava, 1963, 1970) since these cells are not especially rich in protein. Detailed ontogenetic studies have shown, moreover, that not all companion cells are sister cells of sieve elements, and that some cells which are sister cells of sieve elements differentiate not as companion cells but as phloem parenchyma cells containing ergastic substances (Cheadle and Easu, 1958, 1964; Srivastava and Bailey, 1962; Evert, 1963). Thus, distinguishing between companion cells and albuminous cells on the basis of a close ontogenetic relationship with a sieve element, or the lack thereof, may not be as fundamental a difference as earlier workers believed. Since both albuminous cells and companion cells are cytologically similar, and presumably have simlar close physiological associations with sieve elements, Srivastava (1970) has suggested that the term companion cell be applied to both cell types, "but only in a physiological sense with no ontogenetic connotation."

C. Phloem parenchyma cells

Parenchyma cells in the phloem tissue other than companion

cells and albuminous cells are generally called phloem parenchyma cells. They may be further categorized on the basis of the ergastic substances some of them contain, most frequently starch, tannins, or crystals. Phloem parenchyma cells in the axial system may be single elongate cells, called fusiform phloem parenchyma cells, or may form vertical strands of cells, called phloem parenchyma strands, which develop by transverse divisions in derivatives of fusiform cambial initials.

Phloem parenchyma cells typically remain alive after the sieve elements and companion cells collapse, and may undergo continued development in older parts of the phloem.

D. Transfer cells
    In 1884 Fischer recognized certain densely staining parenchyma cells in the phloem of minor leaf veins which he called "Übergangszellen". Recently these cells have received renewed attention, and have been called transfer cells (Gunning et al, 1968; Gunning and Pate, 1969; Pate and Gunning, 1969; O'Brien and Zee, 1971). Transfer cells are specialized parenchyma cells with internal cell wall protuberances. Some transfer cells are sister cells of sieve elements, and are considered to represent companion cells. They are thought to be involved with short distance translocation between mesophyll cells and phloem, and with exchange of solutes between phloem and xylem.

IV. Sclerenchyma cells in phloem

    Sclerenchyma cells, cells with secondary walls which are often lignified, are characteristic components of the phloem of many plants. Two categories of cells are generally recognized, fibers and sclereids. These are differentiated on the basis of form and of origin, but many sclerenchyma cells in phloem do not fit readily into this classification (cf. Esau, 1968).

A. Fibers
    Fibers are narrow elongate cells with pointed ends. They frequently reach lengths of a few to several mm after undergoing symplastic and intrusive growth. Fibers typically begin differentiation early in ontogeny from cells called fiber initials, and are recognizable as fibers even when the phloem tissue in which they are found is still functioning in conduction. Fibers may become transversely septate and multinucleate in later stages of their development.

B. Sclereids
    Sclereids are cells which may be isodiametric, or of irregular twisted shapes, or somewhat elongate, and have very thick walls, often with ramified pits. They typically develop in older,

non-conducting phloem, from cells which were phloem parenchyma
cells.

Typical fibers are easily distinguished from typical
sclereids, but there are some cells which are intermediate in
morphology between the two categories. Such cells are usually
called fiber-sclereids. Some species have sclerenchyma cells
which, on the basis of morphology, closely resemble fibers, but
develop later in ontogeny from phloem parenchyma cells. In at
least one case (Zahur, 1959) cells which are morphologically
sclereids mature in the conducting phloem. Phloem fibers are of
considerable economic importance in some species, and sclerenchyma
in the phloem is of diagnostic and systematic value in taxonomic
studies. However, the relationships between sclerenchyma and
translocation processes in the phloem are not yet evident;
consequently, further details of sclerenchyma in phloem will not be
considered here.

V. Cellular arrangement in phloem tissue

A. Gymnosperms

Most studies of gymnosperm phloem have concentrated on the
secondary phloem of conifers (Fig.2). In this group sieve elements
are sieve cells, ranging from about 1.5 to 5 mm in length. Other
cells in the vertical system include phloem parenchyma cells, both
fusiform phloem parenchyma and phloem parenchyma strands,
albuminous cells, and fibers. Phloem parenchyma cells may exhibit
starch, tannins, oils, or crystals. In the radial system both ray
parenchyma cells and albuminous cells occur (cf. Esau, 1969) and
rays are typically uniseriate or biseriate.

In Pinaceae no phloem fibers occur, but they are present in
other families of conifers. In some cases, (e.g., _Taxodium_, Easu,
1969) there may be a fairly regular radial alternation of fiber,
sieve cell, parenchyma cell, sieve cell, fiber, sieve cell, etc.,
in the derivatives from a single fusiform initial. Adjacent radial
files of cells appear to be coordinated in development, resulting
in the formation of regular tangential bands of similar cell types.
In other cases, (e.g. _Pinus_, Srivastava, 1963) the pattern is much
less regular. A single fusiform initial produces derivatives which
do not differentiate in a regular sequence, and adjacent radial
files of cells do not show close developmental coordination.

Phloem of cycads requires further study (cf. Esau, 1969).
Sieve elements are sieve cells, and are reported to contain starch.
Starch- or crystal-containing phloem parenchyma cells and
albuminous cells are present (Strasburger, 1891). Phloem fibers
are found in some species.

In _Ginkgo_ (cf. Esau, 1969) the phloem is generally similar to
that of conifers. Sieve cells, phloem parenchyma and albuminous
cells, and fibers occur in the vertical system. Some phloem

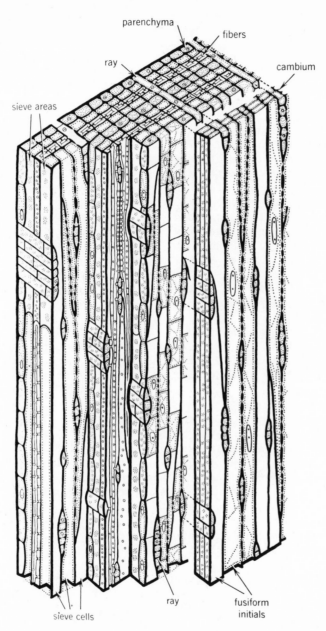

parenchyma
fibers
ray
cambium
sieve areas

ray
fusiform
initials
sieve cells

Fig.2. Block diagram of secondary phloem and cambium of <u>Thuja</u> <u>occidentalis</u>, a conifer. (After K. Esau, 1965. Plant Anatomy, 2nd Ed. John Wiley and Sons, Inc. Original diagram courtesy of I.W. Bailey, drawn by Mrs. J.P. Rogerson under the supervision of L.G. Livingston, redrawn by K. Esau.)

parenchyma cells store starch or starch and tannins, others possess
crystals.   Rays   contain   both albuminous cells and ray parenchyma
cells. There is no regular pattern of radial alternation of cell
types in the vertical system.

The Gnetales are considered to be the most highly evolved
group of gynmosperms.   Phloem structure has been described for
secondary phloem of stems of Ephedra (Strasburger, 1891; Alosi and
Alfieri, 1972; Behnke and Paliwal, 1973), and Gnetum (Strasburger,
1891; Behnke and Paliwal, 1971), and from primary and secondary
phloem of leaf veins of Welwitschia (Evert et al, 1973a, 1973b).
In all cases sieve elements are described as sieve cells. Rays in
secondary phloem of stems are mostly multiseriate and do not
contain albuminous cells. In Gnetum and Ephedra Strasburger and
Alosi and Alfieri recognized both phloem parenchyma cells and
albuminous cells in the vertical system on the basis of light
microscope studies; Behnke and Paliwal were unable to distinguish
between these cell types at the ultrastructural level.   Alosi and
Alfieri (1972) reported the presence of P-protein in Ephedra sieve
elements, but Behnke and Paliwal (1973) were unable to detect
P-protein in either Ephedra or Gnetum.

In Welwitschia leaves Evert et al (1973a, 1973b) found sieve
cells   connected   to   albuminous cells by the typical sieve
element-companion cell connections.   They were unable to find
P-protein in sieve elements at any developmental stages.

B. Angiosperms
    1. Dicotyledons. The great majority of studies on phloem
have been concerned with dicotyledons, and a large amount of
information has accumulated (cf.   Esau, 1969).   Considerable
variation exists from species to species in details of cell
structure, but certain features are common to most or all members
of this taxon.   With only one reported exception, Austrobaileya
(Srivastava, 1970), sieve elements are sieve-tube members.   These
sieve tube members show great diversity in size and shape (Fig.1).
Some form of P-protein is typically present. In some species very
thick nacreous walls develop.

Companion cells are regularly present in the metaphloem and in
the vertical system of the secondary phloem. Their occurrence in
protophloem seems variable.   Several kinds of phloem parenchyma
cells occur, differing both in cytoplasmic contents and in degree
of association with sieve element.   Sclerenchyma is generally
present, at least in older secondary phloem (Fig.3).

    2. Monocotyledons. Sieve elements in monocotyledons are
sieve tube members, with considerable diversity in size and shape.
In all species investigated P-type plastids are present, and in

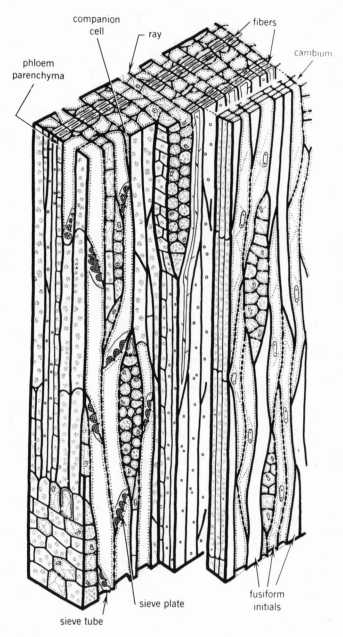

Fig.3. Block diagram of secondary phloem and cambium of Liriodendron tulipifera, a dicotyledon. (After K. Esau, 1965. Plant Anatomy, 2nd Ed. John Wiley and Sons, Inc. Original diagram courtesy of I.W. Bailey, drawn by Mrs. J.P. Rogerson under supervision of L.G. Livingston, redrawn by K. Esau.)

which the protein inclusions frequently have a distinct cuneate shape. In some species starch also occurs in the plastids, in other species it has not been detected. P-protein occurs in sieve elements of some monocotyledons (Currier and Shih, 1969; Behnke, 1969), but has not been detected in others (Evert et al, 1971; Singh and Srivastava, 1972).

In metaphloem there is often a regular alternation of sieve elements and parenchyma cells; in other species the arrangement is much less regular (Cheadle and Uhl, 1948). Some workers have suggested that, in certain monocotyledons, all the parenchyma cells in the interior of the phloem represent companion cells (cf. Esau, 1969); other studies (Parthasarathy and Tomlinson, 1967; Parthasarathy, 1968; Singh and Srivastava, 1972) indicate that both companion cells and phloem parenchyma cells can be distinguished on the basis of differences in content, and in degree of protoplasmic connections with sieve elements.

VI. Longevity

The longevity of sieve elements varies greatly and is related both to the position of the sieve element in the phloem and to the morphological and ecological characteristics of the species involved. Protophloem sieve elements seem to remain functional for only short periods, ranging from perhaps a few days to a few weeks, and metaphloem sieve elements in species with secondary growth probably do not usually remain functional for more than a single growing season, but little attention has been devoted to this subject.

In perennial monocotyledons it is usually assumed that at least some metaphloem sieve elements function throughout the life of the plant. It has been demonstrated (Parthasarathy and Tomlinson, 1967; Parthasarathy, 1974c) that sieve elements and associated parenchyma cells possessing the characteristics of mature, functioning cells can be found in parts of palm stems as much as 50 years old.

In the secondary phloem of several conifers (Esau, 1969; Alfieri and Evert, 1973) most of the sieve elements function for one growing season, but the last few layers of sieve elements formed late in the season overwinter in the mature state, and remain functional at least part-way through a second season.

In the secondary phloem of dicotyledons (Esau, 1969; Davis and Evert, 1968, 1970) the situation is quite variable. In most temperate species studied, sieve elements seem to function during only a single growing season, but exceptions exist. In _Vitis_, sieve elements usually function for two growing seasons, sometimes more (Esau, 1948, 1965). In _Tilia_ sieve elements function for one to five years (Evert, 1962), or even ten years (Holdheide, 1951). In Vitis, dormancy callose is formed at the end of one growing season and removed at the start of the next season. In Tilia, dormancy callose is also deposited and removed, but partly open pores have been found in sieve plates even in winter (Evert and Derr, 1964). In Quercus (Anderson and Evert, 1965), Ulmus (Tucker, 1968), and Acer (Tucker and Evert, 1969) some mature sieve elements overwinter and are reactivated the following spring. In one woody vine, Parthenocissus (Davis and Evert, 1970), sieve elements formed at the end of one growing season overwintered in an immature state, and completed development at the start of the next season. Critical studies of the situation in tropical trees need to be made.

VII. Changes in old phloem

At some point in time, the sieve elements in a given part of the phloem cease to function in conduction, and this part of the phloem becomes known as nonfunctioning or (more appropriately) nonconducting phloem. Sieve elements lose their contents and often collapse or are crushed. Companion cells and albuminous cells typically lose their contents and collapse at the same time that the associated sieve elements do. Phloem parenchyma cells usually remain alive and undergo various changes.

Masses of definitive callose are usually deposited on sieve areas as the sieve element ceases functioning (Esau, 1969), although in some cases (Parthasarathy and Tomlinson, 1967) this apparently does not occur. The contents of sieve elements and their associated companion or albuminous cells disintegrate and disappear. Usually protoplast disintegration occurs very quickly, and we know little about its details. However, a recent study of sieve elements from palm stems estimated to be several decades old (Parathasarathy, 1974c) may be of interest in this respect. It reveals that while sieve elements several years old are basically similar to recently matured ones, the older cells show a progressive decrease in the amount of endoplasmic reticulum, and a gradual degeneration of plastids and mitochondria. It is possible that the sequences of changes observed in long-lived sieve elements over a period of several years are similar to the sequences of changes which occur very rapidly in sieve elements with shorter functional lives.

After sieve elements and companion cells lose their contents they may collapse, as a result of longitudinal stretching (in protophloem) and enlargement of neighbouring cells. Sometimes nonfunctioning sieve elements, instead of collapsing, are occluded by tylosoids, outgrowths from nearby phloem parenchyma cells (Parathasarathy and Tomlinson, 1967). Such tylosoids may develop thick lignified walls. After cell collapse or occlusion the deposits of definitive callose generally disappear.

As sieve elements and associated cells collapse, ray parenchyma and axial phloem parenchyma cells may undergo additional growth (Esau, 1969). Such growth may be restricted only to ray cells, or only to cells in the axial system, or it may occur in both, and is called dilatation growth. Growth may involve only cell enlargement or a combination of cell division of spaces formerly occupied by the collapsed cells, and also compensates for the increased volume involved as the circumference of the stem or root increases.

Some of the living cells in the nonconducting phloem may differentiate as sclereids or fiber-sclereids (Esau, 1969). In many species such ergastic substances as calcium oxalate crystals, tannins, and resins may continue to accumulate.

Eventually, in those species which develop a rhytidome, some of the parenchyma cells in the nonconducting phloem will contribute to the formation of a phellogen. As a periderm develops from the phellogen the living parenchyma cells which are outside the periderm will die, and these pockets of dead phloem will be incorporated into the rhytidome.

References

ABBE, L.B. and A.S. CRAFTS. 1939. Bot. Gaz. 100: 695-722

ALFIERI, F.J. and R.F. EVERT. 1973. Bot. Gaz. 134: 17-25

ALOSI, M.C. and F.J. ALFIERI. 1972. Amer. J. Bot. 59: 818-827

ANDERSON, B.J. and R.F. EVERT. 1965. Amer. J. Bot. 52: 627

BEHNKE, H.-D. 1965. Planta 66: 106-112

BEHNKE, H.-D. 1968. Protoplasma 66: 287-310

BEHNKE, H.-D. 1969. Protoplasma 68: 377-402

BEHNKE, H.-D. 1971a. Taxon 20: 723-730

BEHNKE, H.-D. 1971b. Protoplasma 72: 69-78

BEHNKE, H.-D. 1972. Bot. Rev. 38: 155-197

BENHKE, H.-D. and S. PALIWAL. 1973. Protoplasma 78: 305-319

BOUCK, G.B. and J. CRONSHAW. 1965. J. Cell Biol. 25: 79-96

CHANG, Y.P. 1954. Tappi Monograph Ser. No. 14

CHEADLE, V.I. and K. ESAU. 1958. Univ. Calif. Publs. Bot. 29: 397-510

CHEADLE, V.I. and K. ESAU. 1964. Univ. Calif. Publs. Bot. 36: 143-252

CHEADLE, V.I. and N.W. UHL. 1948. Amer. J. Bot. 35: 578-583

CLARKE, A.E. and B.A. STONE. 1962. Phytochem. 1: 175-188

CRONSHAW, J. and K. ESAU. 1968a. J. Cell Biol. 38: 25-39

CRONSHAW, J. and K. ESAU. 1968b. J. Cell Biol. 38: 292-303

CURRIER, H.B. and C.Y. SHIH. 1968. Amer. J. Bot. 55: 145-152

DAVIS, J.D. and R.F. EVERT. 1968. Bot. Gaz. 129: 1-8

DAVIS, J.D. and R.F. EVERT. 1970. Bot. Gaz. 131: 128-138
DESHPANDE, B.P. AND R.F. EVERT. 1970. J. Ultrastr. Res. 33: 484-494
ENGLEMAN, E.M. 1965a. Ann. Bot. 29: 83-101
ENGLEMAN, E.M. 1965b. Ann. Bot. 29: 103-118
ERVIN, E.L. and R.F. EVERT. 1965. Amer. J. Bot. 52: 627
ERVIN, E.L. and R.F. EVERT. 1967. Bot. Gaz. 128: 138-144
ESAU, K. 1939. Bot. Rev. 5: 373-432
ESAU, K. 1948. Hilgardia 13: 217-296
ESAU, K. 1950. Bot. Rev. 16: 67-114
ESAU, K. 1965. Hilgardia 37: 17-72
ESAU, K. 1969. The Phloem. Handbuch der Pflanzenanatomie, Band 5, Teil 2. Gebruder Borntraeger, Berlin.
ESAU, K. and V. I. CHEADLE. 1958. Proc. Nat. Acad. Sci. U.S. 44: 546-553
ESAU, K. and V.I. CHEADLE. 1959. Proc. Nat. Acad. Sci. U.S. 45: 156-162
ESAU, K. and V.I. CHEADLE. 1965. Univ. Calif. Publs. Bot. 36: 253-344
ESAU, K., V.I. CHEADLE and E.B. RISLEY. 1962. Bot. Gaz. 123: 233-243
ESAU, K. and J. CRONSHAW. 1967. Virology 33: 26-35
ESAU, K. and J. CRONSHAW. 1968. J. Ultrastr. Res. 23: 1-14
EVERT, R.F. 1962. Amer. J. Bot. 49: 659
EVERT, R.F. 1963. Amer. J. Bot. 50: 8-37
EVERT, R.F. and F.J. ALFIERI. 1965. Amer. J. Bot. 52: 1058-1066
EVERT, R.F. and W.F. DERR. 1964. Amer. J. Bot. 51: 552-559
EVERT, R.F., C.J. BORNMAN, V. BUTLER and M. GILLILAND. 1973a. Protoplasma 76: 1-21
EVERT, R.F., C.H. BORNMAN, V. BUTLER and M.G. GILLILAND. 1973b. Protoplasma 76: 23-34
EVERT, R.F., J.D. DAVIS, C.M. TUCKER and F.J. ALFIERI. 1970. Planta 95: 281-296
EVERT, R.F., W. ESCHRICH and S.E. EICHHORN. 1971. Planta 100: 262-267
EVERT, R.F., L. MURMANIS and I.B. SACHS. 1966. Ann. Bot. 30: 563-585
FEINGOLD, D.S., E.F. NEUFELD and W.Z. HASSID. 1958. J. Biol. Chem. 233: 783-788
FISCHER, A. 1884. Untersuchungen uber das Siebrohren-System der Cucurbitaceen. Gebruder Borntraeger, Berlin.
GUNNING, B.E.S. and J.S. PATE. 1969. Protoplasma 68: 107-133
GUNNING, B.E.S., J.S. PATE and L.G. BRIARTY. 1968. J. Cell Biol. 37: C7-C12
HARTIG, T. 1837. Jahr. Fortschr. Forstwiss. und Forstl. Naturk. 1: 125-168
HOHL, H.-R. 1960. Ber. Schweiz. Bot. Ges. 70: 395-439
HOLDHEIDE, W. 1951. In: Handbuch der Mikroskopie in der Technik, by H. Freund. Vol. 5, Part 1: 193-367 Frankfurt a Main, Umschau-Verlag.

HUBER, B.   1939.   Jahrb.   Wiss.   Bot.   88:   175-242
KOLLMANN, R and W.   SCHUMACHER.   1962.   Planta 58:   366-386
KOLLMANN, R and W.   SCHUMACHER.   1963.   Planta 60:   360-389
KOLLMANN, R.   and W.   SCHUMACHER.   1964.   Planta 63:   155-190
MURMANIS, L.   and R.F.   EVERT.   1966.   Amer.   J.   Bot.   53:
    1065-1078
NAGELI, C.W.   1858.   Beitr.   Wiss.   1:   1-156
NORTHCOTE, D.H.   and F.B.P.   WOODING.   1966.   Proc.   Roy.   Soc.   B
    163:   524-537
O'BRIEN, T.P.   and K.V.   THIMANN.   1967.   Protoplasma 63:   443-478
O'BRIEN, T.P.   and S.-Y.   ZEE.   1971.   Aust.   J.   Biol.   Sci.   24:
    207-217
PARAMESWAREN, N.   1971.   Cytobiologie 3:   70-88
PARTHASARATHY, M.V.   1966.   Diss.   Cornell University
PARTHASARATHY, M.V.   1968.   Amer.   J.   Bot.   55:   1140-1168
PARTHASARATHY, M.V.   1974a.   Protoplasma 79:   59-91
PARTHASARATHY, M.V.   1974b.   Protoplasma 79:   93-125
PARTHASARATHY, M.V.   1974c.   Protoplasma 79:   265-315
PARTHASARATHY, M.V.   and P.V.   TOMLINSOM.   1967.   Amer.   J.   Bot.
    54:   1143-1151
PATE, J.S.   and B.E.S.   GUNNING.   1969.   Protoplasma 68:   135-156
PERROT, E.   1899.   Le tissu crible.   Lechevallier, Paris.
SINGH, A.P.   and L.M.   SRIVASTAVA.   1972.   Can.   J.   Bot.   50:
    839-846
SRIVASTAVA, L.M.   1963.   Univ.   Calif.   Publs.   Bot.   36:   1-142
SRIVASTAVA, L.M.   1969.   Amer.   J.   Bot.   56:   354-361
SRIVASTAVA, L.M.   1970.   Can.   J.   Bot.   48:   341-359
SRIVASTAVA, L.M.   and I.W.   BAILEY.   1962.   J.   Arnold Arb.   43:
    234-272
SRIVASTAVA, L.M.   and T.P.   O'BRIEN.   1966.   Protoplasma 61:
    277-293
STRASBURGER, E.   1891.   Ueber den Bau und   dei   Verrichtungen   der
    Leitungsbahnen   in   den Pflanzen.   Histologische Beitrage.   Heft
    3.   Gustav Fischer, Jena.
TAMULEVICH, S.R.   and R.F.   EVERT.   1966.   Planta 69:   319-337
TUCKER, C.M.   1968.   Amer.   J.   Bot.   55:   716
TUCKER, C.M.   and R.F.   EVERT.   1969.   Amer.   J.   Bot.   56:
    275-284
WARK, M.C.   and   T.C.   CHAMBERS.   1965.   Aust.   J.   Bot.   13:
    171-183
WEATHERLEY, P.E.   1962.   Adv.   Sci.   18:   571-577
WOODING, F.B.P.   1966.   Planta 69:   230-243
WOODING, F.B.P.   1967.   Planta 76:   205-208
WOODING, F.B.P.   1968.   Planta 88:   99-110
ZAHUR, M.S.   1959.   Cornell Univ.   Agr.   Exp.   Sta.   Mem.   359
ZIEGLER, H.   1960.   Planta 55:   1-12
ZIMMERMANN, M.H.   1960.   Ann.   Rev.   Plant Phys.   11:   167-190

DISCUSSION

Discussant: M.V. Parthasarathy
            Genetics, Development and Physiology Section
            Cornell University
            Ithaca, New York 14850

The following discussion was preceded by a few remarks from Dr. Parthasarathy.

Parthasarathy: I would like to point out that the reports of the presence of nuclei in certain species of dicotyledons and gymnosperms to which Dr. Lamoureaux referred, are all based on light microscopic work and need to be substantiated at the electron microscope level. Also, the previous reports of P-protein being present in gymnosperms are highly suspect. At the EM level they are all artifacts. A few additional points about the longevity of sieve elements. There seem to be several exceptions to the idea that "the sieve element is short-lived". In palms, for example, Rystoria which lives for 100-120 years, the basal parts of the stem remain intact and, since new roots arise each year, it can be assumed the sieve elements in these parts are conducting. These sieve elements have ER though at a reduced level, connections with the parenchyma cells, and no definitive callose, that is in the main conducting bundles. There is definitive callose in the leaf traces which are not functioning. We have also examined some tree ferns from Costa Rica and the sieve elements in the basal part of these stems are intact which suggest that these plants also have long-lived sieve cells.

Cronshaw: How old do the palms get? Do they live only for 150 years?

Parthasarathy: Oh, no. It's different from species to species. Rystonias and some of the Andean wax palm live for anywhere between 100-200 years; but some others, coconut trees for example, live for 50-60 years.

Cronshaw: So about 200 years is the maximum, then?

Parthasarathy: I would think so. Unfortunately we don't have documented cases. Rystonia is about the only one we know. We are sure it's at least 150 years.

Cronshaw: I find this rather strange, for as you know, the other woody plants live for many thousands of years. If it is a fact that palms only live for 150 years, I find it very interesting that the other plants have the ability to regenerate their phloem and get new phloem every year whereas the palms do not.

21

Parthasarathy: That could be. One other thing that perhaps determines the death of the palm is its inability to transport nutrients and/or photoassimilate after a certain number of years or to get new roots, maybe; I don't know.

Johnson: If these sieve elements live for a very long time, presumably at the end or towards the middle of their life they still have the same sort of cytoplasmic contents which other sieve elements have. Do these individual organelles, plastids or mitochondria or whatever they may be that one can recognize, live for this long length of time? Do they divide or do they perhaps maintain themselves by some sort of patching.

Parthasarathy: I don't know. In palms, in sieve elements estimated to be more than 20-30 years old, I couldn't see anything that was comparable to mitochondria. It's possible that the mitochondria in that particular stage are very fragile and I couldn't preserve them, but there is no doubt that there are plastids, ER, plasmalemma, and protoplasmic connections with the companion cells. The plastids do seem to undergo a lot of degeneration and there is definitely a much reduced amount of ER compared to that in the recently matured metaphloem.

Milburn: In palms, there is a fairly well-defined growing region at the top of the stem where much division occurs; but in sort of small pockets coming down from the top you have areas of stem where cells remain latently developable. So it seems to me that the issue in palms about longevity of sieve elements hinges on the question of when the latest sieve tubes could be formed in the fifty-year-old segment at the very bottom of the tree. Is it possible that the ones that actually conduct could wait fifty years and then develop as a sort of delayed meristematic development?

Parthasarathy: I have examined close to 40-50 palm trunks from top to bottom to see whether I could detect any kind of nucleated cells that could be interpreted as leading to differentiation of new metaphloem but I couldn't find them, so I'm not very sure if anything like that happens.

Aronoff: I am more than naive in this area but I'd like to suggest that the recognition of viability might be tested physically by a very simply-stated methodology, but one which might prove to be rather difficult in practice, namely, the determinaton of the amount of tritium in various portions of a section. In other words, one ought to be able to determine by hydrogen dating, equivalent to that of carbon dating, whether or not parts or cells are indeed of recent origin or such an age that they have essentially lost their tritium content.

Lamoureux: I have looked at some tree ferns, 20-30 years old probably, and I doubt that there is any later differentiation because there is just no place that it would come from. In other words, the phloem tissue or the whole vascular tissue at the base of a tree fern, for instance, has the same number of cell layers that it has at the top and therefore I think at least in those groups, and suspect in the palms also, there is really no evidence for later differentiation. And something at the base of the 50-year-old palm or the 20-year-old tree fern was presumably formed that way in year one and has not changed by the addition of further differentiation of cells to any great extent.

Christ: Are there any other common features of phloem cells which are functional only for one season and, likewise, features that are common to phloem cells which are functional for 20-50 years?

Srivastava: In comparing the structure of sieve elements in dicots and some monocots where sieve elements function for only one year and palms where they seem to live much longer, as far as we know, there are no major differences between the two in the differentiation sequence or in the mature structure.

Parthasarathy: I agree.

Swanson: I was curious about the structure of the magnolia sieve tubes which showed perhaps as much as 90 per cent of the cross-sectional area being occluded by nacreous wall. Is it known that these sieve tubes are functional?

Lamoureux: Apparently these are considered to be functional. These walls develop early, very close to the cambium. The wall is present in this state through most of the functional life of the sieve element. It is absent over the sieve areas, but is present elsewhere, and does, in fact, occupy at least 80 per cent of the volume. But these are extreme cases. Not all nacreous walls are that thick.

Fisher: In the few sieve elements that I have looked at by freeze substitution the walls are much thinner and I suspect that perhaps the swelling that is seen in nacreous walls is an artifact of preparation, in large part due to the aqueous fixative procedures that are used.

Parthasarthy: I was wondering whether anyone has looked at free-hand fresh sections of the material, let it dry, and then examined it again to see if there is a difference in the wall thickness?

Lamoureux: In some Magnoliaceae I have looked at fresh material and again after dehydrating it with ethanol. The shrinkage is not more than, say 30 - 40 per cent from the initial volume.

Fisher: Normally, in the sieve element there is a very high hydrostatic pressure against the wall and as soon as you remove that pressure I think it is quite possible that the wall will hydrate very quickly. So if you look at fresh material you are going to see it. I think the only way you could find out is by freeze substitution.

Johnson: In the freeze-etched material of nymphoides there does not seem to be any obvious difference in the thickness of the walls compared to that in the sectioned (chemically-fixed) material. The freeze-etched sieve elements have not had the pressure released in them because you can see that the walls of the xylem vessels which are under tension inwards are in fact bent in between the regions of wall thickenings and this obvious retention of tension in the xylem leads one to suppose that pressure in the sieve element has not been released.

Dainty: Could I ask Dr. Fisher if he could make the physics of what he is suggesting a little plainer?

Fisher: I do not know if I can but I was thinking, for instance, of some of the colloids that are found in algae that are very sensitive to water potential in their swelling.

Dainty: If I could think this out, I think the water potential would go somewhat more negative as you release the hydrostatic pressure. The other change which would occur if the walls themselves are under tension, is if you release that tension, you may expect maybe a little swelling. But I would not expect from the materials of which the cell wall is built that the release of tension is going to produce so much thickening - at least the change in thickening that you are talking about.

Weatherley: Could I ask Dr. Fisher what happens when you plasmolyze the cell? Does this cause an increase in the wall thickness?

Fisher: I do not know if anybody has looked.

Geiger: Is there a good pattern in the monocots between the phloem of C-3 and C-4 plants, specifically the sieve tubes and the associated companion cells. For instance, in terms of the differences in the vein-loading times between C-3 and C-4 plants; also, some of the posited mechanisms for C-4 plants in terms of transport, what materials may be moving, and so on?

Crookston: We have looked at the phloem parenchyma of C-4

dicots as opposed to C-3 dicots and it appears that when there are
plastids in the phloem parenchyma of C-4 dicots they are more
similar in their structure to the bundle sheath chloroplast than to
those of\ the mesophyll. Whether you can apply physiological
significance to that or not, I do not know. In grasses there do
not appear to be any plastids in the phloem parenchyma.

    Walsh: In Zea I have never seen a sieve element in direct
contact with the bundle sheath cell and thus the mesophyll.
However, there are minor veins in this plant that seem to differ
somewhat from the minor veins of dicotyledons.

    Swanson: Just how much is known about the frequency
distribution and the plasmodesmatal connections among the various
cell types that make up the phloem and between the phloem and the
mesophyll?   The reason, of course, for raising this question is
that one of the current controversies deals with whether the
primary transport pathway from the chloroplast is by way of the
symplast to the loading site or is it by way of the apoplast.
Incidentally, this raises the question of semantics too. What is
mesophyll? As I read the physiological literature I take it that
it means only the palisade and the sponge mesocell, but I believe
the original term referred to all cell types between the upper and
lower epidermis.   Therefore, the vascular system would also be
mesophyll.

    Parthasarthy: As I understand anatomy, I think mesophyll does
not include vascular tissue.

    Lamoureux: The way in which most anatomists talk about it,
mesophyll is the tissue between the epidermal layers derived from
ground tissue rather than from procambium.   This would eliminate
the vascular tissue from consideration.

    Currier: A few questions about terminology.   These things
that pass through the sieve pores - are they to be called
connecting strands, pore contents, or cytoplasmic connections?
Also, the term mictoplasm. I think this could be a useful term if
we modified the meaning from Mark Engleman's original premise to
include the whole substance within the plasmalemma. Finally, the
callose. Is it an artifact?

    Parthasarathy: Engleman's concept of mictoplasm was   the
translucent area in the central part of the sieve element.

    Currier: It seems better to me that we should include
everything within the plasmalemma as mictoplasm and not restrict it
to the aqueous solution.

    Cronshaw: I do not like the term mictoplasm. I like to think

of a cell as a cell surrounded by its plasma membrane and
containing protoplasm which includes the nucleus or cytoplasm that
does not include the nucleus. Both of these terms would include
the vacuoles and whether or not the vacuole is there, does not make
any difference. Referring to the other term, the cell-to-cell
connections or pore contents, I like the term pore contents because
I think that one of the facts we have established over the past few
years is the basic structure of the pore. There is argument about
the size and the various components and the amount of components
that are there, but I think the basic structure has been
established and so when we speak of the material within the pore, I
think it is good to speak of pore contents.

MacRobbie: I would like to know what this stuff in the middle
is because much seems to depend on what we think is happening
there. If we believe that the whole of the centre is actually
heading downwards at a rapid rate, we clearly do not believe that
the gel around the outside is equally heading down at a rapid rate.
Therefore, the question of what separates the stationary gel and
the centre, if we believe the centre is moving, is critical.

Dainty: I was going to ask the same question. We want to
know, if there is no tonoplast, what is the nature of the boundary
between the protoplasm and the vacuole?

Srivastava: We do not really know whether the mature sieve
element, from a differentiation point of view, is in fact the
conducting sieve element, is one that has a fairly thick wall,
plasmalemma, and a few layers of what we call endoplasmic
reticulum. In addition, it has some organelles such as plastids
and mitochondria. Most people believe that during differentiation,
the tonoplast breaks down and, consequently, in the mature element
there is no membrane material that can be distinguished as the
vacuolar membrane internal to the endoplasmic reticulum that lines
the plasmalemma. What else may be present inside the lumen of the
cell depends on what plant you examine. In some plants there are
definitely tubules or fibrils of P-protein type; in others there
may be some other fibrillar material; and in still others, there
may be nothing.

Aikman: Does the term mictoplasm imply that the contents are
shared within the sieve tube, and are not confined to a sieve
element?

Lamoureux: I think this depends on what concept of the pore
contents we have if the pores are open, then in that sense it could
apply to the whole sieve tube.

Cronshaw: I think we have gotten into terminology where we do
not really need to. As I see it, the terminology is fairly clear.

Plant cells differ from animal cells in that they are interconnected and the plasma membrane of one cell is continuous with that of the neighbouring cells via plasmodesmata. The plasma membrane within the whole plant is continuous, especially through the sieve plate pores. The content of one cell is not separated by a semi-permeable membrane from the contents of the next cell. The material that is in the centre in the sieve elements, and everything within the plasma membrane, is continous from cell to cell. The term lumen means everything within, and I think most people use it in that sense. So we have the two good terms we can use, the protoplasm which is everything within the plasma membrane, and lumen when we want to point to the inside and say everything within. I would also like to confirm what Dr. Srivastava said about the absence of a vacuole or vacuolar membrane in the mature sieve element. There is no phase separation, or no semi-permeable membrane within the sieve element, as most of us see it, that separates out an inner from an outer portion.

Currier: I would like to ask the anatomists what they think of the terms peripheral cytoplasm or peripheral protoplasm? Or peripheral mictoplasm? I think in Dr. Esau's book, she talks about the peripheral cytoplasm.

Lamoureux: Unless we can agree on clear definitions of mictoplasm and peripheral cytoplasm since there is no morphological boundary between the two - Dr. Cronshaw's suggestion that we just talk about protoplasm is the best way to settle this - unless the physiologists will insist that they must have a special name for the clear stuff in the centre.

MacRobbie: I think what the physiologists are worrying about is that there is a distinction. If you are diffusing a sucrose molecule in there, there is a big difference between being in the middle of the lumen or the contents, and being in the peripheral cytoplasm. One is full of organelles and membranes and general stuff, stuff that is difficult to diffuse through, and the other is apparently clear.

Milburn: The main difference that I think would be most relevant here, is that there is a sol-gel difference. The exudate that one collects from aphid stylets is presumably not vacuolar content; there is no tonoplast. Therefore, it must be the cytoplasmic sol which is in the centre. The quantity of the exudate that one gets out of the cell is far too great to come from that itself - it comes from a whole sequence of cells, and implies a continuum from cell to cell. I think probably the concept that we use a cytoplasmic or protoplasmic sol is probably the answer for this material.

Eschrich: Is it known whether there is a connection between

the sieve tube systems of the secondary phloem, say in a
dicotyledon or a coniferous plant, and the metaphloem of the same
plant or whether there are connections between the annual
increments of secondary phloem between the younger and the older
parts?   I have never seen these connections in radial sections of,
for example, Tielia, a connection of sieve tubes across the bands
of fibres.   Could it be that in monocotyledons, like palms, the
stem has special bundles which are continous and in which the sieve
tubes stay alive for a long time only inside the stem, whereas the
leaf traces die sooner or later?

Lamoureux: In a dicotyledon or a coniferous tree of any age
there is a single layer of vascular cambium. All the secondary
phloem produced in one growing season is presumably in contact from
the top to the bottom. The critical point is where does the
primary growth in the current year hook on to this continuous layer
of secondary tissue? I think the assumption is generally made that
it must, but as far as the details of that connection go, I am not
sure that that has been studied.

Behnke: In deciduous trees is there not a continuum at all
stages between the primary, or the mostly primary, phloem tissues
in the leaves and the secondary tissues in the stem, via the
special structures called nodal anastomoses?

Srivastava: I think that we must certainly make a distinction
between the pores between 2 sieve elements which are present in the
longitudinal pathway and the connections along lateral walls
between sieve elements and companion cells, and sometimes also
other parenchyma cells. How much lateral transfer occurs in a
conducting sieve element is something that we know very little
about; but on reasonable grounds I would assume that most of the
conduction occurs longitudinally with very limited leakage on the
sides. Whether the sieve area connections on the lateral walls
between sieve element and companion cells function in the same
manner as the pore connections in the end walls is not at all
clear. I would think that they probably do not function in that
way. The plasmalemma along the lateral walls probably exercises a
clear permeability barrier whereas at the plasmalemma that lines
the pores, there is no such barrier, or if there is a barrier then
it is due to the fibrils or some other things that may be there.

Aikman: Dr. Cronshaw made the point that the plasmalemma
defines a continuous phase within the whole plant, that it is
continuous through the sieve pore and through the plasmodesmata.
Although the plasmalemma is continuous through the plasmodesma, is
there not a desmatubule and a tight junction or occlusion of space
between the desmatubule and the plasmalemma?

Behnke: Either there are desmatubules or open ER or there is

some interphase between the ER and the plasmalemma that makes
contact between the protoplasm of the two adjacent cells.

Srivastava: The sieve pores at the end walls between 2 sieve
elements show gradations in size and number which vary from species
to species. The lateral sieve areas between the sieve elements and
the companion cells or their analogous albuminous cells also show a
gradation in complexity or development. They vary from typical
plasmodesmatal connections to far less open pores than are present
between two sieve elements. The connections between the companion
cells and other parenchyma cells in the phloem, through which the
path is established toward mesophyll cells, are almost typically
plasmodesmatal-type connections. Opinions vary on the structure of
plasmodesmata but nobody, to my knowledge, has shown a clear pore
going through a plasmodesmatal connection. What you do see is that
the plasmalemmae of the two cells go through the wall and in
between the plasmalemma tube there is something. As to what it is,
we do not know; but it is certainly not a hollow tube. Whether a
selective permeability is exercised in this manner or whether there
are potential differences across the plasmodesmata - and at least
some people have shown that there are - is not very clear.

Tyree: Dr. Geiger claims that there are very few or no
plasmodesmatal connections between the vascular bundle and the
mesophyll, yet the implication from Cronshaw's remark was that the
sieve cells, companion cells and phloem parenchyma are continuous
with the mesophyll. This is a very important question. I would
like to know what other anatomists have to say about it.

Turgeon: I think when more and more species are looked at we
will probably find that there will be two different types of
connections, especially in the minor veins. There are species
where there are no plasmodesmata, but with transfer cells with
specialized wall ingrowths. In others there may be plasmodesmata.
This applies mostly to the minor veins where you have the very
large companion cells and parenchyma cells, that are apparently
specialized for vein-loading.

Kollmann: There are two fundamentally different types of
intercellular contact in plant cells. In one we have an open
contact between cells by plasmodesmata, sieve pores, and so on; in
the other we have a closed contact as, for instance, in the
transfer cells which are enclosed by a wall labyrinth. The
structural difference between sieve pores and plasmodesmata is not
fundamental. There is a possibility that inside the plasmodesmata
there is a filamentous structure derived from the ER or a
continuous ER tubule, or a solid filament, or as Dr. Robards
proposed, a desmatubule. I feel that this desmatubule is very
similar in structure and function to what we now call P-protein.
If we compare sieve tubes of angiosperms and gymnosperms which may

function similarly, one has P-protein and the other ER in the sieve
pores. It is possible that ER and P-protein have a similar
function, perhaps as a site of enzymatic activity. If this is so,
there is no fundamental difference between plasmodesmata and sieve
pores.

Parthasarathy: You are of course assuming that the sieve
plate pores are indeed plugged with P-protein in its functional
state.

Cronshaw: Dr. Srivastava alluded to, but I do not think,
made his point 100% clear. Sieve elements can be connected to
other sieve elements by lateral sieve plates and I do not think
this has been brought out so far in the discussion. In these
lateral sieve plates the size of the pores is usually smaller than
in the end sieve plates. But that is all. The other basic
structure is very similar. The lateral sieve plates, or lateral
sieve areas as in end wall sieve areas, are double structures.
That is, there is the wall from two adjacent cells. Now in the
lateral sieve areas that Dr. Srivastava mentioned, these are the
connections between sieve elements and companion or other
parenchyma cells. There is a confusion in terminology or no clear
terminology though these are usually referred to as sieve area
connections. There is good terminology from the wood anatomists if
we want to use it, and that is that we would have on the sieve
element side a sieve area which is developed into a pore; and on
the companion cell side we would have a primary pit field, which is
the way we normally refer to thinner areas of the wall that contain
plasmodesmata. So between sieve element and sieve element we have
sieve areas in which we have sieve pores and these are double
structures; or between sieve element and an adjacent parenchyma
type cell we have sieve area primary pit fields and these are
usually referred to as sieve area connections.

Parthasarathy: This term sieve area primary pit field was
proposed in 1958 for such connections, but it did not prove very
popular. I was the only person who used it.

Walsh: That in some instances the wall on the companion cell
side where we have these connections is not a depressed field but
instead is highly thickened.

Dainty: Dr. Srivastava said that we did not know whether
plasmodesmata were open or not. There have been a number of
demonstrations, largely by electrical methods and largely by Dr.
Sovonick, between mesophyll cells in _Elodea_, cortical cells in corn
roots, and across internodal cells in _Nitella_, that there is quite
good electrical contact from cell to cell. We, at Toronto, have
demonstrated the same in _Chara_. It would seem, in our case we used
chloride transfer, that the "conductivity" of a plasmodesmatal pore

is of the order of 30 - 50 times too small; that is, it is more resistant than if it were filled with pure water. But this means it is a very good connection. So there have been clear-cut demonstrations in certain cases, none of which refer to phloem situations yet.

Srivastava: When I say that nobody has demonstrated plasmodesmata as open pores, I mean that under the electron microscope. It does not mean that the things do not conduct.

Structure and Differentiation of Sieve Elements

in Angiosperms and Gymnosperms

L.M. Srivastava

Department of Biological Sciences

Simon Fraser University

Burnaby, B.C., Canada V5A 1S6

I. Introduction

This review covers the literature on structure and development of sieve elements in angiosperms and gymnosperms from 1968 onward. The literature up to that year is well summarized in the book, The Phloem, by Esau (1969). Certain topics on the phloem of angiosperms and gymnosperms such as P-protein, sieve areas, sieve pores and walls, and the parenchyma cells associated with the sieve elements are covered by other authors in this volume. I will concentrate chiefly on studies on living sieve elements and on the changes in protoplast and nucleus during differentiation of the sieve elements. In this account I will be selective rather than comprehensive and will not present an objective summary but a critical review reflecting my own bias. Dr. Eschrich complained not too long ago that "... electron microscopy has confused rather than clarified the structural state of conducting cells ..." (Eschrich, 1970). Rightly so, for there is confusion. But if one takes into account the preparatory artefacts, the vagaries of individual interpretation, and the taxonomic overtoppings with which Mother Nature attempts to confuse us, there emerges a remarkably consistent and meaningful picture of the sieve element structure and differentiation, not only in the angiosperms and gymnosperms but vascular plants as a whole. The brown algae seem to be a case apart as the recent investigations of Dr. Klaus Schmitz in our laboratory show (Schmitz and Srivastava 1974a, b). Finally, I would like to inject one guiding principle. Every now and then one reads that such and such a result was obtained by

light microscopy but under the electron microscope a different
result was obtained. Indeed some authors habitually publish a
paper on light microscopic observations and then a paper on
electron microscopic observations and point out the discrepancies
between the two without offering an explanation. I would like to
emphasize that there is no contradiction between the two methods of
observation. Electron microscopy only extends the range of
observation provided by the light microscope, it does not negate
the observations of light microscopy. Electron microscopy does
require more rigorous and careful preparation of material and,
since one sees only limited amounts of material, a far greater care
in interpretation than is required in light microscopy. This is
where much of the problem lies in phloem research, for
unfortunately, some of the most prolific investigators have not
exercised this caution. The guiding principle, therefore, is,
wherever possible, electron microscopy must be combined with light
microsocpy and should be used to add and extend, not contradict the
results from light microscopy.

II. Studies On Living Sieve Elements

In the past, a number of studies have been carried out on
living sieve elements with 3 main questions in mind: 1. do the
mature sieve elements show protoplasmic streaming; 2. can they be
plasmolysed; and 3. do they accumulate neutral red. The
methodology is relatively simple. Stem pieces are excised, cut
into appropriate shapes for longitudinal (radial and tangential)
sections, and sectioned free hand or on a sliding microtome at
sufficient thickness to include at least one or two layers of
undamaged cells. The tissues are kept flooded with a buffered or
unbuffered sugar solution during sectioning and observation. In
another method, dissected but unexcised and intact phloem bundles
are observed under a suitable optical system (light microscope or
mercury vapour lamp for fluorescence optics), but this method can
be used only for those plants in which long stretches of bundles
can be isolated with relative ease, such as the petiolar bundles of
Heracleum, Pelargonium and Cucurbita.

Most of this work was done before 1955 and is reviewed by
Currier et al (1955), Esau et al (1957) and Esau (1969), but since
the conclusions are important and support the major results from
electron microscopy it is appropriate to summarize them. The few
papers that have appeared subsequent to 1955 are covered in greater
detail.

As a result of the work of Schumacher (1939), Rouschal (1941)
and Currier et al (1955), there seems little doubt that mature
sieve elements, that is, those that lack a nucleus, are
plasmolysable (Fig. 1) and can undergo repeated cycles of
plasmolysis and deplasmolysis if sufficient care is exercised in

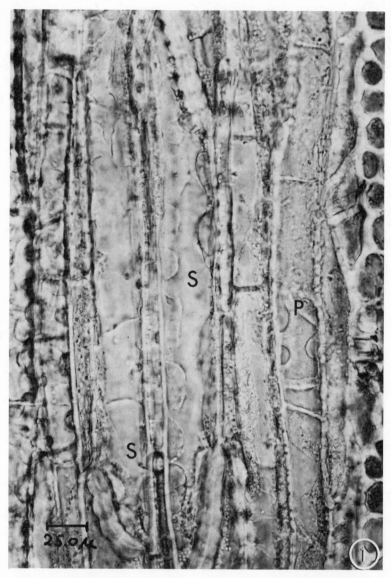

Fig. 1. Tangential section of phloem of <u>Vitis</u> <u>vinifera</u>
showing plasmolysis in the sieve elements (S) and some parenchyma
cells (P). Material collected outdoors on January 6, 1954 in
Davis, California and treated with hypertonic sugar solution.
Figure reproduced by courtesy of the authors, Currier, H.B., Esau,
K. and Cheadle, V.I. and the American·Journal of Botany.

preparation and handling of the material. Moreover, the capacity
of the sieve elements to be plasmolysed or, more precisely, their
sensitivity to injury varies with season. Currier et al (1955)
found that, in <u>Vitis</u>, plasmolysis was more easily and consistently
induced in the protoplasts of mature but dormant sieve elements
than it was in the active elements. The dormant elements differed
from the active elements in that their sieve areas were covered by
masses of callose which lacked any conspicuous connecting strands.
They also found that, whereas in active sieve elements the
protoplast did not separate from the sieve plates at the end walls,
in the dormant sieve elements it separated easily. These are
important observations for they indicate a basic difference in the
behaviour of the plasmalemma in the active vs. dormant sieve
elements. In the active sieve elements the plasmalemma of one
sieve element is continuous across the pores with the plasmalemmae
of the sieve elements above and below, whereas in the dormant sieve
elements, which are heavily callosed, the continuity of plasmalemma
across the pores is broken and each cell is lined individually by
its plasmalemma (Fig. 2).

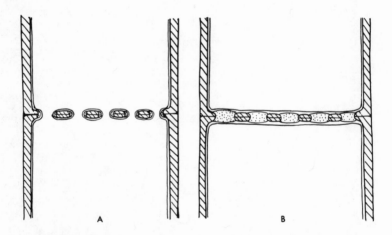

A   B

Fig.2. Drawings showing the relationship between the sieve
area pores, callose, and plasmalemma in the active (A) vs. dormant
(B) sieve elements. The wall is hatched, callose is stippled, and
plasmalemma is drawn as a line.

In agreement with results obtained by Crafts (1932, 1933), and
Rouschal (1941), Currier et al (1955) noted that mature sieve
elements failed to accumulate neutral red, although the young sieve
elements showed a weak reaction and the companion cells gave a
strong positive reaction. Neutral red is a vital stain and is
accumulated selectively in the vacuoles (Bailey, 1930). A lack of
neutral red accumulation is indicative of the lack of a
membrane-delimited vacuole.

Reports on protoplasmic streaming in the sieve elements go far back and have been summarized by Esau (1969). That streaming occurs in the young sieve elements, as in any other parenchyma cell, is beyond question; but from most studies, where critical attention was given to the state of maturity of the sieve element, it seems clear that streaming ceases as the sieve elements mature and is not observed in the enucleate mature cells (Strasburger, 1891; Crafts 1932, 1933; Abbe and Crafts, 1939); a surging motion as a result of osmotic shock, however, has been observed (Bauer, 1953; Currier et al, 1955). In more recent years, streaming of one sort or another has been reported by Thaine (1961, 1962) and Fensom et al (1968). Thaine (1961, 1962) examined free hand sections of fresh petioles of Primula obconica and described transcellular strands, 1-7μ in diameter, extending from cell to cell across the sieve plates, and further reported movement of cytoplasm and particles within the strands. Esau et al (1963) questioned the existence of transcellular strands and considered them to be diffraction artefacts. In subsequent papers, Thaine has tried to prove the existence of transcellular strands and added further details; for instance, that they are delimited by a membrane which is connected to the plasmalemma at the pores and that they are filled with longitudinally oriented fibrils, but these claims are based on chemically-fixed or frozen, not fresh, material (Thaine, 1964; Thaine, Probine and Dyer, 1967, Jarvis and Thaine, 1971; Thaine and de Maria, 1973; Jarvis, Thaine and Leonard, 1973). Fensom et al. (1968) reported streaming within individual sieve elements at rates of 1.5-2.5 cm h^{-1}, occasionally up to 5 cm h^{-1}, in isolated bundles of Heracleum mantegazzianum, but no streaming across the sieve plates. The authors stated that although they could not see the transcellular strands their presence could be inferred by a vibratory motion of particles which were presumably attached to these strands. In a later paper, Lee, Arnold and Fensom (1971) studied "intact functioning sieve tubes" of Heracleum and reported a rapid bouncing motion of plastids and "marker particles" as if they were "attached to or restrained by an invisible network in situ." The marker particles did not move across the sieve plates. The young sieve elements showed protoplasmic streaming along the periphery, probably because they had a vacuole at that time, but "in undamaged mature cells there was evidence neither of a vacuole nor of a transcellular tubule of any size optically detectable." The nearby companion and phloem parenchyma cells showed normal cytoplasmic streaming.

Briefly then, the studies on living sieve elements indicate the presence of a semipermeable membrane around the sieve elements, lack of a vacuole, and lack of cytoplasmic streaming in the mature sieve elements. All these conclusions are supported by the electron microscopic work of the last 10 years. Moreover, the plasmolysability of the sieve elements, in the absence of a vacuole, makes it imperative to visualize one or a series of sieve elements as pressure vessels closed at the two ends. This pressure vessel is easy to imagine for dormant sieve elements with heavily callosed sieve plates at the two ends. For active phloem, one must visualize the plugging of sieve plates at the two ends of the stem block that is sectioned.

III. Studies On Chemically-Fixed Tissues

The electron microscopic studies of chemically-fixed sieve elements have yielded a remarkable wealth of information but are complicated by 2 problems: 1. Conducting sieve elements are under high hydrostatic pressure and the series of sieve elements placed end to end forms a long pressure vessel; the effect of any cut in the series, therefore, is explosive and spreads rapidly through the conduit and is seen in cells far removed from the cut. With care, this effect can be minimized, but cannot be eliminated altogether. 2. Preparatory procedures, particularly fixation with $KMnO_4$, must be viewed with extreme caution. $KMnO_4$ is not strictly a fixative but an electron stain. It does not stabilize soluble proteins nor for that matter neutral lipids, and hence the cytoplasm, nucleoplasm, the internal matrix of plastids and mitochondria, ribosomes, and all other more or less mobile structural components such as microtubules, microfilaments, coated vesicles etc. are not stabilized and are washed out during dehydration and embedment procedures. Its usefulness is as an electron stain for it leaves deposits of MnO_2 and, since everything else is washed out, the more permanent structural components of the cell such as wall, membranes, glycogen, starch etc. stand out in sharp relief. In the absence of the soluble matrix of the cell, however, the spatial relationships of these structures, especially membranes, are often distorted (see also Trump and Ericsson, 1965). Osmium tetroxide, in contrast, cross-links lipids and some proteins especially with double bonds (Stockenius and Mahr, 1965), but it is only the aldehydes: glutaraldehyde, formaldehyde and acrolein, that cross-link proteins and effectively stabilize the otherwise soluble matrix of the cell. Nonetheless, many cell components require special handling for adequate preservation, and soluble carbohydrates and low molecular weight amino acids and proteins that are not part of a macrostructure are almost invariably lost during washing and dehydration.

A. The Young Sieve Element

The young sieve elements show the normal complement of organelles and membrane systems typical of plant parenchyma cells. They have a nucleus and nucleolus, moderate amounts of rough ER of the lamellate type, ribosomes, dictyosomes, mitochondria, plastids, and one or more vacuoles (Fig. 3). Some earlier reports of lack of plastids in the sieve elements of Cucurbita and Tilia were later

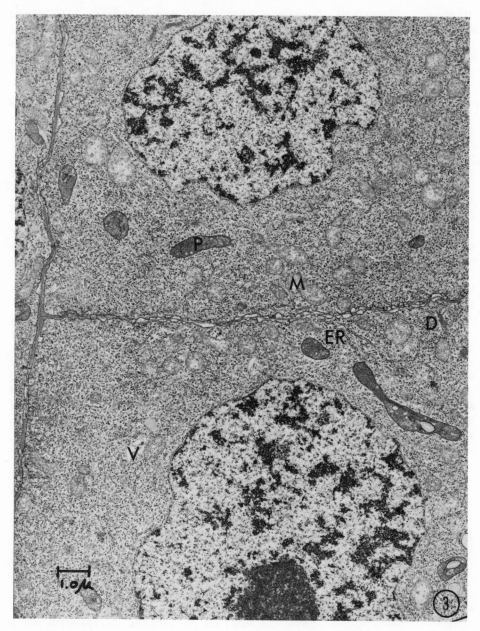

Fig. 3. Parts of 2 young sieve elements in pea (_Pisum sativum_) root. The cells show nuclei, one with nucleolus, plastids (P) with starch, mitochondria (M), ER, dictyosomes (D) and ribosomes in the cytoplasm. Vacuoles (V) are more or less absent in this view.

withdrawn (Esau and Cronshaw, 1968; Evert and Dashpande, 1971) and
another report on lack of plastids in the sieve elements of Smilax
rotundifolia (Ervin and Evert, 1967) is probably incorrect, because
in the electron microscopic study of another species, Smilax
excelsa (Behnke, 1973) plastids are shown clearly.

The young sieve elements also have coated vesicles,
mictrobules, microfilaments, and possibly microbodies.
Microtubules need no introduction. Coated vesicles are known from
a wide variety of plant cells (see Singh and Srivastava, 1973).
They have a uniform size and morphology - approximately 100 mμ in
diameter inclusive of a 20 mμ spoked or alveolate coat - and most
likely arise from the dictyosomes. Despite their wide occurrence,
however, their function remains unknown (but see Bonnett, 1969).
Microfilaments, which occur as bundles of 5-10 nm thick filaments
in the cytoplasm, with their long axis parallel to the long axis of
the cell, are known from various animal and plant cells and slime
molds and have been implicatd in cytoplasmic streaming
(Parthasarathy and Mühlethaler, 1972). They have been reported in
the young sieve elements of palms (Parthasarathy, 1974a) and are
probably present in other species as well. Microbodies are single-
membrane bound packages of different enzymes - enzymes involved
with glycolytic pathway and lipid digestion (Tolbert, 1971;
Trelease et al, 1971) and are reported from a variety of plant
cells (Frederik et al, 1970). They are present in the young sieve
elements (Fig. 8), though in published papers they have not yet
been reported (see Singh and Srivastava, 1972). Still another
structure, spiny vesicle (Newcomb, 1967; Esau and Gill, 1970), has
been reported from the cytoplasm of some phloem parenchyma cells,
but not sieve elements.

B. Sieve Element Differentiation

1. During differentiation, changes occur in nearly all the
organelles and membrane systems. Roughly, there are two rather
distinct but continuous stages in differentiation which may
overlap. The first is a synthetic stage and involves an increase
in the rough ER and polyribosomes, an increase in the dictyosome
activity and wall synthesis, a change in shape and inclusions of
plastids, and synthesis of callose in areas of wall destined to
become sieve pores. In those species that have P-protein, and
these include all the investigated dicotyledons and at least some
monocotyledons, this stage is also characterized by the synthesis
of P-protein. The second stage is marked by a controlled autolysis
of the protoplast, the ER undergoes marked changes in its
morphology and disposition, the nucleus and nucleolus degenerate or
disappear, the plastids lose their internal stroma, the cytoplasm,
ribosomes, dictyosomes, coated vesicles, microtubules,
microfilaments degenerate or disappear, the tonoplast breaks down
or disappears, and the sieve pores are formed by a selective

removal of the wall material. Structures that seem to undergo the
least changes are mitochondria and plasmalemma.   This generalized
overview of sieve element differentiation is valid for nearly all
angiosperms and gymnosperms that have been investigated, though
there are variations among species in respect to the synthetic
activity of one or the other cell component and timing of synthetic
or autolytic activities.   With this overview in mind, let us
examine the changes in cell organelles and membranes in detail.

       2. Dictyosomes: One of the most characteristic features of
sieve element differentiation is the thickening of the wall.
Associated with this thickening is an increase in the number of
dictyosomes and their smooth-surfaced vesicles. Although it is
believed on good grounds that these structures are involved in
synthesis and transfer of non-cellulosic polysaccharides to the
thickening wall, a direct visualization of such transfer in the
sieve elements by electron or light microscopic autoradiography has
not been attempted.   The sieve elements at this stage also have
coated vesicles (Zee, 1969b; Singh and Srivastava, 1972;
Parthasarathy, 1974a) which likewise arise from the dictyosomes,
but their function is unknown.   Microtubules are common in the
parietal cytoplasm and are often arranged normal to the long axis
of the cell (Singh and Srivastava, 1972, Parthasarathy 1974a).

       3. Plastids: The plastids of young sieve elements have the
usual double membrane, a few internal lamellae or thylakoids, and a
fibrillar, relatively electron-dense stroma. Occasional
osmiophilic droplets or plastoglobuli may be present in the stroma,
and in some plants, where membrane-bound osmiophilic inclusions
occur in the plastids of procambial or cambial cells, the plastids
of young sieve elements may also show these inclusions for some
time (Fig. 4). As far as is known, the sieve element plastids do
not show the thylakoid aggregates or grana, typical of
chloroplasts.   They may be called proplastids or leucoplasts,
though the term plastid seems adequate.   Since the presence of
grana is intimately linked to the presence of chlorophyll, some of
the earlier reports of light green plastids in the sieve elements
of Clematis, Viticella, and Foeniculum and of the presence of
chlorophyll in the sieve elements of several aquatics such as
Ceratophyllum, Elodea, Potamogeton, Vallisneria (for references,
see Esau, 1969) must be viewed with caution. At least in Clematis,
Elodea, and Potamogeton, which have been examined electron
microscopically, chloroplasts have not been recorded (Currier and
Shih, 1968; Behnke, 1972) though the neighbouring cells in Elodea
have regular chloroplasts (see Fig. 13 of Currier and Shih, 1968).

       In very young sieve elements or their precursor procambial or
cambial cells, the plastids may be amoeboid or oblong; but soon
they enlarge and round up, and at the same time acquire one or more
different kinds of inclusions, starch (Figs.   3,4), proteinaceous

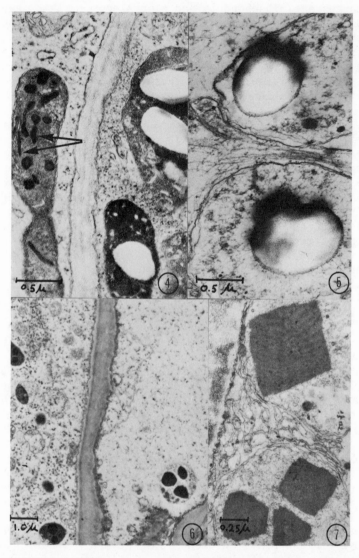

Fig. 4. Plastids in young sieve elements of ash (Fraxinus americana) show membrane-bound osmiophilic inclusions (arrows) and starch. Fig. 5. Plastids in mature sieve elements of ash. Starch grains and a double outer membrane are seen but the plastid stroma is disorganized. Fig. 6. Plastids in young (left) and somewhat older but not yet mature (right) sieve elements of corn (Zea mays). Fig. 7. Plastids in the mature sieve elements of pine (Pinus strobus). This figure is reproduced by courtesy of Protoplasma.

crystalloids (Figs. 6,7), concentrically arranged proteinaceous fibrils, even phytoferritin. Behnke (1972) has made a systematic study of the sieve element plastids and distinguishes 2 basic types and several subtypes. The S-type plastids store starch only and this type occurs in a great number of dicotyledons and has not yet been recorded in monocotyledons. The P-type plastids store protein in form of concentrically-arranged fibrils as in many other dicotyledons (Beta, Tetragonia, members of Caryophyllales) or in form of one or more crystalloids as in most monocotyledons and gymnosperms (see also Singh and Srivastava, 1972; Behnke and Paliwal, 1973; Evert et al, 1973; Parthasarathy, 1974a, b). The presence of starch in P-type plastids is not excluded.

The sieve element starch characteristically stains red or reddish brown with I_2KI, though occasionally blue or violet colouration has been reported (see Esau, 1969). The red colour is associated with low molecular weight polymers of dextrin or amylopectin type and, recently, by a combined digestion with bacterial pullulanase, which is an α-1,6-glycosidase, and α- or β-amylase, Palevitz and Newcomb (1970) confirmed that the sieve element starch was a highly branched polymer of α-glucose with 1,4- and 1,6-linkages. This starch also gives a positive PAS reaction. In electron micrographs, depending upon the fixation procedures, the starch may appear white or varying shades of gray to black. The black colouration is usually seen in $KMnO_4$ fixations and occasionally in glutaraldehyde/OsO_4-fixed material if the osmium has not been washed out thoroughly. The identity of proteinaceous inclusions is based on staining with mercuric bromophenol blue (Mazia et al, 1953) and acid fuchsin. Enzymatic digestion of the crystalloids or concentric fibrils has not been attempted.

In those plants in which proteinaceous crystals are recorded, the crystals appear very early in the sieve element differentiation and indeed some authors have used their appearance and the accompanying thickening of the wall to distinguish the potential sieve elements from their precursors and neighbouring cells in the vascular tissues (Srivastava and O'Brien, 1966; Behnke, 1969; Evert and Deshpande, 1969; Singh and Srivastava, 1972; Behnke and Paliwal, 1973; Esau and Gill, 1973; Evert et al, 1973; Parthasarathy, 1974a, b).

With the rounding up and acquisition of the distinctive inclusions the internal stroma of the plastid loses its electron density and in the mature sieve elements the various inclusions lie in a more or less electron-transparent interior surrounded by the double plastid membrane. This membrane often appears ruptured in electron micrographs and the inclusions scattered in the cell lumen; apparently the rupture is due to osmotic shock during fixation and/ or embedment.

The sieve element plastids differ sharply in their morphology and inclusions from the plastids in the procambial or cambial cells and the neighbouring cells of the vascular and cortical tissue, but the reasons for these differences are obscure. There seems to be a parallel between the sieve-tube starch and floridean starch of red algae or the so-called juvenile starch in the endosperm and other seed tissues (Esau, 1969), but the proteinaceous inclusions of the fibrillar or crystalline type seem to be unique to the sieve elements (but see Hoefert, as cited by Esau, 1971).   These differences among sieve tube plastids of different plants and among sieve tube plastids and plastids of neighbouring cells in the same plant are indicative of biochemical differences, but more work is needed to show if these plastids are indeed unique to the sieve elements and if their stored products bear a relationship to the translocated photosynthate.

4. <u>Callose</u>: The depostion of callose in formation of the future sieve areas is covered elsewhere in this volume (Cronshaw, 1975).

5. <u>Endoplasmic reticulum and ribosomes</u>: In young cells, the ER occurs in the lamellate form, is studded with ribosomes and is dispersed throughout the cell (Evert and Deshpande, 1969; Esau, 1972; Esau and Gill, 1971, 1973; Behnke, 1973). Some authors have commented that the young sieve elements show an increase in the amounts of rough ER and polyribosomes (Singh and Srivastava, 1972; Parthasarathy, 1974b) and others have noted an increased density and/or basophilia of the cytoplasm (Esau and Gill, 1973) (see also Fig. 8). These judgements are based on observations of comparable material of young sieve elements and their precursor cells and are not quantitative. Also, no attempt has been made to feed labelled amino acids and check their incorporation in the sieve elements by histoautoradiography.

From all accounts, the ER undergoes considerable modifications during the sieve element differentiation. It loses its association with the ribosomes and becomes aggregated in orderly stacks and/or convoluted tubules which move to the cell periphery and eventually occupy a parietal position next to the plasmalemma.   These modifications of the ER have been recognized for long, and have been variously referred to as the endomembrane system (Kollmann and Schumacher, 1964), sieve-tube or sieve-element reticulum (Bouck and Cronshaw, 1965; Srivastava and O´Brien, 1966), ER of ordered structures (Wooding, 1967), "gitterartiger Membranstrukturen" (Behnke, 1968), crystalline fibrils and complexes of membranes (Johnson, 1969), though in some other papers they were either missed or confused with slime fibrils. More recently, following Esau and Gill (1971), they are simply referred to as ER, though various descriptive terms such as aggregated or stacked ER,

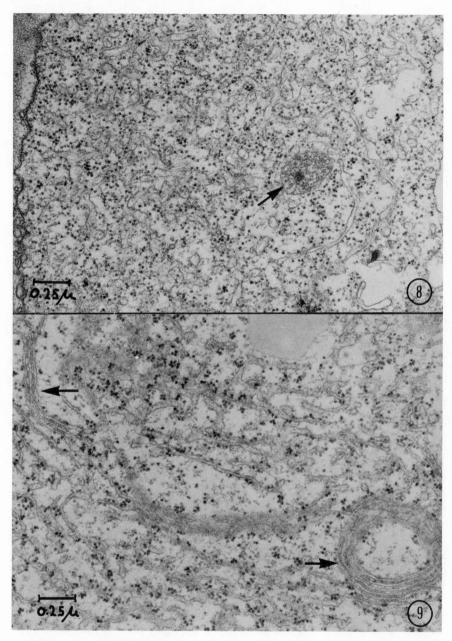

Fig. 8. Part of a young sieve element in a cross-section of
lettuce (Lactuca sativa) hypocotyl. Note the ER and microbody
(arrow.) Fig. 9. Part of a young sieve element in a cross-section
of lettuce hypocotyl. Note the stacking of ER at arrows.

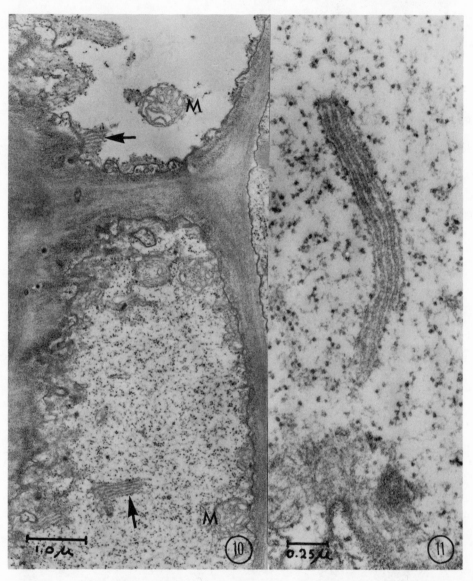

Fig. 10. Parts of 2 sieve elements, one mature (above), the other still with ribosomes and cytoplasm (below) in a longitudinal section of pea root. Note ER stacks (arows) and mitochondria (M). Fig. 11. An ER stack in pea. Part of wall appears at the bottom.

parietal ER, and convoluted or tubular ER have cropped up and are
in danger of being stereotyped.

The aggregation of ER occurs early while the cells have a
nucleus, cytoplasm and ribosomes (Figs. 9, 10). The process of
stacking has been described by several authors and occurs
essentially in the same manner. According to Zee (1969a) and Esau
and Gill (1971), the ER cisternae aggregate in pairs and triplets,
then the stacks increase in depth, the cisternae do not fuse but
are separated by an intercisternal space; simultaneously the
ribosomes are shed from all but the outermost surfaces of the stack
(Fig. 11). These aggregates then move to a parietal position and
lie next to the plasmalemma. By contrast, Parthasarathy (1974b)
reported that in the sieve elements of palms the ER moved to the
periphery, shed its ribosomes, and then aggregated. Evert and
Deshpande (1969) reported that the ribosomes on the cisternal
surface adjacent to the plasmalemma are shed first, those on the
cisternal surface toward the cytoplasm may persist longer.   The
implied difference in timing of aggregation and movement to cell
periphery is probably not important. Most likely, the   two
processes occur simultaneously.   Ribosomes persist on the outer
surfaces of the stacks for some time but eventually they are lost
altogether.

The intracisternal space is often described as clear and the
intercisternal space as well as the space between the outermost
cisternal membrane and the plasmalemma as being filled with an
electron dense material (for details, see Esau and Gill, 1971,
1973). The nature of the intercisternal material is unclear but it
has been suggested that it is P-protein (Zee, 1969b).   Also, the
judgement that the intracisternal space is clear may need
modification, for in many photographs of Esau and Gill (1971,
1973), the intracisternal space harbours some material.

The stacks of ER cisternae may lie parallel or normal or at
various angles to the wall surface within the same cell and the
same stack may show varying orientations in different parts.   The
changing orientation of cisternae in a stack is beautifully
demonstrated in pictures of the same stack taken at different
degrees of tilt in the microscope (Fig. 12A, B). These views also
buttress the concept that the stacks are actually composed of
flattened sacs or discs, much like grana thylakoids.

Behnke (1973) and Behnke and Paliwal (1973) have described the
development of the convoluted ER elements of Smilax and Gnetum,
respectively. According to these authors, the rough lamellate ER,
which was evenly dispersed in the young cell, showed dilation of
intercisternal space and loss of ribosomes at places and eventually
gave rise to smooth tubules of ER which became aggregated and
convoluted into large masses and appeared in a parietal position.

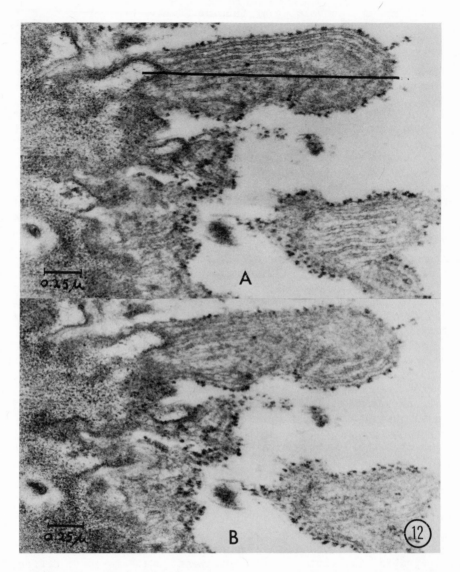

Fig. 12. The same ER stacks in pea in 2 views separated by a 90° tilt. The tilt axis is shown by the straight line. In A, the stack at the top shows the cisternae in sectional view at left and surface view at right; the bottom stack shows them in sectional view. In B, the top and bottom stacks show cisternae more or less in surface view, though in the top stack they appear partly in sectional view at right.

A similar origin of convoluted ER in the sieve elements of Welwitschia was reported by Evert et al, (1973). Esau and Gill (1971) published some excellent photographs of convoluted ER in Phaseolus (Fig. 13).

A parietal ER was distinguished in the mature sieve elements of Cucurbita (Esau and Cronshaw, 1968) and Ulmus (Evert and Deshpande, 1969) and was described as a network closely applied to the plasmalemma and possibly continuous through the sieve plate pores.

The relative amounts of the different forms of ER in mature sieve elements seem to vary among different species and in the same species at different times. For instance, the parietal ER was apparently well developed in the mature sieve elements of Cucurbita (Esau and Cronshaw, 1968) and Ulmus (Evert and Deshpande, 1969). It was present in Mimosa (Esau, 1972), but not in the protophloem sieve elements of Phaseolus (Esau and Gill, 1971) and apparently Allium roots (Esau and Gill, 1973). The stacked forms of ER were encountered only occasionally in Cucurbita (Esau and Cronshaw, 1968), but were extremely common in Phaseolus, where they also showed an unmistakable association with the nuclear envelope, and were more abundant in the late than in the earlier stage of sieve element differentiation (Esau and Gill, 1971). In Mimosa (Esau, 1972) and Allium (Esau and Gill, 1973), in contrast, no marked association with the nuclear envelope was seen; also the stacked forms were less prominent or abundant in older sieve elements than in the younger cells. A possible association of tubular ER with stacked ER was reported in Allium (Esau and Gill, 1973). Evert and Deshpande (1969) noted that the orderly stacks were more common in mature sieve elements of elm than those with an irregular arrangement of cisternae.

How consistent these differences are and their significance, if any, are unknown; but the change from extended and anastomosing lamellae of rough ER with a more or less uniform intercisternal space, typical of young cells, to aggregates of more or less distended but flat cisternae and convoluted tubules of smooth ER of the mature sieve elements (cf. the ER in Figs. 3, 8, 9, 11, 13, 14), must involve, on topographical considerations, a break-up of the lamellae into shorter, closed units, either flat sacs or tubes. At least in some species, this break-up is accompanied by a dilation of the intercisternal space (see Srivastava and O'Brien, 1966; Behnke, 1973; Behnke and Paliwal, 1973); in others, the dilation may not be so evident. Whether the ER in the mature sieve elements should be referred to as the sieve-tube or sieve-element reticulum (Bouck and Cronshaw, 1965; Srivastava and O'Brien, 1966) or simply as ER, with descriptive terms such as stacked, convoluted, or perietal ER, (see Esau and Gill, 1971) is not important. What is important is that this ER differs materially

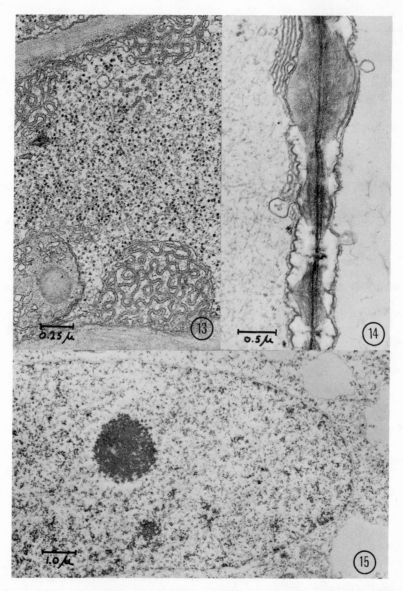

Fig. 13. Convoluted ER in a differentiating sieve element of bean (Phaseolus vulgaris). Figure reproduced by courtesy of the authors Esau, K. and Gill, R.H. and the Journal of Ultrastructure Research.

Fig. 14. Aggregated and parietal ER in mature sieve elements of ash. Fig. 15. Part of nucleus of a young sieve element of corn. The nucleolus shows granular and fibrillar regions.

from the ER in the young cells in aggregation, loss of ribosomes, and an association with the plasmalemma and that the different forms it takes in the mature sieve element are perhaps variations on this basic theme determined by local conditions in the cell or, what is less likely, preparatory procedures. Finally, the total amounts of this ER in mature cells may vary. Some dicotyledons and gymnosperms seem to have it in large amounts, whereas members of the grass family show it only in small amounts.

Several questions about the ER of mature sieve elements remain unanswered: Is it bound to the plasmalemma or is it mobile? We do not know. Is there any significance in the arrangement of the stacks, that is whether they are arranged parallel or normal to the wall surface? Since different arrangements may be seen within the same cell, not only that, but stacked and convoluted forms may occur side by side, it seems unlikely that there is any functional significance to the arrangement or specific form of the ER (See also Esau and Gill, 1973; Behnke, 1973). What is the function of this ER? As reviewed later, it is possible that it has a role in the final autolysis of the cell.

6. <u>Nucleus and nucleolus</u>: It is generally agreed that the mature sieve elements lack a functional nucleus. From the light microscopic literature summarized by Esau (1969) and the treatise by Salmon (1946), who surveyed nuclear degeneration in 52 species of dicotyledons and 2 species of monocotyledons, there seem to be 2 major ways by which this is accomplished. In some species the nuclei enlarge, lose their chromatin and eventually disintegrate; in others, they undergo what is known as pycnotic degeneration (also referred to as necrosis) - they show an increase in condensation of chromatin, lobe and fragment and parts of them may persist in the mature sieve elements. The nucleoli are reported either to disintegrate or in some species persist as the so-called extruded nucleoli.

The electron microscopic investigations of the last 10 years have confirmed the 2 trends of nuclear degeneration and added some details, but they have not been correlated with Feulgen or other histochemical staining at the light microsope level and, hence, the details of the degenerative process, the loss or "extrusion" of nucleolus, and, more important, the early changes in nucleus and nucleolus that herald the sieve element differentiation remain unknown. Let me clarify this last point. In recent years, several authors have studied the nuclear and nucleolar changes in differentiating tracheary cells and have reported a 2-, 4-, or even 8-fold increase in the nuclear DNA and histones (Torrey and Fosket, 1970; Phillips and Torrey, 1973; Lai, 1974). Correlated fine structural studies have revealed that these changes are accompanied by a several-fold increase in the nuclear and nucleolar volume and an elaboration of the nucleolar fine structure into granular, fibrillar, and vacuolar zones, typical of nucleolei engaged in ribosomal RNA synthesis. Finally, what is most interesting, these changes in DNA levels and nuclear and nucleolar volume and fine structure are completed very early, before any overt signs of cell differentiation, such as secondary wall formation, are evident. Indeed, when wall deposition starts the nucleus and nucleolus already show signs of degeneration - aggregation of chromatin into coarse lumps, loss of nucleoplasmic matrix, and the conversion of nucleolus into a dense mass without fine structural detail (Lai, 1974).

Studies correlating changes in DNA, RNA and histones with fine structure of nucleus and nucleolus in the sieve elements are lacking and whatever observations are available deal mainly with nuclear degeneration with only a passing word on nucleolus.

The nuclei of the young sieve elements have the double nuclear envelope with pores and small patches of condensed chromatin which is evenly dispersed throughout the nucleoplasm and at the inner margins of the nuclear envelope except at the pores (Fig. 15). The nucleoplasm itself has a gray fibrillar consistency and shows ribonucleoprotein (RNP) particles. In these particulars, the nucleus of the young sieve elements is no different from that in any other reasonably metabolically active plant or animal cell. The nucleolus of the young sieve elements has not been described and the only references are to its size. For instance, Esau (1972) reported that in the young sieve elements of Mimosa the nucleoli are small and rarely seen and associated this feature "with a reduced nucleolar RNA synthesis". In Allium, in contrast, Esau and Gill (1973) reported a prominent nucleolus in the young sieve elements, but in the slightly older "slowly elongating sieve elements" the nucleoli tended to be smaller than in the adjacent cells. The authors related it to "an expression of reduced meristematic potentiality".

In some of the earlier papers, there were references to disintegration of chromatin (Bouck and Cronshaw, 1965), decrease in stainability of nuclear contents (Northcote and Wooding, 1966) and destruction of nuclear contents (Esau and Cronshaw, 1968) before the breakdown of the nuclear envelope and loss of the nucleus. More recently, Esau (1972) and Esau and Gill (1971, 1972, 1973) have described nuclear degeneration in the stem and leaf phloem of Mimosa and in the root protophloem of Phaseolus, Nicotiana, and Allium. In the 3 dicots the nucleus was lost altogether though there were differences in detail. In Phaseolus and Nicotiana the chromatin disappeared, and thereafter the nuclear region was "invaded by cytoplasmic ribosomes". Although the authors saw no clear indication of breakdown of the nuclear envelope, on the basis of the "obliteration of distinction between the nuclear region and the cytoplasm" concluded that the nuclear envelope had ruptured.

In <u>Mimosa</u> the chromatin persisted for a time and became highly
dense before disappearing.   At the same time the interior of the
nucleus became increasingly electron-lucent. The nuclear envelope
did   not   rupture until after the cytoplasmic ribosomes had
disappeared and there was no reference to invasion of the nuclear
region by cytoplasmic ribosomes. In <u>Allium</u>, in contrast to the 3
dicots, the nucleus persisted in a degenerated condition in the
mature sieve element. As differentiation progressed, the chromatin
moved to the periphery, became granular, and eventually was
converted into an amorphous, alveolate mass with some membranous
material of unknown origin. As in <u>Phaseolus</u>, the authors reported
an entry of cytoplasmic ribosomes into the nuclear region and
concluded that the nuclear envelope "although not obviously
disrupted", did not "maintain the integrity of the nucleus". In
the final degenerated condition, the nuclei were still discrete,
bounded by a "chromatic envelope" and with an electron-lucent
interior.

     In several other monocots, a retention of a necrotic nucleus
in the mature sieve element has not been reported. Instead a
progressive loss of chromatin and structural detail of the
nucleoplasm and a final loss of the nucleus has been reported.
Behnke (1969) reported a gradual reduction of the nucleoplasm and
local dilations of the nuclear envelope prior to the complete
disintegration of the nucleus in several monocots (<u>Asparagus
officinalis</u>, <u>Dioscorea</u> <u>macroura</u>, <u>D.</u> <u>reticulata</u>, <u>Musa</u> <u>velutina</u>,
<u>Scirpus</u> <u>lacustris</u> and <u>Typha</u> <u>latifolia</u>).   Singh and Srivastava
(1972) reported that in the sieve elements of <u>Zea</u> <u>mays</u> the
chromatin disappeared from the interior of the nucleoplasm and
later from its edges near the nuclear envelope.   They reported
lobing of the nucleus. They did not record the final loss of the
nuclear membrane but suspected that it broke down and the nuclear
contents, if any were left by that time, dispersed in the cell.
Parthasarathy (1974b) studied nuclear degeneration in several palms
and reported a lobing of the nucleus, and a blebbing of the outer
membrane of the nuclear envelope. As differentiation progressed,
the chromatin disappeared, the nucleoplasm became electron
transparent and with the rupture of the nuclear envelope was
invaded by the cytoplasm.

     Among the investigated members of gymnosperms, the nuclear
degeneration seems to be of the pycnotic type. Srivastava and
O´Brien (1966) reported that in the sieve elements of <u>Pinus</u> <u>strobus</u>
the nucleoplasm lost its fine structure, the chromatin became
coarse and sparsely distributed, and, ultimately, the nuclei
appeared as clumped dark masses devoid of structural detail.
Retention of the necrotic nuclei in the mature sieve elements of
<u>Pinus</u> <u>strobus</u> was also reported by Murmanis and Evert (1966).
Similar degenerative changes in the nucleus are reported for the
sieve elements of <u>Pinus</u> <u>pinea</u> (Wooding, 1966) and <u>Metasequoia</u>

(Kollmann, 1966, as cited by Behnke and Paliwal, 1973). In the sieve elements of Pinus sylvestris; Parameswaran (1971) described some elongated electron-dense bodies surrounded at places by a double membrane and interpreted them as P-protein bodies. These bodies are most likely necrotic nuclei. More recently, in Gnetum, Behnke and Paliwal (1973) reported that the nuclear matrix disappeared, the chromatin condensed and there was a reduction in nuclear volume leading to pycnotic nuclei. In Welwitschia, Evert et al, (1973) reported that in the young sieve cells the nuclei contained "clumps of chromatin material, some scattered throughout the granular to finely fibrous matrix, and others located near the nuclear envelope." As differentiation progressed, the inner membrane developed protusions into the nuclear matrix, so that in some profiles parts of the nucleus appeared holey. Finally, with maturity, the chromatin became mostly peripheral in distribution but maintained "its coarse granular appearance." The authors also reported that the nucleus retained an intact envelope with pores. Although the authors do not state it, their figures show a progressive condensation of chromatin and loss of its fibrillar detail concomitant with loss of the nucleoplasmic matrix.

In many papers an association of the ER with the nuclear envelope has been reported (Behnke, 1969; Esau and Gill, 1971; Esau, 1972; Behnke and Paliwal, 1973; Parthasarathy 1974b). In contrast, in others this association was not seen (Singh and Srivastava, 1972; Esau and Gill, 1973). Furthermore, several authors have indicated that in the final stages of nuclear degeneration the ruptured nuclear envelope becomes part of the stacked ER (Bouck and Cronshaw, 1965; Northcote and Wooding, 1966) and along with it moves to a parietal position in the cell (Esau and Gill, 1971, 1972; Parthasarathy, 1974b). An association of the nuclear envelope with mitochondria (Evert et al, 1973) and Golgi bodies (Parthasarathy, 1974b) has been reported also, though the significance of these associations is not clear.

The EM studies reviewed above indicate that the chromatin is either lost or condenses into a pycnotic mass and at the same time the nucleoplasm becomes increasingly electron lucent and ultimately structureless. In those forms in which chromatin is lost the nuclear envelope breaks down and probably becomes part of the ER system. By contrast, in species where pycnotic nuclei or nuclear fragments are retained in the mature sieve elements, the nuclei may continue to be bound by the nuclear envelope and may even show pores. As to how functional these pycnotic nuclei are in transcription of m-RNA has not been determined though, from their general appearance, this seems a very remote possibility.

Through all these studies, the changes in the nucleolus and its ultimate fate have been largely ignored. Several authors have noted in passing that the nucleolus disintegrates or disappears

(Bouck and Cronshaw, 1965; Singh and Srivastava, 1972; Esau, 1972; Esau and Gill, 1973; Parthasarthy, 1974b) but the details of this process have not been observed. The breakdown of nucleolus is a common phenomenon during mitosis, but the parallelism between nucleolar breakdown in mitosis and sieve element differentiation cannot be carried far. In mitosis the nucleolar components, granules and fibrils, fall apart but are probably retained in the cell and the nucleolar organizer is often discernible on the nucleolar chromosomes. The disappearance of the nucleolus during sieve element differentiation, in constrast, is more likely related to a cessation of its function and loss of structural detail as in the tracheary cells.

7. <u>Extruded nucleoli</u>: In light microscopic studies of many dicotyledons, it has been reported that the nucleolus is extruded during nuclear degeneration and persists in the mature sieve element (see Esau, 1969). The extruded nucleoli seem to enlarge and develop protrusions on the surface as soon as they leave the nucleus (Esau, 1969) and in several electron microscopic studies extruded nucleoli with orderly sculptings in the form of fibrils and tubules have been shown. A recent paper by Deshpande and Evert (1970) throws doubt on the existence of these extruded nucleoli. These authors re-examined 5 woody dicotyledons and found that the so-called extruded nucleoli were quasi-crystalline aggregates which arose early in the sieve element ontogeny, when nuclei appeared intact and their nucleoli were clearly visible. The authors rejected the possibility that these inclusions were extruded nucleoli, either in part, or in entirety. Are all mature sieve elements enucleate? Scattered throughout the literature are occasional reports of nuclei in mature sieve elements (see Esau, 1969). Many of these reports, as in gymnosperms, deal with degenerated or pycnotic nuclei, others regard it as an exceptional occurrence (Srivastava and Bailey, 1962; Srivastava, 1970). Recently Evert et al, (1970) have questioned the concept that the mature sieve elements lack a functional nucleus. They examined the secondary phloem of 3 species of Taxodiaceae and 13 species of woody dicotyledons and reported nuclei of "normal appearance" in <u>all</u> mature sieve cells of the 3 taxodiaceous species and in <u>some</u> mature sieve tube members in 12 of the 13 wood dicotyledons, but only in 3 of the 12 species: <u>Robinia pseudoacacia</u>, <u>Ulmus americana</u>, and <u>Vitis riparia</u>, did the sieve elements have "apparently normal nuclei." By Feulgen staining, nuclei were detected in mature sieve elements of the 3 taxodiaceous species but only 2 of the 13 woody dicots (<u>Robinia</u> and <u>Ulmus</u>). The authors did not explain why the "nuclei" in the other 10 species of woody dicotyledons did not stain with Feulgen, nor did they explain how they could use Feulgen staining for material killed in FAA or Craf III, both of which contain formaldehyde, without any precautions for binding free aldehyde. Yet, the authors stated that if nuclei disappear in mature sieve elements of any of these 13 species they

"...do not disappear until either immediately before or sometime after the sieve plates are fully perforated, for nuclei were found in sieve elements at all stages of sieve plate perforation ...   ". Perhaps for the same reason, Evert et al, (1973) preferred to call the obviously pycnotic nuclei in <u>Welwitschia</u> as the nuclei of the mature sieve elements and emphasized that they were bound by a double membrane with pores.

8. <u>Mitochondria</u> - In the earlier literature, several authors reported partly degenerated mitochondria in the mature sieve elements, whereas others reported that they were fairly normal (see review in Esau, 1969). In comparison with mitochondria in the neighbouring companion cells or young sieve elements, it seems that mitochondria do undergo some swelling and perhaps loss of cristae but otherwise they remain intact. Whether or not they remain functional in the sense that they have the enzymes associated with oxidation of pyruvate and production of ATP has not been demonstrated cytochemically, although the techniques for electron microscopic localization of several mitochondrial enzymes have been available for some years. In this connection it is interesting to recall that Bauer (1953) and Currier et al, (1955) were unable to detect a reduction of tetrazolium to its coloured formazan in the sieve elements, a reaction which is taken as a rough indicator of mitochondrial respiration. However, those results may have been due to the fact that the tetrazolium reaction is often inconclusive and, if the mitochondrial activity is low, the coloured formazan may be undetectable under the light microscope by unaided eye.

9. <u>The vacuolar membrane and plasmalemma</u> - In conformity with the light microscopic observations, most electron microscopic investigations have also reported that the vacuolar membrane breaks down and in the mature sieve elements there is no precise vacuole. Whereas this observation agrees well with the known inability of protoplasts of mature sieve elements to accumulate neutral red, in several papers Evert and co-workers have raised the possibility that some membrane continues to delimit the parietal cytoplasm from the central cavity (Evert et al, 1969 and earlier references cited therein) at least until after the dispersal of P-protein (Evert and Deshpande, 1969). In the more recent papers on phloem of <u>Welwitschia</u> and some lower vascular plants, however, the vacuolar membrane is reported not to be discernible in the mature sieve elements (Evert et al, 1973; Kruatrachue and Evert, 1974; Warmbrodt and Evert, 1974).

The presence of a membrane in the mature sieve element, in the position one associates with the vacuole, is by itself no indication that it is the vacuolar membrane or that a membrane-delimited vacuole is present. Membrane behaviour in fixation and embedment is not precisely known, but they can break apart and fuse again at sites removed from their original site;

also, there is no a priori reason to believe that the vacuolar membrane, if broken, is completely dissolved during the final autolysis (see below). Furthermore, evaginations of plasmalemma, the so-called plasmalemmasomes, are known from a wide variety of plant cells and, in isolated sections, are often seen as circular profiles with no obvious connection to the plasmalemma. Accumulation of neutral red or some other vital dye remains a crucial test for vacuolar integrity in mature sieve elements, unless selective histological stains at the EM level are developed for the various membranes.

Throughout the above changes in cell protoplast, the membrane system that seems to be the least affected is the plasmalemma. It lines the wall and is continuous across the sieve pores and in good photographs continues to show the typical trilaminate structure of the unit membrane. Of course, the trilaminate structure under the electron microscope by itself is no indication that it has not been altered chemically. Only high resolution microscopy combined with cytochemistry holds any hope that any changes, if they occur, will be revealed.

10. Autolysis of the protoplast: As to what causes the final controlled autolysis of the protoplast and parts of cell wall, whether one or a series of hydrolases are produced, and where they come from are questions for which we have no answer. It seems clear that the enzyme or enzymes, whatever they be, come from the sieve element itself and that the sieve element protoplast is programmed to produce them after certain synthetic activities have been completed. Although occasional microbodies are seen in the young sieve elements, no lysosome-like particles have been observed in the maturing sieve elements (but cf. Zee, 1969c). Attention, therefore, has turned to the endoplasmic reticulum for the production of hydrolyzing enzymes and possibly their storage until use (Esau and Gill, 1972, 1973). In this connection it is pertinent to recall that during xylem vessel differentiation parts of ER with attached ribosomes persist in cells undergoing autolysis and have been implicated in the production of hydrolyzing enzymes (Srivastava and Singh, 1972). The demonstration of acid phosphatase in the ER and ER derived vacuoles of pea (Zee, 1969c) may be relevant in this connection. Most likely, the release of hydrolyzing enzymes affects different organelles and membrane systems differently depending upon their chemical compositions. Some are completely autolyzed, others partially, and still others perhaps not at all. The apparent differences in the relative timing of autolysis of various cell components in different species may have the same explanation. Enzyme histochemistry, after the enzymes have been released, however, is of little value in localizing the site of enzyme production, for they are bound to adhere to membrane surfaces during preparatory procedures.

58

L. M. SRIVASTAVA

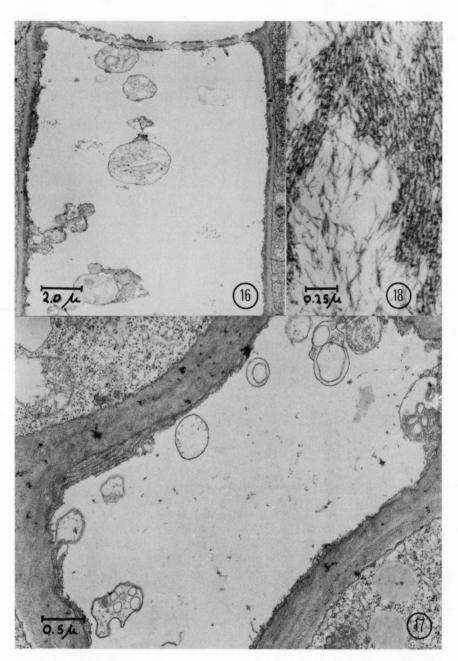

Fig. 16. Parts of 2 mature sieve elements in a longitudinal section of pea root. Fig. 17. Part of a mature sieve element in a x.s. of lettuce hypocotyl. Fig. 18. Tubular P-protein in the mature sieve element of ash. P-protein is absent in Figs. 16 and 17.

In summary then, the young sieve elements in angiosperms and gymnosperms are typical parenchyma cells, but during differentiation nearly all organelles and membrane systems undergo changes. The mature cells have a relatively thick wall lined with plasmalemma and a modified ER which consists of aggregates of flat saccules or convoluted tubules devoid of ribosomes. There are, in addition, mitochondria and plastids, often enmeshed in the ER, and in those species which have P-protein tubular and/or fibrillar, or crystalline forms of P-protein (Figs. 16-18). There is no cytoplasm in the sense of the mobile matrix substance that is present in young cells and there are no ribosomes, dictyosomes, coated vesicles, microtubules, microfilaments, microbodies, and no discernible vacuole. Nucleus and nucleolus degenerate and disappear, but persist in some species as pycnotic but most likely nonfunctional masses. Indeed, this picture of the mature sieve element following aldehyde/OsO$_4$ fixation is so common that there is no problem identifying a sieve element in a cross- or longitudinal section of a plant tissue.

How faithful to the natural state is this picture of the mature sieve element? What about the transcellular strands of Thaine and can this picture be improved by new techniques? First, aldehyde or aldehyde/OsO$_4$ fixation and associated embedment and staining procedures have now been tried by so many different workers on so many different types of plant and animal tissues and the results obtained from electron microscopy have so often been confirmed by cytochemical and biochemical analyses, that there is no doubt whatever that the picture of a cell obtained by a competent microscopist is as faithful a reproduction of the natural state as we are going to get by the available techniques. Admittedly certain cell components are more labile than others, for instance microfilaments, microtubules and possibly P-protein need special handling, and some others, such as soluble carbohydrates and low molecular weight amino acids and proteins, are almost invariably lost. Lipids may be lost if osmium is not used. Moreover, there are problems in embedment, sectioning, staining and during microscopy with contamination, astigmatism, even focussing, but these are minor problems and are taken or should be taken into account during interpretation of results. Finally, the margin of error in estimating size of objects is still of the order of 10-20% unless very special precautions are taken. These handicaps notwithstanding, transmission electron microscopy of chemically fixed material has presented a fairly consistent picture of differentiation and mature structure of the sieve elements, a picture which is in agreement with the 3 main conclusions I outlined earlier from studies of the living sieve elements.

There seems little doubt that the fibrils reported by Thaine and his students in the "transcellular strands" are the slime or P-protein fibrils, but as to membrane-bound transcellular strands of 1-7 µ in diameter, electron microscopy has found no evidence for existence of these structures. The claim that these membranes are flimsy and easily destroyed seems hardly credible, for most other membranes with a lipoproteinaceous framework are easily preserved. Indeed, membranes are among the hardest things to destroy in conventional microscopy. What is more pertinent, in freeze-killed and etched preparations also no membrane bound transcellular strands have been recorded though fibrils have been (Johnson, 1973). Robidoux et al, (1973) used acrolein in dimethyl sulfoxide (DMSO) as a primary fixative in the hope that the rapidity of penetration of the chemicals would preserve the transcellular strands. These authors also found no transcellular strands, though fibrils comparable and perhaps identical to P-protein fibrils were seen extending long distances in the cell.

What are the avenues for future research in structure and development of sieve elements? From a functional point of view, I think one of the chief gaps in our knowledge is the uncertainty that surrounds the structure of the conducting cell. We know the structure of the young and mature cells and some of the stages in differentiation but which of these structures is the one that conducts has not been determined unequivocally. Another avenue of fruitful research is in electron microscopic cytochemistry, particularly with a view to determining how functional the sieve element mitochondria are as a source of ATP and where the enzyme(s) for final autolysis is (are) produced and located. Finally, from the point of view of differentiation and its control, the quantitative cytochemistry of DNA, RNA and proteins in the nucleus and nucleolus correlated with fine structural changes holds great promise.

References
ABBE, L.B. and CRAFTS, A.S. 1939. Bot. Gaz. 100: 695-722
BAILEY, I.W. 1930. Anatomie 10: 651-682
BAUER, L. 1953. Planta 42: 367-451
BEHNKE, H-D. 1968. Protoplasma 66: 287-310
BEHNKE, H-D. Protoplasma 68: 289-314
BEHNKE, H-D. 1972. Bot. Rev. 38: 155-197
BEHNKE, H-D. 1973. Protoplasma 77: 279-289
BEHNKE, H-D. and PALIWAL, G.S. 1973. Protoplasma 78: 305-319
BONNETT, H.T. JR. 1969. J. Cell Biol. 40: 144-159
BOUCK, G.B. and CRONSHAW, J. 1965. J. Cell Biol. 25: 79-96
CRAFTS, A.S. 1932. Plant Physiol. 7: 183-225
CRAFTS, A.S. 1933. Plant Physiol. 8: 81-104
CRONSHAW, J. 1975.
CURRIER, H.B. and SHIH, C.V. 1968. Amer. J. Bot. 55; 145-152
CURRIER, H.B., ESAU, K. and CHEADLE, V.I. 1955. Amer. J. Bot. 42: 68-81
DESHPANDE, B.P. and EVERT, R.F. 1970. J. Ultrastr. Res. 33; 483-494
ERVIS, E.L. and EVERT, R.F. 1967. Bot. Gaz. 128: 138-144

ESAU, K. 1969. Handbuch der Pflanzenanatomie, Vol. V2. Borntraeger, Berlin

ESAU, K. 1971. J. Indian Bot. Soc. Golden Jubilee Vol. 50A: 115-129

ESAU, K. 1972. Ann. Bot. 36: 703-710

ESAU, K. and CRONSHAW, J. 1968a. Can. J. Bot. 46: 877-880

ESAU, K. and CRONSHAW, J. 1968b. J. Ultrastr. Res. 23: 1-14

ESAU, K. and GILL, R.H. 1970. Protoplasma 69: 373-388

ESAU, K. and GILL, R.H. 1971. J. Ultrastr. Res. 34: 144-158

ESAU, K. and GILL, R.H. 1972. J. Ultrastr. Res. 41: 160-175

ESAU, K. and GILL, R.H. 1973. J. Ultrastr. Res. 44: 310-328

ESAU, K., CURRIER, H.B. and CHEADLE, V.I. 1957. Ann. Rev. Plant Physiol. 8: 349-374

ESAU, K., ENGLEMAN, E.M. and BISALPUTRA, T. 1963. Planta 59: 617-623

ESCHRICH, W. 2970. Ann. Rev. Plant Physiol. 21: 193-214

EVERT, R.F. and DESHPANDE, B.P. 1969. Protoplasma 68: 403-432

EVERT, R.F. and DESHPANDE, B.P. 1971. Planta (Berl.) 96: 97-100

EVERT, R.F., TUCKER, C.M., DAVIS, J.D. and DESHPANDE, B.P. 1969. Amer. J. Bot. 56; 999-1017

EVERT, R.F., DAVIS, J.D., TUCKER, C.M. and ALFIERI, F.J. 1970. Planta (Berl.) 95: 281-296

EVERT, R.F., BORNMAN, C.H., BUTLER, V. and GILLIAND, M.G. 1973. Protoplasma 76: 1-21

FENSOM, D.S., CLATTENBURG, R., CHUNG, T., LEE, D.R., ARNOLD, D.C. 1968. Nature 219: 531-532

FREDERICK, S., NEWCOMB, E.H., VIGIL, E.L. and WERGIN, W.P. 1968. Planta (Berl.) 81: 229-252

JARVIS, P. and THAINE, R. 1971. Nature New Biol. 232: 236-237

JARVIS, P., THAINE, R. and LEONARD, W. 1973. J. Exptl. Bot. 24: 905-919

JOHNSON, R.P.C. 1969. Planta (Berl.) 84: 68-80

JOHNSON, R.P.C. 1973. Nature 244: 464-466

KOLLMAN, R. and SCHUMACHER, W. 1964. Planta 63: 155-190

KRUATRACHUE, M. and EVERT, R.F. 1974. Amer. J. Bot. 61: 253-266

LAI, V.H-Y. 1974. M. Se. thesis, Simon Fraser University, Burnaby, B.C.

LEE, D.R., ARNOLD, D.C. and FENSOM, D.S. 1971. J. Exptl. Bot. 22: 25-38

MAZIA, D., BREWER, P.A. and ALFERT, M. 1953. Biol. Bull. 104: 57-67

MURMANIS, L. and EVERT, R.F. 1966. Amer. J. Bot. 53: 1065-1078

NEWCOMB, E.H. 1967. J. Cell Biol. 35: C17-C22

NORTHCOTE, D.H. and WOODING, F.B.P. 1966. Proc. Roy. Soc. (London) Ser. B. 163: 524-537

PALEVITZ, B.A. and NEWCOMB, E.H. 1970. J. Cell Biol. 45: 383-398

PARAMESWARAN, N. 1971. Cytobiologie 3: 70-88

PARTHASARATHY, M.V. 1974a. Protoplasma 79: 59-91

PARTHASARATHY, M.V.   1974b.   Protoplasma 79:   93-125
PARTHASARATHY, M.V.   and MUHLETHALER.   1972.   J.   Ultrastr.   Res.
    38:   46-62
PHILLIPS, R.   and TORREY, J.G.   1973.   Developmental   Biol.   31:
    336-347
ROBIDOUX, J., SANDBORN,   E.B.,   FENSOM,   D.S.   and   CAMERON,   M.L.
    1973.   J.   Exptl.   Bot.   24:   349-59
ROUSCHAL, E.   1941.   Flora 35:   135-200
SALMON, J.   1946.   Recherches cytologiques.   Copyright 1947   by   J.
    Salmon
SCHMITZ, K.   and SRIVASTAVA, L.M.   1974a.   Cytobiologie (in press)
SCHMITZ, K.   and SRIVASTAVA, L.M.   1974b.   Can.   J.   Bot.   (in
    press)
SCHUMACHER, W.   1939.   Jahrb.   wiss.   Bot.   88:   545-553
SINGH, A.P.   and SRIVASTAVA,   L.M.   1972.   Can.   J.   Bot.   50:
    839-846
SINGH, A.P.   and SRIVASTAVA, L.M.   1973.   Protoplasma 76:   61-82
SRIVASTAVA, L.M.   1970.   Can.   J.   Bot.   48:   341-359
SRIVASTAVA, L.M.   and BAILEY, I.W.   1962.   J.   Arnold   Arb.   43:
    234-278
SRIVASTAVA,   L.M.   and O'BRIEN,   T.P.   1966.   Protoplasma   61:
    277-293
SRIVASTAVA, L.M.   and SINGH,   A.P.   1972.   Can.   J.   Bot.   50:
    1795-1804
STOCKENIUS,   W.   and MAHR,   S.C.   1965.   Quantitative   Electron
    Microscopy (ed.   by Bahr, G.F.   and Zeitler, E.H.) The   Williams
    and Wilkins Co., Baltimore, Maryland, pp.   458-469
STRASBURGER, E.   1891.   Histologische Beitrage, heft   III.   Gustav
    Fischer, Jena.
THAINE, R.   1961.   Nature (Lond.) 192:   772
THAINE, R.   1962.   J.   Exptl.   Bot.   13:   152-160
THAINE, R.   1964.   New Phytol.   63:   236-243
THAINE, R.   and DE MARIA, M.E.   1973.   Nature (Lond.) 245:   161-163
THAINE, R., PROBINE, M.C.   and DYER, P.Y.   1967.   J.   Exptl.   Bot.
    18:   110-127
TOLBERT, N.E.   1971.   Ann.   Rev.   Plant Physiol.   22:   45-74
TORREY, J.G.   and FOSKET, D.E.   1970.   Amer.   J.   Bot.   57:
    1072-1080
TRELEASE, R.N., BECKER,   W.M.,   GRUBER,   P.J.   and   NEWCOMB,   E.H.
    1971.   Plant Physiol.   48:   461-475
TRUMP, B.F.   and ERICSSON, J.L.E.   1965.   Quantitative   Electron
    Microscopy   (ed.   by Bahr, G.F.   and Zeitler, E.H.) The Williams
    and Wilkins Co., Baltimore, Maryland, pp.   1245-1323
WARMBRODT, R.D.   and EVERT,   R.F.   1974.   Amer.   J.   Bot.   61:
    267-277
WOODING, F.B.P.   1966.   Planta (Berl.) 69:   230-243
WOODING, F.B.P.   1967.   Planta (Berl.) 76:   205-208
ZEE, S-Y.   1969a.   Aust.   J.   Biol.   Sci.   22:   257-259
ZEE, S-Y.   1969b.   Aust.   J.   Bot.   17:   441-456
ZEE, S-Y.   1969c.   Aust.   J.   Biol.   Sci.   22:   1051-1054

DISCUSSION

Discussant: R.P.C. Johnson
           Dept. of Botany
           Aberdeen University
           Aberdeen, Scotland AB9 2UD

Dr. Johnson was invited to give his paper on "The Distribution of Filaments in Sieve Tubes in Freeze-Etched, Rapidly Frozen Intact Translocating Vascular Bundles". The paper follows.

The Distribution of Filaments in Sieve Tubes in Freeze-Etched, Rapidly-Frozen Intact Translocating Vascular Bundles

R.P.C. Johnson

An outstanding question in the study of translocation in plants with P-protein is "how are the filaments arranged in translocating sieve tubes?"

Weatherley (1972) calculated that if sieve tubes have a low enough resistance to flow for turgor pressure flow alone to account for translocation, then filaments in the sieve pores must be more than 100nm apart; and filaments in the lumen of sieve elements must be bundled together in strands or fibrils.

I have used the freeze-etching method for electron microscopy to examine the _in vivo_ arrangement of filaments in sieve tubes in vascular bundles shown to have been translocating carbon - 14 (see Johnson, 1973). The vascular bundles were frozen rapidly while still intact between leaf and plant. Their temperature was not allowed to rise above minus 75°C until carbon replicas had been made of fracture surfaces through sieve tubes in them.

In this study I have examined a total of 54 sieve plates from 12 petioles on 5 separate plants of _Nymphoides_ _peltata_ (S.G. Gmel.) O. Kuntze. Twenty-five of these sieve plates were frozen, fractured and etched well enough to give useful views. Twenty-two sieve plates showed some filaments in some pores. Thus their occurrence in sieve pores is not an artefact of chemical fixation. Twenty-five pores on 12 sieve plates contained filaments with measurable spacing. Twenty-three pores on 16 sieve plates contained some filaments but with unmeasurable spacing. Five pores

63

Fig. 1. A freeze-etched sieve pore from a translocating vascular bundle. The pore is packed with banded filaments. Banding appears most distinct at arrow. Arrowheads show some filaments arranged across entrance to pore. Fractured plasmalemma appears at m.

in three sieve plates appeared to contain no filaments within the
plane of fracture.   Where a distance between filaments in pores
could be measured it was always less than 100 nm, sometimes much
less.

Filaments were closely packed in some pores (Fig. 1).
Filaments in transverse, and others in longitudinal view appeared
parallel to the sieve plate near some sieve pores (Fig. 1,
arrowheads); possibly these filaments had been displaced.

No sign of callose was seen in most pores, though callose is
obvious in freeze-etched pores treated with glycerol prior to
freezing (Johnson, 1968). Sieve pores of N. peltata fixed with
glutaraldehyde and embedded and sectioned for electron microscopy
contain callose (Johnson, 1968). The absence of callose from sieve
tubes in rapidly frozen, translocating vascular bundles suggests
that they had been disturbed far less than is possible with
conventional chemical fixation; the distribution of filaments seen
in them seems more likely to be as it is when they are
translocating. If the pores are normally so obstructed with
filaments, as most of them appear after mere rapid freezing, then
the turgor pressure-flow hypothesis alone seems inadequate to
account for observed rates of translocation.

Bundles, fibrils or strands of filaments appeared in some
freeze-etched sieve elements (see Johnson, 1973, Fig.2).   Long
lengths and, more often, cross sections of bundles of filaments
appeared in fixed, embedded and sectioned material too. Long
lengths of such bundles would appear only if they happened to lie
in the plane of a thin section or along a fractured surface.
Scanning electron microscopy of sieve elements of N. peltata fixed
with glutaraldehyde, and dried from acetone and Freon 13 by the
critical-point method to avoid damage caused by surface tension,
shows many fibrils or strands near sieve plates (Fig. 2). In the
scanning electron microscope these strands often appear to run from
both sides of a sieve plate to networks of fibrils (Fensom, et al.,
in preparation) in the parietal layer which lines the sieve
elements (Fig. 2). Also, I have found similar strands in cut,
fresh sieve elements mounted in tap water and cine-filmed under
Nomarski differential interference microscopy. The fact that such
strands appear on both sides of sieve plates suggests that they are
not caused by release of turgor pressure as sieve tubes are
prepared. The fact that bundles of filaments occur in
freeze-etched translocating phloem suggests that such strands are
normal in translocating sieve tubes.

The "transcellular strands" described from light microscopy by
Thaine and his co-workers (see Jarvis, Thaine and Leonard, 1973)
are probably bundles of filaments also. Thaine has speculated that
his strands should be surrounded by a membrane or other distinct

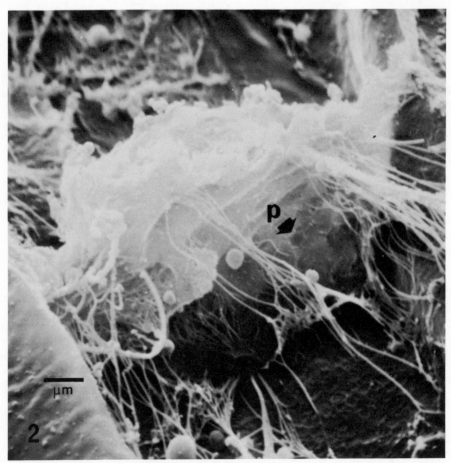

Fig. 2. Scanning electron-micrograph of a sieve plate from a glutaraldehyde fixed, cut, and critical-point dried vascular bundle. Fibrils, or strands, extend from the sieve plate to a network of fibrils lining the sieve elements. A sieve pore appears at P. I am grateful to Mr. Alan Freundlich, who prepared and photographed this specimen, for permission to use it.

boundary. But freeze-etching shows that, at least in <u>N</u>. <u>peltata</u>, the strands have no membranous boundary nor any special arrangement of filaments at their edges; they are merely bundles of filaments (Johnson, 1973).

Filaments are to be expected to aggregate together in parallel in bundles when fluid flows past them as it probably does in the sieve tube. They will tend to aggregate because pressure between them will be reduced by the Bernoulli effect. The low pressure between adjacent filaments will cause them to aggregate side-by-side. Filaments will tend to aggregate in this way whether a flow is driven over them by some mechanism at their surfaces or is caused in some other way, perhaps by turgor-pressure flow. Flow was shown to aggregate filamentous material into strands in a model sieve element (Johnson, 1967). Calculations show that filaments aggregated in bundles in sieve elements will present far less resistance to flow than the same number of filaments distributed uniformly (Weatherley, 1972).

Thus it seems, for a variety of reasons, that filaments of P-protein in sieve tubes are likely to be aggregated in bundles when the sieve tubes are actually translocating. The apparently uniform distribution of filaments sometimes seen in electron micrographs may be a relaxed state formed, perhaps by random thermal motion, after the flow of translocation has stopped.

The possibility remains that filaments are normally bundled in pores also and may have time to disperse to give the apparently uniform distribution I find in freeze-etched sieve pores, even during the fraction of a second taken to freeze them. If this possibility is admitted then one must doubt whether a convincing answer to the question of how filaments are arranged in translocating sieve pores can be obtained by present forms of electron microscopy. A kind of "uncertainty principle" seems to operate; one may prove either that sieve tubes are translocating, or one may examine their structure, but not both together.

In future a greater understanding of how the structure of the sieve tube is related to its function in translocation may be obtained from the study of hydrodynamics and patterns of flow through filaments and pores in working model sieve elements.

References

JARVIS, P., THAINE, R., and LEONARD, J.W. 1973 J. exp. bot. 24: 905 - 919

JOHNSON, R.P.C. 1967. Ph.D. Thesis, University of Aberdeen, Scotland

JOHNSON, R.P.C. 1968. Planta (Berl.) 81: 314-332

JOHNSON, R.P.C.   1973.   Nature (London) 244:   464 - 466
WEATHERLEY, P.E.   1972.   Physiol. Veg.   10:   731 - 742

Hebant: I have recently tried to feed some stem fragments of Sequoiadendron and Metasequoia with tritiated uridine. Preliminary results indicate that at least a large number of nuclei which persist in mature sieve elements are no longer capable of incorporating ^{3}H uridine, whereas the nuclei of the neighbouring parenchyma cells are very strongly labelled. I should also mention that Catesson and Lieberman-Maxe have just demonstrated that cytochrome oxidase is present in mitochondria of mature sieve elements.

Dute: Could both types of nuclear degeneration occur in the same species?

Srivastava: I do not know.   I do not think anybody has described the two processes in the same species.

Hebant: Both pycnotic degeneration and dechromatinization occur in mosses, as well as in ferns.   I have found that protophloem very often shows degeneration of the pycnotic type, whereas the metaphloem sometimes shows both chromatolysis or pycnotic degeneration.

Fisher: I can assure you beyond a shadow of doubt that mature sieve tubes conduct radioactive sugars.

Johnson: How can you do that?

Fisher: Certainly these procedures with plastic embedment have sufficient resolution to tell you which sieve elements are conducting. It is simply a matter of looking at an electron microscope to see what they look like.

Johnson: You then could not argue that you are merely showing the accumulation of these things and not the movement of them?

Fisher: You could, but it is a very weak argument because almost all of the sieve tubes are mature and virtually all of them are labelled.

Swanson: When you find these P-proteins, are they uniformly distributed throughout the lumen, are they considered to be anchored in some way or are they pre-floating, so to speak?

Cronshaw: It is most probable that the P-proteins do not move down the sieve elements. So probably they are anchored in some way, either as a network throughout the sieve elements or in a prorietal position. I would like to ask Dr. Hebant more about Dr. Catesson's paper. Did they show that ATP was produced by the mitochondria in the mature sieve element, or was it just a demonstration of the enzyme cytochrome oxidase?

Hebant: I am not quite sure.

Cronshaw: Well at least someone has shown that there is some enzyme still working in the mitochondria. I would like to comment on the endoplasmic reticulum system. In most of the preparations that I have looked at, the plasma membrane is almost entirely covered by flattened cisternae of the ER that are closely applied to it. And then throughout the cells we have the stacked aggregates scattered.

Srivastava: There are several types of ER described. There is the aggregated or stacked ER, convoluted or tubular ER, parietal ER, and some people have even distinguished between convoluted and tubular ER. Furthermore, there are different amounts of these things in different plants and at different times in the life history of the sieve elements in the same plant. I think that judgements of that nature are very difficult to make, considering the lack of any statistical significance to these observations.

Cronshaw: The parietal ER poses a problem to us in the cytochemical studies in that it is so prolific and covers the plasma membrane so extensively that we have difficulty in establishing when we get reaction product in that area, whether it is in the ER or in the plasma membrane itself.

Johnson: What is the function of this ER considered to be? In some micrographs that we have obtained the endoplasmic reticulum appears to be intimately connected with crystalline structures which have filaments of P-protein near to them. Does it have anything to do with the origin of P-protein?

Cronshaw: In many pictures of the parietal ER there are often dark-staining filaments in between the tightly opposed cisternae. This dark material, I think, is a form of P-protein. The ER at that stage no longer has ribosomes attached to it, so it probably is not a site for translation, but it may be a site of assembly of P-protein.

Aikman: I would like to raise a question about the plasmolysis of sieve elements. If the concentrated solution is only applied to a limited area, then the contents of the sieve tube will be flowing out there and will be replaced by flow from the rest of the functioning active sieve tube. Therefore, one will not expect to observe plasmolysis within this region. It is only perhaps that one might observe the plasmolysis occurring outside this region or, if one exposed the whole length to the concentrated solution, one might expect to see plasmolysis.

Currier: I think that it is possible to plasmolyze an individual sieve element in the middle of a sieve tube without the ends being plugged. And the explanation is, I think, that the distance that water has to travel laterally to cause the plasmolysis is very short and if we believe that there is some resistance in the water going down the length to prevent plasmolysis then I think the plasmolysis would occur first before the replenishment could arrive.

Johnson: I have noticed that if you take a Nitella cell and pipette concentrated sucrose at it, it plasmolyzes at the point where the concentrated sucrose hits the cell.

Behnke: When you plasmolyze, you put the whole section in the plasmolyticum. So you do not have some part of the section inside plasmolyticum and some part outside.

Srivastava: I should add to what Dr. Behnke just said that in the experiments that I have mentioned, hypertonic solution was not applied to a localized part of a section; rather the entire section was immersed in the hypertonic solution.

Turgeon: In these pictures of the sieve elements, one often sees circular membrane profiles which often occur as concentric pairs. I have noticed them in my work, in some of the pictures you have shown Dr. Srivastava, and wonder what they are.

Srivastava: Are you referring to the things that I have called endoplasmic reticulum, or is it something different that you are thinking of?

Turgeon: I do not know. They are not cisternae. They are just round profiles usually in the lumen of the cell and very often close to the sieve plate.

Cronshaw: You are probably thinking of myelin figures. We do not know what they do. I think almost certainly they are an artefact of preparation. The membranous material that is in these myelin figures is probably endoplasmic reticulum. In some preparations of phloem they are particularly striking in that they are very large and elaborate.

Srivastava: If you are talking about the myelin figures, my interpretation has always been that these are phospholipids from various sources in the cell. When the cell is dehydrated, these phospholipids aggregate in a myelin sheath-type pattern. Whether they come exclusively from the ER type membranes or whether they come from loose phospholipids that may be floating around in the cell, I do not know. But, as Dr. Cronshaw said, I think it is very definitely a fixation artefact and in some cells you see it more often than in others.

Johnson: Is it possible that these structures could give rise to the observations of supposed tonoplasts in sieve elements?

Srivastava: No, I doubt that very much, because these things have a very characteristic disposition; not all membranes aggregate in this particular fashion.

Walsh: In discussing the nucleus, we know that in some cases this nucleus will take on a greatly lobed appearance during the "normal course of ontogeny". Yet I do not remember hearing you talk about lobing unless the lobing was a part of pycnotic type. My second question deals with the tonoplast. I have been observing tonoplasts in sieve elements, at least I think these are tonoplasts, and I was just wondering whether anyone else has observed tonoplasts in mature sieve elements or what might be remants of a tonoplast.

Srivastava: As far as lobing of the nucleus is concerned, the detailed paper of Salmon is a very good paper. She describes nuclear lobing in both types of nuclear degeneration. In the EM studies that have been done, the lobing phenomena has been seen, although not always mentioned. As to what lobing means, I do not know. The common assumption is that lobing results in an increased surface contact between the nucleus on the one hand and the cytoplasm on the other. But this still does not explain anything. The lobing occurs in both cases, but I do not know what it's significance is in nuclear degeneration. Referring to the second question about tonoplast, in my judgement the presence of a vacuolar membrane has significance if the cytoplasm, that is, the soluble matrix of the protoplast, is separated from a central cavity which does not show the cytoplasm. That is what the technical definition of vacuolar membrane or tonoplast is. What we are witnessing here is that the cytoplasm, to a great extent also, disappears during differentiation of the cell. Therefore, in the mature cell you have plasmalemma, endoplasmic reticulum, mitochondria, plastids, remnants of ribosomes once in a while, but really no cytoplasmic matrix. And if there is a membrane present somewhere in this cell, it will have to be a very critical study that shows that the membrane is present all around as well as up

and down the cell and delimits the parietal system of membranes and
organelles from whatever is in the centre. This, to my knowledge,
has not been done.

Walsh: I would like to show some slides and whether these are
tonoplasts or not. But they seem to fit Dr. Srivastava's
definition. (At this point Dr. Walsh showed a number of slides of
Lemna minor and Zea mays sieve elements.)

Johnson: This seems to be a convincing demonstration of the
tonoplast. The question is, are these sieve elements
translocating?

Walsh: I do not know and I also do not know whether there is
ever going to be an answer to that type of question.

Schmidt: I have looked at some brown algae including
Macrocystis, and the young sieve elements in these plants are just
packed with vacuoles. The vacuoles seem to be relatively stable
towards fixation. Older sieve elements are very labile and also
are filled with vacuoles which are much bigger and larger but in
other fixations they are disrupted and then a cell looks like a
normal mature sieve element of higher plants.

Walsh: There is a definite correlation between how much care
is taken in the preparation of the tissue and whether or not one
sees a vacuolar tonoplast. With Lemna one can plunge an entire
plant into a vial of fixative. With corn we took a mature leaf at
the end of the dark period, 5 or 6 in the morning, placed it into a
petri dish, allowed some fixation to take place, then made
incisions with a super sharp razor blade, parallel to the
orientation of the vascular bundles. We left it this way for a
time, then made a cross incision. Whenever shorter sieve elements
or a series of sieve tubes composed of shorter sieve elements were
processed, the appearance of these sieve elements was more like
that of adjacent cells in that there was no rupture of plastids,
there was no pulling away of the plasmalemma and that as many as 7
cells would exhibit tonoplast or a vacuole.

Geiger: Some of these slides reminded me of the work that
Cataldo did a few years ago. In slitting longitudinally and fixing
sieve tubes in place, we found that the number of the myelin type
artefacts went up to a great extent. I do not want to make a
definitive statement, but in our experience large vacuole-like
structures were often seen in the sieve tubes that were fixed
somewhat slowly in this way.

Cronshaw: In my judgement, the preservation in these
micrographs was excellent. However, in balancing these slides and

the appearance of the tonoplast against thousands of micrographs of other species, some of which have had equally good fixation, I think we have to think of one or two points as to why possibly these are artefacts. 1. There were tubules in the vacuole. 2. There was some plasmolysis, and whenever we look at fixation and try to make a judgement I think it is much better to make a judgement of the tissue as a whole and not just part of a cell that might be shown in an electron micrograph. 3. The point that Dr. Geiger made is a real one. If you put plant cells in 3% or 6% glutaraldehyde and watch them under a phase microscope, you can see cytoplasmic streaming for about 20 minutes before it stops. In these higher concentrations of glutaraldehyde, fixation is not rapid and there is plenty of time for artefacts to arise. There have been observations of "tonoplasts" in other publications and most of these can be explained as infoldings of the plasma membrane, which is a very common artefact and which can usually be detected by the presence of very fine fibrils, presumably microfibrils pulled out of the wall within the circles of membranes. I have looked for this 3rd artefact in Dr. Walsh's pictures and it was not there. We must take the pictures seriously but we must also weigh the disadvantages of the technique and weight the results in relation to many others already obtained. It warrants further study, particularly with use of more rapidly penetrating fixatives, such as acrolein and studies at the light microscope level of living tissue such as the accumulation of neutral red.

Johnson: This question of the penetration of the fixatives and the speed at which things are locked solid is extremely important. If you consider a sieve tube which is translocating at 40 cm/h and you lock it into its operating configuration in 1/20th of a second, the flow has gone about 5 μm, which is quite an appreciable distance within the sieve element.

Giaquinta: We have also reported on the ultrastructure of functioning phloem in which we used freeze substitution. This technique allows you to look at the entire sieve plate and number of sieve pores at one time and we find that the majority of the pores do not contain any of these filaments and they are not occluded with any type of filamentous proteinaceous material. Also there is no vacuolar membrane. In fact, we only find them if we damage the sieve tubes such as is experienced in low temperature treatment, etc.

Johnson: During this process of freeze substitution, do you freeze the material, stick it in alcohol which will dehydrate it, perhaps put in a fixative, and then warm it up and embed it?

Giaquinta: The initial freezing is done with liquid nitrogen, cooled methylcyclohexane and isopentane, approximating minus 170°C. After that replace the water at minus 80°C in an organic solvent, which is then replaced with a plastic.

Johnson: The plastic has to flow into the sieve element which may remove the filaments from the pore?

Giaquinta: We have shown that when you induce some type of structural damage to the phloem first, then freeze substitute the tissue, you do find these proteinaceous filaments in the pores.

Srivastava: In a few slides, on one side of the sieve pore the filaments were running axially and then all of a sudden on the other side, the axial orientation was lost, as if the filaments had turned.

Johnson: The filaments may have been displaced. Some of the pores certainly seem to have been plugged. But this is after very careful preparation. The filaments are not produced by chemical fixative. They are certainly there but I have not got any evidence that these particular sieve plates are in translocating sieve elements or that they were translocating at the moment they were frozen, although the bundle was. There are a hundred sieve elements in these particular bundles or thereabouts, and I found one sieve plate. I have one replica which had six sieve plates on it.

Tyree: Is there any possibility that what you think are filaments in the sieve pores could arise from the peculiarity of ice crystal formation in confined areas? Has anyone studied ice crystal formation in quartz "capillaries" or in anything that would be confined, like a sieve pore?

Johnson: I tried to investigate this by using the xylem as a set of model sieve pores. Between the spiral thickenings in xylem there are spaces which are about as big as sieve pores, and what I have done is to fill up the vessels with 10% sucrose and freeze-etched them. You do get filamentous structures, but they are not like the filaments that I have found here. They are distinguishable.

Geiger: Even with very careful teasing or cutting out of the phloem, there seems to be some damage done to the sieve tubes. For instance, in bean we find that making longitudinal cuts in a petiole is sufficient to significantly slow or impede the translocation, even without getting near the trace. In this sense the work with Heracleum seems to differ somewhat from the work from the bean. What I am suggesting is that perhaps one of the damages that you see would be this type of filament or P-protein or whatever in the pores when you damage with cyanide or even with rough handling. Is there a possibility that you have not completely stopped transloation but that you have some inhibition just from the longitudinal cutting of the parenchyma area?

Johnson: Definitely. The central vascular bundle has little side bundles at intervals and if one was unlucky enough to hit a sieve element that had been near to one of those, one might have got a pressure-released one. I think perhaps the chances of that are not very great and I have a fair number of sieve plates.

Geiger: Do you actually test that they are all translocating before you freeze them?

Johnson: Not all; two of them, and I cannot tell you offhand which two. But they are both translocating at about 40 cm/h from the leaf where the CO_2 is applied down into the plant and out to the youngest leaf.

Geiger: How many files of sieve elements would you estimate that you have successfully freeze-fractured? Is it a possibility that when you find 5 or 6 plates that these are all in the same file of sieve elements?

Johnson: None at all. The chances of finding a sieve plate are very remote. The chances of finding two in the same sieve tube are almost negligible. And most of these replicas are from different petioles, and there is a fair scattering of the different plants.

Walsh: Are you able to detect callose with freeze-etching?

Johnson: In freeze-etched petioles that have been cut off and had glycerol transpired off of them, you can see very obvious callose and the filaments are very tightly packed in between. But in the uncut, quick frozen specimens, I do not seem to get any callose in the sieve pores.

Behnke: You stated that there might be well over a hundred sieve elements in this vascular bundle that you were freezing, and you do not know exactly whether those that you looked at were actually translocating. Among these one hundred sieve elements, are there developmental sequences in the sieve elements, or are they all at the same stage?

Johnson: There are young sieve elements in these petioles but most of them are what I would call mature.

Swanson: Instead of using 10% sucrose solution in your model system, supposing that you used a more complicated solution something approaching what we might call artificial cytoplasm. Would you get fibrils that might be quite different than in the 10% sucrose solution? I am very much worried that we are looking at artefacts.

Johnson: I have looked at polyethylene oxide which has long filamentous molecules in it and you can see filaments in that. If you freeze very clean water, by looking at the xylem vessels which are presumably highly filtered, you get very large areas of ice crystals and you do not see filamentous structures. But one can obviously produce filaments by using artificial cytoplasm. I am absolutely convinced that most of these filaments, perhaps not the odd one, but most of the ones in the phloem, are genuine P-protein filaments.

Swanson: What convinces you that they are?

Johnson: They have bands on them; they are the right width and in the right place.

Swanson: The fibrils which you see in artificial systems are not banded?

Johnson: Not polyethylene oxide. Presumably if you chose something that produced bands you would have bands on them.

Gorham: Is it possible that, using freon as a cryostating material, you are getting a reaction of the fluoride in the freon? Would you get the same kind of fixation if you used isopentane which would be more neutral? You have a thermal drop from room temperature right down to minus 145° C and there would be time to carry out quite a lot of chemistry.

Johnson: Freon is a standard method of freezing and if you look at freeze-etched ordinary cells they are very comfortingly like the ones you see in sections.

Gorham: But you do not have a very good yardstick to compare to, so we are going in circles.

Tyree: Going back to what you said about freezing 10% sucrose in the vicinity of spiral thickening, to me, this is not an adequate analog of the sieve pores, since you do not have confinement on an entire circle. Would it be possible to infuse some wood and look at the ice crystal formation in a pit between two xylem vessels? This would be a near analog to sieve pores.

Johnson: Yes, it probably would.

Eschrich: When you measure the temperature drop with freon you use a thermocouple. You could not apply it exactly into the sieve tube. You have a 10% sucrose solution. You cut the bundle before. At least in the outer parts there was some stimulation. And we know that a very slight stimulation can cause callose formation and an aggregation of filaments. When we compare this with contraction of muscles or the nerve action in animals, then it must be a matter of picoseconds or milliseconds. I do not think it possible in freezing, even with freon, to get the time so short that you can completely eliminate artefacts caused by slow freezing.

Johnson: I agree entirely. As I pointed out earlier, during the time taken to freeze these bundles the translocation stream could have moved 5 micrometers. If one assumes that the freezing occurred very suddenly, the time I have quoted, .03 seconds, is the time taken to reach minus $50^{\circ}C$ which seemed to me to be a reasonable point to choose, at which one would expect everything to have gone solid. Actually, it would have gone solid a lot sooner that that. But if you take that time, this still gives time for all kinds of things to happen, the relaxments, movements of filaments. I accept that entirely, but how else can one do it?

Watson: To what extent would you claim that the freezing of the cells through the individual sieve tubes that you are looking at, was uniform with respect to time? Would you say that there are ice crystals there?

Johnson: Yes. You get growth of ice crystals and therefore parts of the sieve element will be fluid, while other parts become pure ice.

Watson: This then, may allow for a considerable displacement of structures?

Johnson: Yes. You might get movement of filaments, but I do not believe that you will get movement of filaments into the pores neatly by growth of ice crystals from either direction. That seems to be stretching it too much.

Watson: If they were glued to the edge of the pores would you be prepared to see that sort of displacement?

Johnson: Yes, possibly, if the filaments are around the edge of the pore.

Milburn: It might be a good idea to freeze phloem exudate, which would have randomly mixed proteinaceous material and see if there are the same kind of patterns that you have already demonstrated from the very centre of the lumen.

Johnson: Yes, one should do phloem exudate.

Christ: Some scanning electron micrographs show the filaments or tubules stretching from the sieve plate pore down to the wall. Depending upon how much faith you put in that sort of imagery, it seems to me that by the chemical action of freon or by the freezing process itself, those filaments or tubules are snapped and they snap back to the pore, and enclose it, so that when you finally take the image you see a pore that is filled full of filaments.

P-Proteins

James Cronshaw

Department of Biological Sciences

University of California

Santa Barbara, California 93106

I. Introduction

The characteristic cell of phloem tissue, the sieve element, was discovered in the middle of the 19th century by Hartig, who showed that sieve elements have specialized perforations of the walls arranged in areas now known as sieve areas. In the first study of _Cucurbita_ phloem, Hartig observed characteristic accumulations of material on the sieve areas, especially those of the sieve plates, which became known as slime plugs and the material as slime (Hartig, 1854). Other early workers on the phloem recognized that slime plugs may be extended to form slime strands which are continuous across the entire lumen of the sieve element; that slime bodies were the source of slime in mature sieve elements; and that staining reactions indicated a considerable protein content (Wilhelm, 1880). These observations led to speculation that slime plugs were related to the long-distance transportation of nutrients in the phloem and slime strands were suggested as sites for the movement of insoluble substances through the sieve plates (Wilhelm, 1880).

Several of the early workers suggested that slime plugs resulted from the flow of the contents of sieve elements toward a cut surface (Nageli, 1861; Fischer, 1885), and it was postulated that injury was the primary cause of slime plug formation (Fischer, 1884). Attempts were made to demonstrate this experimentally. Fischer (1884) and LeComte (1889) showed that slime plugs were not formed in scalded plants or in plants that were placed in sugar solution prior to sectioning.

In the century following the discovery of slime, many studies

were made at the light microscope level which confirmed the early
observations. It became established that slime bodies give rise to
slime in mature sieve elements, that slime plugs are often present
on sieve plates, and that slime strands may traverse the lumen of
the sieve element and be continuous through sieve plate pores
(Esau, 1969). In the early studies of phloem tissue at the
electron microscope level slime was observed but not well preserved
by the fixation procedures used. It was not until after the
introduction of glutaraldehyde as a fixative for plant material
(Ledbetter and Porter, 1963) and its use for fixing phloem tissue
(Bouck and Cronshaw, 1965) that the characteristic morphology of
slime became known and the term P-protein introduced (Esau and
Cronshaw, 1967).

The term P-protein, that is the phloem protein, was introduced
to describe proteinaceous substances in phloem tissue that had
previously been named slime (Esau and Cronshaw, 1967). With the
generalized acceptance of the fact that the slime of phloem tissue
was proteinaceous, continued use of the term appeared to be a
misnomer, since plant slimes are compound carbohydrates usually
associated with cell walls (Roelofsen, 1959). In the original
description of P-protein in tobacco and Cucurbita various
morphological forms of P-protein were described and designated by
number, that is, P1, P2, P3, P4 proteins (Cronshaw and Esau, 1967,
1968a). As research on P-proteins continued it became obvious that
many morphological forms existed and that a numerical series of
P-proteins would lead to more confusion than clarity. The present
author has decided to drop the numerical designations and simply
call the substance P-protein, if necessary using a morphological
description in addition. P-protein then, is the proteinaceous
material in the phloem which is sufficiently characteristic when
observed with the electron microscope to warrant a special term.
P-protein body (formerly slime body) is the term used for the
discrete aggregations of P-protein commonly found in immature sieve
elements, in other phloem cells, and in some cases in mature sieve
elements.

II. Isolation and Chemistry of P-protein

It is well known that the contents of the phloem sieve
elements of certain species can be released as phloem exudate by
cutting, and that the exudate contains high amounts of nitrogenous
material. Examination of the cut ends of Cucurbita stems and
petioles has revealed that droplets of exudate emerge only from
mature sieve elements (Eschrich, 1963; Eschrich et al, 1971).
Dispersed in this are P-protein components enabling researchers to
use exudate as a source of P-proteins. Exudate from Cucurbita,
Cucumis, Ricinus and Nicotiana has been examined chemically and by
light microscopy (Thaine, 1964; Thaine et al, 1967; Cronshaw,
1970a; Kollmann et al, 1970; Eschrich et al, 1971; Kleinig et al,

1971a, b; Walker and Thaine, 1971; Weber and Kleinig, 1971; Hall and Baker, 1972; Williamson, 1972; Walker, 1972; Yapa and Spanner, 1972; Cronshaw et al, 1973; Stone and Cronshaw, 1973). Most of the examinations have been made on Cucurbita exudate because of the ease with which it may be obtained from this genus. If droplets of Cucurbita exudate are fixed, sectioned and examined electron microscopically, they are seen to consist of masses of fine filaments. The outside of the droplets is more electron opaque and has infoldings where dense portions of the outer layer are parallel to one another (Fig.1).

Kollmann, Dörr and Kleinig (1970) examined the chemical composition of a trichloracetic acid precipitate of Cucurbita sieve tube exudate and showed that it contained chiefly protein (Table 1). Large amounts of protein in Cucurbita exudate were also found by Walker and Thaine (1971) and Eschrich et al (1971). Repeated removal of a 2 mm slice from a cut Cucurbita stem results in a gradual decline in the amount of exudate but the protein content increases (Eschrich et al, 1971; Table 2).

Table 1. Chemical composition of trichloracetic acid precipitate of 1 ml of sieve tube exudate (Cucurbita).

	mg/ml exudate
Protein	9.8
DNA	<0.02
RNA	<0.01
Phospholipid	<0.0005
Galactolipid	<0.0005
Polysaccharide	None

(From Kollmann, Dörr and Kleinig, 1970)

Kollmann, Kleinig and co-workers have established that the filamentous components in Cucurbita exudate are protein and have partially characterized them (Kollmann et al, 1970; Kleinig et al, 1971a, b; Weber and Kleinig, 1971). They were able to separate five major proteins by polyacrylamide gel electrophoresis and there had molecular weights which approximately doubled from protein to protein (Table 3), implying that four of the proteins represented dimers of the ones with the next lower molecular weight.

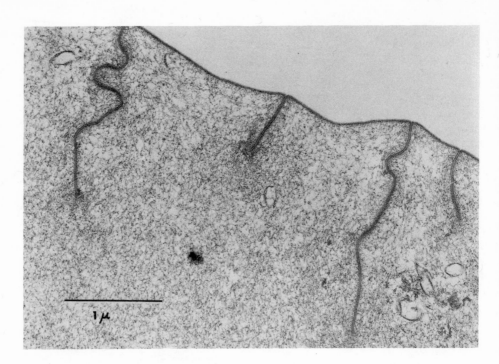

Fig.1. <u>Cucurbita</u> <u>maxima</u>. Section of a portion of a droplet of exudate that had been allowed to gel. The exudate consists of masses of fine fibrils and the gelled droplet has an electron opaque outer layer. There are infoldings where dense portions of the outer layer are parallel to one another.

Table 2. Amounts and protein content of exudate obtained during repeated cutting at 2mm intervals of a stem tip of 40 cm length. A: determined after Lowry et al. (1951); tyrosine value (50 µg/ml) subtracted. B: Biuret determination.

	Cut No.	Exudate (mg)	Protein (mg/ml)
A.	1	12.4	107
	2	8.3	157
	3	3.4	261
	4	1.3	320
B.	1	12.7	81
	2	7.2	117
	3	4.6	186
	4	1.6	221

(From Eschrich et al, 1971)

Table 3. Molecular weights of sieve tube proteins from *Cucurbita* *maxima*

Number of protein	Molecular Weight ±SD
1	15000± 2000
2	28000± 2000
3	59000± 3000
4	116000± 6000
5	220000±10000

(From Weber and Keinig, 1971)

The proteins were all similar in their solubility in dithioerythritol and at low temperature, and in their precipitation with potassium chloride, calcium chloride and vinblastine sulphate. In addition, it was demonstrated that solubilized *Cucurbita* sieve tube proteins could be reversibly aggregated into filamentous structures which resembled natural filaments in negatively stained preparations (Kleinig et al, 1971a).

Exudate from *Cucurbita* was found by Eschrich, Evert and Heyser (1971) to contain at least twelve proteins as shown by dish electrophoresis of a soluble fraction. There were five clearly visible bands and seven other bands that were barely visible. These authors also found some enzyme activity in the soluble fraction of *Cucurbita* exudate, establishing the presence of peroxidases, acid phosphatases, and aldolases. Color tests and assays for the enzymes, ATPase, fructokinase, several dehydrogenases and UDP-glucose: D-fructose-2-glucosyl transferase were negative.

Walker and Thaine (1971) using gel electrophoresis found 17 protein bands from the water soluble fraction of *Cucurbita* exudate in addition to a basic gelling protein. The same method was used to demonstrate six bands from a sample of partially dispersed solids, but at least four of these bands were not represented in the soluble fraction. Later, Walker (1972) reported on the isolation and purification of the gelling protein.

Yapa and Spanner (1972) examined phloem exudate and proteins extracted from whole phloem tissue of *Heracleum* *mantegazzianum* by electron microscopy, isoelectric focusing on polyacrylamide gel, and in sucrose density gradient. The extracted and exudate filamentous proteins were shown to have an isoelectric point of 4.9 indicating that in the alkaline pH of the sieve tube the protein would be negatively charged.

Because the morphology and precipitation properties of P-proteins are similar to those of muscle actin, microtubules and microfilaments, they have been compared with these other protein components known to be involved in biological movement. This comparison has been coupled with speculation that P-proteins may provide the motivating force for nutrient transport in the phloem.

Both muscle actin and P-proteins polymerize with potassium chloride and calcium and can form striated filaments. However, actin has bound nucleotides whereas P-protein fractions do not. The axial peridicities of P-proteins (100-150Å) differ from those of actin (350-360Å) and the molecular weights differ. Moreover the characteristic reacton of actin in binding with heavy meromyosin has not been observed for P-proteins (Williamson, 1972; Palevitz, 1974).

Microfilaments composed of actin-like protein are known to be involved in cytoplasmic streaming in many plants in which the drug cytochalasin B may reversibly depolymerize the filaments and inhibit streaming. Thompson and Thompson (1973) found that cytochalasin B inhibited sucrose transport in Heracleum and interpreted their results as support for a contractile role of P-protein in the phloem. However, Williamson (1972) found no effect of cytochalasin B on ^{14}C transport in Lepidium phloem and we have found no effects on the morphology of P-proteins in tissue slices incubated in cytochalasin B prior to preparation for electron microscopy.

Some tubular P-proteins bear a morphological and some chemical resemblance to cytoplasmic microtubules and have been compared with colchicine-binding microtubule protein. P-proteins precipitated by vinblastine sulphate can be dissolved and reprecipitated reversibly by varying the temperature between 0°C and room temperature as with tubulin from brain extract (Kleinig et al, 1971b). Isolated P-proteins have no colchicine binding activity however, (Kleinig et al, 1972b) and results from this laboratory and others (Wooding, 1969; Sabnis and Hart, 1973) have shown that incubation of tissue slices in colchicine solutions prior to fixation for electron microscopy has no effect on the morphology of P-proteins. Thus the properties of P-proteins appear to differ significantly from those of actin, microfilaments and microtubules although all of these proteins can be broadly categorized together in view of the similarity in morphology and in precipitation properties.

Before direct confirmation by chemical analysis, the proteinaceous nature of P-protein had been deduced from cytochemical tests. Early studies of staining reactions of the phloem indicated that slime was rich in proteins. Mercuric

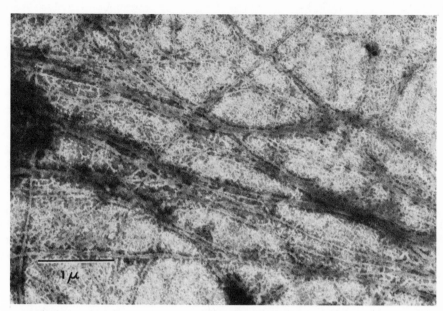

Fig.2. <u>Cucurbita</u> <u>maxima</u>. P-protein fibrils in exudate negatively stained with sodium phosphotungstate.

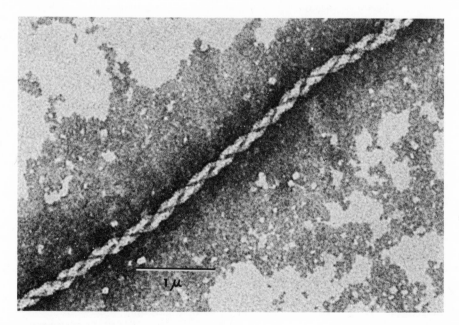

Fig.3. <u>Nicotiana</u> <u>tabacum</u>. Filament from exudate negatively stained with sodium phosphotungstate

bromphenol blue stains the slime bodies of immature sieve elements and the slime plugs in mature sieve elements according to the observations of Engleman (1965), Esau and Cheadle (1965), and Cronshaw and Esau (1967, 1968a), indicating a positive reaction for protein. Other workers have shown that accumulations at sieve plates contain substances such as lipids and ribonucleic acid in addition to protein (Salmon, 1946, 1951; Buvat, 1963). These positive tests are most probably due to the fact that accumulations at the sieve plates consist of both P-protein and other cytoplasmic components.

III. Morphology of P-protein

A. Isolated P-protein
    Early descriptions of P-proteins were made of fixed sectioned material and revealed that P-protein may have several morphological forms; granular, tubular, fibrillar or crystalline (Cronshaw and Esau, 1967, 1968a; Steer and Newcomb, 1969; Wergin and Newcomb, 1970). Recently, several studies have been made of P-proteins isolated in exudate and examined electron microscopically after either negative staining or shadowing. Kollmann and co-workers examined P-protein from Cucurbita maxima and Nicotiana glauca X suaveolens (Kollmann et al, 1970), and found two types of filaments in negatively stained preparations of exudate. One form, which they termed elementary filaments, was nearly 40$\overset{o}{A}$ in diameter and showed a beaded appearance with regular spacing of about 50$\overset{o}{A}$. A second form had a diameter of about 90$\overset{o}{A}$ and it was deduced that it consisted of two helically arranged 40$\overset{o}{A}$ subunits.

    Cronshaw (1970a) and Cronshaw, Gilder and Stone (1973) examined P-proteins in the exudate of Cucurbita, Cucumis, and Nicotiana which had been negatively stained with a sodium phosphotungstate or shadowed with platinum-iridium alloy. In the exudate of Cucurbita and Cucumis the smallest fibrils observed measured 80$\overset{o}{A}$ and the bulk of the filaments were of this type. Also observed were tubules which measured 230$\overset{o}{A}$ in diameter. Typical Cucurbita P-protein fibrils, negatively stained with phosphotungstic acid, are shown in Figure 2 where it can be seen that the 80$\overset{o}{A}$ fibrils form a network and are organized into larger strands.

    From tobacco, exudate can be obtained only if the cut stems or petioles are squeezed so that less material is available for examination. Cronshaw, Gilder and Stone (1973) described filaments in Nicotiana exudate which consisted of two helically wound subfibrils in contrast to the results of Kollmann and co-workers who described filaments in Nicotiana exudate similar to those from Cucurbita. A filament consisting of two helical subfibrils is illustrated in Figure 3. These characteristic fibrils have an overall diameter of about 300$\overset{o}{A}$ and the helical subfibrils a diameter of 150$\overset{o}{A}$ with a pitch of 35$\overset{o}{A}$. The diameter of these

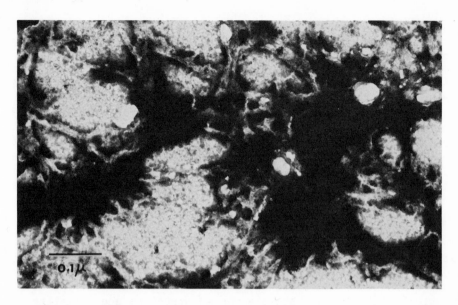

Fig.4. <u>Cucurbita maxima</u>. Isolated P-protein fibrils treated with 50 mM sodium sulphite for 20 minutes and negatively stained with sodium phosphotungstate.

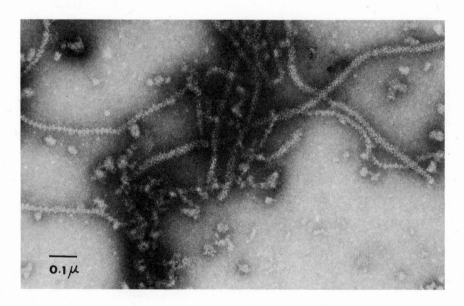

Fig.5. <u>Ricinus communis</u>. Negatively stained filaments in phloem exudate. Some of the filaments have a cross-striate appearance, others are narrower and have projections along their length.

fibrils varies slightly according to the tightness of the windings of the subfibrils. In some special subunits which have a center to center spacing of 70Å can be seen along the helical subfibrils.

The effects of solutions of urea, sodium sulphite, colchicine and potassium chloride on filaments in Cucurbita and Cucumis exudate were examined (Cronshaw, 1970a; and Cronshaw et al, 1973). With both urea and sodium sulphite there was dissolution of some of the smaller fibrillar components at first, and clumping of the remaining fibrils. With extended treatment the fibrils completely dissolved. Treatment with 5% colchicine or 0.6 molar potassium chloride for 30 minutes had no noticeable effect on the P-protein fibrils. The effect of 50 mM sodium sulphite for 20 minutes on Cucurbita fibrils is shown in Figure 4.

Ricinus communis is another species which yields exudate in quantities which are sufficient for chemical analysis and electron microscopical examination. Williamson (1972) examined Ricinus exudate and described filaments of two types, a larger type with a diameter of 200Å and a steeply banded structure and a smaller type with a diameter of 20-80Å. We have examined fibrils in the exudate of Ricinus and observed two main fibrillar components (Stone and Cronshaw, 1973). One type of fibril appears similar to the larger fibrils described by Williamson (1972). This type makes up about 50% of the fibrils. It usually has a cross-hatched appearance and a diameter measuring 200Å (Figure 5). A second type is somewhat less frequently observed, measures 141Å in diameter and consists of a central filament with lateral projections (Figure 5). The projections measure 50 x 140Å and are spaced at intervals of 65-100Å. Some fibrils have characteristics of both these fibril types indicating a structural relationship and possibly an interconvertibility between the two major fibril types (Figure 6). The exudate also contains torus-shaped structures which measure 135-150Å in diameter and other fibril types which are minor components.

B. Fixed and sectioned P-proteins
    P-proteins in fixed sectioned material have been described in many species, in contrast to the smaller number of species for which P-proteins extracted in exudate have been described. The P-proteins vary in morphology and may be granular, tubular, fibrillar or crystalline depending on the species or the maturity of the cell in which they are observed. Tubular P-proteins have been observed in Acer (Northcote and Wooding, 1966; Wooding, 1967), Avena (Esau and Gill, 1973), Beta (Esau et al, 1967), Coleus (Steer and Newcomb, 1969), Coronilla (Palevitz and Newcomb, 1971), Cucurbita (Cronshaw and Esau, 1968; Parthasarathy and Mühlethaler, 1969), Desmodium (Palevitz and Newcomb, 1971), Glycine (Wergin and Newcomb, 1970), Mimosa (Esau, 1972), Nicotiana (Cronshaw and Esau, 1967; Cronshaw and Anderson, 1971; Parthasarathy and Mühlethaler,

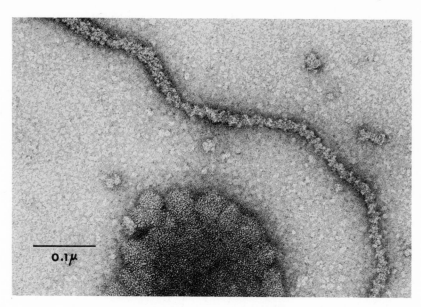

Fig.6. <u>Ricinus</u> <u>communis</u>. Portion of a filament in exudate, negatively stained with uranyl formate. The filament has a cross striate appearance for part of its length and projections for the other part.

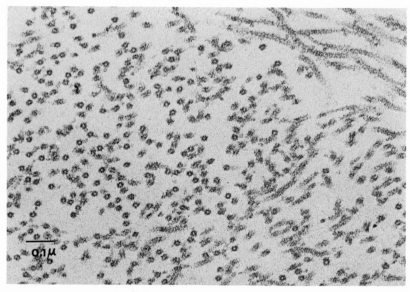

Fig.7. <u>Nicotiana</u> <u>tabacum</u>. Sections of tubular P-protein.

1969; Wooding, 1969), Phaseolus (LaFleche, 1966), Pisum (Zee, 1968), Primula (Tamulevich and Evert, 1966), Tetragonia (Esau and Hoeffert, 1971), Ulmus (Evert, and Deshpande, 1969), and Vicia, (Zee, 1969). In all of these species P-protein also occurs as fibrils. P-protein in the mature sieve elements of most of the species that have been studied is in the form of beaded fibrils and in some species these fibrils are derived from tubular P-proteins. Crystalline forms of P-protein have been described in the papilionaceous legumes and these may show interconversions with tubules fibrils.

Tobacco P-proteins have been studied in detail and tubular and fibrillar forms have been described (Cronshaw and Esau, 1967). Typical P-protein tubules from tobacco are shown in Figure 7. The tubules measure approximately 230Å in diameter have a lumen of approximately 80 Å and may be several microns in length. Transection of P-protein tubules often show spoke-like projections and in some cases there is evidence of a substructure in the wall. Using the Markham reinforcement technique it has been shown that there are most probably six subunits around the wall of the Nicotiana P-protein tubule (Parthasarathy and Mühlethaler, 1969; Cronshaw et al, 1973).

P-protein tubules are often compared with microtubules and it has been demonstrated that in transection microtubules have thirteen equal subunits arranged around the wall (Ledbetter and Porter, 1964). In transection, the electron-lucent lumen of P-proteins is smaller than that of microtubules, an observation which agrees with models of six equal subunits around a P-protein tubule and 13 equal subunits around a microtubule. P-proteins in tobacco also occur in the form of fibrils measuring approximately 150 Å in diameter (Figure 8). These fibrils are striated and the mean center to center distance of the stained bands is about 150Å. In transections the fibrils appear circular in outline with an indication of an electron-lucent center. Beaded fibrils similar to those found in tobacco have been observed in other species such as Coleus, Ricinus, Acer, Aristolochia, and Desmodium. The observations that there appears to be a replacement of the tubular type of P-protein with the fibrillar type during the differentiation of the tobacco sieve element and that P-protein filaments are often observed which have the tubular type of structure for part of their length and the fibrillar type for part of their length, led to the suggestion that P-protein may undergo conformation changes in vivo (Cronshaw and Esau, 1967; Cronshaw, 1970b).

In Cucurbita, P-proteins in fixed section specimens are observed mainly in the form of fine fibrils (Cronshaw and Esau, 1968a). These fine fibrils are approximately 60Å in diameter, are often aggregated into larger bundles, and resemble the fine fibrils

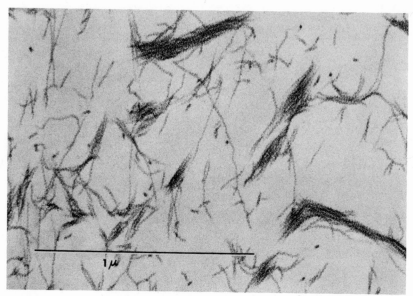

Fig.8.   <u>Nicotiana tabacum</u>.   Sections of beaded fibrillar
P-protein.

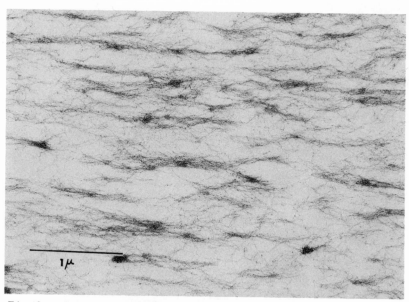

Fig.9. <u>Cucurbita maxima</u>. Section of mature sieve element
with fine fibrils of P-protein.

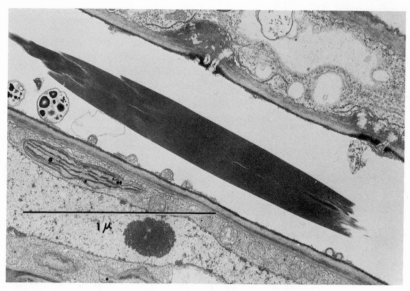

Fig. 10. _Phaseolus vulgaris_. Section of a portion of a mature sieve element showing a typical P-protein crystal.

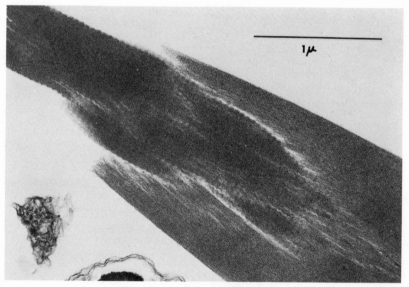

Fig. 11. _Phaseolus valgaris_. Higher magnification view of a portion of a P-protein crystal showing characteristic striations and the insertion of the tail into the main body.

that have been described in Cucurbita exudate (Figure 9).   Tubular
P-proteins   are   also found   in   Cucurbita although in much smaller
quantities; they resemble the tubular P-proteins of tobacco.   Other
P-protein types are found in Cucurbita   P-protein   bodies   where   a
variety   of   structures   ranging from fine fibrils to small tubules
have been described (Cronshaw and Esau, 1968a).

The papilionaceous   legumes   have   characteristic   crystalline
P-proteins   in   addition to having tubular and fibrillar types.   In
transection these crystals are usually square and   in   longitudinal
section   they   are   usually differentiated into a central main body
with a sinuous tapering tail at each end.   Along the entire   length
of   the   crystal find cross-striations about 150Å apart are usually
found (LaFleche, 1966,   Wergin   and   Newcomb,   1970;   Palevitz and
Newcomb,   1971).   A longitudinal section of a typical crystal from
Phaseolus vulgaris is   illustrated   in   Figure   10   and   at   higher
magnification Figure   11   shows the insertion of the tail into the
main body.

In Ricinus communis there is   apparently   a   lack   of   tubular
P-proteins.   In mature sieve elements of this species two distinct
types of P-proteins may be observed.   One   is   a   striated   fibril
about   150Å in diameter with a beaded substructure (Figure 12); the
other is a thicker fibril measuring about 175Å in diameter (Figure
13).   These two types of fibrils obviously correspond with the two
major types described in Ricinus exudate.

C. Effect of drugs on P-proteins
We have examined the effect of solutions   of   cytochalasin B,
colchicine,   vinblastine   sulphate,   and   caffeine   on   isolated P-
proteins and on P-proteins in sections that have been incubated   in
these   reagents   prior   to preparation for electron microscopy.   In
our   preparations   P-proteins   remained   unaffected   by   all   these
treatments and similar results have been found   by   other   workers.
Wooding   (1969)   examined the effect of colchicine on the P-protein
of Nicotiana callus phloem and found   that   pre-incubation   of   the
callus   at   0° C   or   50° C or in solutions of colchicine prior to
fixation had no effect on the individual structures of P-protein or
on the relative proportions of   the   various   morphological   types.
The   number   of   micro-tubules   however,   was reduced by colchicine
treatment.   Sabnis and Hart (1973) obtained   similar   results   when
they incubated Heracleum mantegazzianum in solutions of vinblastine
sulphate   and   colchicine.   Incubation with these alkaloids had no
discernible effects on the ultrastructure of   the   P-protein.   In
this   study   it   was   determined   that   tritiated colchicine freely
entered the tissue and travelled acropetally and basipetally   along
the   vascular   strands.   It is of interest that these authors also
reported that when these alkaloids were applied locally to the stem
of a variety of plant species, they had no effect on   the   movement
of ^{14}C-labelled photosynthates.

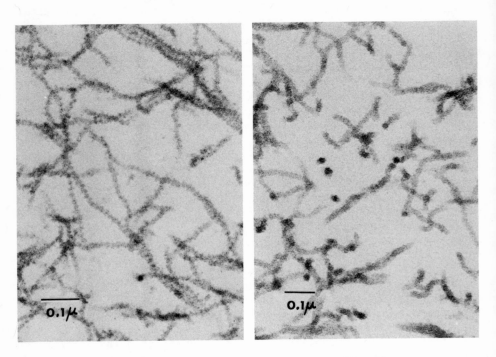

Fig.12   Fig.13

Fig.12 and 13. <u>Ricinus</u> <u>communis</u>. P-proteins from mature sieve elements. Fig. 12. 150 striated fibrils. Fig.13. 175 non-striated fibrils.

IV. Formation of P-proteins

Various organelles have been observed in association with newly forming regions of P-protein and it has been speculated that these organelles may be involved in its origin. Bouck and Cronshaw (1965) observed small regions of granular P-protein in Pisum that were closely associated with cisternae of endoplasmic reticulum, and suggested that endoplasmic reticulum was directly involved in the synthesis of P-protein. Recently we have observed a close association between granular P-protein and the endoplasmic reticulum in very young sieve elements of Nicotiana (Cronshaw and Gilder, unpublished). This observation has been made most frequently in tissue which has been treated by incubation in a 0.1% solution of colchicine prior to fixation, but has also been seen in untreated tissue. There appears to be a specialized region of endoplasmic reticulum which runs through the granular regions of P-protein. This region is devoid of ribosomes and the membranes are straight and parallel to one another (Figure 14). These specialized regions of endoplasmic reticulum may represent sites of assembly of P-proteins. Cronshaw and Esau (1967) observed membrane bound polyribosomes adjacent to developing P-protein bodies in tobacco and it is possible that such polyribosomes are the sites of production of P-protein. Other descriptions of early stages of sieve element differentiation have shown consistently that polyribosomes or groups of ribosomes are associated with developing regions of P-protein.

Newcomb (1967) observed vesicles with spinelike protruberances in pericyclic and procambial cells associated with protophloem sieve elements in the roots of Phaseolus vulgare. Because of the association between these spiny vesicles and regions of P-protein it was suggested that they may play a role in its formation. This association between spiny vesicles and P-protein has also been shown in Cucurbita (Cronshaw and Esau, 1968) Coleus (Steer and Newcomb, 1969) Nicotiana (Esau and Gill, 1970b) and Beta (Esau and Gill, 1970a).

Other observations have shown a close association of apparently active dictyosomes adjacent to developing regions of P-protein (Cronshaw and Esau, 1968a). Again it has been suggested that these organelles are involved in P-protein formation but, as dictyosome derived secretory products are usually transferred through a membrane and it is well established that developing regions of P-protein are not membrane-bound, this latter association may well be fortuitous.

Although a close association has frequenty been demonstrated between developing regions of P-protein and ribosomes, polyribosomes, spiny vesicles, dictyosomes and endoplasmic

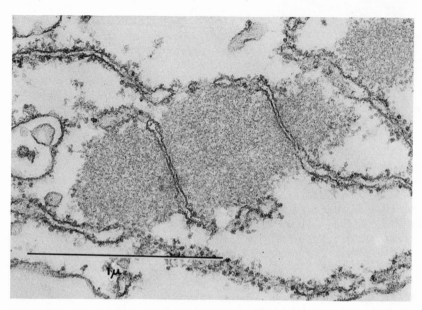

Fig.14. <u>Nicotiana</u> <u>tabacum</u>. Association of P-protein and endoplasmic reticulum in a sieve element at an early stage of differentiation.

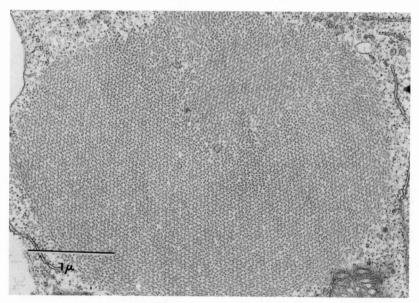

Fig.15. <u>Nicotiana</u> <u>tabacum</u>. Transection of a P-protein body in a differentiating sieve element.

reticulum, there is little evidence to substantiate the suggestion that these organelles are involved in P-protein formation and the answers to questions regarding the origin of P-protein must await further research.

V. P-protein bodies and their dispersal

Ontogenetic studies made on the sieve elements of various dicotyledons have shown that P-protein first appears as small accumulations in the cytoplasm which enlarge to form discrete bodies known as P-protein bodies. Between species, however, there is considerable variation in the morphology of the P-proteins, their arrangement in P-protein bodies, and the changes that occur during sieve element differentiation.

The P-protein bodies of <u>Nicotiana</u> and <u>Coleus</u> are similar and have been studies in detail (Cronshaw and Esau, 1967; Steer and Newcomb, 1969). In the young sieve elements of both of these species P-protein is first apparent as small accumulations of P-protein tubules although in tobacco, P-protein may also be present in very young sieve elements as finely granular material in association with the endoplasmic reticulum (Figure 14). No morphological relationships have been observed between the finely granular material and the small groups of P-protein tubules. Usually there is only one P-protein body per cell and this increases in size by the addition of more tubules until it reaches a size approximately equal to that of the nucleus or in some cases even larger. The P-protein bodies have no bounding membrane, are ellipsoidal in shape, and are elongated in the same direction as the developing sieve elements. Within them, the P-protein tubules are arranged roughly parallel to one another with their long axes in the same direction as the major axis of the P-protein body. Thus, a transection of a P-protein body shows most of the tubules in transection (Figure 15). In some P-protein bodies the tubules are arranged in crystalline aggregates having a hexagonal close-packed arrangement (Steer and Newcomb, 1969; Cronshaw and Anderson, 1971). The individual tubules have stained spoke-like projections which appear to connect with adjacent tubules and these connections may be responsible for the structural integrity of the bodies (Figure 16).

At an intermediate stage of differentiation of the sieve elements the P-protein bodies disaggregate. This occurs at the stage of cell differentiation when degenerative changes are apparent in the nucleus, the endoplasmic reticulum is reorganizing, and the sieve plate pores are developing. The disaggregation of the P-protein bodies starts with individual tubules gradually moving apart and becoming dispersed in the surrounding cytoplasm. At about this stage of sieve element differentiation beaded fibrils are apparent in the cytoplasm adjacent to the dispersing P-protein

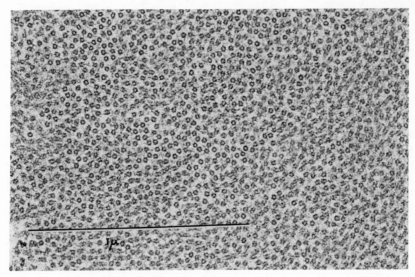

Fig.16. _Nicotiana_ _tabacum_. Higher magnification view of a portion of the P-protein body shown in Figure 15. The tubular nature of the P-protein can be clearly seen and also fine fibrils which interconnect some of the tubules.

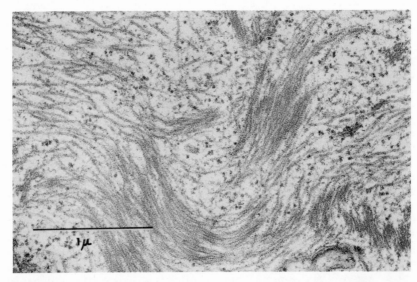

Fig.17. _Nicotiana_ _tabacum_. Portion of a sieve element at a comparatively late stage of differentiation. Two forms of P-protein are visible in the dispersing P-protein body, a tubular form and a beaded fibrillar form.

tubules and appear to be derived from them (Figure 17). As dispersion proceeds, the P-protein tubules and fibrils become evenly distributed throughout the lumen of the sieve element.

The P-protein bodies of Cucurbita maxima, studied in detail by Cronshaw and Esau (1968a,b), are first apparent in differentiating sieve elements as small areas of granular or finely fibrillar material interspersed with ribosomes, dictyosomes and endoplasmic reticulum. These small aggregates gradually increase in size until numerous P-protein bodies are formed. Other organelles become excluded from the developing aggregates during this process. The P-protein bodies are arranged in the peripheral cytoplasm and two types may be distinguished, a larger type and a smaller type (Figure 18). The bulk of the P-protein is in the larger type which contains mainly fibrillar material and may have less dense regions that have been described as vacuoles. The smaller type of P-protein body is composed of tubules similar in size and morphology to the P-protein tubules of tobacco. These tubules may be regularly arranged in crystalline aggregates with quadrangular packing.

The fine fibrils of P-protein in the larger bodies may be modified to show a range of structures some of which have an electron lucent core and resemble tubules. These tubules are much smaller in diameter than those in the small P-protein bodies. From the variability of structures found it has been suggested that there is a developmental sequence from the fine fibrils initially formed to the small tubules (Cronshaw and Esau, 1968a).

At an intermediate stage of differentiation of the sieve element the P-protein bodies begin to disaggregate (Figure 19). The first evidence of this is a swelling of the large P-protein bodies as the individual filaments move apart from one another. The bodies fuse and eventually, as dispersal proceeds, the lumen of the sieve elements becomes filled with the products of disaggregation. The small P-protein bodies often remain intact as crystalline aggregates of tubules dispersed, together with the fine fibrils, in the lumen of the mature sieve element. The disaggregation of P-protein bodies in Cucurbita is not a consistent phenomenon and in certain sieve elements both types of P-protein bodies fail to disperse.

Studies of Ricinus (Cronshaw and Stone, unpublished) have shown the P-proteins in this species to be either granular or fibrillar with no clearly defined tubular P-proteins. Ricinus sieve elements at early stages of differentiation show a finely fibrous or granular P-protein in the cytoplasm intermixed with ribosomes and vesicles of various sizes. During differentiation the amount of fibrillar material in these P-protein bodies increases and adjacent bodies may fuse to form larger ones. In

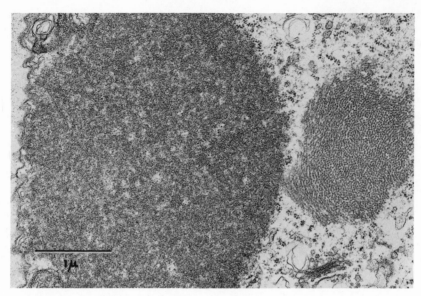

Fig.18. _Cucurbita_ _maxima_. Section through P-protein bodies in a differentiating sieve element. The large body contains fibrillar P-protein, the small body consists of P-protein tubules.

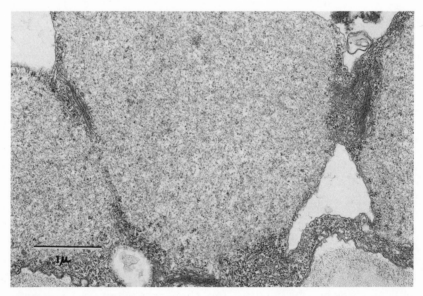

Fig.19. _Cucurbita_ _maxima_. Longitudinal section of a sieve element at an intermediate state of differentiation showing disaggregating P-protein bodies.

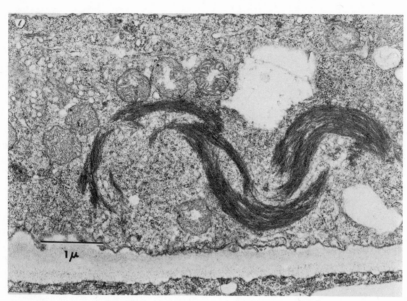

Fig.20. _Ricinus_ _communis_. P-protein bodies in a sieve element at an early stage of differentiation. The bodies consist of P-protein fibrils in loose aggregates.

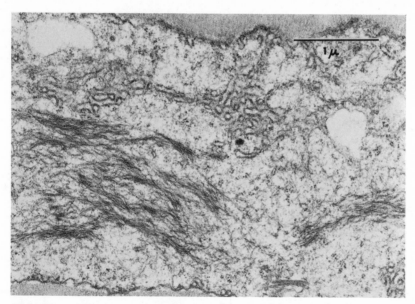

Fig.21. _Ricinus_ _communis_. Longitudinal section of a portion of a differentiating sieve element showing a dispersing P-protein body.

some cases this results in the formation of a single large fibrillar P-protein body which may occupy most of the length of the cell.   The P-protein bodies are loose aggregates and do not have the compact form of those in <u>Nicotiana</u> and <u>Cucurbita</u> (Figure 20). The fibrils in developing P-protein bodies have a diameter of about 150Å, and usually appear striated. As differentiation proceeds larger fibrils of about 175Å are observed in the bodies, which often lack the striated appearance. At later stages of differentiation the P-protein bodies disperse throughout the lumen of the sieve element (Figures 21 and 22). In mature sieve elements two distinct types of P-protein are observed, 150Å striated fibrils with a beaded substructure and thicker non-striated 175Å fibrils.

The large, refractive, crystalline bodies of the papilionaceous legumes, observed by Strasberger (1891) in <u>Robinia,</u> have been examined at the electron microscope level in several species and have been variously named irregular bodies (Bouck and Cronshaw, 1965), slime bodies (Wark and Chamber, 1965), crystalline inclusions (Zee, 1968), flagellar inclusions (LaFleche, 1966), and mucoid bodies (Gerola, Lombardo and Catarra, 1969). These bodies are now known to be P-protein bodies and Newcomb and co-workers have made detailed studies of them and other P-proteins in several species of papilionaceous legumes.

In soybean, Wergin and Newcomb (1970) described finely granular P-protein which accumulated in the young sieve elements amongst ribosomes, vesiculate endoplasmic reticulum and numerous dictyosomes. Within these accumulations, scattered bundles of tubules appeared which then aggregated and aligned themselves longitudinally in the sieve element. These tubules were converted into an electron opaque crystalline body which continued to grow by aggregation and transformation of additional tubules until it attained a large size and the characteristic shape: square in cross-section and terminated by sinuous tails. As the sieve element matured this crystal dispersed into a mass of fine striated fibrils that filled the lumen of the mature sieve element.

Palevitz and Newcomb (1971) studied the primary sieve elements of several papilionaceous legumes and in particular <u>Desmodium</u> <u>canadense</u>. In this species the crystalline inclusions were first seen in small thin crystals in the cytoplasm of young sieve elements. Numerous tubules appeared in the cytoplasm near to the developing crystalline P-protein bodies and appeared to be concerned with their growth. These tubules seem to be synthesized and accumulate in the vicinity of the crystals. Apparently the tubules are incorporated into the crystals since they are clearly continuous with the crystals and can be traced directly into the central bodies. The tubules have an overall diameter of 157 and usually have a parallel linear orientation. At later stages of differentiation the crystals undergo a limited form of dispersal

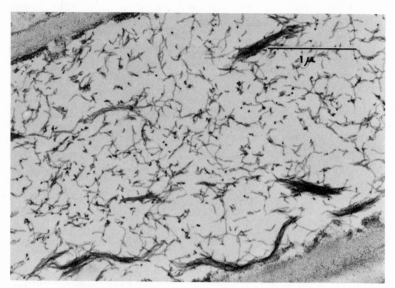

Fig.22. <u>Ricinus communis</u>. Longitudinal section of a portion
of a mature sieve element. The P-protein fibrils have dispersed
and are almost uniformly distributed throughout the lumen of the
cell.

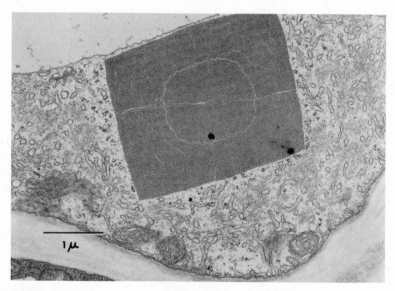

Fig.23. <u>Phaseolus vulgaris</u>. Sieve element at an early stage
of differentiation. Granular masses of P-protein are associated
with cisternae of endoplasmic reticulum. A P-protein crystal is
shown in cross section.

during which they expand, become less electron opaque, and lose their crystallinity. This occurs by a loosening and then a separation of fine filaments starting at the periphery of the central body. At maturity the sieve element lumen is filled with fine fibrils resulting from the dispersal of the P-protein components. The mass of tubules produced appear to be in excess of the amount incorporated into the crystal body, and the tubules that remain in the cytoplasm are converted into striated fibrils as in the case of tobacco. In some preparations four forms of P-protein were observed simultaneously in sieve elements: crystalline inclusions, fine filaments derived from the crystals, tubules and striated fibrils derived from these tubules. In other species, e.g. Coronilla varia, tubules similar in appearance to those in Desmodium were observed, but no direct association with the developing crystalline inclusions was found.

LaFleche (1966) in an ontogenetic study of Phaseolus vulgaris sieve elements described large crystalline inclusions and masses of tubules. We have recently examined the P-proteins of this species and have found granular, tubular, fibrillar and crystalline P-proteins. At early stages of differentiation, granular masses appear in the cytoplasm associated with the endoplasmic reticulum (Figure 23). Small crystalline aggregates and eventually large crystals with the characteristic shape develop in the cytoplasm, usually in regions rich in the granular P-protein (Figure 23). Aggregates of tubular P-protein develop into P-protein bodies, these are usually distinct and appear to be unrelated to the crystalline P-protein (Figure 24). As the sieve elements mature the crystals may or may not disperse as the sieve elements mature although an association of fibrillar P-protein and endoplasmic reticulum may persist in the mature sieve element.

Detailed ontogenetic studies have been made of only a small number of species mostly of dicotyledonous plants, but from numerous electron microscopic observations of those and other species it appears that P-proteins are universally present in the phloem of dicotyledons. In addition to being present in the sieve elements, P-proteins and P-protein bodies may be present in the companion cells and phloem parenchyma cells of most species. In other taxa our information on P-protein is much less complete. In general P-proteins are absent in gymnosperms and the lower vascular plants (Esau, 1969; Neuberger and Evert, 1974) with the possible exceptions of Selaginella kraussiana (Burr and Evert, 1973) and Isoetes muricata (Kruatrachue and Evert, 1974). In the monocotyledons, P-protein is absent from some species and present in others. P-protein has been observed in Dioscorea (Behnke and Dörr, 1969; and Behnke, 1968); Elodea (Currier and Shih, 1968); Avena and Secale (O'Brien and Thimann, 1967); Tradescantia (Heyser, 1971) and in the phloem of several palm species (Parthasarathy, 1974a). P-protein is absent in the leaf sieve elements of corn

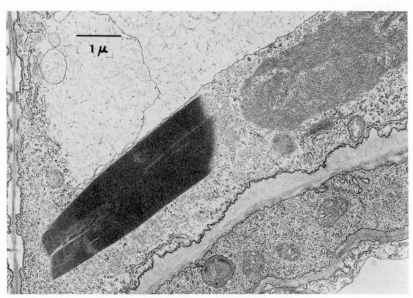

Fig.24. _Phaseolus vulgaris_. Longitudinal section of a sieve element at an early stage of differentiation. A P-protein crystal and a body consisting of tubular P-protein are developing in the cytoplasm.

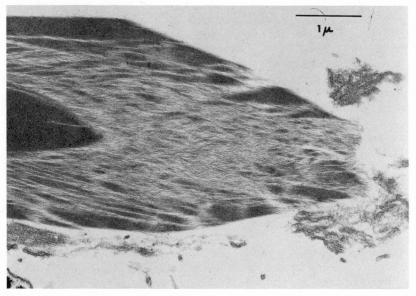

Fig.25. _Phaseolus vulgaris_. A P-protein crystal in an almost mature sieve element appears to be dispersing into fine fibrils.

(Singh and Srivastava, 1972) and <u>Hordeum</u> (Buvat, 1968; Evert et al, 1971) and absent in some metaphloem sieve elements of <u>Triticum</u> (Kuo et al, 1972). Recently, Parthasarathy (1974a, b, c) made an extensive study of the phloem of twelve species of palm. P-protein filaments 70-120 Å were present in the differentiating sieve elements of some species although the amount of P-protein was variable. In other species there was a total absence of P-protein. Tubular P-protein or P-protein bodies with compacted tubules were apparently absent in all species. When present, P-protein filaments were usually helical or twisted and often had a fibrillar structure with transverse striations. The largest amount of P-protein observed was in the metaphloem sieve elements of <u>Prestoea montana</u>.

VI. P-protein in mature sieve elements

One of the most controversial aspects of the structure of the phloem sieve element protoplast is the distribution of P-protein in the mature sieve element. As sieve elements mature technical difficulties arise in preserving them for microscopical investigation as they begin to accumulate osmotically active substances and develop a high hydrostatic pressure. It has been estimated that mature sieve elements have an osmotic pressure which may be as high as 30 atmosheres (Weatherley, 1962) and cutting or alteration of the membrane permeability properties by using chemical fixatives releases the hydrostatic pressure causing a surge of images of mature sieve elements. For this reason interpretation of images of mature sieve elements obtained by light electron microscopy after various preparation procedures is difficult and has led to controversy about their structure.

Several investigators have attempted either to reduce the hydrostatic pressure of the sieve elements before fixation or to fix tissue in other ways which would prevent surging. Various treatments have been tried such as immersion in boiling water (Fischer, 1885, 1886), pre-treatment of phloem with glycerine before excision (Rouschal, 1941), fixing portions of plants before excising the phloem (Engelman, 1965), fixing whole plant (Cronshaw and Anderson, 1969), rapidly freezing plants in liquid nitrogen before fixation of whole plants (Anderson and Cronshaw, 1970) and starving prior to fixation (Eschrich et al, 1971). We have also tried incubating whole plants in sucrose solutions prior to fixation for electron microscopy.

There is little agreement in the results obtained using these experimental procedures. Some investigators, e.g. Wooding (1969) and Anderson and Cronshaw (1970) have described an even distribution of P-protein throughout the lumen of mature sieve elements. It was suggested that P-protein is distributed as a more or less even network throughout the lumen passing through the sieve plate pores so that the network is continuous from one sieve element to the next. This distribution is especially apparent in plants wilted prior to fixation (Anderson and Cronshaw, 1970). In preparations showing an even distribution of P-protein other cytoplasmic components, cisternae of endoplasmic reticulum, plastids and mitochondria, are usually organized in the parietal position, close to the plasma membrane (Figure 28).

Further evidence of an even distribution of P-protein comes from studies of sieve elements which have been induced to differentiate in isolated nodules in tobacco callus (Wooding, 1969) or tobacco pith cultures (Cronshaw and Anderson, 1971). These nodular sieve elements are in a closed system and are not part of a continuous transport system such as that found in intact plants. One would expect sieve elements in nodules to be fixed without the artifacts associated with the release of hydrostatic pressure. In mature sieve elements of these tissue culture nodules there tends to be an even distribution of P-protein throughout the lumen of the cell, although in some cases, probably where hydrostatic pressure had been developed, P-protein may plug the sieve plates (Cronshaw and Anderson, 1971).

Other investigators have described a parietal distribution of P-protein. Evert and co-workers (1973) tried to minimize surging of sieve elements in Cucurbita maxima hypocotyls by reducing the supply of nutrients. The cotyledons and first foliage leaves of 16-day old Cucurbita maxima plants were removed two days prior to fixing portions of the hypocotyls. In the majority of the sieve elements of tissue fixed in this conditon the P-protein was entirely parietal in distribution. This parietal distribution of P-protein continued through the sieve plate pores and the pores were either unplugged or merely lined with P-proteins.

Strands which may traverse the sieve element lumen and be continuous from sieve element to sieve element through the sieve plates pores have been observed in many preparations of phloem tissue examined by light microscopy. These strands were recognized by early workers on the phloem and have been described in detail by Crafts (1932) and Thaine and co-workers (Thaine, 1964; Thaine et al 1967). The concept of strands was refined by Evert and co-workers (Evert et al, 1969) who described strands confined to the parietal position. Later work by Evert, Eschrich, and Eichhorn (1973) showed no transcellular strands, and most workers are agreed that at the electron microscope level there appears to be no evidence for strands other than accumulations of P-protein and other organelles formed during release of P-protein and other organelles formed during release of hydrostatic pressure. Robidoux, Sandborn, Fensom and Cameron (1973) recently described strands of P-protein in Heracleum phloem fixed in a mixture of glutaraldehyde, DMSO and acrolein. These authors state that this fixation mixture reduces damage in the lumen of the sieve tube although there appears to be

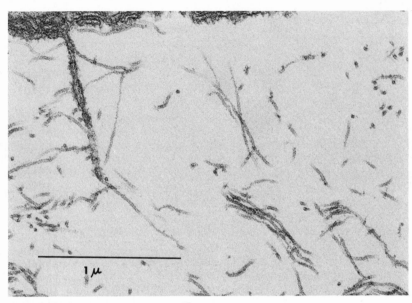

Fig.26. <u>Nicotiana</u> <u>tabacum</u>. P-protein filaments in a mature sieve element some of which show structural characteristics of more than one type of P-protein.

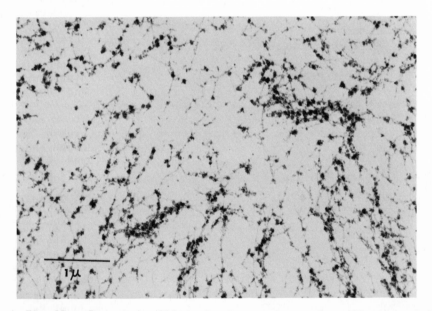

Fig.27. P-protein filaments in a mature sieve element stained to show the localization of ATPase activity. Reaction product is associated with the P-protein.

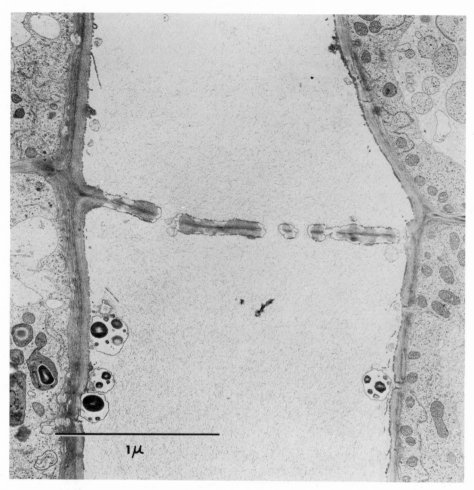

Fig.28. <u>Nicotiana</u> <u>tabacum</u>. Longitudinal section of portions of mature sieve elements and adjacent cells from a wilted plant. The P-protein fibrils are distributed almost uniformly in the lumens of the sieve elements.

no evidence of this as judged by the published micrographs.

In order to determine whether or not P-protein is stationary in the sieve elements we have examined mature sieve elements of various ages in the stems of tobacco plants and have found no differences in the distribution of P-protein. As mature sieve elements no longer contain the cell components necessary for protein synthesis it seems likely that P-protein does not move in the assimilate stream. Alternatively it is possible that it may be synthesized in adjacent cells and continually replaced in the sieve element. In this case either the P-protein would have to move out of the sieve elements at the sites of nutrient utilization or it would accumulate in sieve elements in this region. There is no evidence that either of these events take place.

The amount of P-protein in sieve elements in different regions of the plant does vary. In general sieve elements in roots both during differentiation and when mature contain much less P-protein than do sieve elements in stems.

VII. Conformation changes

Observations of the apparent replacement of one type of P-protein by another as differentiation of the sieve element proceeds, and of P-protein filaments which show structural characteristics of more than one type of P-protein (Figure 26) have led to the suggestion that P-proteins may undergo conformation changes and these have been related to the substructure of P-proteins, expecially that of P-protein tubules (Parsatharathy and Mühlethaler, 1969; Cronshaw, 1970b; Cronshaw et al, 1973). High magnification images of transections of P-protein tubules have shown that there are most probably six subunits around the tubule (Parthasarathy and Mühlethaler, 1969; Cronshaw et al, 1973). Negatively stained preparations of Nicotiana exudate reveal P-protein tubules that are comprised of two helical subunits (Cronshaw et al, 1973). In some fixed sectioned preparations P-protein filaments have been observed with two helical subunits (Cronshaw, 1970b). Parthasarathy and Mühlethaler (1969) suggested that P-protein tubules were composed either of a single helical strand of subunits with either six or twelve subunits per turn of the screw or of two helical strands of subunits with 12 subunits per turn of the screw. These authors suggested that a loosening and unwinding of the helical subfibrils and a stretching of the tubules would produce apparently striated fibrils of the type observed. Cronshaw et al (1973) described a model based on those of Parthasarathy and Mühlethaler suggesting that P-protein tubules are constructed of two helical subfibrils composed of near spherical subunits with six subunits per turn. This later model is consistent with Parthasarathy and Mühlethaler´s suggestion that a loosening and unwinding of the helical subfibrils and a stretching of the tubules would produce striated structures and would represent a conformation change of the P-protein filaments.

This conformation change would account for the abundance of striated fibrils which are observed near and in sieve plate pores of sieve elements when no precautions have been taken to prevent surging of the contents of the sieve element prior to fixation for electron microscopy. Under those circumstances it would be expected that as the release of hydrostatic pressure took place the P-protein would be forced into and through the sieve plate pores and stretched. This view is substantiated by the observations of Cronshaw and Anderson (1969) and Wooding (1969) who found little striated P-protein associated with the pores of sieve elements grown in tissue culture.

Observation of conformational changes such as that described for tubular P-proteins in tobacco and of morphological variation has given rise to the idea that P-proteins in general may consist of the same or basically similar subunits and that their morphology may be sensitive to external conditions. The various conformations of P-protein probably reflect a dynamic equilibrium amongst several states with the position of the equilibrium determined by environmental conditions. Cronshaw, Gilder and Stone (1973) suggested that the various conformations of P-protein may represent responses to changes in the sieve element conditions, especially those that occur during differentiation, and this may have functional significance. The structural equilibrium may be sensitive to such conditions as the levels of cations, ATP or carbohydrate.

VIII. Enzyme activity

Cytochemical studies of the distribution of ATPase activity in the phloem cells of <u>Nicotiana</u> and <u>Cucurbita</u> have shown that P-proteins may have an ATPase activity associated with them (Cronshaw and Gilder, 1972; Gilder and Cronshaw, 1973a, b, 1973). This ATPase activity was not apparent at early stages of cell differentiation during the formation of P-protein bodies, but appeared as differentaition and P-protein dispersal proceeded. The largest amounts of reaction product and ATPase activity were usually seen in mature sieve elements which were presumably functional in the long distance transport of nutrients (Figure 27).

It has not been determined whether ATPase activity is a property of the P-protein subunits themselves or is the property of another protein associated with them. The activity appears at about the same stage of cell differentaion as the conformation change of P-protein tubules to beaded fibrils occurs and it has been speculated (Gilder and Cronshaw, 1973b) that in <u>Nicotiana</u> ATPase activity may be located in or on the P-protein or its

subunits in such a way that it could be either activated or
unmasked during the dispersal and conformation change. The
loosening, stretching and unwinding of the tubule which has been
postulated to explain the conformation changes could present the
active site to the liquid phase in the sieve element. It was also
speculated that an alternative site for the ATPase is in the
spoke-like projections which connect adjacent tubules in P-protein
bodies. These intertubule connections are broken during dispersal
of the P-protein filaments and again may present the active site to
the liquid phase. It is possible that in one of these ways the
conformation change may regulate an ATPase activity of the
P-protein and may be the way in which changes in the environment of
the sieve element provide for a regulation of the function of the
P-protein. This regulation could be through the levels of ATP in
the phloem or possibly through levels of divalent of monovalent
cations or inorganic phosphate which move with the nutrient stream.
Calcium sensitivity of P-proteins has previously been reported by
Kleinig et al (1971b), and the polymerization of tubulin into
microtubules is known to be extremely sensitive to low
concentrations of calcium. Calcium is present in very low
concentration in sieve elements and may well regulate P-protein
conformation and ATPase activity.

IX. Functions of P-protein

In the recent literature there has been a great deal of
speculation concerning the functions of P-proteins. The function
of P-protein in plugging sieve plate pores on injury has been well
documented (Esau, 1969; Crafts and Crisp, 1970). This rapid
plugging prevents the continued loss of nutrients that would take
place on wounding in response to the hydrostatic pressure generated
by phloem loading. Other functions of P-proteins are far less
certain. The P-protein provides a large surface area in the
functioning sieve element and may well serve to distribute
metabolically active molecules.

Recently there have been several suggestions that P-proteins
may be directly functional in the long distance transport of
nutrients in the phloem, possibly by a contractile mechanism
(MacRobbie, 1971). Demonstration of an ATPase activity associated
with the P-proteins in mature sieve elements has given credence to
this speculation but obviously definitive statements on the
functioning of P-proteins must await further research.

References
ANDERSON, R. and CRONSHAW, J. 1970. Planta 91: 173-180
BEHNKE, H.D. 1968. Protoplasma 66: 287-310
BEHNKE, H.D. 1969. Protoplasma 68: 377-402
BEHNKE, H.D. 1973. Protoplasma 77: 279-289
BEHNKE, H.D. and DORR, I. 1967. Planta 74: 18-44
BOUCK, G.B. and CRONSHAW, J. 1965. J. Cell Biol. 25: 79-96

BURR, F.A. and EVERT, R.F. 1973. Protoplasma 78: 81-97
BUVAT, R. 1963. Compt. Rend. Acad. Sci. 257: 733-735
BUVAT, R. 1968. Compt. Rend. Acad. Sci. 267: 406-408
CRAFTS, A.S. 1932. Plant Physiol. 7: 183-225
CRAFTS, A.S. and CRISP, C.E. 1971. Phloem transport in Plants.
    W. Freeman, San Francisco.
CRONSHAW, J. 1970a. Roy Microscop. Soc. Micro 70 Conf.
    Catalogue, pp. 199-200
CRONSHAW, J. 1970b. In: Septieme Congres International de
    Microscopie Electronique, Grenoble. pp. 429-430
CRONSHAW, J. and ANDERSON, R. 1969. J. Ultrastruct. Res. 27:
    134-148
CRONSHAW, J. and ANDERSON, R. 1971. J. Ultrastruct. Res. 34:
    244-259
CRONSHAW, J. and ESAU, K. 1967. J. Cell Biol. 34: 801-815
CRONSHAW, J. and ESAU, K. 1968a. J. Cell Biol. 38: 25-39
CRONSHAW, J. and ESAU, K. 1968b. J. Cell Biol. 38: 292-303
CRONSHAW, J. and GILDER, J. 1972. In: 30th Annual Proceedings
    of the Electron Microscopy Society of America, Los Angeles.
    C.J. Areneaux, Ed. Claitor's Publishing Division, Baton Rouge.
    230.
CRONSHAW, J., GILDER, J. and STONE, D. 1973. J. Ultrastruct.
    Res. 45: 192-205
CURRIER, H.B. and SHIH, C.Y. 1968. Amer. J. Bot. 55: 145-152
ENGELMAN, E.M. 1965. Ann. Bot. 29: 83-101
ESAU, K. 1969. The phloem Handbuch der Pflanzenanatomie.
    Histologie. Band V, Teil 2. Gebruder Borntraeger,
    Berlin-Stuttgart.
ESAU, K. 1972. Ann. Bot. 36: 703-710
ESAU, K. and CHEADLE, V.I. 1965. Univ. Calif. Publs. Bot.
    36: 253-344
ESAU, K. and CRONSHAW, J. 1967. Virology 33: 26-35
ESAU, K. and GILL, R.H. 1970a. J. Ultrastruct. Res. 31:
    444-455
ESAU, K. and GILL, R.H. 1970b. Protoplasma 69: 373-388
ESAU, K. and GILL, R.H. 1973. J. Ultrastruct. Res. 44:
    310-328
ESAU, K. and HOEFFERT, L.L. 1971. Protoplasma 72: 237-253
ESAU, K., CRONSHAW, J. and HOEFFERT, L.L. 1967. J. Cell Biol.
    32: 71-87
ESCHRICH, W. 1963. Planta (Berl.) 59: 243-261
ESCHRICH, W., EVERT, R.F. and HEYSER, W. 1971. Planta (Berl.)
    100: 208-221
EVERT, R.F. and DESHPANDE, B.F. 1969. Protoplasma 68: 403-432
EVERT, R.F., ESCHRICH, W. and EICHHORN, S.E. 1971. Planta
    (Berl.) 100: 262-267
EVERT, R.F., ESCHRICH, W. and EICHHORN, S.E. 1973. Planta
    (Berl.) 109: 193-210
EVERT, R.F., TUCKER, C.M., DAVIS, J.D. and DESHPandE, B.P. Amer.
    J. Bot. 56: 999-1017

FISCHER, A. 1884. Untersuchungen uber das Siebrohren-System der Cucurbitaceen. Gebruder Borntraeger, Berlin.

FISCHER, A. 1885. Ber. Beut. Bot. Ges. 3: 230-239

FISCHER, A. 1886. Ber. Sachs. Ges. (Akad.) Wiss. 38: 291-336

GEROLA, F.M., LOMBARDO, G. and CATARO, A. 1969. Protoplasma 67: 319-326

GILDER, J. and CRONSHAW, J. 1973a. Planta (Berl.) 110: 189-204

GILDER, J. and CRONSHAW, J. 1973b. J. Ultrastruct. Res. 44: 388-404

GILDER, J. and CRONSHAW, J. 1974. J. Cell Biol. 60: 221-235

HALL, S.M. and BAKER, D.A. 1972. Planta (Berl.) 106: 131-140

HARTIG, T. 1854. Bot. Z. 12: 51-54

HEPLER, P.K. and PALEVITZ, B.A. 1974. Ann. Rev. Plant Physiol. 25: 309-362

HEYSER, W. 1971. Cytobiologie 4: 186-197

KLEINIG, H., DORR, I. and KOLLMANN, R. 1971a. Protoplasma (Wien) 73: 293-302

KLEINIG, H., DORR, I., WEBER, C. and KOLLMANN, R. 1971b. Nature (Lond.), New Biol. 229: 152-153

KOLLMANN, R., DORR, I. and KLEINIG, H. 1970. Planta (Berl.) 95: 86-94

KRUATRACHUE, M. and EVERT, R.F. 1974. Amer. J. Bot. 61: 253-266

KUO, J., O'BRIEN, T.P. and ZEE, S.Y. 1972. Aust. J. Biol. Sci. 25: 721-737

LAFLECHE, D. 1966. J. Microscop. 5: 493-510

LECOMTE, H. 1889. Ann. Sci. Nat. Bot. 10: 193-324

LEDBETTER, M.C. and PORTER, K. R. 1963. J. Cell. Biol. 19: 239-250

LEDBETTER, M.C. and PORTER, K.R. 1964. Science 144: 872-874

LOWRY, O.H., ROSEBROUGH, N.J., FARR, A.L. and RANDALL, R.J. 1951. J. Biol. Chem. 193:265-272

MACROBBIE, E.A.C. 1971. Biol. Rev. 46: 429-481

NAGELI, C.W. 1861. Konigl. Bayer. Akad. Wiss. 1: 212-238

NEUBERGER, D.S. and EVERT, R.F. 1974. Amer. J. Bot. 61: 360-374

NEWCOMB, E.H. 1967. J. Cell Biol. 35: C17-C22

NORTHCOTE, D.H. and WOODING, F.B.P. 1966. Proc. Roy. Soc. Ser. b 163: 524

O'BRIEN, T.P. and THIMANN, K.V. 1967. Protoplasma 63: 443-478

PALEVITZ, B.A. and NEWCOMB, E.H. 1971. Protoplasma (Wien) 72: 399-426

PARTHASARATHY, M.V. 1974a. Protoplasma 79: 59-91

PARTHASARATHY, M.V. 1974b. Protoplasma 79: 93-125

PARTHASARATHY, M.V. 1974c. Protoplasma 79: 265-315

PARTHASARATHY, M.V. and MUHLETHALER, K. 1969. Cytobiologie 1: 17-36

ROBIDOUX, J., SANDBORN, E.B., FENSOM, D.S. and CAMERON, M.L. 1973. J. Exp. Bot. 24: 349-359

ROELOFSEN, P. 1959. The Plant cell wall. Handbuch der

Pflanzenanatomie Band 3, Teil .
ROUSCHAL, E.   1941.   Flora 35:   135-200
SABNIS, D.D.   and HART, J.W.   1973.   Planta (Berl.) 109:   127-133
SALMON, J.   1946.   Recherches cytologiques.   p.   235.   Copyright
   1947 by J. Salmon.
SALMON, J.   1951.   Compt. Rend. Acad. Sci. 233:   495-496
SINGH, A.P.   and SRIVASTAVA, L.M.   1972.   Can.   J.   Bot.   50:
   839-846
STEER, M.W.   and NEWCOMB, E.H.   1969.   J.   Cell Sci.   4:   155-169
STONE, D.L.   and CRONSHAW, J.   1973.   Planta (Berl.) 113:   191-206
STRASBURGER, E.   1891.   Ueber den Bau und die Verrichtungen der
   Leitungsbahnen in den Pflancen: Histologische Beitrage. Heft
   111. Gustav Fischer, Jena
TAMULEVICH, S.R.   and EVERT, R.F.   1966.   Planta 69:   319-337
THAINE, R.   1964.   New Phytol.   63:   236-243
THAINE, R., PROBINE, M.C.   and DYER, P.Y.   1967.   J.   Exp.   Bot.
   18:   110-127
THOMPSON, R.G.   and THOMPSON, A.D.   1973.   Can.   J.   Bot.   51:
   933-936
WALKER, T.S.   and THAINE, R.   1971.   Ann. Bot. 35:   773-790
WALKER, R.S.   1972.   Biophys. Acta 257:   433-444
WARK, M.C.   and CHAMBERS, T.C.   1965.   Aust.   J.   Bot.   13:
   171-183
WEATHERLEY, P.E.   1962.   Advan. Sci. 18:   571-577
WEBER, C.   and KLEINIG, H.   1971.   Planta (Berl.) 99:   179-182
WERGIN, W.P.   and NEWCOMB, E.H.   1970.   Protoplasma 71:   365-388
WILHELM, K.   1880.   Beitrage zur Kenntnis des Siebrohrenapparates
   dicoyler Pflanzen. Wilhelm Engelmann, Leipzig.
WILLIAMSON, R.E.   1972.   Planta (Berl.) 106:   149-157
WOODING, F.B.P.   1967.   Protoplasma 64:   315-324
WOODING, F.B.P.   1969.   Planta (Berl.) 85:   284-298
YAPA, P.A.   and SPANNER, D.C.   1972.   Planta (Berl.) 106:   369-373
ZEE, S.Y.   1968.   Aust. J. Bot. 16:   419-426
ZEE, S.Y.   1969.   Aust. J. Biol. Sci. 22:   1573-1576

DISCUSSION

Discussant: Hans Kleinig
               Institut fur Biologie II
               Lehrstuhl fur Zellbiologie
               Universitat Freiburg i. Br.
               Germany

Srivastava: Dr. Kleinig was invited to give his paper on
"Biochemistry of Phloem Proteins". The paper follows.

Biochemistry of Phloem Proteins

Hans Kleinig

In recent years, biochemical investigations of proteins from
phloem exudate and phloem tissue were done nearly exclusively with
only three plant species, Cucurbita maxima, Cucurbita pepo, and
Heracleum mantegazzianum.   In the following, the results of these
investigations will be briefly summarized.

1. Cucurbita maxima

Kollman et al. (1970), isolated filamentous structures from
the exudate of C. maxima and showed the proteinaceous nature of
these filaments. The proteins were soluble in buffers containing
SH-group protecting agents like mercaptoethanol or dithioerythritol.   It was demonstrated that the proteins could be
reversibly aggregated into filamentous structures (Kleinig et al.,
1971a). Using polyacrylamide gel electrophoresis, several major
and minor protein bands were found in the exudate fractions
(Kleinig et al., 1971a; Weber and Kleinig, 1971; Eschrich et al.,
1971; Beyenbach et al., 1974; Weber et al., 1974). The two major
protein components, which comprised more than 40% of total protein
each, were separated by Sephadex and ion exchange chromatography
and further analyzed by analytical ultracentrifugation, amino acid
analysis, determination of IEP and others (Kleinig et al., 1971a;
Weber and Kleinig, 1971; Beyenbach et al., 1974).   Both proteins
showed a relatively high lysine content and were highly basic with
an IEP of above 9.5. The larger protein had a MW of 116,000 D,
showed the tendency to form higher aggregates, and formed a gel in
the absence of SH- protecting agents. This protein could not be
split into smaller units either by 8 M urea or by 6 M guanidinium
hydrochloride. The smaller protein turned out to be a dimmer of MW
60,000 D, which could be reversibly split into monomers of MW

117

30,000 D under highly reducing conditions. At least 10 minor
protein components were present in the exudate. No glycoproteins
could be detected using the PAS staining on polyacrylamide gels.

No enzyme activity can be attributed as yet to the major
proteins, although several enzymes seem to be present in the
exudate , the best characterized being the phosphodiesterase,
described by Heyser et al. (1974). Several authors failed to
detect any ATPase activity in the exudate (Kleinig et al., 1971a;
Eschrich et al., 1971; Beyenbach et al., 1974; Weber et al., 1974).
Using a lead phosphate precipitation technique, Gilder and Cronshaw
(1973) found ATPase activity associated with the P-protein in situ.
Weber et al., (1974), however, pointed out that P-protein may have
a tendency to adsorb particulate phosphate materials which might be
mistakenly identified as ATPase activity.

No binding of ATP, GTP, CTP, colchicine (Kleinig et al.,
1971a) or of α -naphthaleneacetic acid (Kleinig, unpublished) to
isolated P-protein of C. maxima could be detected. Colchicine
also appeared to have no morphological effects on the filaments
(Cronshaw et al., 1973). The proteins, however, were found to be
precipitable by vinblastine (Kleinig et al., 1971b).

Recently, it was hypothesized by Weber et al., (1974), that
P-proteins are not specifically synthesized during phloem
differentiation but are of cytopathological origin deriving from
ribosomal material during a general degeneration of cellular
components. More evidence for this interesting view should be
accumulated, before one should reject the general interpretation
made by others for Cucurbita (Cronshaw and Esau, 1967) and other
plant species (e.g. Behnke, 1969a, 1969b, 1973; Cronshaw and Esau,
1967; Esau, 1971; Heyser, 1971; Northcote and Wooding, 1966) that
the close association of the protein filaments with ribosomes
during sieve element differentiation, is the expression of
P-protein synthesis by ribosomes in a translation mechanism.

2. Cucurbita pepo

Walker and Thaine (1971) found several protein bands using
DEAE cellulose chromatography and polyacrylamide gel
electrophoresis from the exudate of C. pepo. A gelling protein
that constituted up to 4% of the total protein in the exudate, was
isolated and further analyzed by analytical ultracentrifugation and
amino acid analysis (Walker, 1972). A high proportion of basic
amino acids was found (18%) and a MW of 136,000 D was determined.
Seemingly, this protein shows properties similar to the gelling
protein from C. maxima (Beyenbach et al., 1974) although the
latter occurs in a higher concentration in C. maxima exudate.

3. Heracleum mantegazzianum

Yapa and Spanner (1972) determined the IEP of the main proteins of exudate and whole phloem tissue extracts from H. mantegazzianum by isoelectric focusing on polyacrylamide gels and sucrose gradients. The IEP was found to be 4.9. Using phloem tissue extracts, Hart and Sabnis (1973) showed that P-protein was unable to bind colchicine in vitro. In addition, it was demonstrated by the same authors that colchicine and vinblastine had no effect on the ultrastructure of sieve elements from isolated vascular strands (Sabnis and Hart, 1973).

Considering the results of the biochemical analyses reported above, it seems that in the two Cucurbita species, at least some similarities of the proteins may exist. The proteins from Heracleum, however, seem to have quite different properties. Actin-like proteins could not be demonstrated in the exudates of either species. If one assumes a uniform translocation mechanism in igher plants and a function of P-proteins in translocation, whatever this function might be, major diversities of the properties of these proteins would not be expected. Therefore, it seems to be a very meaningful, and for any theoretical consideration of translocation, a very fundamental question whether these kind of P-proteins known from C. maxima with their characteristic properties, are universally distributed throughout the plant kingdom or not. From the numerous morphological observations, such a universality would be suggested.

References

BEHNKE, H.D. 1969. Protoplasma 68, 289
BEHNEK, H.D. 1969. Ibid. 68, 377
BEHNEK, H.D. 1973. Ibid. 77, 279
BEYENBACH, J., WEBER, C., KLEINIG, H. in press. Planta
CRONSHAW, J., ESAU, K. 1967. J. Cell. Biol. 34, 801
CRONSHAW, J., ESAU, K. 1968. J. Cell. Biol. 38, 292
CRONSHAW, J., GILDER, J., STONE, D. 1973. J. Ultrastruct. Res. 45, 192
ESAU, K. 1971. Protoplsma 73, 225
ESCHRICH, W., EVERT, R.F., HEYSER, W. 1971. Planta 100, 208
GILDER, J., CRONSHAW, J. 1973. Planta 110, 189
HART, J.W., SABNIS, D.D. 1973. Planta 109, 147
HEYSER, W. 1971. Cytobiol. 4, 186
HEYSER, W., ESCHRICH, W., HUTTERMANN, A., EVERT, R.F., BURCHARD, R., FRITZ, E., HEYSER, R. 1974. Z. Pflanzenphysiol. 71, 413
KLEINIG, H., DORR, I., WEBER, C., KOLLMANN, R. 1971a. Nature New Biol. 229, 152
KLEINIG, H., DORR, I., KOLLMANN, R. 1971b. Protoplasma 73, 293
KOLLMANN, R., DORR, I., KLEINIG, H. 1970. Planta 95, 86
NORTHCOTE, D.H., WOODING, F.B.P. 1966. Proc. Roy. Soc. B 163, 524

SABNIS, D.D., HART, J.W. 1973. Planta 109, 127
WALKER, T.S. 1972. Biochim. Biophys. Acta 257, 433
WALKER, T.S., THAINE, R 1971. Ann. Bot. 35, 773
WEBER, C., KLEINIG, H. 1971. Planta 99, 179
WEBER, C., FRANKE, W.W., KARTENBECK, J. 1974. Exp. Cell Res. 87, 79
YAPA, P.A.J., SPANNER, D.C. 1972. Planta 106, 369

Kollmann: Dr. Cronshaw, what do you think about the origin of the P-protein within the cell? Is it synthesized by the ribosomes, or does it represent the protein of the disintegrated cytoplasmic material?

Cronshaw: From general knowledge of biochemistry, the site of translation is probably the ribosome and most probably, the ribosomes on the ER, though we do not know this in the sieve elements; there is so much ER in the regions of formation of P-protein. We also have to think in terms of self assembly of the macromolecular structures.

Kollmann: The self assembly process does not rule out synthesis of units by the ribosomes?

Cronshaw: No, the protein subunits, the polypeptides, are most likely synthesized on the ribosomes. But then from the polypeptides to the macromolecular structures could be a self assembly process.

Kleinig: I agree. Self assembly processes are well known for such structural proteins.

Walsh: There are reports in the literature on the occurrence of P-protein in cells other than sieve elements. Have you investigated the ontogeny of the P-proteins in these cells, and if so, considering that these cells are perhaps less liable to damage than the sieve elements, do the results contradict your observations on the sieve elements?

Cronshaw: Yes. In the associated parenchyma cells of the sieve elements, we also find P-protein. P-protein bodies are formed, but then they do not go through the dispersion stages.

Dorr: I was interested in your pictures of negative stained tubules. Can you be sure that these are P-protein tubules? Can it be that they are parietal microtubules which are frequent in sieve tubes, or perhaps, even microtubules from developing xylem elements? I do not know exactly how you collect the exudate - whether you separate phloem from xylem.

Cronshaw: No. We collect the exudate as droplets, but usually there is no contribution from the xylem and we always blot the ends first to remove most of what comes out of the parenchyma cells. We are not sure these are P-protein tubules. We do not see them too frequently, but then the sectioned material of Cucurbita also shows fibrils and tubules. As for the parietal microtubules, they are present during the early stages of differentiation, but they are absent in the mature sieve elements.

Johnson: Is it usual that the filamentous material is evenly distributed in the sieve elements, and would it be possible to measure the volume of the P-protein bodies, and to calculate what this volume of P-protein material would occupy, if it was uniformly distributed?

Cronshaw: Yes, there is fairly even distribution of P-protein in the cell. In the Ricinus picture, one that did not have a sieve plate in it, there was a fairly even distribution of P-protein in that cell. Yes, we could make calculations. We know for instance, in tobacco, that there is usually only one P-protein body. We know it's size and the size of the filaments. The one parameter that we would have to make a guess at, is the spacing and separation.

Weatherley: If there is a continuous network filling the whole lumen of a mature sieve tube, what is it's fate in relation to fluid moving down the sieve tube? At the rate of about 1 second, from sieve plate to sieve plate, it is a very rapid flow. Do you think this network is anchored, that it is not being swept down as a sort of suspension from sieve element to sieve element? If so, where does it end up and what happens to it? Therefore, it must be anchored. I do not know the dimensions of this, but I would imagine the fluid drag by such a continuous network filling the whole lumen would be enormous, not only putting a pressure flow quite out of court, but probably any other form of fluid movement.

Cronshaw: I agree with you that there is a much stronger case of anchoring of P-protein than there is for even distribution of P-protein.

Eschrich: Have these P-proteins been found in xylem exudates? We have fairly good evidence from electron micrographs that protein coming out in the sieve tube droplet comes only from one and the last cut sieve element.

Mittler: It may not represent the normal movement of proteins in the intact plant but, if my calculations are correct, the data of Eschrich and his collegues that was presented, would indicate a protein concentration ranging from 1% to 30% weight per volume. If that is correct, then we have a big soup protein in the exudate. This amount of protein must occupy quite a sizeable volume in a 30%

concentrated protein solution, perhaps 15, 20% of the volume of the phloem element.

Cronshaw: I would like to emphasize that <u>Cucurbita</u> is a very special case. It has more P-protein than any other plant that I know.

Eschrich: The cucurbits have an extremely high content of P-protein in their own sieve tubes.

Mittler: Has anybody obtained this protein through aphid stylets?

Kollmann: I realize <u>Cucurbita</u> is an unusual situation, but if the aphids were ingesting sap of the composition that approached anywhere near that of the exudation, even in other plants, they would not just be getting (what we traditionally thought) free amino acids in their diet, but they would be getting proteins. From the long association of aphids and plants, I would imagine that the aphids would be able to utilize these proteins. I do not see why they should waste them. We have some data showing that aphids, fed on artificial diets, can digest dipeptides as well as some unconfirmed evidence that they are able to cope with unhydrolysed proteins.

MacRobbie: This 1% protein solution is very dense, but even at that level, spread throughout the lumen, is a very fine collection of fibrils. If we space 1%, and something a good deal less in other plants, then this is relatively loose and there are fewer problems for mass flow of one kind or another. Even if we go up to the higher figures in Cucurbits of 30 mg/ml, this is still not a serious problem. Regarding the aphid, it may well be that if there is a P-protein network anchored, this leaves the aphid intact, whereas if we come along and chop at it, then we do break it up and therefore get the network out.

Spanner: Dr. Cronshaw mentioned that in the lower plants it has been shown that P-protein did not exist. I would like to refer to Dr. Evert and his co-workers' work with <u>Selaginella</u>. I know that Dr. Evert has interpreted the material in pores as not being P-protein. The same is true of Isoetes, because it is formed - at least in <u>Selaginella</u> it appears to be formed - in vacuoles and not in the cytoplasm. I would like to ask Dr. Cronshaw what he thinks about this material. I would also like to ask whether he sees any affinity between P-protein and the fibrillar material in the plastids; for instance, in the centrospermae of <u>Tetragonia</u>, which, in not very carefully prepared sections, appears to come out and mix itself with the P-protein.

Cronshaw: We have seen similar fibrillar material in vacuoles in maize and yet there appears to be no P-protein in maize that corresponds with P-proteins in other plants and I have not seen this material out of vacuoles. In Dr. Evert's work, this material was confined, always membrane bound. I would not regard this as P-protein. I believe Dr. Evert states that in each of the lower plants that he has examined, and also in the more recent paper on gymnosperms, there is no P-protein. In answer to your second question, I do not think it is P-protein in the plastids. I think it is a specific plastid inclusion and it may be liberated from the plastid probably as an artefact of preparation. If it is liberated, then it is there in very small quantities compared with the other P-protein in the cells.

Kleinig: When we speak of P-proteins, we mean proteins which are structurally observed under the electron microscope. However, when we do biochemistry, then this term becomes confusing because there are a lot of other proteins that may be there.

Srivastava: I have been bothered by the numbers of things that one sees in the electron micrographs which appear sometimes as fibrils sometimes as aggregations, etc., and the source of these things is not always clear. As to "P-protein", unless certain fibrils or tubules can be shown to have a certain type of origin which pertains to what has been described for P-protein, I do not think you can conclude that these are P-protein simply by looking at a few fibrils. They could be. There is a substantial amount of autolysis that goes on in these cells; many membranes and organelles are destroyed, and this debris must go somewhere. Either it becomes completely dissolved, or parts of it persist in the cell. As for the vacuolar contents, vacuoles do have certain substances which occasionally precipitate. Phenolic compounds are a classic example, but there may be other substances which also precipitate and take the form of fibrils. To conclude that the fibrillar stuff that one sees in the vacuole sometimes is P-protein, is rather dangerous.

Behnke: I agree that there are no real P-proteins in the lower vascule plants and that plastids in normal development do not break down and release their proteins. I also think that P-protein should be called P-protein only if we know it's origin. In our survey of more than 300 angiosperms, we found that the tubular type of P-protein is much less common than the filamentous type, so that the P-protein bodies that are seen in <u>Nicotiana</u> or <u>Cucurbita</u>, are not encountered as often as the so-called aggregates of filamentous proteins.

Johnson: There seem to be two possible functions for filaments in sieve tubes: 1) they are part of some sort of pump and 2) they help to block the sieve pores when the sieve elements are damaged. Considering the second possibility, and assuming that the filaments in plastids come out if the sieve element is damaged, is there any known relation between the concentration of P-protein and the concentration of filaments in plastids?

Behnke: There is no known relation. Also, these filaments in the plastids are only found in certain kinds of plants, Centrospermae Pinacae, and in some other families. Other protein-containing plastids mostly have crystalline structures, which disintegrate very rarely, if at all, if the plastids break down.

Spanner: Dr. MacRobbie's contention that 1% P-protein in the lumen of sieve tubes would not constitute a hydrodynamic difficulty, could be seriously questioned. If the P-protein is concentrated, as it appears very likely, into half the sieve element, you could at once, get 2%, and if one thinks of something like 2% agar gel, it would increase the resistance to flow to movement of water, quite appreciably.

Geiger: Would Dr. Johnson make some estimate of the spacing and the arrangement of the P-protein in freeze fractured material, not just near the plate but its general distribution throughout?

Johnson: I really have no comment to make about freeze-etching. It is extremely difficult to find filaments in the lumen. One tends to find them when they are together in bundles. When one does find the odd filament, it tends to be by itself.

Swanson: I am still very much intrigued by the distribution of P-protein, in Ricinus in particular, and if this is the true in vivo distribution in a translocating system, it seems that there is the possibility of being able to prove definitively that there is no flow in the sieve tubes, for how can that distribution be stabilized against flow?

Kleinig: There are examples from Heliozoa where there is flow and there are many structural proteins, microtubules.

Cronshaw:? In many animal cells we do have lots of structural proteins around and lots of flow. In answer to Dr. Spanner's point about the gelatin, we have examined some sections of gelatin and there are no filaments in there. All that you can see at high magnification, is just a granularity in the background. Therefore, the 2% protein system is much more diffusely spread whereas the P-protein system is aggregated into much larger macromolecular structure with space around it.

Weatherley: Referring to Dr. MacRobbie's point, I do not think there is any difficulty in blocking the system up with 1%, if it is distributed in the right way. Dr. Spanner's point about the gelatin is not quite fair. There may be only 1 or 2% of protein, but what blocks gelatin from mass flow, is all the oriented water around the protein middles. The whole system is rigid with water. It is an elastic solid really, through which only diffusion could occur. We have done the simple thing of impregnating a porous pot with gelatin and exerted hydrostatic pressure and you cannot push water through at all.

Spanner: I agree that the comparison is not entirely fair, but I did quote, I think, agar and not gelatin and I think agar does not imbibe and hold the water quite in the same way as gelatin does.

Cronshaw: It was agar we examined, not gelatin.

Milburn: Dr. Dixon tried to get Ricinus to hold aphids. They managed to persist there for a few days but then got sick and died, presumably because Ricinus contains at least 3 highly toxic principles - an alkaloid ricinine, an allergenic protein ricin, and also one or two other things that have carcinogenic properties about which less is known. Ricinus is a rather free exuder and many of the protein analyses done on Ricinus have been on the considerable amount of exudate that comes out over a protracted period of time. If the protein only comes from the one sieve element adjacent to the cut, then there has got to be a lot there, as I presume all the remaining material coming from other sieve elements in the file would merely dilute the protein. The protein in cucurbits seems to be exceptional in that it polymerizes under the influence of oxygen. If you collect the exudate from cucurbits in a tube, it will remain like that for weeks, but if you break the tube open, it gels immediately. If the proteins are involved in sealing, the sealing seems to be reversible. For example, you can squeeze a Ricinus stem on the outside and stop any exudate when you cut, but one hour later, if you make a cut in the same zone, you will get exudation. This seems like induced blockage, which is reversible, and the system is almost certainly protein. Referring to aphids, it may seem that P-protein is there in excessive amounts to carry out a very simple function - to limit the quantity of sap that the probing parasite can get out, to as small an amount as possible, either by sealing or by oxidative gelling; unless of course, it has got other defense mechanisms such as poisons, etc. This may be the reason why exudate is so difficult to obtain from many plants, under most circumstances.

Crookston: Is there any taxonomic pattern to the distribution of P-protein within the plant kingdom?

Behnke: There seem to be no systematic implications in the distribution of P-proteins other than that they occur only in angiosperms. Among these, the absence of P-protein has been recorded so far only in the Poales, the grasses, such as maize and wheat.

Walsh: No P-protein was observed in Lemna minor, either.

Aronoff: Dr. Kleinig, to complete the correlation between the EM and your biochemistry, have you studied the aggregation properties of the purified proteins, and ascertained whether they are able to aggregate in a similar way or whether you require oxidation? Also, you described the reversibility of one of the systems, but did not indicate how you did your reduction.

Kleinig: The reversibility was induced by removing the reducing conditions, and dialysing out, for example, mercaptoethanol. It seems to turn out that both the dimerization and the gelling process are at least partly due to the formation of S-S bond. Only the 160,000 protein exhibits gelling properties - not the 60,000 or 30,000 ones. We have not yet studied the aggregation of the isolated and purified proteins into larger structures discernible under the electron microscope.

Kollmann: We have been able to reaggregate the proteins of the first and the second peak after pretreatment with vinblastin, potassium and calcium, and also form the reversible filament and membranes and tubules.

Kleinig: I meant under careful ionic conditions; not with vinblastine. With vinblastine you can precipitate all these things.

Kollmann: Walker and Thaine do not find the gelling factor in agreement with that reported by Dr. Kleinig.

Guiaquinta: This fraction that has the high isoelectric point of close to 10 - does it represent the major fraction of P-protein, which would be positively charged at the physiological pH?

Kleinig: The two fractions comprise about 40% of the total and both are highly basic; therefore, it is actually about 80% of the total protein.

Giaquinta: Would you comment on the proposed similarity between ribosomal proteins and P-proteins?

Kleinig: It has been hypothesized that P-protein is formed, not by, but from, ribosomal material. This is quite a strange hypothesis, and I think more evidence is needed before it can be discussed.

Geiger: If, in view of species differences in the P-proteins, would it not be possible to devise a grafting experiment to determine if the P-protein is in fact, normally moving. If the two plants could be different enough to be easily distinguished in their P-protein, and would also take a graft in which phloem conduction would be possible between the two parts, it would seem that one could look at the sieve tubes and determine whether the P-protein is in fact, the native type or the type that has moved from the other part of the graft.

Giaquinta: Are there any biophysical or biochemical differences in the P-protein in various physiological regions in the plant source sink?

Kleinig: ATPase is not evenly distributed in the plant. Dr. Spanner recently published in Planta, on the localization of ATPase activity in mature sieve elements, and discussed some possible artefacts due to diffusion of the lead precipitate.

Spanner: There is a considerable amount of controversy as to whether the ATPase can be located satisfactorily by this test. It seems that this is a point on which one must be fairly guarded.

Cronshaw: The two main criticisms that have been raised of the lead precipitation technique for the localization of nucleoside phosphatase activity are: 1) lead can itself cause hydrolysis of ATP so that we can have phosphate cleaved and reaction product deposited due to the action of lead directly on the ATP without any enzyme being there at all; 2) lead can inhibit the action of ATPase so that if we have lead in the solution, plus an ATPase, we do not get any enzymatic activity. If you put these two things together, you are heading for artefacts in your preparations. There is a whole series of papers in the literature describing the technique, some say things are valid, and others say things are not valid. That is why we did the experiments we did. We had as many controls as we could think of and we did the biochemistry to ascertain the percentage of activity that remained after fixation. Compared with animal cells, our percentage activity of the APTase was about 13%. That figure of remaining activity is very high indeed and we thought it was sufficient to give us localization. But the main control that we did was with the other subtrates. We got good hydrolysis with ATP, UTP, GTP and ITP. We got good hydrolysis with ADP and 5 AMP, but we did not get hydrolysis with 2 AMP or 3 AMP. We did not get hydrolysis of p-nitrophenylphosphate or b-glycerophosphate at the pH that we used.

Sieve Element Cell Walls

James Cronshaw

Department of Biological Sciences

University of California, Santa Barbara

Santa Barbara, California

A.   Terminology

Sieve elements typically have non-lignified cellulosic walls similar to those of parenchyma cells. The walls are often thickened and in most instances this thickening is described as "nacreous" thickening or the "nacre" wall. The term "nacre" was introduced because of the refractve properties of this wall and its characteristic lustre when viewed with the light microscope. Cell walls are usually classified as either primary or secondary but in the case of the nacreous wall of the sieve element there is no general agreement as to the use of these terms (Esau, 1969). The term primary wall is used for the wall which is first formed around a developing cell and which surrounds the protoplast during growth in surface area. The term secondary wall is used for wall layers that are laid down after growth in surface area has ceased. Although it has not been formally accepted that the nacreous wall is a secondary wall, it is best considered as such in order to clearly distinguish between the major wall layers. Between the primary walls of contiguous cells is the intercellular substance.

The part of the sieve element wall which bears a group of performations through which the contents of contiguous sieve elements are in contact is termed the sieve area. If the sieve areas on all walls are of equal specialization as in the gymnosperms and vascular cryptograms, the sieve element is termed a sieve cell. In the angiosperms the end walls are termed sieve plates as they have one or more sieve areas that are more specialised than those on the laterial walls. Sieve areas on lateral walls are termed lateral sieve areas.

The terms sieve areas and sieve plates usually refer to double structures, that is, similar structures in the walls of two contiguous cells. Longitudinal series of sieve elements connected by sieve plates are termed sieve tubes.

In parenchyma cells there are thinner areas in the wall through which plasmodesmata pass. These areas are known as pit fields or primary pit fields. Where sieve elements are connected with parenchyma cells the walls may be developed into sieve areas on the sieve element side only; the parenchyma cell walls having primary pit fields.

B. Basic structure of the wall

The shape of sieve elements, especially in relation to their taxonomic position, has been extensively reviewed by Esau, (1969). The length and width of sieve elements vary greatly both within the same plant and between species and genera. The degree of thickening of the walls of sieve elements also varies greatly between species and genera. In some cases the wall thickening is so extensive that the lumen of the cell is almost occluded (Esau, 1969).

Irrespective of the shape or degree of thickening of the sieve element wall, the basic structure is that of cellulose microfibrils embedded in an amorphous matrix. Few studies have been made of the chemistry and submicroscopic structure of sieve element walls themselves. Much of the information available is the result of comparison with other cell types that have been intensively studied.

Cellulose consists primarily of β-D-glucose residues linked together by β-1,4 linkages into long chains with a degree of polymerization of 8,000-12,000. The glucan chains are arranged strictly parallel to one another over at least part of their length and are regularly spaced into a crystal lattice. These crystallites are more than 600Å long and about 50Å wide. The crystalline and non-crystalline regions are organized into microfibrils which vary in size depending on species. They are usually extremely long and vary in width according to species although the exact measurements are in dispute at the moment (Preston, 1971). Some investigators have described elementary fibrils which are subdivisions of microfibrils. Usually the microfibrils are described as between about 80Å and 250Å, being approximately half as thick as wide (Northcote, 1972). The crystalline part of the microfibrils forms a central core and is built of β-D-glucose residues only. Around the central core is a region of cellulose molecules which contains chains of sugar residues other than glucose in variable amounts. These prevent close packing of the chains into a crystalline structure (Preston and Cronshaw, 1958). The mechanical properties of cell walls are to a large extent determined by the structure, size, and orientation of the microfibrils.

The matrix material of the wall is composed of linear and branched polysaccharide polymers, many of which are heteropolymers, together with water and a small amount of protein. The structure of the polysaccharides varies from plant to plant although there are families of polysaccharides for which basic structures can be described. These are the xylans and arabinoxylans; gluco- and galactomannans; arabinogalactans and galacturonorhamnans (Northcote, 1972). Another wall matrix component present in localized regions is callose. Chemical studies have shown that callose is an unbranched polymer of D-glucose residues united by β-1,3 linkages; and with a degree of polymerization of approximately 100 (Kessler, 1958). In contrast to cellulose, callose is isotropic and identifiable in most preparations at the electron microscope level by its electron- lucent appearance. With the light microscope callose is identified by its specific staining reaction with aniline blue and resorcinol blue. Its density (1.62) is higher than that of cellulose (1.55).

C.   Primary walls

Primary walls must be capable of deformation as developing cells increase their surface area. Sieve element primary walls have been shown by staining reactions to be rich in matrix materials and to contain some cellulose (Srivastava, 1969). Muhlethaler (1950, 1953) carried out electron microscopic studies on the microfibrils of the primary walls of sieve elements in Avena and Zea coleoptiles. The orientation of the microfibrils and the pattern of growth was shown to be similar to that in parenchyma cells. The microfibrils of young sieve elements were oriented predominantly in a transverse direction, though some on the outside of the cells had an axial orientation. Corner thickenings were observed with an axial orientation of microfibrils similar to those found in parenchyma cells. In older cells, axially-oriented microfibrils on the outside surface were more numerous and the whole network of microfibrils was denser. There were numerous primary pit fields distributed over the whole surface of the cells. The normal orientation of microfibrils was interrupted at the primary pit fields, which were thinner regions in the wall. There were numerous small openings in the primary pit fields where plasmodesmata had traversed the wall in vivo. From these observations it is apparent that the primary walls of the sieve elements develop by a multi-net mechanism similar to that of parenchyma cells.

The distribution of microfibrils in developing sieve plates of Cucurbita sieve elements was studied by Frey-Wyssling and Muller

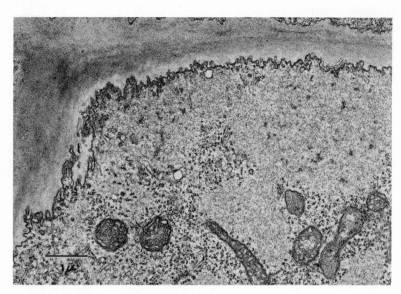

Fig.1. <u>Cucurbita</u> <u>maxima</u>. Portion of a sieve element at an early stage of differentiation. The cell wall cytoplasmic interface is highly convoluted.

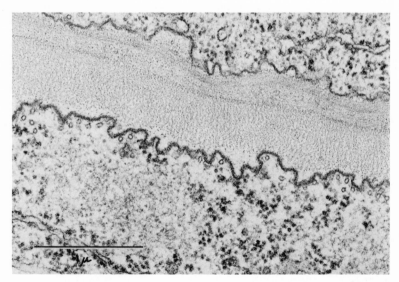

Fig.2. <u>Cucurbita</u> <u>maxima</u>. Portions of a sieve element at an early stage of differentiation and companion cell showing transections of peripheral microtubules.

(1957).   They showed that there were numerous pores in young cells scattered throughout a thin network of microfibrils. As the wall thickened the number of microfibrils increased, and the pores became rounded with the microfibrils curved around them.

In sieve elements that have been processed for electron microscopy after glutaraldehyde-osmium fixation, the orientation of microfibrils is visible in the walls, and the fine structure of the cytoplasm is preserved.   The cytoplasmic components show many of the structural features that have been associated with wall development in other cell types (Cronshaw, 1965).   In sieve elements at early stages of differentiation where wall development is proceeding the plasma membrane often has a highly convoluted form indicative of intense surface activity and most probably, fusion of vesicles (Fig. 1). Often the plasma membrane stains more heavily in the sieve element than in surrounding cells (Bouck and Cronshaw, 1965). It may be stained asymmetrically, the outer electron opaque layer appearing thicker than the inner one. Microtubules are present in the peripheral cytoplasm of cells with developing walls and, as in other plant cells, these are oriented parallel to the microfibrils that are being deposited (Fig. 2).

D.   Secondary Wall

The secondary or nacreous wall of sieve elements has long been known from staining reactions and studies by polarization optics to have a high cellulose content (Esau, 1969; Srivastava, 1969). Esau and Cheadle (1958) surveyed the occurrence and characteristics of the nacreous wall in the secondary phloem of 142 species in 121 genera of 58 families of dicotyledons. Forty-five of these species had a nacreous wall layer detectable by light microscopy and within these species there was a wide range of variation in thickness of wall.   Few of these species have been examined with the electron microscope.

We have examined the sieve element secondary wall of Cucurbita maxima, Nicotiana tabacum, Phaseolus vulgaris, Beta vulgaris, Pisum sativum and Zea mays. In all of these species the mature wall consists of a fairly uniform layer of predominantly transversely oriented microfibrils, (Figs. 3 and 4) with the exception of an inner layer which often is much more electron opaque and shows a distinct striate pattern when cut in oblique section (Figs. 5 and 6).   In such oblique sections the striae are often distinct, are parallel to one another, and may appear to be continuous with membranes in the cytoplasm both being perpendicular to the long axis of the cell (Esau and Cronshaw, 1968).   From similar observations Spanner and Jones (1970) concluded that there are extensions of the plasma membrane into the wall forming microvilli similar to those of some animal cells, a finding that they thought was relevant to potassium exchange in electroosmosis.   Oblique

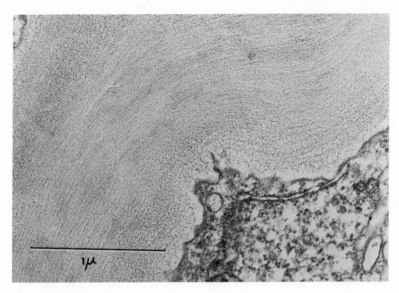

Fig.3. <u>Nicotiana</u> <u>tabacum</u>. Section of a portion of a sieve element showing the transverse orientation of microfibrils in the nacreous wall.

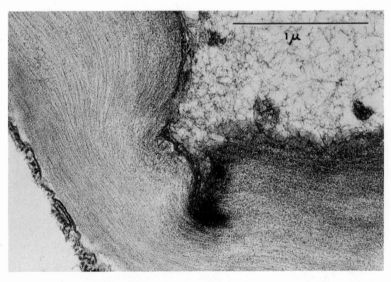

Fig.4. <u>Cucurbita</u> <u>maxima</u>. Section of a portion of a sieve element showing the transverse orientation of microfibrils in the nacreaous wall.

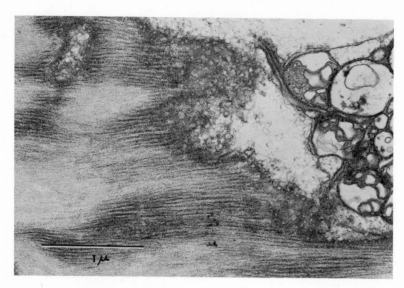

Fig.5. _Cucurbita_ _maxima_. Oblique section of a portion of a sieve element. Distinct striae are visible in the wall.

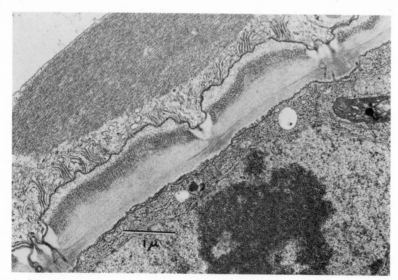

Fig.6. _Phaseolus_ _vulgaris_. Longitudinal section of a portion of a differentiating sieve element and of a companion cell. The sieve element has an electron opaque layer of nacreous wall and cisternae of endoplasmic reticulum oriented approximately at right angles to the plasma membrane.

sections are difficult to interpret, however, and it is most probable that the membranes are part of the peripheral endoplasmic reticulum system which, in well-oriented longitudinal sections can be seen to be closely applied to the plasma membrane but distinct from it (Fig.6).

The secondary walls of sieve cells of gymnosperms show distinct lamellae and have a characteristic helical structure (Abbe and Crafts, 1939; Esau and Cheadle, 1958). Detailed studies have been made of Pinus strobus sieve elements by Srivastava (1969) and Chafe and Doohan (1972) and these authors differed in the interpretation of their results. Srivastava described the secondary walls of the sieve elements as thick and lamellate and composed predominantly of cellulose with lesser amounts of polyuronides. The lamellae were composed of two intersecting parallel sets of microfibrils and these were envisaged as making angles to both the horizontal and vertical axes of the cells, and so not to lie in the plane of the lamellae but at an angle to it. Chafe and Doohan (1972) disagreed with this interpretation and described the thickened sieve cell wall as composed of alternate lamellae in which microfibrils describe either S or Z helices around the cell. It was judged that the orientation was of moderately low helical pitch, perhaps 50° to 60° or more from the cell axis although there was considerable variation in the orientation of the microfibrils to the cell axis both within and between walls. The wall structure proposed by Chafe and Doohan seems more reasonable since it is comparable to the structures of the walls of other cell types such as wood fibres, tracheids and phloem fibres.

E. Sieve areas and sieve area pores

The sieve areas that have been most frequently studied with the electron microscope are those of the sieve plates of angiosperms. Sieve plates are termed simple if they have only one sieve area, compound if they have two or more. Compound sieve plates may be scalariform if the sieve areas are in one row or reticulate if they form a netlike pattern. The sieve plates connect one sieve element to another and have large pores at maturity. These pores are lined with the plasma membrane which is continuous from cell to cell (Figs. 7 and 8). Surrounding the plasma membrane is a cylinder of callose of varying thickness into which the plasma membrane may infold (Fig. 8). The microfibrillar network can be clearly seen in electron micrographs of fixed, sectioned sieve plates. It ends at the electron-lucent callose cylinder, which is devoid or almost devoid of microfibrils. The primary walls in sieve plates as in other regions of the cell often stain more intensely than the intercellular substance and the secondary wall.

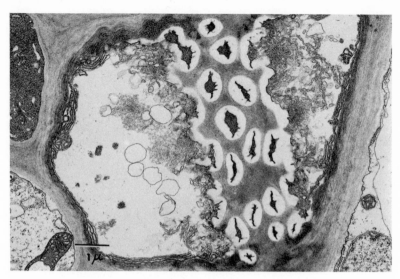

Fig.7. <u>Phaseolus</u> <u>vulgaris</u>. Transverse section of a sieve element showing a sieve plate consisting of a network of microfibrils containing plugged pores surrounded by callose cylinders.

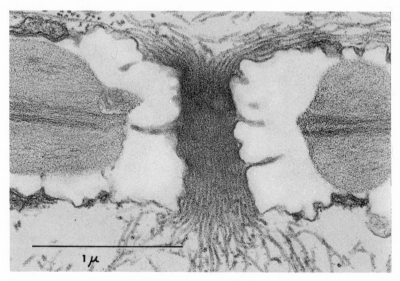

Fig.8. <u>Nicotiana</u> <u>tabacum</u>. Median longitudinal section through a sieve plate pore. The plasma membrane is continuous from cell to cell through the pore which is plugged with P-protein and is surrounded by a callose cylinder.

The amount of callose observed on mature sieve plates varies according to preparation procedures and the degree of injury to the sieve elements. The deposition of callose in response to wounding is well known as is also the deposition of callose in some sieve elements prior to dormancy. The literature on the properties of callose and the factors influencing its deposition has been extensively reviewed (Esau, 1969; Eschrich, 1970; Crafts and Crisp, 1971).

The sieve areas of gymnosperms have pores which are joined in the middle of the wall by a median cavity (Kollmann, 1964; Kollmann and Schumacher, 1963; Srivastava and O'Brien, 1966; Evert et al, 1973). As in the angiosperms, the plasma membrane is continuous from cell to cell, lining the pores. Usually there are callose cylinders around the pores, although the amount of callose has been reported as small or, in some cases, absent (Srivastava and O'Brien, 1966; Evert et al, 1973). Membranous material has been reported in the median cavity of all species so far examined.

F. Sieve area pore contents

The pores of mature sieve plates and the nature of their contents have been the subject of intense study in view of their importance in determining the mechanism of the function of sieve elements (Crafts and Crisp, 1971). The previous paper discussed the controversy over the distribution of P-protein in mature sieve elements, and the problems of interpretation of electron micrographs of mature sieve elements. These problems must be especially considered in relation to the contents of sieve-plate pores. When sieve plates were first examined by electron microscopy, fixation was with osmium tetroxide and the pores were described as plugged with osmiophilic contents (Hepton et al, 1955; Schumacher and Kollmann, 1959; Kollmann, 1960). Later, potassium permanganate was introduced as a fixative. Esau and Cheadle (1961, 1965) using this method of fixation, demonstrated unplugged pores in Cucurbita. Potassium permanganate, however, is unsuitable for use as a fixative for electron microscopy of sieve-plate pores since it dissolves away many cytoplasmic components.

After the introduction of glutaraldehyde and acrolein for the fixation of phloem tissue (Bouck and Cronshaw, 1965), which yielded excellent fixation, plugged pores were described for many species (see Cronshaw and Anderson, 1969). The validity of these observations of plugged pores was questioned when Esau, Cronshaw and Hoeffert (1967) described pores in the sieve elements of sugar beet plants that were plugged with virus particles. Sugar beet yellows is known to move in the nutrient stream and the results were interpreted to mean that virus particles were forced into the pores, plugging them during preparation of the material for electron microscopy. It was suggested that sieve plate pores were unplugged in vivo but could be plugged either by virus particles or by P-protein filaments as a results of the release of hydrostatic pressure. These observations conflicted with previous demonstratons of plugged pores and a series of experiments was initiated to try to determine unequivocally the nature of the contents of the sieve plate pores (Cronshaw and Anderson, 1969; Anderson and Cronshaw, 1969, 1970b).

When tobacco plants are cut prior to fixation in glutaraldehyde, formaldehyde-glutaraldehyde, or acrolein, the pores are universally found to be plugged with P-protein. Similarly, when whole plants are fixed in glutaraldehyde or formaldehyde-glutaraldehyde the pores are universally found to be plugged. After fixation of whole plants with the rapidly penetrating fixative acrolein, however, some pores lack dense P-protein plugs. Some unplugged pores are also observed when whole plants are rapidly frozen in liquid nitrogen and then transferred to chemical fixatives (Cronshaw and Anderson, 1969). Unplugged pores can also be observed when very thin slces of tobacco plants are rapidly cut into chemical fixatives and processed for electron microscopy.

In view of the uncertainty of judging surging artefacts in sieve elements Anderson and Cronshaw (1969) attempted to determine the precise effects of the release of hydrostatic pressure. Observations were made near to the cut ends of specimens that had been excised into fixative. The plastids were usually disrupted and the starch grains liberated. It was found that sieve plates had a dense plug of material on the "upstream" side of the sieve plates which consisted of P-protein and other organelles. On the "downstream" side of the sieve plate strands had been formed usually of aggregates of P-protein although in some cases strands consisted of cisternae of endoplasmic reticulum. Most of the pores were plugged with P-protein. In some instances, however, starch grains larger than the diameter of the pores were observed to block them on the "upstream" side. Pores blocked by starch grains in this way were devoid of P-protein plugs (Fig. 9). It was argued that starch grains may occlude the pores and may prevent their plugging with P-protein on the release of hydrostatic pressure. Siddiqui and Spanner (1970) interpreted these results differently. These authors argued that starch grains coming from the peripherally located plastids would arrrive at the sieve plte later than the P-protein coming from the central region of the lumen. They suggested that perhaps large starch grains were responsible for the removal of P-protein plugs originally present in pores. However, large starch grains which occlude empty pores often have substantial amounts of P-protein on the "upstream" side and this observation seems to negate Spanner's argument. Also, in some cases, medium sized starch grains are found within the pores together with P-protein.

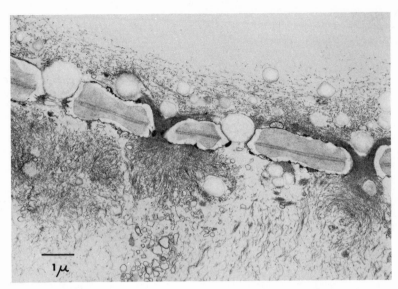

Fig.9. <u>Nicotiana</u> <u>tabacum</u>. Longitudinal section of a portion of a sieve plate from near the cut surface of a specimen cut prior to fixation. The sieve plate pores are either plugged with P-protein or blocked on the "upstream" side by starch grains.

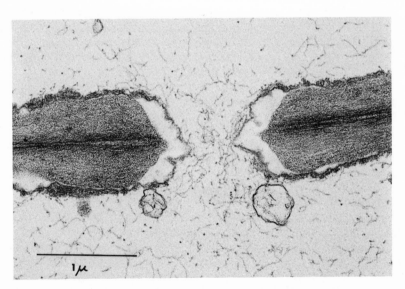

Fig.10. <u>Nicotiana</u> <u>tabacum</u>. Median longitudinal section through a sieve plate pore of a plant wilted prior to fixation. The pore is unplugged and the callose cylinder narrow.

In an attempt to minimize the effects of hydrostatic pressure
in the sieve element, tobacco and bean plants were wilted prior to
fixation (Anderson and Cronshaw, 1970).   Electron microscopical
examination of these plants showed that in some plants the pores
were unplugged. Where unplugged pores were observed most of the
pores within the plant were in the unplugged condition. In other
plants the pores were plugged and again this condition was uniform
throughout the plant. In the plants with unplugged pores there was
little evidence of rapid flow in the sieve element due to the
release of hydrostatic pressure. The P-protein appeared to be
almost evenly distributed in a network in the lumen of the sieve
elements. This network was continuous through the sieve-plate
pores (Fig.10).   The callose cylinder surrounding the pores in
wilted plants in which the pores were unplugged was comparatively
small and this was taken as an indication of lack of injury to the
plant (Fig. 10).

In tissue culture sieve elements can be induced to
differentiate in isolated nodules, and in this situation they form
a closed system, not part of a continuous phloem system as in whole
plants. Anderson and Cronshaw (1969) examined sieve elements
differentiated in tobacco pith cultures. Again, both plugged and
unplugged pores were observed. The isolated sieve elements in the
nodules, although not functional in long distance transport, may
have developed sufficient hydrostatic pressure to cause the
plugging that was observed.

The results show that unplugged pores can be demonstrated, in
some cases by fixation with rapidly penetrating fixatives, in other
cases by fixation after experimental manipulation. It seems most
probable that this is the condition of the pores in vivo in view of
the fact that pores are always plugged when observed in specimens
cut from the plant and fixed in chemical fixatives. None of the
individual experiments in themselves are definitive, but taken
together they present a strong case for unplugged sieve area pores.
These results have been substantiated by work from other
laboratories. Johnson (1968, 1973) studied sieve area pores in
phloem cells prepared for electron microscopy by freeze fracturing
and found that there was a uniform network of P-protein continuous
from cell to cell through the pores with no dense plugging.
Unplugged pores have also been observed by Shih and Currier (1969)
in acrolein-glutaraldehyde fixed intact cotton plants and by
Currier and Shih (1968) in Elodea.   Behnke (1971) observed
unplugged pores in Aristolochia phloem although P-protein
filaments, tubular endoplasmic reticulum and sometimes a
filamentous component derived from plastid crystalloids extended
through the sieve plate pores.   Evert et al (1973) observed a
peripheral distribution of P-protein in the pores of Cucurbita
plants which had been starved prior to fixation and again the pores
were unplugged (Evert et al, 1973).

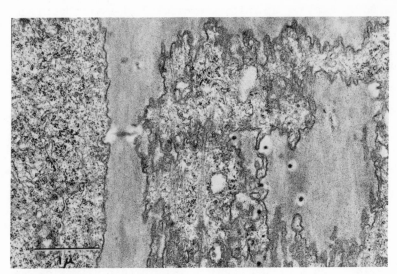

Fig.11. <u>Nicotiana</u> <u>tabacum</u>.   Oblique   section   through
developing   sieve   plate   of   a   sieve element at an early stage
differentiation.   Plasmodesmata   can   be   seen   in   transverse
longitudinal section.   Callose deposition has been initiated aro
the plasmodesmata which are future pore sites.

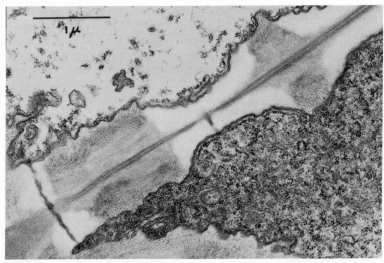

Fig.12. <u>Cucurbita</u> <u>maxima</u>. Longitudinal section of portions
two immature sieve elements with a differentiating sieve pl
between them.   Callose platelets occur at sites of future pores
endoplasmic   reticulum   cisternae   are   localized in contact with
plasma membrane at the sites of the callose platelets.

G. Development of sieve area pores

In the angiosperms the development of sieve area pores in sieve plates has been studied in Cucurbita (Esau, Cheadle and Risley, 1962; Evert, Murmanis and Sachs, 1966), Acer pseudoplatanus (Northcote and Wooding, 1966), Nictiana tabacum (Anderson and Cronshaw, 1970), Saxifraga sarmentosa (Deshpande, 1974) and several species of palm (Parthasarathy, 1974). The sequence of events during pore development appears to be similar in all cases. The sieve area pores develop from plasmodesmata which are distributed in the developing sieve plate, each plasmodesma eventually forming a pore (Fig.11). Early in the development of sieve area pores, cisternae of endoplasmic reticulum become closely applied to the plasma membrane at the pore sites (Figs. 12 and 13). Rings of callose are deposited in the wall around the plasmodesmata just outside the plasma membrane. Deposition of callose continues until two callose platelets are formed at opposite sides of each plasmodesma in the developing sieve plate (Fig. 12). These callose platelets increase in size to form cone-shaped pads which eventually fuse. Continued deposition of callose results in cylinders of callose surrounding the plasmodesmata. Wall material appears to be displaced or removed from the site of the pore by the developing callose masses. The plasmodesmata enlarge in the region of the primary walls and intercellular substance to form median nodules. The plasmodesmatal cylinders then widen from the median nodules with dissolution of callose until finally large pores are formed. The plasma membrane remains continuous from cell to cell through the pores. In mature cells residual callose is nearly always observed as thin walled cylinders around the pores.

Much less information is available on the development of sieve areas in gymnosperms (Esau, 1969). Recently Evert, Bornman, Butler and Gilliland (1973) have described the development of sieve areas in leaf veins of Welwitschia. In this species the development of the pores is not associated with callose platelets and the plasmodesmatal canals widen more or less uniformly along their length. Median cavities which connect the plasmodesmatal canals are formed by the union of median nodules.

H. Sieve element connections to other cells

Adjacent sieve elements are connected to one another by lateral sieve areas. They also have connections with other parenchymatal cells and these connections develop from primary pit fields. The plasmodesmata in the primary pit fields have the usual structure consisting of a tube of plasma membrane continuous from cell to cell which contains a central tubule. During differentiation of the sieve element and companion cells, the plasmodesmata often become branched on the companion cell side (Fig.14). On the sieve element side, callose is often deposited

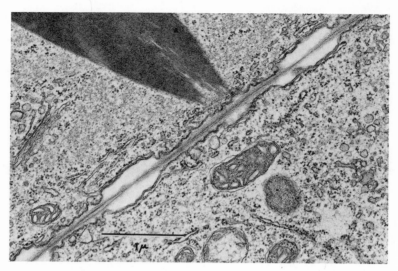

Fig.13. _Phaseolus_ _vulgaris_. Longitudinal section of portions of two sieve elements with a sieve plate at a stage of differentiation similar to the plate shown in Fig.12.

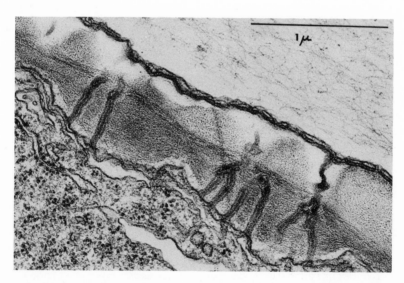

Fig.14. _Cucurbita_ _maxima_. Longitudinal section of portions of a sieve element and a companion cell. Plasmodesmatal connections between the cells are branched on the companion cell side.

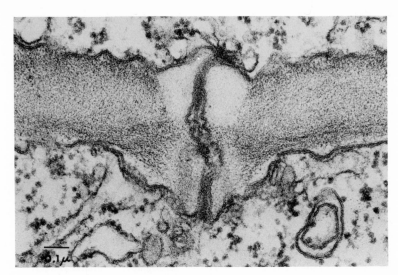

Fig.15. <u>Cucurbita</u> <u>maxima</u>. Plasmodesmatal connection between a differentiating sieve element and a companion cell. Callose is associated with the plasmodesma on the sieve element side.

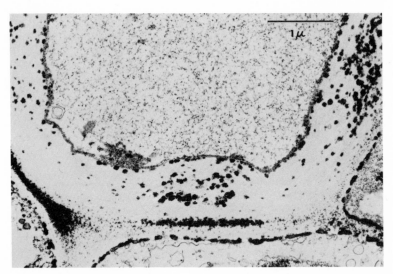

Fig.16. <u>Cucurbita</u> <u>maxima</u>. Transverse section of portions of a sieve element and adjacent cells, stained to show the localization of ATPase. Reaction product is localized in the walls between the sieve element and contiguous cells indicating regions of ATPase activity.

around the plasmodesmata which may become modified to form po
canals. Branched plasmodesmata are characteristic of t
connections between sieve elements and companion cells but al
occur between sieve elements and any contiguous parenchyma cell
In some cases the central part of the plasmodesma may beco
modified and enlarged to form a structure similar to that of t
median nodule of the developing sieve plate pore (Fig. 15).

I. Wall ATPase

In a cytochemical study of the distribution of ATPase activi
in the phloem of <u>Cucurbita</u>, Gilder and Cronshaw (1973a) observ
specific regions of reaction product in the walls indicating ATPa
activity. In the phloem of minor veins reaction product w
observed in the wall between sieve elements and companion cells a
between sieve elements and parenchyma cells (Fig. 16). ATPa
activity was also evident within plasmodesmatal connections betwe
sieve elements and contiguous cells. Localized wall ATPa
activity was not observed in petiole phloem; it was suggested th
the wall ATPase in minor veins may possibly indicate an apopla
pathway for the movement of carbohydrates.

References

ABBE, L.B. and CRAFTS, A.S. 1939. Bot. Gaz. 100: 695-722
ANDERSON, R. and CRONSHAW, J. 1969. J. Ultr. Res. 29: 50-5
ANDERSON, R. and CRONSHAW, J. 1970a. Planta 91: 173-180
ANDERSON, R. and CRONSHAW, J. 1970b. J. Ultr. Res. 3
    458-471
BEHNKE, H.D. 1971. J. Ultr. Res. 36: 493-498
BOUCK, G.B. and CRONSHAW, J. 1965. J. Cell Biol. 25: 79-96
CHAFE, S.C. and DOOHAN, M.E. 1972. Protoplasma 75: 67-78
CRAFTS, A.S. and CRISP, C.E. 1971. Phloem Transport in Plant
    W.F. Freeman and Co., San Francisco
CRONSHAW, J. 1965. In: Cellular ultrastructure of woody plant
    W.A. Cote, Jr. (Ed.). Syracuse University Press, Syracus
    New York
CRONSHAW, J. and ANDERSON, R. 1969. J. Ultr. Res. 2
    134-148
CURRIER, H.B. and SHIH, C.Y. 1968. Amer. J. Bot. 55: 145-1
DESHPANDE, B.P. 1974. Ann. Bot. 38: 151-158
ESAU, K. 1969. The phloem. Handbuch der Pflanzenanatomi
    Histologie. Band V, Teil 2. Gebruder Borntraege
    Berlin-Stuttgart.
ESAU, K. and CHEADLE, V.I. 1958. Proc. Nat. Acad. Sci. U.
    44: 546-553
ESAU, K. and CHEADLE, V.I. 1961. Proc. Nat. Acad. Sci. U.
    47: 1716-1726
ESAU, K. and CHEADLE, V.I. 1965. Univ. Calif. Publs. Bo
    36: 253-344

ESAU, K. and CRONSHAW, J. 1968. J. Ultr. Res. 23: 1-14
ESAU, K., CHEADLE, V.I. and RISELY, E. B. 1962. Bot. Gaz.
    123: 233-243
ESAU, K., CRONSHAW, J. and HOEFFERT, L.L. 1967. J. Cell Biol.
    32: 71-87
ESCHRICH, W. 1970. Ann. Rev. Plant Physiol. 21: 193-214
EVERT, R.F., BORNMAN, C.H., BUTLER, V. and GILLILand, M.G. 1973.
    Protoplasma 76: 23-34
EVERT, R.F., ESCHRICH, W. and EICHHORN, S.E. 1973. Planta
    (Berl.) 109: 193-210
EVERT, R.F., MURMANIS, L. and SACHS, I.B. 1966. Ann. Bot. 30:
    563-585
FREY-WYSSLING, A. and MULLER, H.R. 1957. J. Ultr. Res. 1:
    38-48
GILDER, J. and CRONSHAW, J. 1973a. Planta (Berl.) 110: 189-204
GILDER, J. and CRONSHAW, J. 1973b. J. Ultr. Res. 44: 388-404
HEPTON, C.E.L., PRESTON, R.D. and RIPLEY, G.W. 1955. Nature 176:
    868-870
JOHNSON, R.P.C. 1968. Planta (Berl.) 81: 314-332
JOHNSON, R.P.C. 1973. Nature 244: 464-466
KESSLER, G. 1958. Ber. Schweiz. Bot. Ges. 68: 5-43
KOLLMANN, R. 1960. Planta 55: 67-107
KOLLMANN, R. 1964. Phytomorph. 14: 247-264
KOLLMANN, R. and SCHUMACHER, W. 1963. Planta 60: 360-389
MUHLETHALER, K. 1950. Ber. Schweiz. Bot. Ges. 60: 614-628
MUHLETHALER, K. 1953. Z. Zellforsch. U. Mikroskop. Anat. 38:
    299-327
NEUBERGER, D.S. and EVERT, R.F. 1974. Amer. J. Bot. 61:
    360-374
NORTHCOTE, D.H. 1972. Ann. Rev. Plant Physiol. 23: 113-132
NORTHCOTE, D.H. and WOODING, F.B.P. 1966. Proc. Roy. Soc. B
    163: 524-537
PARTHASARATHY, M.V. 1974. Protoplasma 79: 93-125
PRESTON, R.D. 1971. J. Microscop. 93: 7-13
PRESTON, R.D. and CRONSHAW, J. 1958. Nature 181: 248-250
SCHUMACHER, W.R. and KOLLMANN, R. 1959. Ber. Deut. Bot. Ges.
    72: 176-179
SHIH, C.Y. and CURRIER, H.B. 1969. Amer. J. Bot. 56: 464-472
SIDDIQUI, A.W. and SPANNER, D.C. 1970. Planta (Berl.) 91:
    181-189
SPANNER, D.C. and JONES, R.L. 1970. Planta (Berl.) 92: 64-72
SRIVASTAVA, L.M. 1969. Amer. J. Bot. 56: 354-361
SRIVASTAVA, L.M. and O'BRIEN, T.P. 1966. Protoplasma 61:
    277-293

DISCUSSION

Discussant: L.M. Srivastava
              Department of Biological Sciences
              Simon Fraser University
              Burnaby, B.C., Canada V5A 1S6

Srivastava: In companion cell sieve element connections, I presume that you start out with the regular plasmodesmata type connections. How do the branched plasmodesmata arise on the companion cell side?

Cronshaw: That is a very difficult question. I do not have any results from my own work. It is said that, as the wall grows, it increases in surface area and at that time the plasmodesma is pulled apart and develops this branched structure.

Aronoff: You mentioned that callose was a β-1,3-glucan. I believe also, that it is a pyranoside rather than a furanoside. Have you made a space-filling model to attempt to ascertain why this particular kind of substance which, to the best of my knowledge, does not exist elsewhere in nature, would be located there as compared for example, to amylose which is also unbranched and possibly as insoluble as callose? I do not know anything about the solubility properties of the 1-3 glucans as compared to the 1-4 glucans.

Cronshaw: No. I also know nothing of the difference in solubility of the 1-3 vs. the 1-4 glucans. The 1-3 linkage does not form close- packed crystalline structures and so may be more amenable to rapid attack by an enzyme. It has to be manipulated in the way the plant manipulates it, lays it down and takes it back out again. Most of the 1-4's do pack into crystal structures, which I think would be more difficult for enzymatic digestion.

Currier: Callose seems to have permeability properties which are rather unique. It certainly is much less permeable to solutes than cellulose and the ordinary primary wall material. Heslop-Harrison concluded that from his pollen work where there is a shell of callose around the developing microspore. I have often wondered why nature has selected this 1-3 glucan to exist in regions where you might expect transfer of materials from one cell to another. It always forms between the plasmalemma and the wall, and the development of the sieve pores is a good example to illustrate how the formation of callose can prevent the deposition of wall substance in that region. There are other instances of this too, in nature, where the plasmalemma is separated from the wall and growth stops in that area.

Eschrich: I suggest that the insolubility of callose may be due to the 1-3 combination. There are no free original OH groups left and this prevents attack of most of the solvents. Only special enzymes can dissolve such molecules.

(Dr. Walsh presented some slides of Zea mays and Lemna minor) Walsh: My conclusion is that in Lemna minor, it appears as though there is no callose involved in the process of sieve pore development).

Behnke: Dr. Walsh, how did you manage to get so much contrast in your callose in Zea mays? Callose normally has no contrast at all. How did you fix your material?

Walsh: I used 10% glutaraldehyde or glutaraldehyde-paraformaldehyde, modified Karnovsky's on the corn, and less than 1% modified Karnovsky's on the Lemna.

Behnke: You commented that there was little or no information on development of sieve pores in monocotyledons. We did these developmental studies in 1969 and obtained quite the same views as Dr. Cronshaw has showed us.

Spanner: I would like to make some observations on plugged or unplugged pores. Dr. Cronshaw remarked that many people have found some pores unplugged. I would counter that by the remark that all people have found more pores plugged. I think that is a fair statement and I would maintain that the plugged state is one which has to be taken seriously in any theorizing on mechanism. It is no embarassment to me if some pores are unplugged. It is a very acute embarassment to the proponent of the pressure flow theory, if many pores are plugged. I have another observation about virus particles in plasmodesmata. Are these particles there because of surge artefacts also? From our own work, and various other lines of argument, I am convinced that the plugged condition is one which must be taken seriously.

(Ed. note: Dr. Spanner described his technique of making twin cuts 1mm apart on intact stolons of Saxifraga and other materials and gave a critique of Dr. Cronshaw's work on Nicotiana in which intact seedlings were prechilled by a suitable liquid cooled to liquid nitrogen temperature before fixation in ice-cold glutaraldehyde.)

(Ed. note: Dr. Fisher questioned the use of wilted, plasmolyzed, or otherwise starved plants for study of the sieve element ultrastructure, and presented his own work on freeze-substituted and thin-sectioned, sieve elements of soybean, which, in neighbouring cross sections, had been shown to be labelled with ^{14}C translocate. Briefly, he reported that the lamina of the sieve element reticulum are composed of parallel arrays of 100 nm fibrils spaced about 250 nm apart, and that there is absolutely nothing in the centre of the translocating sieve elements except sucrose, and possibly plastids and mitochondria. There are no transcellular strands and there is the typically square and non-dispersed P- protein crystal. He also reported that in the sieve plates of functional sieve elements, there is not much in the pores. There is some callose lining the pores and part of the tail of the P- protein crystal usually is draped across the sieve plate, and usually passes through one of the pores, but that was it. According to him, at a higher magnification of some of the pores, one could see the unit membrane around the sieve pore, and perhaps a little bit of stuff in the pore, but certainly not a great deal. Dr. Fisher emphasized that the 100 nm fibrils are organized into sieve element reticulum, but was uncommitted as to whether they were P-protein. He also thought that a sieve element which had scattered fibrils in the lumen, did not fit the criterion of good fixation).

Srivastava: The picture that emerges of the sieve element structure from freeze-substitution, shows some very important points of resemblance to the chemically-fixed sieve elements, particularly in relation to the aggregations of the ER and, at least in some of the pictures that were shown, structure of the pores.

Ho: Dr. Fisher, in one of your autoradiographs there were only a few sieve tubes with silver grains. Does it mean that in the same vascular bundles, some sieve tubes are not functioning?

Fisher: There is a problem in that not all of the sieve elements appear to be labelled. I counted the percentage of labelled sieve elements in each of these experiments and summed it up - approximately 80% of the sieve elements are labelled. In longitudinal sections, I cannot be sure that I am looking at a sieve tube that is functional, whereas I know I am looking at a functional sieve element in cross sections, because I can match the ones that I see in electron microscope with the ones that are seen in the autoradiograph.

Swanson: Dr. Fisher, were you also able to do the converse, where you could match unlabelled sieve tubes with plugged sieve plates?

Fisher: No.

Walsh: Can one observe rapid freezing under the microscope?

Fisher: I see no way at all of getting it to occur as rapidly as you could fix a free half millimetre span. I do not think you would be able to do it.

Cronshaw: The way the cytologists judge the quality of their fixation is by comparison with other cells, with animal material, all kinds of plant material, and with other techniques. We fit information together from many techniques and make judgements as to which technique to use for a certain thing. As far as the cytology of sieve elements at early stages of differentiation is concerned, I think there are not many arguments as to the results that have been obtained. The arguments start when we get to this very specialized cell, the sieve element. It could be that with the freeze-subsitution technique you are able to get some information which we would consider a gross level; for instance, on the distribution of what is in the sieve element lumen, or on the blocking or non-blocking of the sieve plate pore. I think you would have to be very careful when you go down to the level of interpreting structure of the endoplasmic reticulum or the fibrils. For instance, it is generally agreed amongst cytologists that most of the work that has been done with freeze-etch, without prior fixation in glutaraldehyde, will not stand up to critical analysis, and will have to be repeated. There have been many examples shown of the movement of particles in membranes during the freeze-fracturing technique.

Fisher: The judgements I am trying to make are not down to a few angstroms. We are talking about the very basics of sieve element structure - whether the pores are plugged, or whether there is loose slime in the sieve element lumen.

Eschrich: Every technique should be tried. When you fix the wilted leaf, surely there will be no turgor as in parenchyma cells. According to our experiments with cucurbits, we found that the concentration of sieve tube sap rises under water stress. That means that surging must be much greater when you have a wilted plant than when you have a watered plant. A new method should be evaluated - whether it is just so much technique or whether any real result can be expected from it.

Gorham: Some years ago we examined the translocation in wilted tissue. In cucurbits, there is very good translocation of ^{14}C in highly wilted material, confirming what Dr. Eschrich says. Dr. Fisher, in those autoradiographs at an advancing front, are we looking at 40-minute translocation, or is it that the front is way down beyond?

Fisher: Way down beyond.

Companion Cells and Transfer Cells

H.-Dietmar Behnke

Lehrstuhl fur Zellenlehre der Universität Heidelberg

Im Neuenheimer Feld 230, D-69 Heidelberg, Germany

Phloem tissue as a whole, according to the introduction of this term by Nageli (1858), consists of sieve elements, sclerenchymatic elements and parenchymatic elements. Companion cells and transfer cells, the two cell types to be considered more in detail in the following, belong to the latter group. Parenchymatic elements of phloem as a general term, however, comprise a wide variety of nucleate cells which are referred to as related or unrelated to the enucleate sieve elements. By definition of the phloem those are characteristic constituents in seed plants and ferns, but they are also present in the leptom of bryophytes.

There is a noticeable gradation in cell structure and a stepwise increase in degree of association in a sieve element in going from unrelated to related phloem-parenchyma cells, to more specific albuminous cells, and to most specialized, ontogenetically-related, companion cells. All of these, except the unrelated, are thought to be physiologically and functionally associated with the sieve element. Close spatial relationship of a nucleate cell, including frequent plasmatic connections, seems to be predestined to assist the enucleate, and in many respects highly differentiated, sieve elements in their task of translocation.

Although there are a lot of common features among all the related parenchyma cells that probably predominate over the distinguishing ones, some extremely specialized cells - e.g. companion cells and albuminous cells - are always separated from the whole and from each other, in part by their structural peculiarities or only by tradition. With the description of the companion cell, we try to portray the most prominent of the specialized parenchyma cells as a representative to the whole

153

group, while at the end of the following part, in a comparison of its different and common characters with the albuminous cells, the validity of its exceptional status will be reconsidered.

A second part will be devoted to transfer cells, a structurally unique type of parenchyma cells found in different plant tissues, and in the phloem tissue, to be considered here, of close physiological relationship to sieve elements.

Companion Cells

Companion cells are characteristic components of, and restricted to, the phloem tissue of angiosperms. Therefore, they can be used as a helpful criterion (one among others) to distinguish angiosperm sieve-tube members from gymnosperm sieve cells (see Esau, 1969: p. 124). In general, companion cells are present in primary as well as secondary phloem tissues, although there are some reports of their absence from the protophloem of several species. Actually, as could be demonstrated by Resch (1961), the identification of companion cells within the protophloem may be obscured by an inconsistency in their occurrence and the inability to trace their derivation from a mother cell common to the respective sieve element.

One of the most important characteristics of a companion cell is this close ontogenetic relationship to the sieve element. A procambial or cambial initial gives rise to a phloem mother-cell that, in its turn, divides unequally and results in the formation of a larger sized sieve-tube member and a narrow companion cell. (These relative size proportions are reversed only in the minor veins, the endings of a vascular bundle in leaves; see e.g. Morretes, 1962; Esau, 1965.) The uneven sister-cells, though structurally and functionally related throughout all their lives, add to the size differences a peculiar differentiation of their protoplasmic contents. Sieve-tube members undergo changes in protoplasmic and wall structure, which culminate in the formation of an enucleate, but highly hydrated protoplasm lined by a plasmalemma and containing the characteristic P-protein, all within a set of open-pore connected conduits. Companion cells, on the other hand, retain the integrity of all the different cell organelles and are distinguished by an increase of density of their protoplasts and by especially large nuclei.

In cross sections of phloem tissue of many plant species a regular pattern of alternating sieve elements and companion cells reflects their derivation from a common mother cell by an unequal division. This is maintained throughout their life-time and is therefore readily demonstrated (e.g. Scirpus, Fig. 2), while in many others this ontogenetic relationship during tissue differentiation has been obscured frequently by uneven growth of

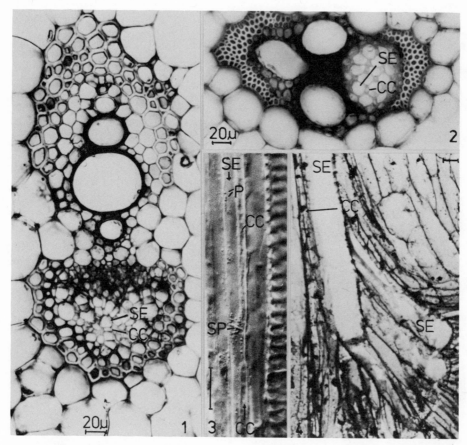

Figs. 1 - 4. Companion cells (CC) and sieve elements (SE) in light-microscopic transverse (1,2) and longitudinal sections (3,4) of phloem of <u>Musa</u> <u>velutina</u> (1), <u>Scirpus</u> <u>lacustris</u> (2), <u>Montia</u> <u>perfoliata</u> (3), <u>Dioscorea</u> <u>reticulata</u> (4), SP = sieve plate, P = plastids. Originals by the author.

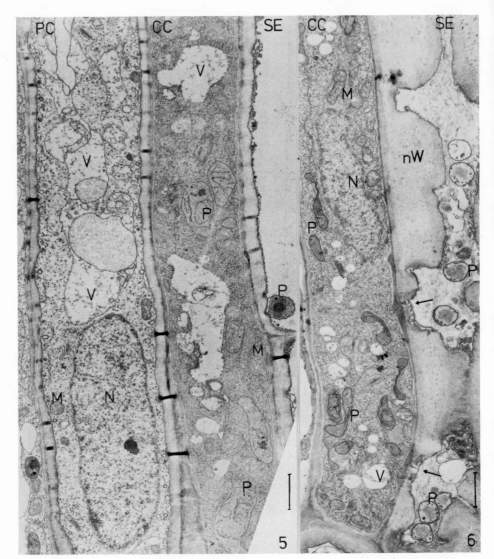

Fig. 5. <u>Decaryia</u> <u>madagascariensis</u>, longitudinal section of phloem: Ultrastructure of companion cell (CC), adjacent to sieve element (SE), and compared to phloem-parenchyma cell (PC). Fig. 6. <u>Annona</u> <u>cherimola</u>: Companion cell (CC) with dense cytoplasm and, abundant organelles, connected to sieve element (SE) by specific plasmatic bridges (arrows). M = mitochondria, N = nucleus, P = plastids, V = vacuoles, nW = nacreous wall. Originals by the author.

the   adjacent   cells   and   therefore   become   less   evident.
Nevertheless, sieve elements and companion cells in general can be
distinguished by size and spatial relations (e.g. Musa, Fig. 1).
In longitudinal sections, however, the demonstration of companion
cells often is made more difficult by several influences during the
differentiation of the phloem tissue from the procambial or cambial
to the meta or secondary condition. One of those is to be sought
in the development of the companion cell itself.   Besides the
simple one sieve element - one companion cell relation (Fig. 3)
there may be a longitudinal series of companion cells (Fig. 4),
resulting from various divisions of the original phloem mother-cell
derivative.   Although   there may be multifold variations of
companion cells in size, shape, number, and in their association to
a sieve element, a rough calculation has demonstrated that in meta-
and secondary phloem tissues each sieve element is accompanied by
at least one companion cell (for details, see Esau, 1969).

     The idea that this close ontogenetic and spatial relationship
is reflecting an intimate physiological and functional association
between the two sister cells is strongly supported by the
observation that both cell types cease to function and obliterate
at the same time.

     Zahur (1959), in a light microscopic survey on the phloem of
423 woody dicotyledons, has recorded the presence and
characteristics of companion cells in almost all species. Our own
comparative   electron   microscopic   investigations   of   some   330
angiosperms, initiated to document the distribution and taxonomic
signifance of different plastid-types of sieve elements (Behnke
1972,   1975),   also   furnished   a   great   deal of data to a general
characterization of companion-cell ultrastructure.

     The extraordinary density of its cytoplasm, which had been
emphasized in light microscopic observations, was confirmed in most
of the studied species and helped to identify companion cells and
clearly distinguish them from phloem-parenchyma cells, even in
low-power micrographs.

     In Fig. 5, a longitudinal section through meta-phloem of
Decaryia madagascariensis, the differences in density of cytoplasm
between a companion cell (CC; next to a sieve element, SE) and a
phloem-parenchyma   cell   (PC) are quite obvious.   Cytoplasmic
organelles like plastids (P) are equally present in both cell
types, although in companion cells plastids and mitochondria are
more frequently encountered (Figs. 5, 6).   Most evident is the
abundance of ribosomes which, together with the surrounding
cytoplasmic matrix, are responsible for the quoted high density of
companion-cell protoplasts (Fig. 13). Vacuoles, delimited against
the cytoplasm by a tonoplast membrane, are fairly frequent in
companion cells (Figs. 5-7; 10, 11) but of smaller size compared

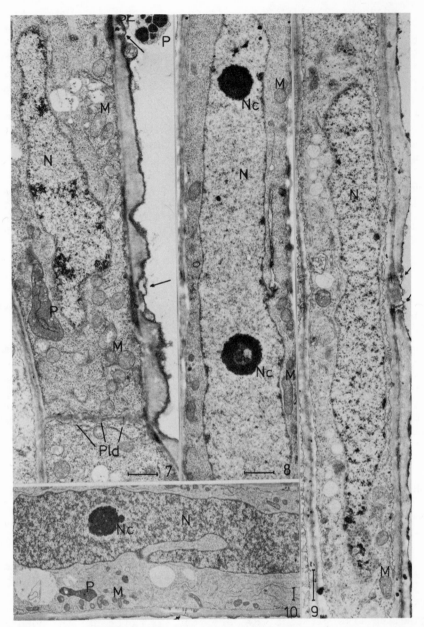

Figs. 7 - 10. Electron micrographs of companion cells of
Lupinus polyphyllus (7), Amaranthus retroflexus (8), Persea
americana (9), Achlys triphylla (10) documenting large nuclei:
(N), plastids (P), and mitochondria (M). Nc = nucleoli, Pld =
plasmodesmata, SE = sieve element. Arrows point to sieve-pore pit
field connection. Originals by the author.

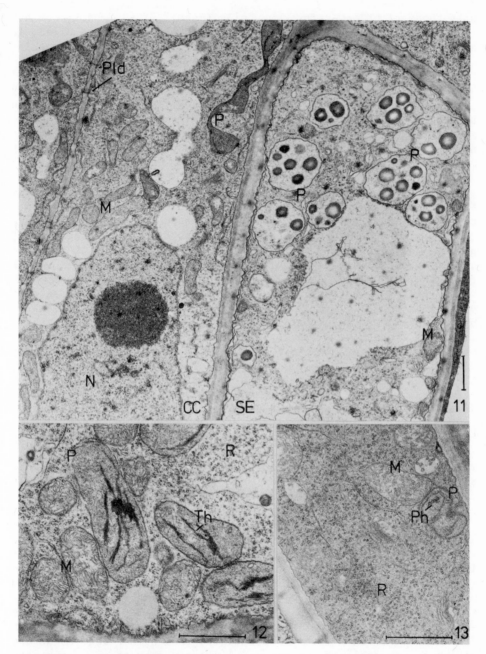

Figs. 11 (<u>Lindera</u> <u>praecox</u>), 12 (<u>Saponaria</u> <u>officinalis</u>), 13 (<u>Decaryia</u> <u>madagascariensis</u>): Cytoplasm of companion cells rich in ribosomes (R), mitochondria (M), and plastids (P). Th = thyladoids, Ph = phytoferritin. Originals by the author.

with those of phloem-parenchyma cells (Figs. 5, 6). Last, but not least, the nucleus, as the most prominent of the organelles, is another factor contributing to the special features of a companion cell.

Nuclei of companion cells in general are particularly elongated (Figs. 8, 10) and also occupy a large proportion of the breadth of the cell (Figs. 8-10). Their matrix is rather dense, and they are usually of a regular and even outline. Sometimes nuclei are extremely lobed (Fig. 9), probably reflecting a high metabolic activity.

One of the distinctive features attributed by light microscopy to the selective characterization of companion cells is the absence of starch in their plastids. Fine structural studies have demonstrated poorly differentiated plastids which, because of absence of detectable amounts of chlorophylls, are to be filed among photosynthetically inactive leucoplasts. Their shape varies from the elongated amoeboid (preferentially in young companion cells, Fig. 11, but see also Fig. 6) to ovoid or stretched elliptical forms (Fig. 12). Their internal structure is dominated by an electron-opaque matrix (Figs. 5-7; 11; 12:P). This matrix is always crossed by a few thylakoids which aggregate randomly to build small grana stacks (Fig. 12:P). Plastoglobuli are fairly common (Fig. 6). In some species phytoferritin is present within the matrix of companion-cell plastids (Fig. 13:Ph). Starch, a common storage product of phloem-parenchyma cells, has been identified in companion-cell plastids in only a few examples (e.g. in Cucurbita by Esau and Cronshaw, 1968). Its presence may be an indication of an altered function of these cells as discussed by Esau (1971). In Larix (Sauter and Braun, 1968) albuminous cells were shown to store large amounts of starch during a period when the associated sieve elements were immature. This starch was completely removed during differentiation of the sieve elements.

Mitochondria are frequent in companion cells, they are elongated, in contrast to the prevalent spherical sieve-element mitochondria, and have well developed invaginations of their internal envelope (Figs. 7, 11-13). Their abundance supports the idea of a high metabolic activity in companion cells.

Eventually, the plasmatic connections between sieve elements and their associated cells are considered another major indication for their functional accord. Calculated from number and structure of these connections, companion cells again show the most intimate contact with sieve elements.

Whereas normal plasmodesmata connect companion cells with each other (Fig. 7) and with adjacent phloem-parenchyma cells (Fig. 11), the plasmatic connections between companion cells (or

equivalent contiguous nucleate cells) and sieve elements are of a
very peculiar structure, not found elsewhere within plant tissues,
and probably reflecting the difficulties in uniting the normal
protoplast of a parenchymatic element with the hydrated protoplast
of a sieve element.

The existence of plasmatic connections between sieve elements
and their companion cells was recognized by many of the last
century's early light microscopists, their special composition of a
sieve area on the sieve-element side and a primary pit-field on the
companion-cell side having already been considered by Hill (1908),
but with their characteristic structure revealed only by the
electron microscope.

A first micrograph of a median section giving us an idea of
their micromorphology was published by Eschrich (1963). Subsequent
observations made by many authors resulted in our current knowledge
of their composition and finer plasmatic contents. Several fine
plasmatic channels, each 50 to 60 nm wide and actually comparable
to plasmodesmatal halves or branches, occur in the
companion-cell-half of the wall between the two sister cells (Figs.
14-19). A single sieve-pore half, the diameter of which is
variable from plasmodesmatal size to regular sieve-pore size (up to
0.2 μm), penetrates the sieve-element side (Figs. 15-18), meets
with the plasmodesmatal halves in the area of the middle lamella,
and unites with them to form a continuous plasmatic connection.
The restriction of the plasmodesmatal part to the companion-cell
half of the wall is nicely documented by micrographs recording a
difference, most likely artificial, in contrast to the two wall
parts (Fig. 16).

Most common is the formation of a "median cavity", an enlarged
connecting area between plasmodesmatal branches and sieve-pore
half, in the region of the middle lamella (Figs. 15, 18, 19:Mc).
This median cavity, similar to those reported in normal
parenchyma-cell plasmodesmata (Krull, 1960), justifies the term
plasmodesmatal "branches" instead of "halves" of single
plasmodesmata. The number of these branches probably varies
greatly (between 5 and 27 are reported) but, similar to the
morphology of the entire connection and which only median sections
disclose (cf. Fig. 16), their actual number can only be
calculated from sections glancing to the wall surface of the
companion-cell side. Undefined oblique sections, therefore, giving
longitudinal (Fig. 17) or cross-sectioned (Fig. 14) aspects of
the branches, can only result in the rough impression that there
are more than five, more than ten, etc.

Each plasmodesmatal branch is lined against the cell wall by
plasmalemma (Fig.17: Plm) and penetrated by a membrane-like
structure (Fig. 17, arrow). Cisternal profiles of endoplasmic

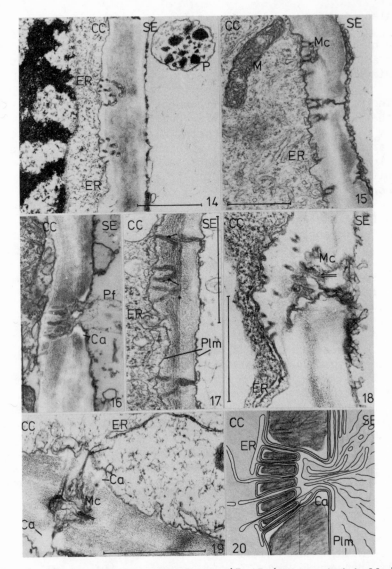

Figs. 14 (<u>Sprekelia</u> <u>formosissima</u>), 15 (<u>Achlys</u> <u>triphylla</u>), 17 (<u>Ulmus</u> <u>rubra</u>), 16 (<u>Dioscorea</u> <u>macroura</u>), 18 (<u>Tinantia</u> <u>fugax</u>), 19 (<u>Anoectochilus</u> sp.): Sieve-pore pit-field connections between companion cells (CC) and sieve-element (SE). Ca = callose, ER = endoplasmic reticulum, Mc = median cavity, Plm = plasmalemma; arrows point to membrane-like structures within the plasmodesmata, double arrows to membranes in median cavities. Originals by the author.

Fig. 20. Schematic interpretation of sieve-pore pit-field connection and its plasmatic contents.

reticulum (ER) coming close to the entrance of plasmodesmatal branches (Figs. 14, 15, 17-19:ER) and ER-like membrane-delimited spaces inside the median cavities (Fig. 18: double arrow), suggest a participation of ER in the formation of the internal structures of the branches or a penetration of ER itself. Both possibilities are also discussed for the micro-morphological composition of normal plasmodesmata (Robards, 1971). The plasmalemma lining the plasmodesmatal branches also delimit the median cavity and the sieve-pore half against the cell wall and is continuous with the cell´s own plasmalemmas on each side of the plasmatic connection (Fig. 17:Plm). On the sieve-element side ER likewise lies next to the sieve-pore half, most probably participating in its internal structure (Fig. 16). Occasionally, protein filaments are to be found inside the sieve-pore half (Fig. 16:Pf). Callose formation is regularly encountered next to the sieve-element side of plasmatic connections but also seems to be present at the companion-cell side (Figs. 16, 18: Ca).

Since none of the published micrographs nor any of the pictures recently taken from so many species yielded a satisfactory document of the sieve element (the companion cell connection as a whole, the reason having been discussed earlier), a drawing composed of the main structural features of Figs. 14 - 19 and comparable to one published previously by Kollman (1973), may be used to interpret its structure as seen in a median section (Fig. 20).

Summing up the structural data given in the previous chapters, it may be confirmed finally, that we draw a picture of a nucleate cell extraordinarily differentiated and closely associated to the enucleate sieve element, both implicating a tight physiological, as well as structural relationship. This portrait conforms with previous observations by many authors and has been reviewed by Esau (1969). Only some aspects, which have not been touched in our survey, as well as the results of some recent publications, will be added.

A comparison between plastids of the different phloem cells in Cucurbita was done by Esau and Cronshaw (1968). They noticed a rough structural similarity between sieve-element and companion-cell plastids, especially during early stages of differentiation, while plastids of phloem-parenchyma cells differed much from both. Wooding and Northcote (1965) reported companion-cell plastids of Acer pseudoplatanus to be closely sheathed with endoplasmic reticulum. Since this association between the two organelles was found to be permanent, the authors suggested a participation of ER in the rapid translocation of sucrose through the plasmatic connections and through companion-cell cytoplasm and a temporary storage of this material in the plastids. Behnke (1973), opposing reports and theories on the absence of plastids in companion cells by ontogeny, emphasized the general presence of plastids in an extended record on plastids in sieve-elements and companion cells.

Incidentally, P-protein has been found in companion-cells and ascribed a factor in enhancing the extreme density of their protoplasts (cf. Esau, 1969).

The degeneration of the companion-cells, dependent on the degeneration of its sister sieve-element, has been followed by the electron microscope in Mimosa (Esau, 1973). These include: the inflection of nuclear envelopes preceding disorganization of the nuclear contents, the swelling of mitochondria and plastids while fragmentation and disappearance of internal membranes proceed, and loss of identity of all cell organelles.

Next to the ontogenetic relation and the quoted structural features, the coincidence in the degeneration of sieve-elements and companion cells has always been looked upon as a special proof of the interdependence of the two cell types and the priviledged position of companion cells among related parenchymatic elements of phloem. Reviewing the type of association of albuminous cells with gymnosperm sieve cells, which lack true companion cells, it must be emphasized that albuminous cells show the same characteristics as companion cells, except for the ontogenetic relation. Albuminous cells contain a dense protoplasm including large nuclei, and, although arising from a separate precursor, are connected to sieve cells by combinations of sieve-pore and primary pit-field halves (cf. lit. in Esau, 1969). Finally, they degenerate at the same time when sieve cells cease to function. The latter could be demonstrated by several transverse and radial, as well as tangential longitudinal sections, in Metasequoia (Figs. 21-24, from Kollman, 1966).

This data raises the question of the validity of the ontogenetic relationship as a distinctive factor among the nucleate associates of sieve elements. Srivastava (1970) in an investigation on the secondary phloem of Austrobaileya first raised this question and proposed to deal with the physiological relationship as the most important character that might or might not be accompanied by an ontogenetic relationship. The adoption of a new term, such as "associate cell", or a new definition of the term "companion cell", not including ontogenetic relation (proposed by Srivastava, 1970), would help to name those cells where the alignment to one of the presently defined phloem cells is not possible, e.g. in Austrobaileya or Gnetum (Paliwal and Behnke, 1973). A new term or a new definition should apply to all gradations of associated parenchymatic elements: companion cells, albuminous cells and related parenchyma cells as well.

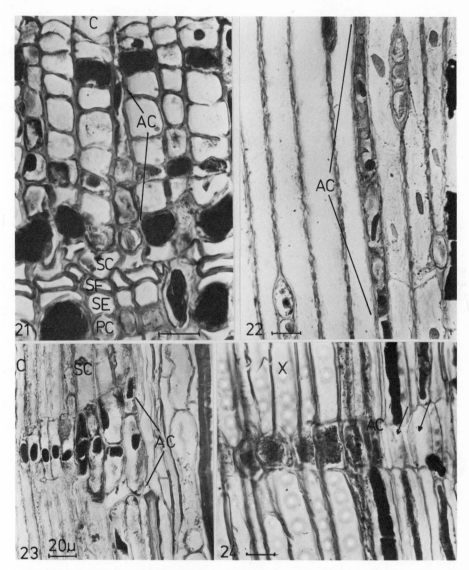

Figs. 21 - 24. Light microscopic transverse (21), tangential
(22), and radial (23,24) longitudinal sections through phloem of
Metasequoia glyptostroboides. Degeneration of albuminous cells
(AC) in association to sieve cells (SC). C = cambium, PC =
parenchyma cell, SF = bast fiber, arrows point to degenerated
albuminous cells. Originals kindly provided by R. Kollmann (from
Kollmann, 1966).

Transfer Cells

Transfer cells, a term that was first introduced by Gunning, et al (1968) in the description of some minor vein-phloem parenchyma cells, now more generally applies to plant cells that are involved in, and by structural modifications made specialists for, short- distance translocation. It is to the special credit of Gunning, Pate and co-workers, after the rediscovery of Fischer's (1884) "Ubergangszellen", to have pursued the occurence of these cells throughout the plant body as well as the entire plant kingdom, to have assigned to them a special function, and given them an appropriate name.

Comprehensive reviews on the different types of transfer cells, their distribution and possible function, were given by Gunning and Pate (1969, 1974) and Pate and Gunning (1972) and may be consulted for a broad survey of their various anatomical and physiological possibilities in plants. Vascular transfer cells (i.e. phloem and xylem transfer cells), the type of our present specific interest, are predominantly associated with nodes of stems including hypocotyls (Gunning et al, 1970; Pate et al, 1970); and with roots only when pathologically altered by nematodes (Jones and Northcote, 1972) or by nodule formation (Pate, et al 1969); and with leaf veins (Gunning and Pate, 1969). Vascular transfer cells are also found in haustoria of seed plant parasites, e.g. in Cuscuta in contact with the host sieve-elements (Dörr, 1972), and in Orobanche attached to the xylem elements of the host plant (see discussion to follow by Dr. I. Dörr).

The description of structure, function, and developmental aspects of transfer cells in the following, will be confined mainly to their occurrence in leaf minor veins for two reasons: (1) their functional role has been discussed very extensively and their distribution within plant kingdom screened thoroughly; (2) correlated structural and functional evidence of transfer cells associated with phloem is most pertinent to the general theme of this symposium.

The Structure of Transfer Cells

The general structure of transfer cells is dominated by cell-wall in-growths which are already visible in routine preparations for light microscopy (cf. Fig. 28). The application of histochemistry, and transmission and scanning electron-microscopy, revealed the composition, the finer structure, and three-dimensional arrangement of the wall in-growths: They form part of the secondary cell wall and are composed of randomly oriented microfibrils, most likely of cellulosic nature. Experiments with lanthanum hydroxide colloid penetrating the wall in-growths of transfer cells, but not the lignified parts of cell

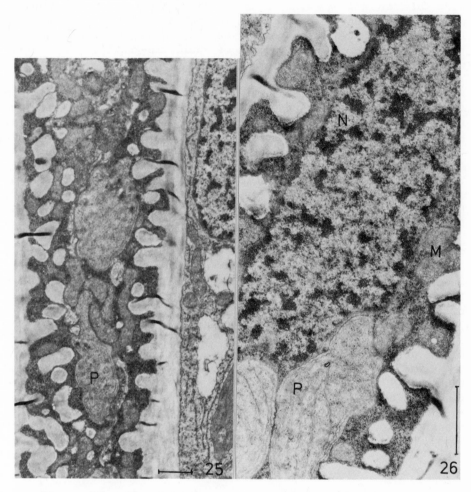

Figs. 25 - 26. Cytoplasmic contents and wall ingrowths of A-type transfer cells in longitudinal sections of Armeria pseudarmeria. N = nucleus, M = mitochondria, P = plastids. Originals by the author.

walls (Gunning and Pate, 1969; Zee and O'Brien, 1971). It has demonstrated the existence of interconnected channels between the microfibrils, wide enough for a passage of solutes and water. The size of wall in-growths is not uniform, their shape not regular, and their frequency and distribution along the cell walls not constant, although some transfer cell types are characterized by a polar arrangement and restriction of wall in-growths to certain wall parts.

The particular structure of transfer cells within leaf minor veins motivated Gunning and Pate (1969) to distinguish between four different types which they name A- to D-type cells consecutively. Morphological characters, including specific association to certain vascular cells and, in part, specific restriction of wall-ingrowth formation to wall parts adjacent to these cells, were the primary parameters used for the definition of these types. Different functional characters are presupposed and will be discussed below. Our description of the four types follows that given by Pate and Gunning (1969) and Gunning and Pate (1974).

A-type transfer cells are conspicuous by their close association to sieve elements. They are easily distinguishable from the other three types by their extremely dense cytoplasm and the arrangement of wall in-growths that are almost evenly distributed around the entire cell wall (Figs. 25, 28, 29:A). The presence of specific plasmatic connections with the sieve element, in addition to other structural features (ribosome-rich cytoplasm, well-organized mitochondria, dense nucleus; compare Figs. 25, 26 to Figs. 5 - 13,) have led Gunning et al. (1968), and Pate and Gunning (1969), to the view that A-type cells are modified companion cells; but their ontogenetic relationships remain to be resolved. Chloroplasts with starch granules as found in A-type transfer cells (Fig. 26 and Gunning et al, 1968, Pate and Gunning 1969), are also more common in minor vein companion cells (cf. Figs. in Falk, 1964; Esau 1965) and rarely present in companion cells of stem vascular bundles (Figs. 5, 6, 11 - 13).

B-type transfer cells, on the other hand, are looked upon as modified phloem-parenchyma cells. The obvious differences in protoplast structure and density, as compared to A-type cells, reflect the differences recorded between companion cells and normal phloem-parenchyma cells (see Fig. 5 and description thereof). B-type cells do not elaborate plasmatic connections to sieve elements but, when adjacent to them, selectively provide their common wall parts with wall in-growths. The obvious polarity in the formation of wall in-growths, documented by their restriction to wall parts abutting on sieve elements, as well as to those shared with sieve-element associated A-type cells, is a major morphological character of all B-type transfer cells.

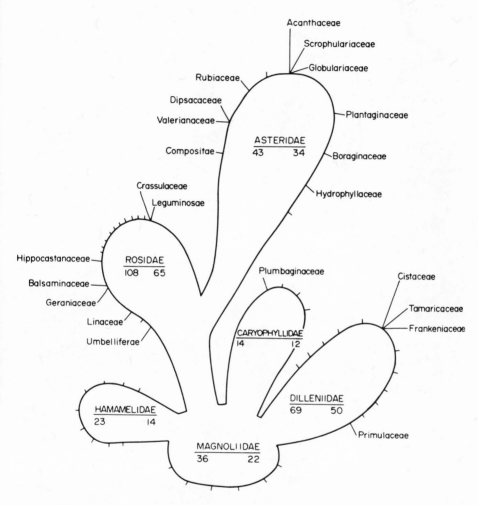

Fig. 27. Distribution of families bearing minor-vein
transfer cells in angiosperm system after Cronquist (1968). From
Gunning & Pate (1974).

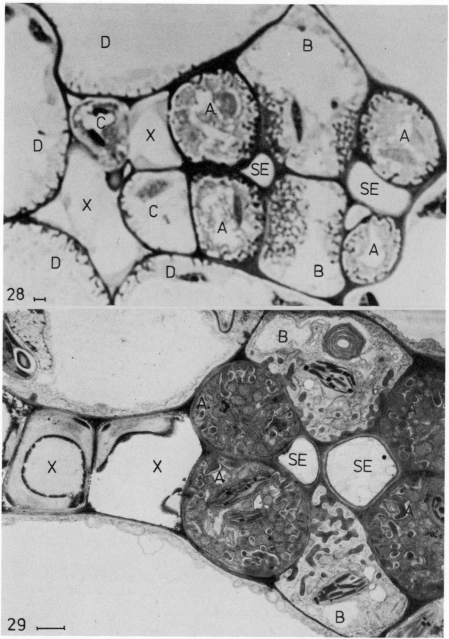

Figs. 28 - 29. Light and electron micrographs of cross sections through minor veins demonstrating structure and arrangement of A- to D-type transfer cells in _Anacyclus pyrethrum_ (Fig. 28, after Pate & Gunning, 1969), and _Senecio vulgaris_ (Fig. 29, after Gunning & Pate, 1974). x = xylem, SE = sieve element.

C- and D-type transfer cells, both located in the immediate neighbourhood of vessels and tracheids, also show a polar distribution of their wall in-growths, being confined to wall parts contiguous to these xylem elements (Fig.28). While C-type cells, modified xylem-parenchyma cells, most likely are of the same type present as xylem transfer cells in stem, D-type cells, as derived from bundle sheath cells, exclusively follow minor veins (Gunning and Pate, 1974).

The systematic search for transfer cells in the recent classes Lycopodiatae to Liliatae, i.e. including pteridophytes, gymnosperms, and angiosperms, carried out by Gunning, Pate and co-workers, covers more than 1000 species. But, with the exception of the monocotyledonous family Zosteraceae, only dicotyledons, and preferentially their herbaceous genera, develop minor-vein transfer cells (Pate and Gunning, 1969). Fig. 27, adapted from Gunning and Pate (1974), lists those 22 families (from a total of 197 investigated), which were proven to have at least one species with at least one type of minor-vein transfer cell and gives their distribution within the six subclasses recognized by Cronquist (1968). The number of families investigated in each subclass (right hand numbers beneath subclass names) is plotted against the total number of families identified in the Cronquist system (left hand numbers).

Pate and Gunning (1969) also recorded the distribution of the four types of minor vein transfer cells within the taxa examined, and found the combination of types to be species-specific. There is no indication of a genus- or family-specificity with this character. Their percentage calculation confirmed earlier suppositions: the A-type transfer cell is by far the most common in minor veins. Ninety-seven per cent of the positive genera contain A-type cells, either exclusively (more than 50%) or in combinations with the other types. Only 38% have B-type cells, a large number of them in various combinations, 3% as a single type. There are no genera reported to incorporate C- and/or D-type cells exclusively. In spite of the wide variation in the equipment of minor veins with morphologically-different transfer cells, a general functional accord is surmised for all, differences being admitted only for quantitative relations.

The Function of Transfer Cells

Wall in-growths, the morphological character common to all types of transfer cells, multiply the surface area of the plasmalemma. Estimations by Gunning, et al (1970) have demonstrated that walls bearing in-growths are able to increase the plasmalemma area up to the twenty-fold (in minor-vein transfer cells to 27-fold), as compared to equivalent walls without in-growths. The plasmalemma is one of the main obstacles limiting transport processes from the apoplast to the symplast, and vice versa. Its amplification not only faciliates this transport, but also can be taken as an indication of an effective transmembrane transport in the respective cells.

Surface area of plasmalemma, however, is not the only factor implicated in a consideration on how transport functions in transfer cells. Simple diffusion phenomena as moving forces are easily exempted, since wall in-growths considerably lengthen the diffusion path of substances secreted or absorbed by the plasmalemma. Being aware of these limitations, Gunning and Pate (1974) discuss the electro-osmotic as well as the "standing-gradient osmotic flow" hypotheses, as driving forces. For electro-osmosis the small channels between microfibrils of wall in-growths, proven to exist by the lanthanum experiments, could be fitted with fixed anionic charges establishing a potential difference to the protoplast and thus enabling the flow of solutes and water. The "standing-gradient osmotic flow hypothesis", adapted from attempted explanations for transport phenomen in animal tissues (Diamond and Bossert, 1967; Oschman and Berridge, 1971) and transferred by Gunning and Pate (1974) to the transfer-cell situation, starts with the assumption of the presence of solute pumps inside or next to the plasmalemma and is applicable to both absorptive and secretory situations. These pumps actively and unidirectionally drive solutes from or to the apoplast (i.e. wall-in-growths lumina) to or from the surrounding symplast, producing an osmotic gradient which allows water to follow. The surplus (or deficit) of solutes and water created could be compensated by their export (or import) within the symplast (e.g. by plasmodesmata). The localization of ATPases next to wall in-growths (Maier and Maier, 1972) support ideas on an active transmembrane transport of solutes.

Returning now to the particular situation in minor-vein transfer cells, their physiology is concerned primarily with the loading of veins, both with solutes arriving and reabsorbed from the xylem, and with assimilates produced by the photosynthetically-active mesophyll cells. Whereas the latter arrive in transfer cells by symplasmic connections as well as by leakage to, and subsequent absorption from, the apoplast, solutes ascending with the transpiration stream accumulate only by a selective absorption at the membrane barrier between apo- and symplast. Their removal to the minor-vein sieve elements is by plasmatic connections.

Therefore, regardless of some possible specific functions of the different types of minor-vein transfer cells, the number of cells bearing wall in-growths and, consecutively, the amount of augmentation of contact area between apo- and symplast, is of decisive influence. The morphological differences between species

bearing and others missing minor-vein transfer cells reflect
quantitative rather than functional variations. The amplification
of plasmalemma surface area in herbaceous dicotyledons give those
plants a larger capacity for an apoplast-symplast transport.   They
are known to have a high assimilate production and transpiration
rate and have developed the most advanced sieve elements for a
rapid translocation. Gunning et al (1974), performing quantitative
calculations on transfer-cell structure of minor veins, e.g. their
plasmalemma-surface amplification and frequency of plasmodesmatal
connections with sieve elements, tried to include the major
arguments in favour and against such a model of transfer cell,
retrieving solutes from the xylem and loading the vein sieve
elements.

Aspects of Transfer Cell Development

    The initiation of wall in-growth formation in transfer cells
coincides with the onset of their absorptive or secretory function.
Experiments impeding the normal development of a tissue also stop
the formation of the in-growths and associated amplification of
plasmalemma, e.g. low light intensities applied to leaves
(Gunning, et al. 1968) or seedlings grown in darkness or under
CO_2-deficiency (Pate, et al. 1970).   These close correlations
between function and specific structural differentiation, similar
to those in other plant tissues, however, do not account for the
species- and/or tissue-specific morphogenetic processes responsible
for a given type of transfer cell.   The entire complex of the
morphogenesis of transfer cells so far has only been scratched on
its surface.

    A possible explanation for the mode of wall-ingrowth formation
has recently been contributed by Schnepf (1974) who tries to
elucidate   the function of microtubules in the formation of
secondary walls in plants. He studied the deposition of very
restricted wall parts against the turgor pressure. Unlike other
examples (e.g. Sphagnum, Cobaea), which confirmed his theory of a
mere cytoskeletal function of microtubules (by "lifting off the
plasmalemma they form distinct extracytoplasmic spaces to allow the
deposition of plastic wall material against the turgor pressure",
(Schnepf, 1974).   Transfer cells in septal nectaries of Aloe and
Gasteria do not elaborate microtubule-plasmalemma relationships
during the formation of wall in-growths. The irregular deposition
of wall material does not necessarily include the aid of
microtubules. However, if there are species-specific morphological
characters of wall in-growths in some of the transfer cell types,
another morphogenetic factor must be involved in these cases.

Acknowledgement
    B.E.S. GUNNING and R. KOLLMANN kindly contributed original
micrographs for inclusion in this review. I wish to express my

gratitude for their permission to publish these pictures as Figs.
27 - 29 and 21 - 24, respectively. I am also indepted to Mrs. L.
POP for skillful assistance in the unpublished part of my own
research and in preparation of the micrographs for this review.

References

BEHNKE, H-D. 1972. Bot. Rev. 38: 155-197
BEHNKE, H-D. 1973. Planta 110: 321-328
BEHNKE, H-D. 1974. Ann. Mo. Bot. Gard. (in press)
CRONQUIST, A. 1968. The Evolution and Classification of Flowering
    Plants - Nelson, London.
DIAMOND, J.M. and BOSSERT, W.H. 1967. J. Gen. Physiol. 50:
    2061-2083
DORR, I. 1972. Protoplasma 75: 167-184
ESAU, K. 1965. Proc. Am. Phil. Soc. 111: 219-233
ESAU, K. 1969. Encyclop. Plant Anatomy Vol. V,2. Borntrager,
    Berlin, and Stuttgart.
ESAU, K. 1971. J. Ind. Bot. Soc. 50 A: 115-129
ESAU, K. 1973. Ann. Bot. 37: 625-632
ESAU, K. and CRONSHAW, J. 1968. Can. J. Bot. 46: 877-880
ESCHRICH, W. 1963. Planta 59: 243-261
FALK, H. 1964. Planta 60: 558-567
FISCHER, A. 1884. Untersuchungen uber das Siebrohren-System der
    Cucurbitaceen. Berlin, Borntrager
GUNNING, B.E.S. and PATE, J.S. 1969. Protoplasma 68: 107-133
GUNNING, B.E.S. and PATE, J.S. 1974. In: Dynamic Aspects of
    Plant Ultrastructure, A.W. Robards (eds.). McGraw-Hill,
    Maidenhead (U.K.). pp. 441-480
GUNNING, B.E.S., J.S. PATE and L.G. BRIARTY. 1968. J. Cell
    Biol. 37: C7-C12
GUNNING, B.E.S., J.S. PATE and L.W. GREEN. 1970. Protoplasma
    71: 147-171
GUNNING, B.E.S, J.S. PATE, F.R. MICHIN and J. MARKS. 1974.
    Symposium 28 on Soc. Exp. Biol. (in press).
HILL, A.W. 1908. Ann. Bot. 22: 245-290
JONES, M.G.K. and D.H. NORTHCOTE. 1972. Protoplasma 75:
    381-395
KOLLMAN, R. 1966. Zur funktionellen Morphologie des
    Coniferen-Phloems. Habilitationsschrift Bonn.
KOLLMANN, R. 1973. In: Grundlagen der Cytologie, G.C. Hirsch,
    H. Ruska and R. Sitte, (eds.). Fischer, Stuttgart. pp.
    479-504
KRULL, R. 1960. Planta 55: 598-628
MAIER, K. and U. MAIER. 1972. Protoplasma 75: 91-112
MORRETES, B.L. DE. 1962. Amer. J. Bot. 49: 560-567
NAGELI, C.W. 1858. Beitr. Wiss. Bot. 1: 1-156
OSCHMAN, J.L. and M.J. BERRIDGE. 1971. Fed. Broc. 30: 49-56
PALIWAL, G.S. and H.-D. BEHNKE. 1973. Phytomorphology 23:
    183-193

PATE, J.S. and B.E.S. GUNNING. 1969. Protoplasma 68: 135-156
PATE, J.S. and B.E.S. GUNNING. 1972. Ann. Rev. Plant Physiol.
    23: 173-196
PATE, J.S., B.S.E. GUNNING and F.F. MILLIKEN. 1970. Protoplasma
    71: 313-334
RESCH, A. 1961. Z. Bot. 49: 82-95
ROBARDS, A.W. 1971. Protoplasma 72: 315-323
SAUTER, J.J. and H.J. BRAUN. 1968. Z. Pflanzenphysiol. 59:
    420-438
SCHNEPF, E. 1974. Port. Acta Biol. ser. A (in press)
SRIVASTAVA, L.M. 1970. Can. J. Bot. 48: 341-360
WOODING, F.B.P. and D.H. NORTHCOTE. 1965. Amer J. Bot. 52:
    526-531
ZAHUR, M.S. 1959. Cornell Univ. Agr. Exp. Sta. Mem. 358
ZEE, S-Y. and T.P. O'BRIEN. 1971. Aust. J. Biol. Sci. 24:
    35-49

DISCUSSION

Discussant: Inge Dörr
          Botanisches Institut der Universität Lehrstuhl II
          D-2300 Kiel, West Germany

Dr. Dörr was invited to give her paper on "Development of Transfer Cells in Higher Parasitic Plants". The paper follows.

Development of Transfer Cells in Higher Parasitic Plants

Inge Dörr

We have investigated higher parasitic plants in order to gain information about assimilate transport. These are very unusual plants to study such fundamental physiological processes, but we thought that the unusual characteristics of plant parasitism may help us in getting new results.

Our first object was Cuscuta odorata, a holoparasite belonging to Convolvulaceae. Biochemical investigation showed that there is absolutely no chlorophyll in this plant. The parasite grows as long, pale shoots and it gains all water and nutrition from the host plant by sending haustoria into the foreign tissue. The tip of the Cuscuta haustorium very soon spreads into single cell strands, the so-called "searching hyphae", which transverse the foreign tissue dissolving the walls of the host cells.

The searching hyphae resemble growing pollen tubes. When a searching hypha reaches a host sieve tube, it forms a special contact cell. This contact cell looks like a hand grasping the conducting element of the host. Walls of the sieve tubes and companion cells are spared from dissolution, so that the conducting system of the host stays undamaged for a while. Finger-like protrusions of the contact hypha grow between the different cell types of the host phloem and are closely attached to the sieve element (Fig. 1). The contact cell shows typical characteristics of a transfer cell. It develops a conspicuous wall labyrinth which is polarized and appears only at those parts which are adjacent to the host sieve tube. In this respect, the parasitic contact cell is to be compared with the B-type transfer cells in leaf minor veins described by Pate and Gunning (1969). The wall labyrinth in the contact cell of Cuscuta enlarges the inner surface up to 20 times. The wall ingrowths consist of spongy wall material and short fibrillar structures are arranged irregularly.

177

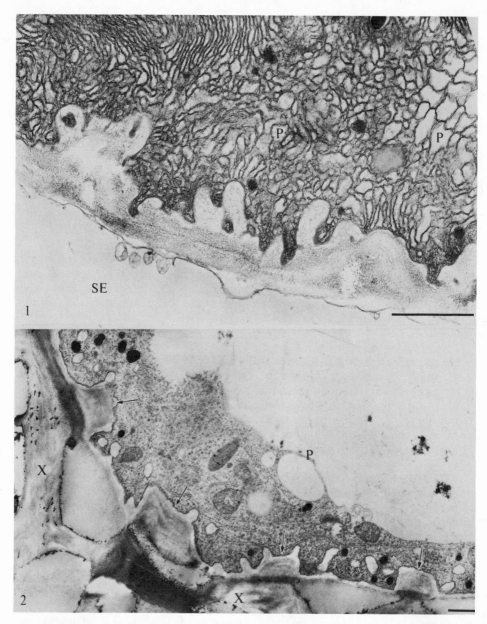

Fig. 1. Longitudinal section through a sieve tube (SE) of
Pelargonium zonale with a closely attached parasitic contact cell
of Cuscuta (P). This contact hypha shows a polar developed wall
labyrinth and a striking smooth surfaced ER. Fig. 2. Haustorial
cell of Orobanche crenata (P) adjacent to an xylem element (x) of
the host (Vicia faba). Beside small labyrinthine structures within
the parasitic cell there are recognizable xylem wall thickenings
(arrows).

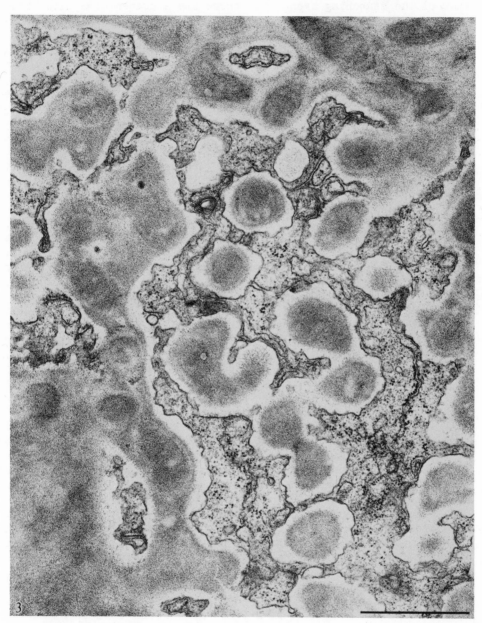

Fig. 3. Tangential section through an haustorial cell of
Orobanche within the xylem of the host. Wall ingrowths (here in
magnification) are cut transversely.

The fine structure of the contact cell differs very much from
that of the searching hypha, but there are a number of similarities
to the development of sieve tubes. The nucleus disintegrates, the
tonoplast is dissolved, ribosomes disappear, mitochondria and
plastids look very much like those known from sieve tubes. The ER
appears only in its smooth form.   Most instructive is the
appearance of typical plasmatic filaments or tubular structures,
the so-called P-protein, which are well known from the sieve tube
cytoplasm.

We can state that the parasitic contact cell described here is
a transfer cell, but this transfer cell does not show a protoplast
normally seen in this type of cell. The protoplast here is highly
specialized, and obviously concerned with a high rate of assimilate
transport.

The second parasitic plant we investigated was Orobanche.
This brownish holoparasitic plant grows on roots of different host
plants. It develops slowly, subterraneously for several months;
then the flower stalk appears above ground and grows quickly. The
haustorium of Orobanche crenata enters the root tissue like a cone.
The walls of the host cells in front of the haustorial tip are
dissolved by the growing parasitic cells.   When the vascular
cylinder is reached, the haustorial cells spread among the
conducting tissues of the host. We did not recognize any transfer
cells within the phloem part of the host root, but as for Cuscuta,
walls of sieve tubes are not attacked by the parasite, and they can
be detected well preserved, but completely surrounded by Orobanche
cells.

In the xylem tissue, single parasite strands grow between the
host cells. The former compact haustorium showed outer front walls
of striking thickness. Single cells that leave this compact
formation of haustorial cells, do not show this striking wall any
longer.   Instead, they develop a conspicuous wall labyrinth (Fig.
3) at the wall part adjacent to the host xylem element.   This
parasitic transfer cell shows a highly structured protoplast. The
ribosomes are often arranged as polysomes. The mitochondria, which
are often strikingly long and branched, accumulate near the wall
labyrinth, and sometimes are situated between the trabeculae.

We note that sometimes the parasitic hyphae penetrate the
xylem elements of the host and resemble tyloses, but they are
recognized as labyrinthine parasitic cells by the structures at the
wall. Normally the parasitic cell stays outside and is closely
attached to the host xylem. This transfer cell obviously is not a
final stage of development. Within the spongy wall material of the
labyrinth, one can detect the development of real xylem
thickenings.   They are recognizable by their different wall

structure, and are usually arranged correspondingly to those of the host xylem. They extend into the cell lumen, while the labyrinthine structures are reduced (Fig. 2). In an intermediate stage of development, wall labyrinth is merely visible on the pit membranes - a region of obviously, predominant passage of substances. Finally, trabeculae disappear completely, the protoplast disintegrates, and a typical xylem element is formed.

Generally, in transfer cells, it is well known that wall labyrinth is not developed when the cell is suppressed in its special function (Gunning et al., 1968). Moreover, wall labyrinth can be reduced if the transfer cell stops to function (Schnepf, 1965; Schwab, Simmons and Scala, 1969; Fahn and Rachmilevitz, 1970).

We have detected a further characteristic of vascular transfer cells of the parasitic haustorium. Obviously, these transfer cells can develop into true conducting elements. This is convincingly shown by the special cytoplasmatic state of the modified cells. The contact cell of the haustorium of Cuscuta functions as a transfer cell (wall labyrinth is present), as well as a sieve element (the cytoplasm shows the typical characteristics, Dörr, 1972). Transfer cells can directly develop into xylem elements, as we have seen in Orobanche. Both structural changes might reflect an increased functional stress.

These observations are not merely restricted to parasitic transfer cells. A thickening of the whole wall of older transfer cells in which the labyrinth is breaking down, was described by Yeung and Peterson (1974). It is possible that the change of transfer cells to conducting elements occurs rather frequently in the plant kingdom.

References

DORR, I. 1972. Protoplasma 75: 167-184
FAHN, A. and RACHMILEVITZ, T. 1970. New Research in Plant Anatomy, Suppl. J. Linn. Soc. pp. 51-56. N.K.B. Robson, D.F. Cutler and M. Gregory, eds. London: Academic Press
GUNNING, B.E.S., PATE, J.S. and BRIARTY, L.G. 1968. J. Cell Biol. 37: C7 - C12
PATE, J.S. and GUNNING, B.E.S. 1969. Protoplasma 68: 135-156
SCHNEPF, E. 1965. Ber. dtsch. bot. Ges. 78: 478-483
SCHWAB, D.W., SIMMONS, E. and SCALA, J. 1969. Amer. J. Bot. 56: 88-100
YEUNG, E.C. and PETERSON, R.L. 1974. Protoplasma 80: 155-174

Tyree: It was stated that the protoplasts of all plant cells are connected through plasmodesmata. This is not entirely a

correct statement for two reasons, the first being that
translocation persists in <u>Cucurbita</u> even when the leaves are well
beyond the wilting point. This suggests that the osmotic
concentration of the solutes in the vascular bundle, the phloem
tissue, must be considerably higher than that in the surrounding
tissue. I do not see how high concentrations of solutes can
persist, if there is one complete symplasmic system in the plant.
The second reason resulting from Geiger and co-workers' plasmolytic
studies, where the osmotic concentration in veins of sugar beet
leaf in the sink, are almost twice the concentration of the
surrounding mesophyll. Again, I cannot see how this can arise if
there is only one symplasmic system. The implication seems to be
that we have in a plant, two independent symplasmic systems - one
being vascular tissue, the phloem, sieve cell, phloem parenchyma
and maybe bundle sheath complex - and the other, everything else.
Possibly there are even more than two symplasmic systems. Are
there frequent plasmodesmatal connections between the vascular
symplasm and the rest of the symplasmic network? If there are
frequent connections of this sort, how does one reconcile that with
the two observations I have just made?

Behnke: I do not believe that there are two independent
symplasmic systems. All the living cells are connected with each
other by plasmodesmata. There is one exception that has always
been given - that the guard cells are not connected to the
subsidiary cells, but here too, some investigators have shown
plasmodesmata between these two types of cells. There seem to be
two independent systems of translocation within the plant - the
symplast and the apoplast, and in going from the apoplast to the
symplast, one has to go through the plasmalemma barrier. Possibly
this passage is facilitated by the wall ingrowths. In connections
between two independent species, as in host parasite interactions,
there is no symplasmic connection, but there is this membrane
barrier, and the transfer cells are thought to facilitate the
transport of substances from the apoplast to the symplast. How
translocation occurs in the whole symplasm is something for the
physiologists to consider, but there are supposed to be
concentration gradients.

Tyree: At a physiological level, it is meaningful to talk of
symplasmic system only if you have free and rapid communication of
solutes between connected cells. This cannot possibly exist. If
it can be shown that one group of cells in a plant, which are
supposedly in one complete symplasmic system, has a substantially
higher solute concentration than the neighbouring cells, and if
they show connections under the electron microscope, surely they
are non-functional.

MacRobbie: Possibly we do not have a qualitative, but a
quantitative, problem. We can have a continuous symplasm and

concentration differences between different parts of the symplasm, provided we have very vigorous pumps localized in parts of the symplasm which are pumping solutes into that part from the apoplasm, and provided we do not have too many connections between that part of the symplasm and the apoplast. If we have very vigorous pumping of sucrose into the sieve elements, through the transfer cells, then the mere fact that there are connections between the vascular symplasm and the mesophyll symplasm does not necessarily pose a large problem. If, on the other hand, we find that there are very extensive connections between the two, and not necessarily a very large area of apoplast - symplast contact in the vascular system, which would give us tremendous loading, then perhaps we do have a problem. Whether all these connections are, in fact, as freely permeable to everything as we sometimes think, is questionable.

Lamouroux: In any first year botany lab when one tries a plasmolysis experiment, one finds that in the same piece of tissue, some cells will plasmolyse and adjacent cells will not, and I suggest that this indicates that in any tissue, neighbouring cells may have quite different osmotic concentrations.

Cronshaw: There is the possibility that plasmodesmata can be either open or closed - like valves if the central tubule has a dumbbell shape.

Parthasarathy: Plant pathologists routinely use Cuscuta for transmitting virus, as well as mycoplasma, from one plant to another.

Aikman: If the mesophyll cells are in good contact with the upper end of the phloem, then the concentration of sucrose would build up within the mesophyll cells to the same high level as it is within the upper end of the phloem. That is, if it is not being used up within the symplasm of the mesophyll cells. We are still left with the qualitative problem.

MacRobbie: Two things have been suggested about the system which, if we accept them, are relevant. One, the possibility that the mesophyll is in fact fairly leaky, and two, the enormous amplification of surface area in the transfer cells. If, in fact, we had leaky mesophyll cells and transfer cells with a very high rate of active transport of sucrose out of the apoplast, we have got the possibility of concentration differences, provided that resistance to flow in the sieve element is downwards, to carry it away, and is low compared with the resistance back into the mesophyll cells. It seems that the system will work as it appears to, but it involves active uptake at one end and perhaps some degree of leakiness in the other.

Tyree: Dr. MacRobbie's defense of a single symplasmic system, namely that the plant is working very hard to overcome backward leaks, is the best argument I can think of for an evolutionary trend towards dual independent symplasmic arrangement. If this were going on for eons, the plant would eventually lose all the connecting plasmodesmata.

Johnson: The apoplast is a continuous system, and from the mesophyll cells to the transfer cells, you get movement through the cell wall of sugars. Apparently at the same time, there is a transpiration stream going on in the opposite direction. Why is it that substances are not wafted in the wrong direction by the transpiration stream? Is there some sort of compartmentation in the apoplast in this cell wall system?

Currier: Dr. Johnson's point is well taken. If too many solutes get into the cell wall system, they are going to be swept along with the transpiration stream. I am not aware though, that the apoplast does much sugar conducting, other than possibly in sugar cane.

MacRobbie: The active transport is essential in any system, and in some plants, for example sugar cane and sugar beet, there is evidence of leakiness, but in other plants, there was argument about the degree of leakiness. The leakiness would help, but is not essential. The active transport system is essential and has to be fairly powerful.

Cronshaw: The elaboration of the plasma membrane most probably is to provide a greater surface area so that the pumps can work better and pump something into the transfer cells, but we have to think not only of sugar transport, but also of ion transport. These may not necessarily be at the same sites. What has been discussed so far is just sugar transport. If the transfer cells pump sugar into the symplast, then there must be a symplastic loading of the sieve elements. Otherwise, the sieve elements must be loaded from the apoplast through pumps in the sieve element plasma membrane. We have the same two possibilities for ion loading into the sieve elements. It is quite possible that the transfer cell elaboration may be for an ion pathway and may have nothing to do with the transfer of carbohydrates.

Turgeon: Dr. Johnson asked if there is compartmentation. The preferred pathway for water, once it gets out of the minor vein xylem, is to travel upwards and downwards towards the epidermis and then laterally. Therefore, there may be in fact, a compartmentation to some degree to keep a solute from flowing away from the phloem.

Swanson: I would like to see a table showing the relative

frequency of plasmodesmata between mesophyll cells and mesophyll cells; mesophyll cells and phloem parenchyma; phloem parenchyma and sieve cells; sieve cells and companion cells; companion cells and phloem parenchyma, and transfer cells, for quite a few different species. This might be somewhat helpful in evaluating part of the apoplast, symplast transport problem.

Behnke: Gunning and Pate have given a percentage distribution on plasmodesmata in different types of walls in minor veins. They counted a total of 251 plasmodesmata and found: (1) from the A type transfer cells to the sieve element, about 39% of all the plasmodesmata in Vicia faba, and about 40% in a member of Compositae; (2) from A type to A type transfer cells, about 1 - 9%; (3) from A type to unspecialized phloem parenchyma cells, more than 25% in Vicia, and none in the Compositae; (4) from A type to B type, none recorded; (5) from A type to xylem parenchyma cells, about 2% (this must be at the ground level); (6) and from A type to bundle sheath cells, (it might be C or D type cells) 1 to 9% also. Therefore, the most important connections are from A type cells to sieve element, or from A type cells to unspecialized phloem parenchyma cells. Some other figures from B type to unspecialized parenchyma cells show below 5%, but they are not so important.

Walsh: There is very little literature on the ultrastructure of minor veins, whether they be in monocotyledons or dicotyledons, but it appears as though there may be at least two types of mechanisms involved in the movement of sugars from mesophyll to vascular tissue. No one has demonstrated the occurrence of transfer cells in monoctyledon leaves, although we have seen them in spikelets and in other regions of the plant. It appears that transfer cells are primarily associated with the leaves of the dicotyledons. There are numerous plasmodesmata between the bundle sheath cells and the mesophyll cells in Zea mays. However, there are more plasmodesmata between adjacent bundle sheath cells. Recently, O'Brien and Carr reported about suberitic lamellae associated with the vascular bundles of the leaves in some monocotyledons, in which case I can envision a way of preventing leakage, or of preventing leakage loss of whatever it is you are interested in retaining.

Crookston: Dr. Walsh, have you observed plasmodesmata between the bundle sheath and the phloem tissue in Zea mays? In all the C4 grasses that we have looked at, we have found numerous plasmodesmata in outer walls of the sheath cells where they come in contact with the mesophyll, but we have failed to observe any plasmodesmata at all, between the bundle sheath and the phloem cells which it surrounds.

Walsh: Yes, the plasmodesmata are present.

Millburn: <u>Cuscuta</u> does contain chlorophyll, admittedly in a lower concentration than in most green plants, and can, and does, carry on photosynthesis as proved by C14 studies.

Srivastava: I would like to ask Dr. Behnke or Dr. Dörr about the wall thickenings in these transfer cells. Dr. Dörr mentioned that these thickenings do shrink - come and go, more or less. The same kind of phenomenon has been reported in some of the insectivorous plants, for example the Venus fly-trap, where the authors fed slices of lean beef to these plants and monitored the fate of the protein that was fed over 24 to 36 hours, and at the same time watched the cytology of the cells and noted some very remarkable changes in the wall ingrowths. What is the character of these walls? Is it cellulose or what?

Behnke: As far as I know, it is cellulose, but possibly it is put together differently.

Johnson: The ingrowths on these walls always seem to have a white lining around them. Is this a cavity or callose, or evidence of shrinkage?

Behnke: Gunning and Pate, and others, think this is a kind of free space with some solutes inside. It is not callose, but there is a report of callose in these sites. Gunning and Pate could not detect callose at all in this place.

Johnson: We have observed that at the point of contact of mesophyll cells in <u>Nymphoides</u>, there is a white layer which looks exactly like these white layers between the plasmalemma and the cell wall. By an unfortunate accident in one preparation, we got uronylacetate crystals in this place which seems to suggest that is a space.

Cronshaw: I agree with Dr. Johnson. The electron-lucent appearance of that region is very different from the electron-lucent appearance of callose. I think this is a space, but I do not think it is a real space in the living cell. What we have, is that these wall ingrowths are highly hydrated, and in preparation procedures they have shrunk, and we have this artefact of a space between them and the plasma membrane.

Srivastava: On the basis of cytochemical staining, these walls are low in cellulose and very high in polyuronides and pectins.

Phloem Tissue and Sieve Elements in Algae, Mosses and Ferns

H.-Dietmar Behnke

Lehrstuhl fuer Zellenlehre der Universität Heidelberg

Im Neuenheimer Feld 230, D-69 Heidelberg, Germany

During the phylogenetic development of plants from unicellular aquatic algae to highly specialized seed plants, the acquisition of a particular vascular tissue has become one of the major bases for life on the land. Parallel to the development of organs like stem, leaves, and roots, the differentiation of specific tissues like epidermis, rhizodermis, sclerenchyma, phloem and xylem adjust the plant to the altered surroundings. As we will see, most of the tissues, like the vascular tissue, were subject to phylogenetic changes during the evolution from primitive to more advanced land plants. Although there was a need for the development of cells specialized in long-distance transport in the larger brown algae, e.g. the giant kelps, differentiation of organized tissues for food translocation, closely associated to those facilitating the distribution of water, were initiated only in bryophytes. In the three parts to follow we will review the structure of the respective cells and/or tissues of algae, mosses, and ferns.

Algae

The brown algal taxon Laminariales has long been known to include many species whose giant thalli differentiate some cells particularly adapted to long-distance transport of assimilates. Therefore, investigations on translocation phenomena in algae have been confined almost exclusively to a few genera, comprising Laminaria, Macrocystis, Nereocystis, and Pelagophycus. Except for a very limited number of positive results, the search for specific assimilate conducting cells has failed, as far as pursued at all in the other classes unlike brown algae. A notable exception is the description of elongated conducting cells in the phylloids and cauloids of some red algae by Hartmann and Eschrich (1969). These authors applied leucine-^{14}C to young phylloids of Delesseria and Cystoclonium, traced their distribution by autoradiography, and in

187

a light-microscopic histological study of the phylloid, revealed elongated cells that probably are interconnected by pit-like structures. In identifying these cells with long-distance translocation they verified a previous supposition on the presence of specific conduits in the Florideae made by Feldmann and Feldmann (1946).

In Laminariales the structure of the conducting elements is not uniform. There are at least three morphologically different cell types to be recognized, largely depending upon their location in the various parts of the algal thalli and the different species studied. Two of them occur in the medulla (Fig. 1), the innermost part of phylloids and cauloids, which represent leaf- and stem-like differentiations in algae. The medulla consists of a network of elongated cells, named hyphae by Oliver (1887), which are claimed to be the cellular units for long-distance translocation (Fig. 2: H, from Steinbiss and Schmitz, 1974). Some of them are preferentially longitudinally-oriented and are often enlarged where they are connected to each other by transverse walls; they are called "trumpet hyphae" (Fig. 2: TH). The third type, qualified as true sieve tubes by several authors, occurs in the perimedullar region of only some of the Laminariales, not in Laminaria itself. All three types, because of pittings in their transverse end walls which are similar to true sieve plates, were likewise compared with and, by analogy, also called sieve elements. Many other terms have been proposed for these conducting elements (for a review see Esau, 1969), but despite their morphological and topographical diversities, histologically, they all lack the close association to a nucleate cell. Consequently, sieve elements of algae are rather isolated conducting cells, unlike higher plant sieve elements which are organized within a specific tissue, e.g. the leptom of bryophytes or the phloem of ferns and seed plants.

The perimedullary sieve tubes, as found to be present in Macrocystis, Nereocystis, and Pelagophycus, are characterized by especially well-developed sieve plates in their end walls. Light microscopic investigations on Macrocystis pyrifera revealed sieve pores lined by callose and of a diameter similar to angiosperms (Ziegler, 1963). Up to 244 pores, each measuring 2.4 to 6.0 µm in diameter have been recorded by Parker and Huber (1965) to be present within the simple and transverse verse sieve plate. Largely due to inadequate fixations for electron microscopy the protoplasmic contents of the premedullary sieve-tubes are almost unknown. Occasionally mitochondria, plastids, and dictyosomes have been reported to be present (Parker and Huber, 1965), but a developmental study has not yet been carried out nor have the presence of nuclei been clarified.

Translocation of ^{14}C-labelled assimilates and of fluorescent dyes, although occasionally shown to be effective in Macrocystis

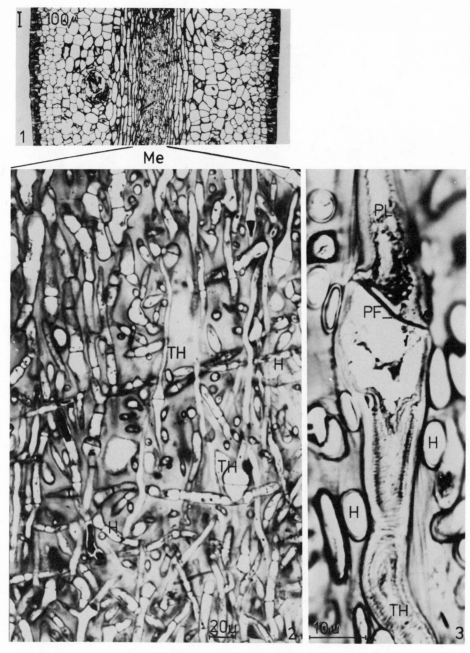

Figs. 1 - 3. _Laminaria hyperborea_. Arrangement and structure of "hyphae" (H) and "trumpet hyphae" (TH) in the medulla (ME). Pit-fields connect two trumpet hyphae in 3 at PF. Originals kindly provided by Steinbiss from Steinbiss & Schmitz (1974).

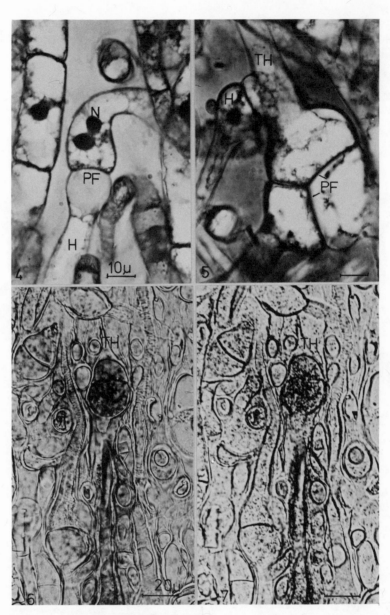

Figs. 4 - 5. <u>Laminaria</u> <u>hyperborea</u>. Hyphae within the medula connected by pit-fields (PF) to either other hyphae (H) or to trumpet hyphae (TH). N = nuclei. Originals by Steinbiss, from Steinbiss & Schmitz (1974).

Figs. 6 - 7. <u>Laminaria</u> <u>hyperborea</u>. Autoradiography of C^{14}-labelled assimilates in trumpet hyphae (TH), focused on tissue (6), and on grains (7). Originals from Steinbiss (1972).

(Parker, 1963, 1964, 1966) and _Nereocystis_ (Nicholson and Briggs, 1972), has not yet been confined to a single perimedullar sieve-tube. Such a cellular localization, however, has been accomplished for the medullary trumpet hyphae of _Laminaria_.

Steinbiss (1972) and Steinbiss and Schmitz (1973) offered ^{14}C-labelled sodium hydrogen carbonate to old phylloids of _Laminaria hyperborea_ and followed the translocation of the radioactive assimilate to the growing parts of the thallus. By histoautoradiography (including freeze-substitution of the material) they were able to localize the assimilates within the medulla of the phylloid. Moreover, a strict limitation of labelled assimilates to the longitudinally oriented trumpet hyphae, predominantly to their younger elements (Figs. 6,7; from Steinbiss, 1972), and the absence of radioactive material in (the approximately) horizontally-connecting ordinary hyphal cells, suggest a physiological differentiation between the two cell types found in the medulla.

An additional developmental study on the phylloid of _Laminaria hyperborea_ by the same authors (Steinbiss and Schmitz, 1974) contributed further details on the structure of ordinary hyphae and trumpet hyphae: Hyphae transverse the medulla in various directions and by a series of cylindrical cells connect the trumpet hyphae to the cells of the inner cortex. During development of the phylloid the inner cortex in its turn gives rise to additional hyphae crossing the medulla. Hyphae contain nuclei and are frequently connected by pit-fields to trumpet hyphae (Figs. 4,5; from Steinbiss and Schmitz, 1974). Trumpet hyphae with prominent pitted transverse walls most likely undergo an ontogenetic development, and there is accumulating evidence that young trumpet hyphae are derived from parenchymatic cells of the inner cortex which first pass through a hyphal stage. Further support to this idea is found in the observations of Went et al, (1973b) who found transverse filaments (=hyphae) connecting the medulla with the cortex in the cauloid of _Laminari digitata_. Numerous plasmodesmata connect the transverse filaments to the cortical cells.

It is obvious that in the central part of the medulla are located the most elongated trumpet hyphae (up to 1 mm), displaying the largest deposition of callose at their end walls and heaviest formation of secondary wall material (Fig. 3, from Steinbiss and Schmitz, 1974), both implying that they are in a non-functional state. An interesting new hypothesis, casting some light on the possible derivation of the "trumpet" from a normal hypha, has recently been proposed by Went and Tammes (1973). They gave some evidence that "trumpets" are formed only during turgor release, while longitudinal cell walls generally swell and only the transverse end walls resist.

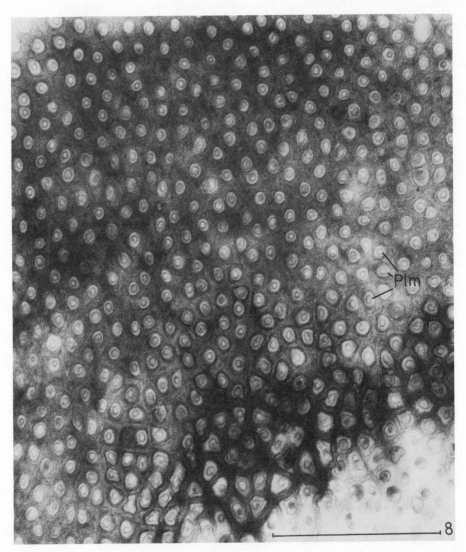

Fig. 8. Electron micrograph of transverse section of pit-field in end wall of _Laminaria_ trumpet hyphae. Remarkable density of plasmodesmata, all lined by callose and plasmalemma (Plm). Original from Ziegler after Ziegler & Ruck (1967).

Electron microscopy of the trumpet hyphae has so far only yielded some details on the structure of the pitted end walls. Exact information on the plasmatic structure of either hyphae or trumpet hyphae are not yet available. An electron microscopic study on the ontogeny of these cells in Laminaria is under way (Schmitz, pers. comm.) and will add to the knowledge of their fine structure.

The transverse end walls between two trumpet hyphae of Laminaria have been shown to be composed of plasmatic connections similar to seed plant plasmodesmata. Ziegler and Ruck (1967) disclosed their fine structure as being lined by plasmalemma, and although often narrowed down by callose deposits, considerable space is left for a plasmatic strand (Fig. 8, from Ziegler and Ruck, 1967). However, density calculations yielded higher ratios per unit area than in any other wall bearing plasmatic connections. Went et al (1973a) were also interested in the ultrastructure of these plasmatic connections because of the possible induction of an experimental fluid flow through Laminaria digitata trumpet hyphae (Went and Tammes, 1972). In their preparations callose was absent and consequently a larger open area within plasmodesmata was thus, according to their interpretation, more capable of rapid fluid flow.

Further fine structural investigations using more adequate fixations to yield more satisfactory results are needed for a clarification of cytoplasmic and connecting strand details in trumpet hyphae as well as "true sieve tubes" of Laminariales.

Mosses

According to Haberlandt's (1886) early investigations of the gametophyte stem of Polytrichum juniperinum, the conducting tissues of mosses consist of a "hadrom" (for water transport) and a "leptom" (for assimilate transport). The leptom - not phloem by Naegeli's (1858) terminology, since sclerenchymatic elements are missing - is composed of leptoids, very similar to sieve elements, and of parenchyma cells. In Polytrichum, the most intensely studied genus, a leptom was identified in both generations, gametophyte and sporophyte. The same is true for other genera studied, whereas in some taxa a leptom occurs only in the sporophyte and in still others it is absent altogether. Hebant (1964a, 1966a, 1970a) who performed most of the studies noted consequently interpretes the diverse appearance of conducting tissues in the mosses of today as the result of an evolutionary process leading to a reduction in the more advanced taxa. Mosses which, on the basis of some characteristics, are among the most primitive, still contain conducting tissues in both generations. The haploid generation is the first to lose the leptoids, as exemplified by some Funariales of intermediate advancement (Hebant, 1966a).

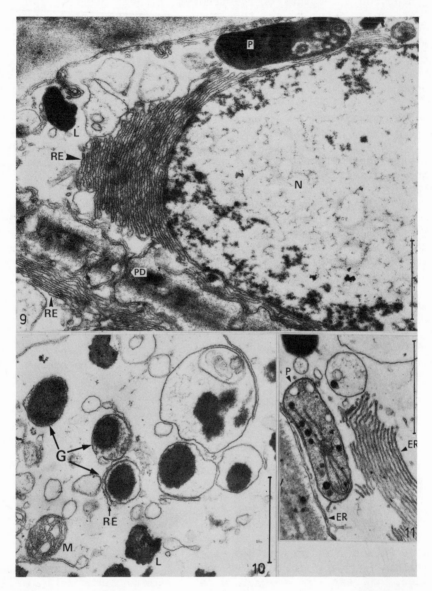

Fig. 9. <u>Polytrichum</u> <u>commune</u>. Two sieve elements connected by plasmodesmata (Pd) and containing endoplasmatic reticulum (RE), plastid (P), lipid droplets (L), and degenerating nucleus (N). Original after Hebant (1969a).

Fig. 10. <u>Polytrichum</u> <u>commune</u>. Refraction spherules (G) in sieve element. Original after Hebant (1969b).

Fig. 11. <u>Polytrichadelphus</u> <u>magellanicus</u>. Plastid (P) and ER in a mature sieve element. Original after Hebant (1974).

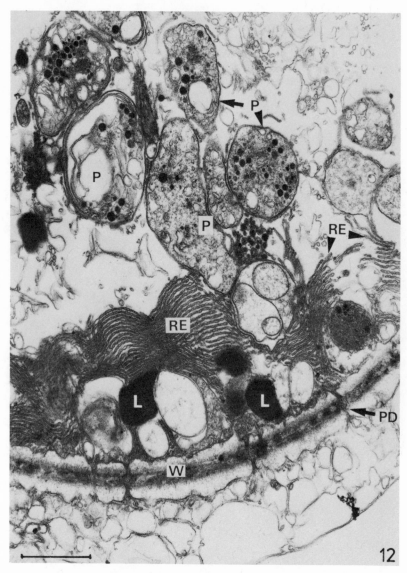

Fig. 12. _Polytrichum alpinum_. Longitudinal section of differentiated sieve element and end wall penetrated by plasmodesmata (Pld). Original from Hebant (1973b).

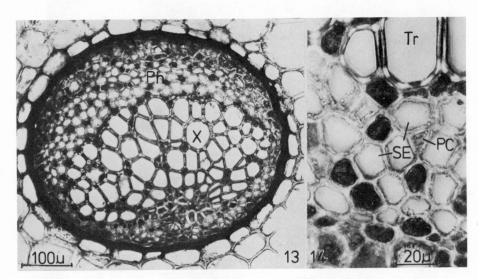

Figs. 13 - 14. <u>Dryopteris</u> <u>filix-mas</u>. Cross section of
vascular bundle (13) and phloem (14) from rhizome. PC = parenchyma
cell, PH = phloem, SE = sieve element, Tr = tracheid, x = xylem.
Unpublished originals by the author.

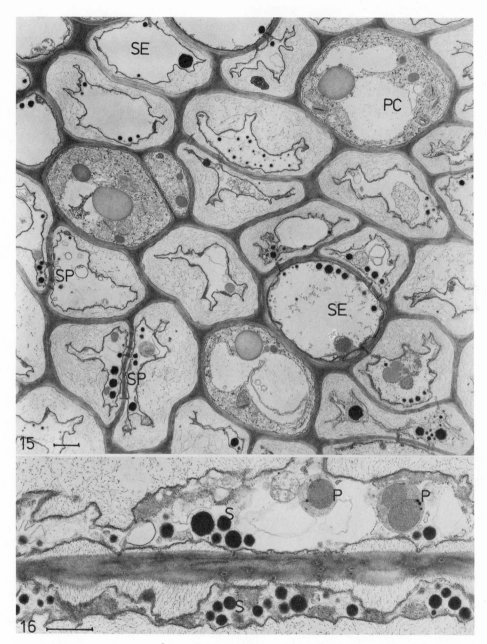

Figs. 15 - 16. <u>Platycerium</u> <u>bifurcatum</u>. EM transverse section of phloem (15) and longitudinal section of sieve elements (16). P = plastids, PC = parenchyma cell, S = spherules, SE = sieve elements, SP - sieve pores. Originals from Evert & Eichhorn (1974b).

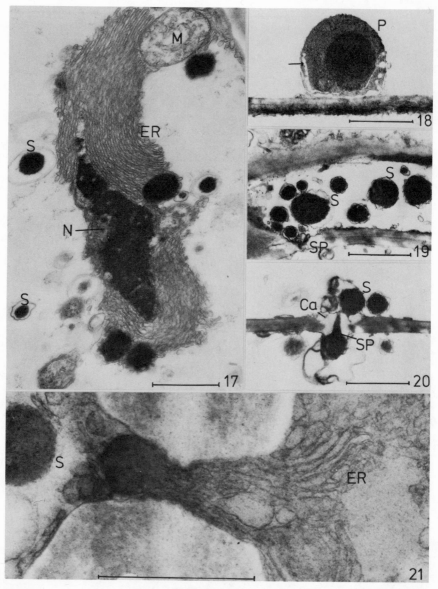

Fig. 17. <u>Polypodium</u> <u>vulgare</u>. Degenerating nucleus (N) of
sieve element associated by stacks of ER. M = mitochondrium, S =
spherules. Original from Maxe (1966).

Figs. 18 - 20. <u>Salvinia</u> <u>natans</u>. Plastids (P), spherules
(S), and sieve pores (SP, narrowed down by callose, Ca). Originals
by the author.

Fig. 21. <u>Polypodium</u> <u>vulgare</u>. Sieve pore, partly filled with
ER. Original from Maxe (1964).

There seems to be no anatomical difference between leptoids of the gametophyte and those of the sporophyte. Their structure, as elucidated with the light microscope, is largely comparable to that of fern sieve elements. Therefore, leptoids henceforth will be named as sieve elements (of bryophytes). Except for some early contributions by the classical 19th-century microscopists, our present knowledge of the ontogeny and light microscopic structure of moss sieve elements is based largely on Hebant's extensive investigations of different Polytrichales (1964b, 1966b, 1967a, 1967b, 1973a). During tissue differentiation sieve elements of bryophytes are to be distinguished from contiguous parenchyma cells by their elongated shapes with slightly oblique end walls. They contain a dense cytoplasm plus a large nucleus with prominent nucleoli. Mitochondria as well as plastids may also be identified. The ontogenetic development of the sieve elements includes a thickening of their longitudinal walls, a loss of stainable contents in the cytoplasm, and a final degeneration of the nucleus. Refractive spherules, known to be present in fern sieve elements, have also been found in some polytricheaceous mosses. Mitochondria and plastids can be recognized at all stages of sieve-element development. All the above cited morphological and cytoplasmic features confirm the concept of an extensive similarity between moss and fern sieve elements, as again emphasized by Hebant (1967a, 1968, 1970a).

At the time when K. Esau wrote her book on "The Phloem" (1969), no electron microscopic study was available on the fine structure of the sieve elements of bryophytes. It is thus again to the special credit of C. Hebant to have carried out successful ultrastructural investigations of the bryophyte leptom. Therefore, in the following I shall only give a short synopsis of the common fine structural characteristics of moss sieve elements, and C. Hebant will allude to some particular aspects in his subsequent presentation.

A transverse section through the leptom of Polytrichum nicely demonstrates its composition of two different types of cells: sieve elements (i.e. leptoids) and parenchyma cells (cf. Hebant, 1969a). The sieve elements are easily discernable by their light plasmatic contents which are dominated by opaque refractive spherules. End walls between two contiguous sieve elements are fitted with numerous plasmodesmata. Refractive spherules are quite similar to those long reported to be present in fern sieve elements (compare Fig. 10 with Figs. 15 - 20); they are surrounded by a unit membrane. Parenchyma cells are distinguishable from sieve elements by their different plasmatic contents, including, among others, starch-storing chloroplasts. Plastids of bryophyte sieve elements, on the other hand, are poorly developed leucoplasts containing only some thylakoids as well as osmiophilic globules (Figs. 9, 11, 12; compare also Hebant 1969a, 1973b, 1974).

     In bryophytes the protoplasts of sieve elements undergo a
ontogenetic differentiation similar to the respective conducting
cells in vascular plants.   Plastids at the beginning of their
development contain several thylakoids, but are less organized into
granal   stacks   than   in   neighbouring parenchyma cells.   They
gradually break down their internal membranes and change from an
elongated to a more or less spherical outline (Hebant, 1970b,
1973b, 1973c, 1974).   The endoplasmatic reticulum seems to become
especially abundant during sieve-element differentiation.   It is
present as flattened saccules or often as stacks of cisternae,
frequently found close to degenerating nuclei or next to end walls
(cf. Figs. 9, 12; see Hebant 1973b, 1973c, 1974).   Whereas
mitochondria persist during all stages, ribosomes seem to be absent
at the end of cell differentiation.   The great number of highly
active dictyosomes present in developing sieve elements is probably
related to the formation of the walls (Hebant,   1970b);   there are
none   recorded   within   the   differentiated   sieve   elements.
Eventually, the nuclei also pass through degenerative changes which
at the end render them inactive (as proven by their inability to
further incorporate radioactive precursors of RNA-synthesis).   In
terms of their structure, however, they do not dissolve (Hebant,
1973b).   Nonetheless sieve-element nuclei of bryophytes also show a
gradual decline of their contents (Fig.   9) and/or sizes,   often
recalling chromatolytic processes of angiosperm sieve elements
(Hebant, 1969, 1973b), but sometimes being likened to a pycnosis of
gymnosperm sieve cells (Hebant, 1974).

     Summing up: the fine structural aspects of sieve-element
protoplasts in bryophytes have much in common with sieve elements
of vascular plants and, as we will see, particularly with those of
ferns.   The protoplasts of two adjacent sieve elements, however,
are not connected by sieve areas or by single sieve pores but by a
large   number   of   plasmodesmata   similar to, though not as
concentrated as in, brown algal conducting cells.   Tests for
callose are positive in the end walls (Hebant, 1973c).   Details on
the contents of the plasmodesmata are still missing.   In some
sections plasmodesmata appear to have median cavities (Hebant,
1969a), in others they seem to be widened towards an initiation of
narrow sieve pores (Hebant, 1969a, 1974).

     In ending the section on mosses, it should be added that the
leaf veins of diverse Polytrichales, contain conducting cells which
have been given the particular name "deuters" but which, according
to fine-structural investigations (Hebant, 1972), may display some
of the characteristics of the sieve elements of stem leptom.
Finally, it must be mentioned that the liverworts also contain
conducting   tissues,   but their anatomical structure is only
incompletely known.

Ferns

Under this heading we will preferentially be dealing with the phloem structure of the Filicinae and only at the end report on some recent electron-microscopic research on species from taxonomic groups considered to stand at the base of the vascular plants. All the diverse taxa, including the Filicinae, which often are referred to as "the vascular cryptogams" have developed a typical phloem tissue, i.e. sieve elements related to phloem-parenchyma cells and associated with more or less pronounced sclerenchymatic elements. Lamoureux (1961) has made an extensive comparative study on the phloem of vascular cryptogams, recording histological, anatomical, and cytological data for the specific groups included, and Esau (1969) in a painstaking review collected all the details known from a widespread literature. Consequently it will be the aim of the present compilation to essentially review only those mostly-ultrastructural investigations that have accumulated since.

Of the nearly ten ultrastructural studies of Filicinae phloem available, all species investigated are exclusively of the family Polypodiaceae. Therefore, Dryopteris filixmas might well serve to introduce the topography of its phloem tissue: A cross section through the rhizome displays the typical periphloematic concentric vascular bundle which is almost isomorphic to a bicollateral condition (Fig. 13). The phloem is mainly composed of sieve elements (Fig. 14 : SE) and phloem-parenchyma cells (Fig. 14:PC). Additional information on the organization of the phloem of a polypodiaceous fern may be taken from an electron-microcopic transverse section (Fig. 15: Platycerium, from Evert and Eichhorn, 1974b). This allows a first impression on the cytoplasmic differences between sieve elements and parenchyma cells. Phloem-parenchyma cells have been examined by Maxe (1965) in Polypodium and by Warmbrodt and Evert (1974b) in Lycopodium. We will not deal further with their structure.

Sieve elements of ferns in general are very elongated cells with slightly, to very extremely, oblique end walls, resembling coniferous sieve cells in many respects. This also applies to the structure and distribution of sieve areas along the lateral as well as end walls. Although among different species their size varies, the pores of lateral and end wall sieve-areas are of equivalent differentiation.

One of the outstanding cytoplasmic characteristics present within all fern sieve elements are refractive spherules identified previously by the classical microscopists (among the earliest reports were those of Dippel, 1864 and Bary, 1877). They are highly refractive in unstained sections, respond readily to cytoplasmic stains and fixations, and probably are of proteinaceous nature. Maxe (1964), in the very first ultrastructural study of a

fern sieve-element, depicts the refractive spherules of Polypodium
as electron-opaque globules surrounded by a single unit membrane
(Fig. 17: S, from Maxe, 1966). Nearly identical views have since
been obtained from refractive spherules of Nephrolepis (Hebant,
1970b), Microsorium (Sakai and Lamoureux, 1973), Platycerium (Figs.
15, 16) and Microgramma (Evert and Eichhorn, 1974b), Phlebodium
(Evert and Eichhorn 1974a), Salvinia (Figs. 19, 20), Marsilea and
Regnellidium (Behnke, unpubl.). In an ontogenetic study of
Polypodium sieve elements Liberman-Maxe (1971) traced back their
origin and obtained evidence for a close relationship between
refractive spherules and endoplasmic reticulum. Whereas Sakai and
Lamoureux (1973) came to the same conclusion from their
developmental study of Microsorium, Evert and Eichhorn (1974b)
implicate dictyosomes in the formation of the spherules. The
latter authors also record a final fusion of the spherule membrane
with the plasmalemma and discuss a transport of glycoprotein to the
cell walls that may contain hydroxyprolin-rich proteins. Their
micrographs, however, show a difference in size and contrast
between vesicles fusing with the plasmalemma and those containing
the highly refractive material and might well suggest two
independent structures.

The nucleus of fern sieve-elements, as is generally admitted,
degenerates as the cell matures. Hebant (1969b), who carried out a
comparative light-microscopic study of proto- and metaphloem sieve
elements of 26 species belonging to the Filicinae, recorded a
pycnotic degeneration to be most frequent, although a chromatolysis
was found with some species. The nuclei of only a single species,
Ceratopteris cornuata, though considerably denatured, seem to
persist throughout the life-time of the sieve element. Prior to
this comparative report Maxe (1966), in a correlated light- and
electron-microscopic investigation, followed up the different steps
in nuclear degeneration of Polypodium sieve elements: a close
association of stacks of ER to the nuclear envelope and the
concentration of the chromatic material at some places of the
nuclear matrix. These are the major visible features of the
nuclear transformation in the microscope and resemble a pycnosis in
every detail. In late stages of dissolution the highly contrasted
nuclear remnants are largely surrounded by massive aggregations of
ER cisternae (Fig. 17, from Maxe, 1966). Recently Evert and
Eichhorn (1974a) examined nuclear degeneration in sieve elements of
four polypodeaceous species. Stacking of ER alongside the nuclear
envelope was also found to be present in Platycerium and
Phlebodium, but in these species the type of degeneration was
assigned to a kind of chromatolysis. The parts of broken nuclei
finally released somewhat resemble pycnotic nuclei, being still
surrounded by ER (see Evert and Eichhorn, 1974a: Figs. 12 and
16). The species chosen by these authors also contain prominent
crystalloids in their nuclei, which are eventually liberated from
the nuclei, and persist up to the obliteration of the sieve

element. Judging from light-microscopic reports (e.g. Hebant, 1969b) protein crystalloids are fairly common in fern nuclei.

Plastids and mitochondria are equally present in young and mature fern sieve-elements. However, there is a marked variation among plastid structures of the different species studied: Poorly structured plastids with only some plastoglobuli, short thylakoids, and rare starch accumulations inside a light matrix are found in Polypodium vulgare (Liberman-Maxe, 1971). Platycerium (Fig.   16, from Evert and Eichhorn, 1974b) sieve-element plastids contain prominent inclusions of higher contrast than the surrounding matrix. Salvinia (Fig. 18) obviously develops similar plastids, whereas Evert (1974) reports on the presence of extensive thylakoids, partly organized into grana, within plastids of Microgramma and Polypodium schraderi.

Ribosomes are lost during sieve-element ontogeny. Likewise, the granular ER which is evenly distributed throughout the cytoplasm of young cells (Liberman-Maxe, 1971; Evert, 1974) changes to the agranular form in mature elements, predominantly found as aggregates of stacked cisternae of the type associated to the nuclear envelope. Dictyosomes are present and still active during early stages of sieve-element differentiation, they most probably contribute to the formation of the cell wall (Liberman-Maxe, 1971, especially to the nacreous wall parts when present.

Nacreous walls have hitherto been depicted with the EM only in three fern species: Nephrolepis (Hebant, 1970b), Polypodium schraderi (Evert and Eichhorn, 1974b). In Polypodium schraderi, Evert (1974) assigns a role to both dictyosomes and ER in the formation of the typical wall thickenings. Vesicles or other compartments of these organelles discharge their contents into the wall region which, in ferns, (as discussed by Evert, 1974) have a lysosomal-like or autophagic activity. This suggestion, however, as is admitted by Evert, needs to be proven by the detection of hydrolytic enzymes.

The plasmatic connections between contiguous sieve elements contain open sieve pores, much like those known in angiosperms. While their average size far exceeds the diameter of plasmodesmata, in Polypodium pore diameters of 0.3 - 0.5 μm (Fig. 21) equal measurements taken from many angiosperm sieve pores. Again, Maxe was the first to examine a fern sieve pore. Fig. 21 is taken from her 1964 paper. She found ER cisternae in contact with the pore and, in part, penetrating the pore canal which is completely lined by plasmalemma. Evert (1974) likewise showed ER aggregations next to sieve pores. In some stages of development and, dependent upon fixation procedures, sieve pores are considerably narrowed down by callose deposits (Fig. 20: Ca). Refractive spherules often accumulate at the pore site (Figs. 19 - 21).

204 H.-D. BEHNKE

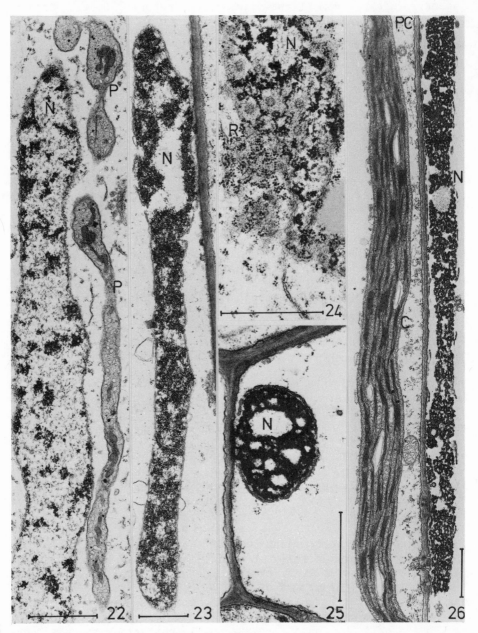

Figs. 22 - 26. _Isoetes_ _muricata_. Stages of the nuclear (N)
degeneration in sieve elements. Large parts of the nucleus persist
in mature cells (26). Originals from Kruatrachue & Evert (1974).
P = plastids, PC = parenchyma cells, R = ribosomes, C =
chloroplast.

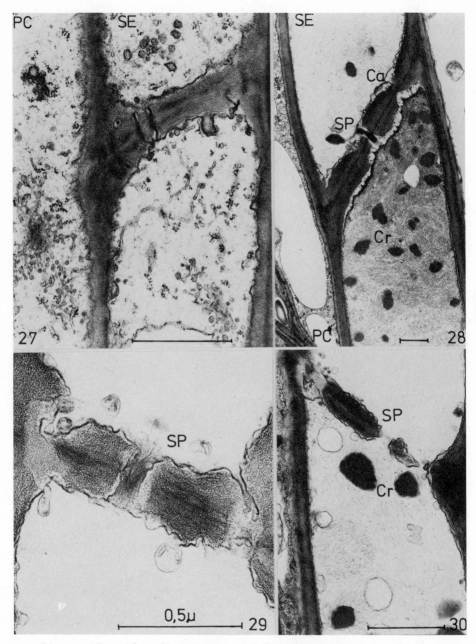

Figs. 27 - 30. Sieve pores of <u>Isoetes</u> <u>muricata</u>. Open pores (SP) are crossed by filamentous material that is derived from crystalloids (Cr). Pores develop from plasmodesmata (27), they may be closed by callose deposits (28; Ca). PC = parenchyma cell, SE = sieve element. Originals from Kruatrachue & Evert (1974).

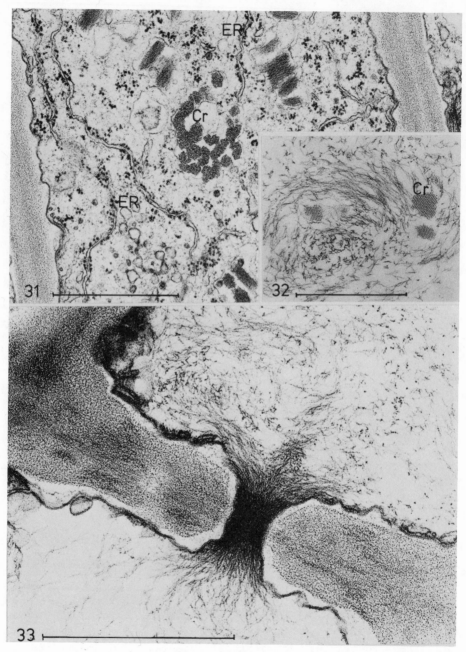

Figs. 31 - 33. Selaginella kraussiana. Protein crystalloids
(Cr) inside ER (31), released and originating filaments (32), and
inside sieve pore (33). Originals from Burr & Evert (1973).

Our knowledge of the phloem of the taxa considered as the most primitive vascular plants has been improved considerably by the recent pioneering ultrastructural investigations of Evert and co-workers, who examined sieve elements and/or phloem parenchyma cells in Psilotum, Lycopodium, Selaginella, and Isoetes.

Because of the equal size of pores in lateral and end walls, sieve elements of Lycopodium and Selaginella are considered as sieve cells; those of Isoetes show plasmodesmata in lateral- and either plasmodesmata or sieve pores in end walls. Although in general much resembling fern sieve-cells, some protoplasmic features are different and have to be reviewed in greater detail.

Contrary to some light microscopic reports, Lycopodium, Selaginella, and Isoetes do not contain refractive spherules. It is suggested by Burr and Evert (1973) and Evert (1974) that in light-microscopic studies plastids or some multivesicular bodies may have been mistaken for these. Spherules with some different structure have been identified in Psilotum nudum (Sakai and Lamoureux, 1973).

Except for Lycopodium (Warmbrodt and Evert, 1974a) the nucleus obviously dissolves only incompletely. Kruatrachue and Evert (1974) followed up single steps in the nuclear degeneration of Isoetes sieve elements (Fig. 22 - 26). They found large proportions of the nuclei associated with membrane parts to be present in mature sieve cells (Fig. 26). Only small remnants of the nucleus persist in Psilotum (Evert, 1974) and Selaginella (Burr and Evert, 1973).

Plastids, mitochondria, ER, and dictyosomes are of similar organization, as reported for fern sieve cells. Plastids of Isoetes elaborate similar inclusions in their matrix as portrayed for Platycerium (Fig. 17) and Salvinia (Fig. 18), whereas those of Selaginella and Lycopodium contain thylakoids in various arrangements.

As regards the plasmatic connections in the end walls, sieve pores of considerable diameters (0.2-0.8 µm) develop from single unbranched plasmodesmata (cf. Figs. 27 - 30, from Kruatrachue and Evert, 1974, and Fig. 33, from Burr and Evert, 1973). They are generally lined by plasmalemma and may be occluded by callose (Fig. 28:Ca), but in Isoetes (Fig. 29) and Selaginella (Fig. 33) their filamentous contents bear a strange similarity to the structure of an angiosperm sieve pore crossed by protein filaments. P-protein, however, (and this is the ubiquitous result of all the phloem research), is absent from vascular cryptogams. Nevertheless, the filaments demonstrated in the sieve pores of Isoetes and Selaginella are proteinaceous, but derived from crystalloids originating in a quite different way. In <u>Selaginella</u> protein crystalloids originate inside ER-derived small vacuoles (Fig. 31, from Burr and Evert, 1973). In Isoetes, on the other hand, filaments and crystalloids are deposited within dilated cisternae of ER and, as multivesicular bodies, fuse with the plasmalemma, discharging their proteinaceous contents. Some cisternae may release their contents earlier and thus contribute to the material inside sieve pores (Figs. 28 - 30, from Kruatrachue and Evert, 1974).

Summary

Despite the morphological variety of cells serving as conduits for assimilate translocation, a number of common characteristics may be recorded: degeneration of nuclei, dissolution of tonoplast (although not actually mentioned), loss of dictyosomes and ribosomes being among the most prominent. These are changes in the cytoplasm which are well-known to occur also in the sieve elements of seed plants. The gradual change from a primary pit-field having an extraordinary dense arrangement of plasmodesmata to sieve areas with true sieve pores suggests a phylogenetic line which eventually ends in the angiosperms displaying the largest and probably most effective sieve pores. According to this view the structure of sieve areas in gymnosperms would implicate a side line of development. Their more complex structure of several plasmodesmatal branches uniting in a median cavity cannot easily be derived from single pore openings as present in ferns but are likely to be traced back to plasmodesmata with median cavities in the end walls of bryophyte sieve-elements.

Acknowledgements

R.F. Evert and his co-workers, as well as C. Hebant, M. Liberman-Maxe, H.H. Steinbiss, and H. Ziegler kindly contributed original micrographs for inclusion in this review. I am indebted to them for their permission to publish these pictures as Figs. 15-16 and 22-33, 9-12, 17 and 21, 1-7, and 8, respectively.

References

BARY, A. de. 1877. Vergleichende Anatomie der Vegetationsorgane der Phanerogamen und Farne. Engelmann, Leipzig.
BURR, F.A. and R.F. EVERT. 1973. Protoplasma 78: 81-97 DIPPEL, L. 1864. Ueber die Zusammensetzung des Gefaessbuendels der Kryptogamen. Ber. Dt. Naturf. u. Aerzte, 142-148
ESAU, K. 1969. The Phloem. In Zimmerman, W., P. Ozenda, and H.D. Wulff (eds.), Encyclop. Plant Anatomy, Vol. V2. Borntraeger, Berlin and Stuttgart
EVERT, R.F. 1974. In Yearbook of science and technology. McGraw-Hill, Maidenhead (UK)
EVERT, R.F. and S.E. EICHHORN. 1974a. Planta 119: 301-318

EVERT, R.F.   and S.E.   EICHHORN.   1974b.   Planta 119:   319-334

FELDMANN, J.   and G.   FELDMANN.   1964.   Recherches   sur   l'appareil
    conducteur des Floridees.   Rev.   Cytol.   Cytophysio.   8:159-209

HABERLANDT, G.F.J.   1886.   Beitraege zur Anatomie und   Physiologie
    der Laubmoose.   Jahrb.   wiss.   Bot.   17:   359-498

HARTMANN, T.   and W.   ESCHRICH.   1969.   Planta 85:   303-312

HEBANT, C.   1964a.   Nat.   Monspel., ser.   Bot.   16:   79-86

HEBANT, C.   1964b.   C.R.   Acad.   Sci.   258:   3339-3341

HEBANT, C.   1966a.   C.R.   Acad.   Sci.   263:   1065-1068

HEBANT, C.   1966b.   C.R.   Acad.   Sci.   262:   2585-2588

HEBANT, C.   1967a.   C.R.   Acad.   Sci.   264:   901-903

HEBANT, C.   1967b.   Nat.   Monspel., ser.   Bot.   18:   293-297

HEBANT, C.   1968.   C.R.   Acad.   Sci.   266:   2190-2192

HEBANT, C.   1969a.   C.R.   Acad.   Sci.   269:   2530-2533

HEBANT, C.   1969b.   Nat.   Monspel., ser.   Bot.   20:135-196

HEBANT, C.   1970a.   Phytomorphol.   20:390-410

HEBANT, C.   1970b.   C.R.   Acad.   Sci.   271:   1361-1363

HEBANT, C.   1972.   Nova Hedwigia 23:   735-766

HEBANT, C.   1973a.   J.   Hattori Bot.   Lab.   37:   211-227

HEBANT, C.   1973b.   C.R.   Acad.   Sci.   277:   1445-1447

HEBANT, C.   1973c.   C.R.   Acad.   Sci.   276:   3131-3134

HEBANT, C.   1974.   Protoplasma 81:   373-382

KRUATRACHUE, M.   and R.F.   EVERT.   1974.   Amer.   J.   Bot.   61:
    253-266

LAMOUREUX, C.H.   1961.   Diss.   Univ.   of California

LIBERMAN-MAXE, M.   1971.   J.   Microscopie 12:   271-288

MAXE, M.   1964.   C.R.   Acad.   Sci.   258:   5701-5704

MAXE, M.   1965.   C.R.   Acad.   Sci.   260:   5609-5612

MAXE, M.   1966.   C.R.   Acad.   Sci.   262:   2211-2214

NAEGELI, C.W.   1858.   Beitr.   wiss.   Bot.   1:   1-156

NICHOLSON, N.L.   and W.R.   BRIGGS.   1972.   Amer.   J.   Bot.   59:
    97-106

OLIVER, F.W.   1887.   Ann.   Bot.   1:   95-117

PARKER, B.C.   1963.   Science 140:   891-892

PARKER, B.C.   1965.   J.   Phycol.   1:   41-46

PARKER, B.C.   1966.   J.   Phycol.   2:   38-41

PARKER, B.C.   and J.   HUBER.   1965.   J.   Phycol.   1:   172-179

SAKAI, W.S.   and C.H.   LAMOUREUX.   1973.   Protoplasma 77:   221-229

STEINBISS, H-H.   1972.   Zur Entwicklung und Anatommie des Phylloids
    von Laminaria hyperborea unter besonderer Beruecksichtigung
    moeglicher Assimilatleitbahnen.   Diplomarbeit, Bonn

STEINBISS, H-H.   and K.   SCHMITZ.   1973.   Planta 112:   253-263

STEINBISS, H-H.   and K.   SCHMITZ.   1974.   Helgol.   wiss.
    Meeresunt.   26:   (in press)

WARMBRODT, R.D.   and R.F.   EVERT.   1974a.   Amer.   J.   Bot.   61:
    267-277

WARMBRODT, R.D.   and R.F.   EVERT.   1974b.   Amer.   J.   Bot.   (in
    press)

WENT, J.L.   VAN, A.C.   VAN AELST, and P.M.L.   TAMMES.   1973a.   Acta
    bot.   Neerl.   22:   120-123

WENT, J.L. VAN, A.C. VAN AELST, and P.M.L. TAMMES. 1973b. Acta
    bot. Neerl. 22: 77-78
WENT, J.L. VAN, and P.M.L. TAMMES. 1972. Acta Bot. Neerl. 21:
    321-326
WENT, J.L. VAN, and P.M.L. TAMMES. 1973. Acta Bot. Neerl. 22:
    112-119
ZIEGLER, H. 1963. Protoplasma 57: 786-799
ZIEGLER, H. and I. RUCK. 1967. Planta 73: 62-73

DISCUSSION

Discussant: Charles Hébant
             Laboratoire de Paleobotanique
             Université des Sciences et Techniques du Languedoc
             34060 Montpellier, Cedex, France

    Dr. Hébant was invited to give his paper on "The Phloem
(Leptome) of Bryophytes". The paper follows.

              The Phloem (Leptome) of Bryophytes

                       Charles Hébant

    In   this   review,   I   shall   try   to   summarize   which
characteristics, either general or specialized, make the leptome of
Bryophytes particularly interesting for the study of translocation
mechanisms in land plants. Particular attention will be given to
the structural variability encountered in leptome,   as   this
variability is responsible for the contradictions found in some of
the papers recently published in this field.   This account is
essentially based on a comparative study of the leptome in the
gametophytes of a wide range of species of polytrichaceous mosses
belonging to eight genera (Atrichum, Dawsonia, Dendroligotrichum,
Oligotrichum, Pogonatum, Polytrichadelphus, Polytrichum, and
Psilopilum), but observations on the seta (part of sporophyte) of a
few species, have also been made.

                1. Light and electron microscopy

    The study of underdeveloped specimens has led some recent
authors to serious misinterpretations as to the normal structure of
certain Polytrichaceos mosses.   Correlated light and electron
microscope studies of the leptome of Bryophytes have shown, that in
the gametophytes of certain polytrichaceous mosses, this tissue can
exhibit a level of specialization strikingly similar to that of the
phloem in the true vascular plants. This refers mainly to (a) the
functional   association   of   the   food-conducting   elements
(leptoids)with parenchyma cells. As in the vascular plants, these
latter cells may show increased enzyme activities (e.g.
peroxidases;   phosphatases;   cytochrome   oxidase;   succinate
dehydrogenase), and some of them in Polytrichum commune (Hébant,
1972, Fig.   23) show an ultrastructure comparable to that of the
parenchyma cells associated with sieve cells in certain
Pteridophytes (Polypodium vulgare, Maxe, 1965; Lycopodium

211

lucidulum, Warmbrodt and Evert, 1974). (b) Secondly, the cytological organization of the best developed leptoids can hardly be distinguished from the sieve elements of the true vascular plants. Their ultrastructure has already been described by Dr. Behnke. Their "pores" are mostly enlarged plasmodesmata. However, in some forms, they may reach a development comparable to that of the small pores which have been observed in the protophloem sieve elements of diverse vascular plants. ER frequently shows relations with these enlarged plasmodesmata or small "pores" - several tubules of ER sometimes entering one single "pore".

Such a high level differentiation has already been observed in the fully developed gametophytes of such mosses as Dendroligotrichum dendroides, Polytrichadelphus magellanicus, Polytrichum alpinum or Polytrichum commune. It will certainly be found in a number of other Polytrichales.

That the leptome showing this type of organization is functional, is clearly shown by experiments with labelled compounds (Eschrich and Steiner, 1967; Collins and Oechel, 1974), and also by observations on exudation (Fig. 1 - A,B).

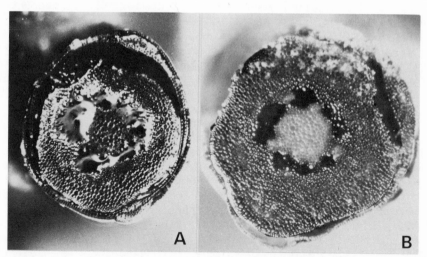

Fig. 1. Exudation in Polytrichum commune gametophytes. In Fig. 1A, a clear exudation is seen in the leptome zone around the central strand of water conducting elements (hydroids). In Fig. 1B, exudation is rather weak, but it´s precise origin can be seen: exudate originates not only from the leptoids sensu stricto, but also from the many cells of specialized conducting parenchyma which occur in association with them, and are related to the leaf traces (compare with Fig. 1 in Hebant, 1972, which gives the histological localization of these cells).

However, it must be emphasized that the level of specialization of the leptome in Bryophytes is not always so high.

2. Structural variability in Bryophyte leptome

Important variations are seen in the phloem of vascular plants, but it is not certain that they affect the functional organization of this tissue fundamentally. For instance, in all the Tracheophytes investigated up to now, the sieve elements have always been found clearly distinguishable from their neighbouring parenchyma cells.

In mosses, things are different. Here structural variability reaches a really unusual level, and elements showing structures intermediate between those of leptoids and of parenchyma cells are frequently seen. The occurrence of such elements is as follows: (a) they are seen in the gametophytic phase of the most highly developed species: (i) in their juvenile forms (ii) in underdeveloped specimens (for instance, in specimens grown in an unfavourable and too wet environment) (iii) in leaf nerves, leaf traces, and where the leaf traces merge with the central conducting system. Typical zones of transition, and gradients of differentiation are seen at these levels (Hébant, 1972). (b) they are also normally met in the many Polytrichales with a lower level of specialization. In the gametophytes of such mosses, a true phloem-like organization is usually not established. The leptoids can be poorly differentiated, to such an extent that they are no longer easily distinguishable from the adjacent parenchyma cells. A large variety of "intermediate" types of organization exists. In the most regressed forms, the only distinction which can be made, is between the cells of specialized conducting parenchyma of the internal cortex, and the ordinary parenchymatous elements of the external cortex. This latter type of organization prefigures that found in the leafy stems of the more "advanced" Arthrodonate mosses, where no leptoids have been found up to now. (In these mosses, leptoids are seen in the seta of a few species only).

In any case, structural variability is puzzling. For instance, the leptome cells (and sometimes nearly all cells of the "internal cortex") of the gametophytes of certain Polytrichales (e.g. Atrichum ligulatum, Atrichum undulatum, Polytrichum juniperinum, Polytrichum piliferum, ...) can exhibit thick walls very similar in appearance to the nacreous walls of sieve elements of certain vascular plants; but, in view of several other characteristics, the leptome of these mosses does not appear remarkably differentiated.

The tissue which, in the seta of the Polytrichales and certain other mosses, corresponds to the leptome of the gametophyte, must also be mentioned here. In the few species which have been investigated with the electron microscope (e.g. <u>Dawsonia</u> <u>superba</u>, <u>Polytrichum</u> <u>commune</u>, <u>Polytrichum</u> <u>formosum</u>), <u>Seta</u> leptoids did not show an exceptional level of specialization. They are never intermixed with specialized parenchyma cells, and although they present some sieve element characteristics in their ultrastructure (existence of modified plastids, occurrence of specialized ER), the plasmodesmata in their contact end walls have never been seen, in the specimens studies, to be enlarged to such an extent as to reach the structure of small "pores", as this may occur in the corresponding gametophytes. Further study of this tissue is needed.

Note: No liverwort is known with certainty to possess leptoids. The recent report of the occurrence of leptoids in <u>Pallavicinia</u> <u>wallisii</u> by Winkler (1969) needs substantiation. The cells he described as such, from the study of herbarium specimens, are probably ordinary elongated parenchyma cells.

3. Concluding remarks

What is the interest of these observations for the comparative histology and physiology of translocation in primitive land plants? (a) Leptoid structure: As in higher plants, it remains questionable as to what extent the observed ultrastructures are artefactual. And if they are not artefactual, their true functional significance is not completely understood. Nevertheless, the remarkable similarity, or identity, of organization of the leptoids when in the most highly developed form, to that of the sieve elements of the Tracheophytes, is really striking. With reference to such structural features as the lack of P-protein; the retention of a degenerated nucleus; the organization of the "pores" and their relations with the ER; the resemblance of the leptoids to the protophloem elements of certain angiosperms (e.g. Parthasarathy, 1974a,b), and to the sieve elements of gymnosperms, is particularly impressive. The resemblance of leptoids to protophloem elements is not surprising as both are involved in short, or relatively short, distance translocation. Extension of knowledge on the phloem of Pteridophytes will perhaps make possible further comparisons between Bryophytes and vascular plants. (b) From a more specialized phylogenetic point of view, it must be pointed out that while the occurrence of food-conducting cells in certain algae is evidently the result of convergent evolution, the situation in Bryophytes may be different. The presence of leptoids in these plants could be the consequence of another parallel development. However, Bryophytes and Tracheophytes share such an impressive ensemble of common characteristics (biological, morphological, physiological, biochemical) that the occurrence in some of the most primitive contemporary mosses of conducting tissues very similar to those of primitive Tracheophytes could also be relictual. This constitutes a reasonable alternative hypothesis which must not be entirely rejected when comparative studies are made on conducting tissues of land plants. With ther persisting degenerated nuclei; small "pores"; extensive ER, and lack of P-protein, and also because of their structural variability, the leptoids of mosses may well represent a model of archaic sieve element. (c) From a physiological point of view, the structural variability found in the leptome of Bryophytes is particularly interesting. It could provide a new material for the experimental study of translocation mechanisms. From which level of specialization do the elements of the leptome become as efficient as the sieve elements of the phloem in vascular plants? Observations on exudation in Polytrichum commune, in fact show that not only the leptoids sensu stricto, but also various cells of specialized conducting parenchyma associated with them, in which leptoid characteristics are more or less well developed, are capable of exudation (Fig. 1B). However, the leptoids sensu stricto seem to be more efficient - in wilted plants, exudation, if any, originates preferentially from them.

References
COLLINS, N.J. AND OECHEL, W.C. 1974. Can. Journ. Bot. 52: 355-363
ESCHRICH, E. AND STEINER, M. 1967. Planta 74: 330-349
HÉBANT, C. 1972. Nova Hedwigia 23: 735-766, 11 plates
MAXE, M. 1965. C.R. Acad. Sci (Paris) 260: 5609-5612, 4 plates
PARTHASARATHY, M.V. 1974. Protoplasma 79: 93-125 and 265-315
WARMBRODT, R.D. AND EVERT, R. 1974. Amer. Jour. Bot. 61: 437-443
WINKLER, S. 1969. Osterr. Bot. Z. 117: 348-364

(Ed. Note: Dr. K. Schmitz´s talk on "Sieve Elements of Brown Algae" is summarized briefly as follows.)

Schmitz: This work is still in progress and should not be taken as definitive. We have looked at Laminaria groenlandica; Alaria marginata; Nereocystis lutkeana, and Macrocysti integrifolia. I will present slides only from Laminania. In a hand section, already one can see that the sieve elements are elongated cells with blown up ends. Therefore, they have been termed "trumpet cells" or "trumpet hyphae". The sieve elements are linked together and form sieve tubes; they are also cross-linked with each other. We have seen up to 6 sieve elements come together and form one very complicated and composite sieve area. From our autoradiographic studies, we have good reason to believe that these are conducting sieve elements. The young sieve elements have thin walls; a nucleus; some cytoplasm; lots of vacuoles, and mitochondria and plastids. The plastids start out as normal brown algal type plastids but during sieve demand differentiation, acquire various shapes - cup or doughnut shapes - and at high magnifications, usually shows six lamellae, three stacks of

membranes forming the thylakoids. Some dark staining bodies, of
unknown origin and nature, occur close to the plastids. During
differentiation, the stroma of the mitochondria becomes electron
light, but otherwise, the mitochondria show the typical tubular
invaginations of the inner membrane. The nucleus is lobed, and
shows some condensed chromatin, and a prominent nucleolus. The
nuclear envelope shows normal pores. There is rough ER; some
smooth ER; ribosomes in the cytoplasm, and coated vesicles. The
nucleus is usually surrounded by dictyosomes, and the sieve
elements acquire a thick secondary wall along the longitudinal, as
well as the cross, walls. In a section through the sieve area, the
plasmalemma is continuous through the pores, and there is
cytoplasmic connection between adjacent sieve elements. Whereas
some pores are open, others contain some particulate or granular
material - what it is, I do not know. In some larger pores, the ER
may be seen close to, or even penetrating the pore, but we have not
been able to follow it through the pore from one cell to another.
The dimension of these pores in Laminaria, is in the range of 0.04
- 0.067 μm. There is an electron light space between the pore wall
and the plasmalemma lining it, but I do not think that it is callose
as I have never obtained a positive staining for callose under the
light microscope in these sieve elements, though by the same
techniques it is easy to demonstrate callose in the old sieve
elements. This light space is a fixation artefact, probably caused
by shrinking of the cell wall. In the brown algae, the cell wall
is mainly composed of alginic acid with some cellulose it in. The
alginic acid is very labile - it shrinks and swells up very easily.
In sections parallel or at a slight angle to the cross wall, the
cellulose fibrils are seen to be laid concentrically around the
pores. In Laminaria and all the other species mentioned, as the
sieve elements age, the secondary wall becomes thicker and appears
to be irregularly laid down in rings resembling the secondary
thickenings in xylem elements. The protoplast finally degenerates;
the cytoplasm becomes very electron light; the ribosomes and
dictyosomes disappear, and cell organelles, like plastids and
mitochondria lose their internal matrix where their membranes
separate from each other and often vesiculate. The nucleoplasm
also becomes electron light and the membranes of the nuclear
envelope separate from each other, but the nucleolus persists in
Laminaria. During these changes, the sieve areas at either end
become solidly blocked with callose, and eventually the whole lumen
of the cell may be filled with callose. I have not presented the
ontogeny of sieve elements, but we have followed it in Laminaria;
Alaria; Nereocystis, and Macrocystis, and as far as the ontogeny
and the main structural features are concerned, they are the same
in the 4 genera, though there are some differences. It seems that
there is a phylogenetic line in the Laminariales, starting with
sieve elements like those in Laminaria which have very small, but
numerous pores in the end walls, and leading to those which have
large, but relatively only few pores at the end walls as in

<u>Macrocystis</u>. It seems that in this phylogenetic line, we have sieve elements which retain their nucleus until the cells become nonfunctional as in <u>Laminaria</u>, and sieve elements which seem to lose their nucleus while they are still functional, as in <u>Macrocystis</u>.

Milburn: Dr. Schmitz, would you explain exactly where you made the measurements for pores? It strikes me that in dealing with shrinkage problesm, the "sieve plate" is rather dense and may not shrink so much as the tubular lining of the pores.

Schmitz: The tubule formed by the plasmalemma, is usually in the range of 300 nm, which is quite constant, but the width of the pore, from rim to rim, varies from 0.04 - 0.06 or 0.07 µm.

Srivastava: Because of the concentric arrangements of fibrils around the pore, the pore pulls apart, enlarging rather than shrinking, and therefore, a space is created between the wall and the plasmalemma.

Lobban: Dr. Schmitz, could you give us some information on rates of transport in <u>Laminariales</u>, and how this might relate to pore size?

Schmitz: We have looked at 14 species of <u>Laminariales</u> and in all of them, we have found translocation of 14 labelled photoassimilates, which suggests that all of them have a conducting system for the photoassimilates. <u>Laminaria</u> shows a translocation velocity of about 10 cm/h and has very fine, but numerous pores. In <u>Macrocystis</u> <u>pyrifera</u>, Parker reported translocation velocities of 60 to 75 cm/h. In <u>Macrocystis</u> <u>integrifolia</u>, we find velocities between 50 to 60 cm/h, but one has to be very careful in giving this data, as translocation velocities vary quite a bit and are influenced by so many factors, that one can only give a range. <u>Alaria</u> and <u>Nereocystis</u> range in between. In <u>Alaria</u>, velocities range between 15 to 35 cm/h, occasionally a bit more. <u>Alaria</u> fits very well in this line between <u>Laminaria</u> and <u>Macrocystis</u> in terms of the width of the pores. In <u>Alaria</u>, the pore diameter varies between 0.2 - 0.8 µm, and in <u>Nereocystis</u> it ranges from 0.2 - 1.2 µm. As Dr. Behnke mentioned, the pore size in <u>Macrocystis</u> is much higher - up to 4, 5, or even 6 µm.

Kollmann: Have you tested whether these young cells, which contain nuclei and vacuoles, show exudation?

Schmitz: It all depends on what you call exudation. In <u>Laminaria</u> you may get about one ml of exudate, but then it stops. You may cut again and get a tiny little bit. This may be the extruded content of the injured cells. I do not call it exudate. In <u>Nereocystis</u> it exudes better, and in <u>Macrocystis</u> it really flows. If you cut Macrocyctis at the base of the stipe and leave the whole plant in sea water, just raise the cut face a bit above the sea level and blot off the first few droplets. Then you are able to collect for about 15 minutes, and it really flows.   The flow depends very much on the condition of the plant; on temperature, and on various other factors.   After a while, the exudate starts to gel, and it may then be necessary to make a new cut, which will again flow for about 10 minutes.

Dute: Dr. Hébant, have you noticed refractive spharules in tissues of moss other than the leptome, and are they typically present in the intermediary cells?

Hébant: No, they disappear very quickly.   You find them mostly in the sieve elements, not in intermediate cells.

Behnke: Dr. Hébant, is it a possibility that the intermediate cells are not yet fully mature leptoids or sieve elements? We sometimes have these kinds of cells in the nodes of vascular plants.

Hébant: It does not seem that they differentiate any further. I have looked at stems 1 - 2 years old, and have found a similar type of organization.

Srivastava: Dr. Behnke, are there any parenchyma cell associations with the sieve elements in ferns along the lines of companion cells or albuminous cells?

Behnke: I did not find records of plasmatic connections between these two cells, although they are always associated spatially with each other.

Dute: There is an old report about the roots of Equisetum, apparently in which there are true companion cells - that is, cells drived ontogenetically from the same mother cell as the sieve element. I am studying Equisetum, but I have not gotten far enough to see whether there is a true ontogenetic relationship.

Hébant: The cells associated with sieve elements in the Pteridophytes show fantastic enzyme activities; phosphatase, and respiratory activities.   Therefore, there is a very strong physiological association.

Walsh: Dr. Hébant, do you find ribosomes in cells which you referred to as acting analogous to companion cells? In higher vascular plants, companion cells are usually very densely occluded with ribosomes.

Hébant: I find ribosomes in parenchyma cells associated with sieve elements in mosses, but I have never found such a high density as in the vascular plants.

Lee: Dr. Behnke, why, or what, is this material passing through the pores. These filamentous structures look very similar to that which have been found in higher plants. How do you distinguish between this filamentous proteinaceous material and P-protein?

Behnke: I do not call it "P-protein" because P-protein as defined by Esau and Cronshaw, has a different ontogeny than these things have. Maybe Dr. Kruatrachue can answer the question.

Kruatrachue: We decided not to call it P-protein because of the difference in ontogeny between P-protein of higher vascular plants and of this substance. In <u>Selaginella</u> it originates in the intracisternal space of the ER. Both are membrane bound, as opposed to P-protein of higher vascular plants that derives from the slime body which is not membrane bound.

Aronoff: The ontogenetic criterion may not be an adequate basis for distinguishing between polymers. I suggest that biophysical characterization be made before a decision is rendered as to whether these compounds are in fact, identical. We have been told of electrophoresis properties, but there are other parameters which could be used.

Kollmann: I agree, but there would be a real difference if this protein originates within a non-plasmatic phase (vacuole), while the P-protein always originates within the plasmatic phase. It may be that it is synthesized in the plasmatic phase, but secreted into the non-plasmatic phase.

Cronshaw: We should not call this material "P-protein" at this stage. We should leave this question open until there has been more research, as there obviously is this difference in ontogeny and perhaps difference in functions. Also, this generalization that there is no P-protein in gymnosperms and lower vascular plants should be left tentaive. In the original definition of P-protein, we did not include the criterion of ontogeny. What was said, was that these materials are proteinaceous, therefore the term "slime" appears to be a misnomer. We would like some other term to include the fact that those things have a characteristic morphology in the electron microscope. These were the two characteristics we included in the paper in Virology. In the subsequent paper on the ontogeny of the tubular and fibrillar components of <u>Nicotiana</u> cells, we described the ontogeny, but ontogeny was not one of the original criteria.

Walsh: What is the life span of the common garden variety of bryophyte, and do these "sieve elements" in bryophytes exhibit definitive callose?

Hébant: The life span of bryophytes varies greatly, from a few weeks to several years - up to 20 years or more for some species. The Polytrichales are more or less perennial and can live up to 10 or more years. It is very easy to detect callose in the sieve elements in all cells of conducting parenchyma in bryophytes.

Lobban: It is very important to remember that the sieve tubes in algae have had quite a different evolutionary history from the phloem of higher plants. We have to always keep this in mind, particularly when we are thinking in terms of function.

Geiger: Could it be that the spherules serve the function of plugging the pores or openings, or plasmodesmata? It would be difficult to conceive of them as functioning as pumps or boosters, or whatever.

Lamoureux: In microscopy observations, prepared in the usual ways, the spherules have been reported to be attached to the pores in the sieve areas. With some of the better EM fixation, it is obvious that this is not the case, though I have recently seen a SEM photo of phloem of Pteridium, in which each pore was plugged by a refractive spherule. These plants frequently have callose, in addition to refractive spherules, and you can see deposits of callose which build up more or less in relation to the method of preparation. Therefore, the location of spherules in the pores is a preparatory artefact.

Tyree: There are a number of ferns, commonly referred to in North America as resurrection ferns, that can survive tremendous dessication. It has been demonstrated in one species in the East Mediterranean, that photosynthesis can occur down to 20% of the plant's original water content, and many of them can survive for months at 20 or 8% of the original water content. Has anyone examined these "resurrection ferns", and found any definitive differences in the sieve tube structure that could explain this abnormal resistance?

Hébant: I have not studied the "resurrection ferns", but in some tropical ferns that have to go through a period of drought, the sieve elements have extremely thick nacreous walls. In bryophytes, the species which show nacreous walls in the sieve elements, or in the leptome sensu strieto, are species which most often have to go through periods of drought. In the Mediterranean region, some sites are extremely dry for 3 - 4 months, and the mosses that are found in these sites show the best nacreous walls.

Srivastava: We have looked at the "resurrection" plant,

<u>Selaginella</u>, and that plant can certainly be hydrated and dehydrated through several cycles. In the parenchyma cells in the dehydrated state, there are hardly any vacuoles; the cytoplasm shows very little lamellate ER; the ribosomes are scattered throughout the general matrix, and as you hydrate the plant, they appear at the same time vacuoles begin to appear. You can watch this change repeatedly. We have not looked at the sieve elements.

Lamoureux: I have looked at a few desert ferns and would agree with Dr. Hebant that in these plants there are very thick nacreous walls. I have made no attempt to study them when they were dried, and when they were unfolded and green. I would like to add one word of caution - there are many other ferns which show nacreous walls in the sieve elements, which are in no sense desert plants. They, in fact, occur in tropical rain forests.

Troughton: Dr. Schmitz, we have never been able to obtain "trumpet hyphae" in brown algae after freeze drying. Do you think that the trumpet shape is an artefact?

Schmitz: If you look at the ontogeny of the sieve elements in brown algae, you find that they start out as cylindrical cells, but as they age, they become trumpet-shaped. They are pulled out with the extension growth of the plant, stipe as well as the phylloid. They start out with a length of about 40 µm and end up at a length of more than 1,000 µm. At the same time, they acquire a thick secondary wall, and eventually, because of these two processes, the lumen becomes very narrow and nearly disappears. There is no space for the nucleus, for example, and it moves toward one of the ends near the cross wall. In fresh hand sections, you can see that the older sieve elements are very nicely trumpet-shaped, but the younger they are, they are less trumpet-shaped. Tammes and van Went say that the trumpet shape is an artefact, but I cannot believe it. All the anatomical literature also states that they are trumpet-shaped.

Srivastava: Dr. Schmitz, you may like to indicate that the cross wall remains the same diameter.

Schmitz: Yes, there is no change in diameter of the cross wall, but there is a tremendous change in the thickness of the wall as well as in the length of the cell.

Troughton: Initially, we hand-cut the specimens from the stipe and looked at them with a light microscope and saw characteristic "trumpet hyphae", but when we freeze-dried the material and looked at it again under the EM, we could not find the characteristic trumpet shape. With the electron microscopic techniques, do you find the trumpet shape consistently in old cells?

Schmitz: We have done freeze substitution as well as freeze drying, and have found trumpet-shaped sieve elements.

Spanner: Dr. Schmitz, how do you reconcile the view that the cross walls remain the same size - unshrunken - with those callose-like cylinders, which attribute to shrinkage artefact?

Srivastava: I think we are talking about two different things, Dr. Spanner. One is the change in the shape of the cell as the cell ages. This change in shape from a cylindrical cell to a cell with bulbous ends, is brought about by the fact that the cell continues to elongate as the thallus elongates, but near the cross walls, the diameter remains fixed. That is why with aging, the cell eventually becomes very stretched, and narrow in the middle with bulbous ends. The other point is the increase in size of the pores during preparation of the material for electron microscopy. That is brought about because the celluloe fibers run concentrically around the pores, and consequently, since these walls are highly hydrated, when you dehydrate them, the pore tends to enlarge. That is why a space is created between the plasmalemma and the wall itself. We have seen this kind of retraction of the wall from the plasmalemma in parenchyma cells by watching the whole process of fixation and subsequent dehydration, under the light microscope.

Milburn: If one observes the cut end of the stipe of Macrocystis, as exudation procees, from being a cleanly cut cylinder, it becomes depressed like a dish. This must clearly indicate that as the sieve tube system empties, there is a volume displacement within the stipe as a whole. The exudation is very rapid at first, and then falls away rapidly. If you make a second cut, a considerable distance from the initial cut, you get almost no exudate - the quantity is very much reduced. This indicates that your initial cut goes a very long way to draining the turgor pressure from the whole system. I have some preliminary information that the turgor pressure of the system in Macrocystis, about 10 metres tall, is of the order of two bars - not very much. Therefore, I am very interested to know what the pore diameters might be. In connection with the lining of the pores, the protoplasmic lining would normally be under turgor pressure, and this thing is undoubtedly elastic. It appears that it is more likely to shrink away from the wall inwards, rather than the plate to have shrunk away, as has been suggested. It also seems that there must be some trumpeting effect when one releases the turgor from the sieve tube-like hyphae.

Cronshaw: Regarding the shrinkage of this wall, cellulose microfibrils, and the microfibrils of the Laminariales are primarily cellulose. Cellulose microfibrils have a tensile strength that is approximately equal to steel. If you wrap microfibrils round and round as we have, or as has been suggested, around these pores, I can see no way in which we could have this binding stretched out or shrunk. It is more probable that Dr. Milburn is correct - that the elastic protoplasmic parts shrink inwards. Within the Laminariales, we have on other 1,3 linked glucan, laminarin. When tests are made for callose, does it stain?

Eschrich: No.

Cronshaw: There is a possibility that this clear area could be laminarin. I do not know where laminarin is, in these plants.

Schmitz: Laminarin is probably found within the protoplast, not outside the plasmalemma. Referring to cellulosic fibrils, these are somehow linked together, and there is lots of space in between. We have not measured the distance from fibril to fibril, but is there not the possibility that, as you dehydrate, the fibrils come closer together, not that the fibrils are being stretched, thus enlarging the pore?

Cronshaw: No, if the microfibrils are wound round and round the pore, as they are in the first part of the pit border that is laid down, you could not have slippage of those microfibrils past one another.

Srivastava: I disagree with Dr. Cronshaw for the following reason. The length of the cellulose fibril will not change. That will stay constant, but if the filler material is of a high degree of hydration then as you remove water, you are liable to bring two cellulose fibrils closer together, and to have sort of an appressing effect. This is something that you can see very clearly in some of the nacre walls. When the nacre walls are dehydrated, the fibrils do come closer together. Referring to laminarin, we do not know whether we have laminarin in our material. We do know that in young sieve elements, we do not get callose staining with resorcin blue or aniline blue, but in older elements we do. It is on that basis that we conclude that in the young sieve elements, the space that we see is not due to callose, but that it is an artificial space created during fixation and dehydration procedures.

Johnson: One of the features of this so-called artefact space, is the extreme regularity of it. We have seen a similar space in the protuberances into the transfer cell. Again, there is the plasmalemma which is uniformly spaced from the wall. I cannot see what should keep it such a carefully spaced distance, unless there is something filling the space, or unless there are very fine filaments anchoring it at intervals, which do not show up.

Cronshaw: There is a very good explanation, put forward by Dr. Srivastava, that the wall shrinks. One would expect from the wall characteristics which are fairly uniform, that the shrinkage would be uniform, and therefore would have a uniform space in these examples that you mentioned.

Walsh: Dr. Hebant and Dr. Schmitz, why do you call these, pores and not plasmodesmata? is it a size, or a structural discrepancy?

Schmitz: I call them sieve pores for two reasons. First, I do not want to exclude Laminaria from the other Laminariales. There is no doubt that Macrocystis has sieve pores. We have Alaria, which has smaller sieve pores, and finally we have Laminaria which has the smallest pores. Second, if you define the plasmodesma in terms of the plasmodesma being traversed by the plasmalemma, and that some sort of a desmatubule is always found within it, then these are pores and probably not plasmodesmata.

Lobban: Going back to Dr. Milburn's comment, I do not believe he realized, that with a cut you cannot drain such an enormous system as that of macrocystis. It is simply far too big for that. Also, if you make a second cut, you will usually find further exudation. Crafts found this in 1939, and I believe Dr. Schmitz has found this as well.

Milburn: If you make a series of cuts rather rapidly, one after another, before the sealing becomes significant, the second cut that you make, a distance away from the first one, will give you significantly less exudate, suggesting that you have reduced the turgor pressure in the whole column - this goes for a distance of several metres.

Schmitz: I never have tried to make a rapid series of cuts because I wanted to have time to collect the exudate, before making the next cut, but I have checked where I found good exudate. To get a good exudate, you have to cut at the very bottom. If you cut at the top, and Macrocystis is a plant which grows at the top, you do not get exudate.

Lobban: I have been looking at which blades in a Macrocystis plant export, and which ones import, and the ones at the top do not export, therefore you have material flowing from the bottom of the plant towards the top. Dr. Milburn, which side of the cut did you look at - particularly when you are getting towards the apex?

Milburn: I cut the stipe at the junction where the horizontal floating thallus, lying along the sea surface, plunges down to the hold-fast - about 10 metres below, in my case. The flow comes out from both surfaces. When you cut near the hold-fast, it behaves exactly as if you have cut through a turgid cell of infinite length, in the middle. You seem to get exudate of comparable volume from both ends.

Sieve Element Structure in Relation to Function

Rainer Kollman

Botanisches Institut der Universität Kiel

Lehrstuhl II (Zellbiologie)

Dusternbrooker Weg 17, D-23 Kiel, W-Germany

I. Introduction

Translocation in the phloem of higher plants represents a complex physiological process with several partial reactions working together. Uptake and release of substances, energy transfer as well as storage of assimilates, are involved; or there may be a requirement in some cases for the mechanism of long-distance transport. Various cell types are engaged in the whole process.

The function of the different categories of phloem cells is reflected by their cytoplasmic fine structure (for a review of the literature, see Esau 1969; Kollman 1973). Some modifications of the fundamental type of a plant cell are realized. The parenchyma cells of the axial system of the phloem tissue and those of the ray system represent storage cells, with large vacuoles and many starch-containing leucoplasts. A special organization of the ER and the arrangement of the plasmodesmata in these cells point to their additional function in short distance transport. The companion cells and albuminous cells are specialized parenchyma cells. The high metabolic activity of these cells is indicated by their peculiar fine structural features, such as dense protoplasts lacking predominant central vacuoles, an abundance of ribosomes; rough ER; large, often-lobed nuclei; and leucoplasts which generally do not store considerable amounts of starch. From this latter group of cells we distinguish the transfer cells. Their enlarged cell surface points to the particular function of the plasmalemma in the uptake and/or secretion of substances from and/or into the extracellular free space. (For a more detailed

225

description, see Behnke´s report on companion cells and transfer cells).

Though we have learned to correlate structure and function of the parenchymatic phloem cells, the interpretation of the main conducting units, the sieve elements, still remains problematical. This uncertainty arises from the peculiarities of these cells. (1) Sieve elements undergo fine structural changes during their development which are unique among plant and animal cells. (2) Due to their altered cytoplasmic fine structure and their high internal pressure, mature sieve elements are extremely sensitive to any kind of manipulation. This accounts, at least in part, for the differences in the results of various anatomical and physiological investigations. Sieve elements may be the most thoroughly investigated plant cells; but the exact arrangements of the peculiar fine structures and their functional significance under physiological conditions may still remain open.

Our present consideration will deal with three aspects of sieve element structure, which may be essential to any discussion on the transport mechanism. We will first summarize the basic cytological features of sieve-element differentiation which are generally accepted; some functional aspects will be indicated. Secondly, we will point out those peculiarities of fine structural differentiation, the functional interpretation of which seems to be problematical. Finally, we shall try to show some ways which may lead to a better understanding of the sieve element structure. In this connection, the results of some recent biochemical and cytochemical investigations will be reported critically. Some hypothetical interpretations of the findings will conclude this report.

II. Overview of Fine Structural Differentiation of the Sieve Elements

(Selected literature only will be cited; for detailed references, see Esau 1969 and Kollman 1964, 1973)

The main cytoplasmic changes including the development of the intercellular bridges will be considered here (for cell wall differentiation we refer to Dr. Cronshaw´s report).

A. Sieve pore formation

Of special significance with regard to the function of the sieve elements, is the formation of the sieve pores in the sieve areas and sieve plates. These cytoplasmic bridges originate by enlargement of plasmodesmata. ER and callose may be involved in the removal and/or exchange of wall material (Esau et al, 1962; Northcote and Wooding, 1966; Deshpande, 1974); the precise mechanism is unknown. The completed pore, in angiosperm sieve tubes, is lined by a continuous plasmalemma and traversed by

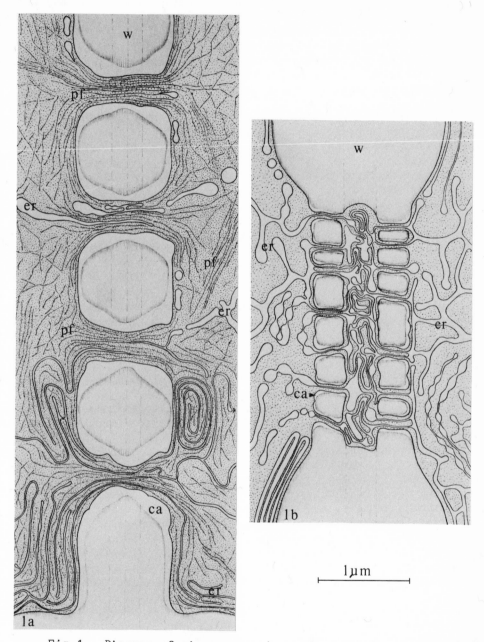

Fig.1.   Diagram of sieve pores in angiosperm sieve tubes (1a) and   conifer   sieve   cells.   (1b).   w=cell wall,   ca=callose, pf=filamentous P-protein, er=sieve-tube reticulum; (for further explanation of various angiosperm plants (1a) and of Metasequoia glyptostroboides (1b), see Kollmann (1973)).

protein filaments and scattered tubular protrusions of the ER, thus
directly connecting the two adjacent sieve element protoplasts
(Behnke, 1971). The narrower pore channels of conifer sieve cells
are branched with a joint median cavity in the region of the
primary wall. No protein filaments, but distinct ER structures,
predominate within these cell connections (Kollman und Schumacher
1962, 1963; Kollman 1964; Srivastava and O'Brien 1966; Wooding
1966; Parameswaran 1971; Evert et al, 1973b). There is no doubt
that the enlargement of cell connections between sieve elements is
related to an increased intercellular transport. (If we compare
the principle of cell contact in sieve elements with that in
transfer cells, we may discern the two different ways for an
increased intercellular transport indicating two different
mechanisms (Kollman und Dorr, 1969).

B. Disintegration of cytoplasmic components

The most striking event during sieve element development is
the disintegration of some main cytoplasmic components. Mature
sieve elements are devoid of normal organized nuclei, ribosomes,
dictyosomes, and tonoplasts.

Disorganization of the nucleus has often been described in
higher and lower vascular plants. Two different ways can be
distinguished: so-called chromatolysis (Resch 1954) and pycnotic
degeneration. The former mechanism appears to occur mainly in
angiosperm plants (Behnke 1969a; Esau and Gill 1972; Parthasarathy
1974b), while the latter, with some exceptions (Esau and Gill
1973), may be typical for gymnosperms (Maxe 1966; Wooding 1966;
Evert et al, 1973a). The nuclear envelope becomes incorporated in
the modified endomembrane system of mature sieve elements (Behnke
1969a; Parthasarathy 1974b). The exact time of the nuclear
degeneration, especially with regard to a certain stage of cell
differentiation, has not been determined so far. There are some
indications, however, that main processes of cell synthesis
requiring metabolic activity, such as cell wall development, pore
formation, and the production of protein filaments, are advanced at
the time when the nucleus breaks down (Parthasarathy 1974b). In
some plants disorganized nuclear material persists for a longer
time in fully developed sieve elements (Evert et al, 1970, 1973a;
Esau and Gill 1973). Whether or not these remnants of altered
nuclei have any physiological function remains to be proved.

As a consequence of the breakdown of the nucleus, ribosomes,
which are abundant in young cells, disappear during maturation of
the sieve elements. From this very important alteration of the
protoplast we have to conclude that, in mature sieve elements, no
further protein synthesis occurs. This statement brings out some
problems as to any transport mechanism controlled by the metabolism
of the mature sieve elements. However, we have to bear in mind
that young sieve elements evidently are able to act in

translocation (Kollman 1968), and that the conducting elements are
structurally   and   physiologically   associated   with   their
metabollically active companion cells and albuminous cells. Most
likely, the bulk of enzymatic activity identified within the phloem
exudate may be contributed by these associates of the conducting
elements (Lehmann 1973).

    Disappearance of the dictyosomes and microtubules indicates
maturity of the sieve elements. No further synthesis and delivery
of wall material takes place, and the process of translocation is
not related with these cell structures.

    Distintegration   of   the   tonoplast   characterizes   some
culminating   points   of   sieve element differentiation.   This
phenomenon means that cytoplasmic and vacuolar sap fuse as a result
of subsequent alteration of the protoplast. Most likely due to the
high amount of osmotic substances occuring within the sieve-cell
cytoplasm, the latter becomes more and more hydrated. Dilution of
the protoplast proceeds to a certain degree of biophysical adaption
of the plasmatic phase and the non-plasmatic compartment (see
"mictoplasm", a term proposed by Engleman 1965). The boundary
between both phases breaks down and fusion occurs.   The often-
described extension of the parietal protoplast preceeding the
breakdown of the tonoplast may be interpretated according to this
concept (Kollman and Schumacher 1964; Evert and Deshpande 1969).

C. Changes in fine structure of cell organelles and endomembranes
    Modifications of the other cytoplasmic components, such as
endoplasmic reticulum, mitochondria and plastids hardly can be
interpreted. Alteration of the rough ER into a smooth, parietally-
distributed, and stacked "sieve-tube reticulum" (Bouck and Cronshaw
1965; Wooding 1967a; Behnke 1968; Esau and Cronshaw 1968) coincides
with the loss of ribosomes and demonstrates some essential
biophysical changes of the protoplast. The latter becomes devoid
of lipid components during cell maturation.   Suggestions that
stacking of the ER may represent a sequestering of the membrane
system in an inactive form should only be accepted with caution.
The occurrence of a highly developed ER system in mature conifer
sieve cells (Kollman and Schumacher 1964; Wooding 1966; Neuberger
and Evert 1974), as well as in the assimilate-conducting elements
of the haustoria of higher parasitic plants (Dörr 1972; Dörr and
Kollman 1974), points to a peculiar role of the endomembrane system
in the process of translocation. Its parietal distribution at the
lateral walls and at the sieve areas leads us to suggest that the
ER may have some function in the exchange of substances between the
conducting elements and the surrounding tissues.

    Plastids   and   mitochondria   are   retained   during   cell
differentiation. Both organelles usually show sparse internal
structures but have real functions in the living, enucleated sieve

elements. Recent cytochemical investigations prove the existence
of some enzymes of the electron-transport chain (cytochrome c;
cytochrome oxidase) in the mitochondria of mature sieve cells and
sieve tubes (Catesson and Liberman-Maxe 1974). The leucoplasts
accumulate starch-like polysaccharides and/or proteins. (In the
sieve-element starch 1,6-linkages of the highly-branched glucan
chain predominate (Palevitz and Newcomb 1970). The different
storage products of the sieve-element plastids have some taxonomic
significance (Behnke 1972). The obvious lability of the
mitochondria and especially of the plastids during the various
procedures of preparation and fixation emphasizes the difficulties
in understanding the peculiar structure of the sieve-element
protoplast as a whole.

D. P-protein development

Protein filaments and tubules - the so-called P-protein
(according to Esau and Cronshaw, 1967), are specific components of
the sieve tubes in angiosperms. They originate within the
cytoplasm during cell development. We always identify these
typical structures as the main components of mature and so-called
functioning conducting elements. In gymnosperm sieve cells
P-protein has not been shown so far.

The origin of the protein structures takes place at very early
stages of sieve element differentiation. Several small but
distinct areas of granular and fine filamentous material can be
recognized most often in close association with polysomes (Buvat
1963; Behnke 1969b; Zee 1969; Wergin and Newcomb 1970). The latter
most likely are involved in the synthesis of the electron dense
protein material. Other cell organelles are separated from these
aggregations, but often surround them. Thus we cannot exclude at
present that especially ER and dictyosomes with their forms of
vesicles may participate in the process of formation of the
peculiar protein structures (Bouck and Cronshaw 1965; Cronshaw and
Esau 1968; Steer and Newcomb 1969; Esau and Gill 1970).

The first steps of protein formation are followed soon by a
subsequent organization of the more or less amorphous material.
The structure of organized protein varies in the sieve elements of
the different plants. Generally, for instance in Nicotiana
(Cronshaw and Esau, 1967), tubules of about 20nm (15-25nm) in
diameter, and of undetermined length, are formed within large
compact bodies. The tubules often show regular, nearly hexagonal,
arrangement. Spoke-like extensions often seem to connect
laterally-adjacent tubules (Cronshaw and Esau 1967; Parthasarathy
and Mühlethaler 1969). In some plants, such as Cucurbita (Cronshaw
and Esau 1968), filaments are formed in separated bodies in
addition to tubular structures. Tubules and filaments may be
interpreted as different forms of protein assembly. This is shown
evidently by further changes of the protein structures as cell

development proceeds. At that stage of differentiation, which is marked by the disorganization of the nucleus and tonoplast, the protein bodies disaggregate and the tubules or filaments disperse throughout the whole cell lumen. From that time on they also penetrate the developed sieve pores as more or less aggregated strands. In the sieve elements of a few plants only, for instance Nuphar (Behnke 1969b), tubular structures persist within the mature sieve tubes. As a rule, however, the tubules undergo some further modifications during the process of dispersal. They are transformed into solid filaments, which often show striation of a strong periodicity (Northcote and Wooding 1966; Wooding 1967b; Cronshaw and Esau 1967; Steer and Newcomb 1969). In mature sieve elements then we find only filaments with a diameter of about 10nm (6-15nm).

Finally, another form of protein aggregation should be mentioned here. From light microscopic investigations it is well known that, in the sieve elements of most papilionaceous plants, spindle- shaped crystalline inclusions persist for a long time. As Wergin and Newcomb (1970) and Palevitz and Newcomb (1971) have shown, this crystalline protein represents a specific intermediate stage, temporarily established when the tubular protein complex is transformed into filamentous structures.

As to the process of transformation of the tubular structures into filaments, Parthasarathy and Mühlethaler (1969) have proposed a model. According to it, sieve-tube tubules are made of either one or two helically-arranged filaments, each filament representing a linear strand of globular subunits. Polymerization and depolymerization, in combination with stretching and loosening of the helices, lead to the different structural aspects.

The P-protein of the sieve tubes is the most problematical component in many respects. (1) The mechanism of its formation has not been cleared up entirely. (2) The distribution of the peculiar structures in undisturbed functioning sieve elements is a matter of controversy. (3) Their function, especially with regard to the mechanism of translocation, is absolutely obscure. (4) Finally, we always have to bear in mind that these typical structures obviously are lacking in the sieve cells of conifers and of most lower vascular plants. In the following, therefore, we shall be concerned with those essential questions.

III. Some Concepts of P-Protein Formation; Its Distribution and Function Within Mature Sieve Tubes

As to the mechanism of formation of the protein structures, several ways may be discussed at present: A. De novo synthesis and unspecific assembly process. Structural protein units are synthesized by the ribosomes. The process of assembly of the

initially produced material into tubules or filaments is influenced
by various environmental factors, such as ion composition and lipid
content of the cytoplasm, pH, hydrostatic pressure and certain
metabolic compounds. (See discussion in Wergin and Newcomb, 1970,
and Parthsarathy, 1974a).

B. Reorganization of protein structures from disintegrated
cytoplasmic material

According to this concept, P-protein is the product of
reorganization of protein material derived from disintegrated
cytoplasmic structures such as membranes, ribosomes and others.
Assembly of the free membrane proteins or ribosomal proteins takes
place according to changes in the environmental conditions as
mentioned before. There are some indications that filamentous
structures may arise from cytoplasmic membranes in various kinds of
cells (for literature, see Franke 1971). Moreover, results of
recent biochemical investigations such as gel-electrophoresis,
amino acid analysis, and immune reactions were interpreted by Weber
et al, (1974) with regard to some relationships between sieve-tube
proteins and ribosomal proteins.

According to these concepts, the P-protein arises from
unspecific protein units under certain physiological conditions of
the sieve elements by a self-assembling mechanism. Thus the
peculiar structures would only reflect a certain metabolic state of
the conducting elements without any functional significance. More
detailed morphological and biochemical studies are necessary to
prove such a hypothesis.

C. De novo synthesis of P-protein with regard to
particular function

This alternative concept points out that P-protein may
represent a specific cell structure which is synthesized during
cell development and may have a considerable influence on the
process of translocation. However, any discussion at this point
has to consider that the peculiar protein structures are restricted
to angiosperm sieve tubes. Nevertheless the main hypothesis being
discussed so far should be examined more intensively.

1. Moving protein filaments

As is well known, actin-like proteins responsible for cell
movements have been identified in many plant and animal cells. A
recent hypothesis points to the possibility that the protein
filaments of sieve tubes represent actin-like proteins. Such a
concept was stimulated first by some morphological data. Indeed,
the substructural features of isolated protein filaments resemble
the characteristics of actin-filaments as described for diverse
animal and plant cells (Kollman et al, 1970). Some chemical
in-vitro reactions of the sieve-tube protein, such as G-F

transition in presence of Ca^{++} and K^+ ions and precipitation behaviour with vinblastine, seemed to support the concept of a relationship between actin-like proteins and sieve-tube proteins (Kleinig et al, 1971b). Some other characteristics of actin and/or tubulin microtubular protein, however, could not be identified with the purified sieve-tube proteins. Under in-vitro conditions, the latter were unable to participate in any nucleotide and and colchicine-binding, and they did not show any ATPase activity in presence of divalent cations such as Ca^{++} and/or Mg^{++} (Kleinig et al, 1971b).

Some observations on the behaviour of the protein structures in intact sieve elements under functional conditions, after treatment with colchicine, vinblastine, and cytochalasin B, emphasized the dissimilarity between sieve-tube protein and actin-like proteins (Williamson 1972; Sabnis and Hart 1973). In contradiction to this statement, inhibition by cytochalasin B of assimilate transport in phloem strands of Heracleum has been reported. (Thompson and Thompson 1973). As Sabnis and Hart (1974) pointed out in their recent paper, the use of cytochalasin B to implicate microfilaments in the transport mechanism may be of doubtful value on account of the questionable specificity of this drug. Furthermore, Sabnis and Hart (1974) once again failed to detect any actomyosin-like proteins in the phloem extract of Heracleum. Thus, there is no convincing evidence so far for any existence of "contractile" proteins as a structural basis for protoplasmic streaming in mature sieve elements.

2. Surface interactions
We shall not discuss the various theories on transport mechanism (refer to the papers by Dr. MacRobbie and Dr. Spanner), but we may point to one essential biochemical and biophysical property of the sieve-tube protein with regard to certain motive forces generated by surface interactions. It may be suggested that the protein filaments form an extended frame work with charged surfaces. The electrochemical properties of the protein thus should be of special significance. Isoelectric focusing of the sieve tube proteins obtained from Heracleum phloem resulted in an isoelectric point (IEP) of 4.9 (Yapa and Spanner 1972). At the alkaline pH of the sieve-tube sap the proteins would therefore be negatively charged. This finding was interpreted by Yapa and Spanner (1972) as supporting their electroosmotic theory of translocation. However, the results with Heracleum cannot be generalized. In Cucurbita, for instance, the main proteins of sieve-tube exudate have been identified as being highly basic with an IEP above 9.5 (Beyenbach et al, 1974). The discrepancy between the two results cannot be explained at present; thus the electrochemical nature of the sieve-tube proteins remains an open question.

3. Protein filaments as sites of enzymatic activity.

It has often been suggested that the tubular and filamentous
protein structures within the sieve tubes might represent sites of
enzymatic activity at which energy is released for the transport
mechanism. Association of phosphatase activity with the sieve-tube
content was indicated from earlier histochemical investigations
(Braun and Sauter, 1964; Kuo, 1964; Lester and Evert, 1965).
Biochemical tests for ATPase activity in the exudate and extract of
phloem tissue showed different results. Positive reactions have
been obtained with Robinia (Wanner 1953) and Heracleum (Sabnis and
Hart 1974); with Cucurbita exudate, both positive (Lehmann,
unpublished results) and negative results (Eschrich et al, 1971)
have   been reported.   The enzyme activity was stimulated by
monovalent cations, especially by K^+, and inhibited by divalent
ions such as Mg^{++} and Ca^{++} (Sabnis and Hart 1974). As mentioned
earlier, purified sieve-tube proteins from Cucurbita exudate did
not display ATPase activity under the various conditions employed
(Kleinig et al, 1971b). Whether or not this was due to calcium- or
magnesium-sensitivity cannot be decided at present.

Recently Gilder and Cronshaw (1973 a, b; 1974) succeeded in
localizing cytochemically at the electron microscope level ATPase
activity with the protein structures of Cucurbita and Nicotiana
sieve tubes.   Significantly, the enzyme activity occurred in
association with the dispersed protein filaments of mature and
functioning sieve tubes; undispersed protein complexes of immature
cells did not show any positive reaction. Similar results were
obtained with other nucleoside triphosphatases.

Provided that artefacts were excluded, the identification of
protein filaments as sites of ATPase activity and energy release
would be of great importance with regard to the function of the
sieve tubes.   Further biochemical investigations now have to
establish such a concept. We shall return to this point later on.

D. Arrangement of protein structures within undisturbed
functioning sieve elements

Whatever the results of biochemical and cytochemical
investigations of specific sieve-tube contents might be, there is
one most outstanding question which has not been answered
unequivocally so far: how are the diverse fine structures arranged
in the undisturbed functioning conducting elements? As we have
mentioned earlier, any interpretation of the sieve-element fine
structure has to consider some artefacts which in particular might
be due to the phenomenon of dislocation of the cell contents during
preparation and fixation. It is open to question whether we will
really be able to overcome this main difficulty. (For details, see
Srivastava´s   paper).   However, from the results of numerous
different techniques carried out under various precautions, a
nearly adequate picture of the functioning sieve elements has been
established. (See also Johnson, 1973). We already have referred

to it in the first part of the present paper. Let us return once again to the structural aspects.

There are specific subcellular structures which extend throughout the conducting elements. In angiosperm sieve tubes protein filaments, and in conifer sieve cells ER tubules, have been discerned. These fine structures may interact with the various translocates along the pathway and especially in the region of the sieve pores. The fine structure of the conifer sieve areas, generally accepted so far, rules out any discussion of so-called open sieve pores. (Nevertheless, conifer sieve cells pierced by aphid stylets showed exudation (Kollman and Dörr, 1966), a phenomenon which has often been quoted as an indication of the high permeability of the sieve pores. The question is whether the protein filaments and ER tubules act only as barriers along the pathway or whether these structures have some special motive function in phloem transport. Before answering this question, more detailed information is required on the molecular structure and, especially, on the biochemistry of these peculiar sieve-tube contents. In a final part of this lecture we will point to some experimental ways.

IV. Recent Approaches in Biochemical Characterization of the
Sieve-Element Contents.

As previously indicated, the problem of transport mechanism now is focused on a thorough interpretation of those particular fine structures which connect the sieve elements. Some preliminary results have already been reported. Further investigations will be determined by the different techniques employed.

With regard to cytochemical work, various angiosperm and gymnosperm plants should be tested. The most important question is whether the ER tubules in the sieve cells of conifers display ATPase activity as shown by the protein filaments in the sieve tubes of Cucurbita and Nicotiana.

As to biochemical studies, the problem of isolation of cell contents always exists. Plants showing the phenomenon of phloem exudation will be preferred; results obtained from analysis of phloem extracts are unreliable and should be interpreted cautiously. Phloem exudate consists mainly of soluble components and those fine structures which can pass through the sieve pores. With Cucurbita the characteristic sieve-tube filaments have been identified electron-microscopically as the main structural component of the phloem exudate (Kollmann et al, 1970). Other cell structures were retained at the sieve plates. Chemical analysis has proved the pure protein fraction of the exudate. Thus Cucurbita phloem represents a rather good object for biochemical studies on the sieve-tube protein.

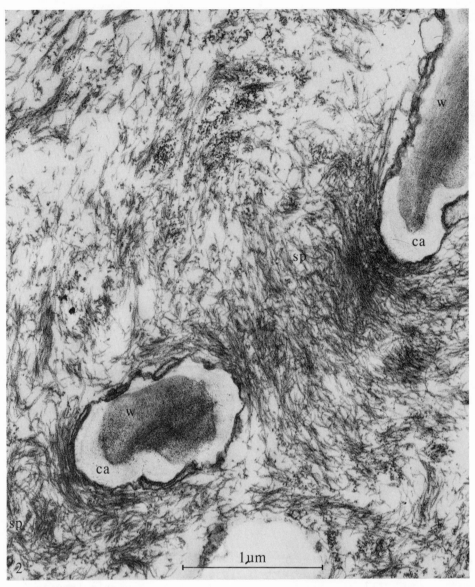

Fig.2. Typical aspect of a mature angiosperm sieve tube.
Arrangement of filamentous P-protein in the sieve pores (sp).
w=wall of the sieve plate, ca=callose. Cannabis sativa. Original
by Dr. Inge Dorr.

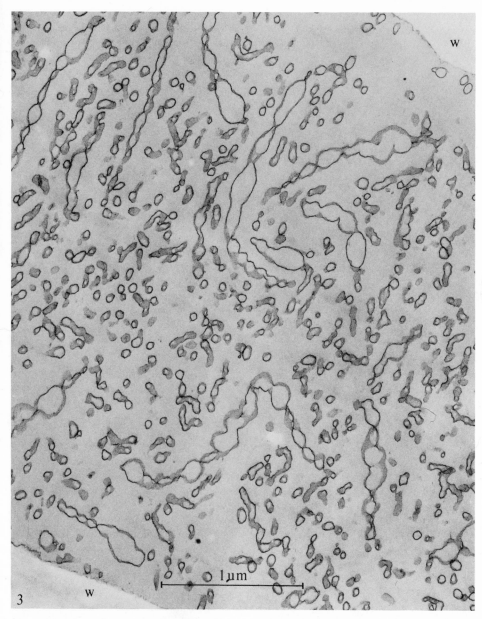

Fig.3. Typical aspect of a mature gymnosperm sieve cell. Network of smooth tubular ER extends throughout the cell lumen. w=cell wall. _Metasequoia glyptostroboides_. Original by the author.

0,5 μm

4

Fig.4. Filamentous P-protein negatively stained in the exudate of <u>Cucurbita</u> phloem. Original by Dr. Inge Dörr.

Biochemical characteriztion of the phloem proteins has been attempted in the meantime by several investigators using Cucurbita exudate and Heracleum-phloem extract; (for references, see the paper on P-protein by Dr. Cronshaw). In detail, we will only refer to the recent investigations of Dr. Kleinig's team (Beyenback et al, 1974). It has been shown that the phloem exudate of Cucurbita consisted mainly of two basic proteins and several minor, possibly acidic, protein components. The basic proteins comprised about 40% each of the total protein. Their molecular weights were 30,000 and 116,000 as determined by various methods such as gel filtration, polyacrylamide gel electrophoresis and analytical ultracentrifugation. The smaller protein with the molecular weight 30,000 existed in vitro only under highly reducing conditions. It dimerized very easily into a stable 60,000 component. The larger protein with the molecular weight 116,000 displayed gelling- and association properties when SH-protecting agents were omitted. Dimerization and association of the proteins most likely were due to formation of intermolecular disulfide bonds. Both proteins had an exceptionally high IEP (above 9.5) which corresponded to a high amount of lysine. The minor, non-basic proteins of the exudate have not been characterized so far.

Previous investigations have shown that the purified phloem proteins solubilized in thiol reagents assembled as filamentous structures in the presence of some monovalent and divalent cations such as K^+ and Ca^{++}. With the alkaloid vinblastine, protein lamellae as well as filaments and tubular structures were formed (Kleinig et al, 1971 a, b). It has been suggested that the in vitro filaments correspond to the protein filaments found in the fresh exudate and in the intact cells. We now have to determine whether the purified proteins could form the P-protein structures alone or whether some additional minor factors are required.

A main problem arising from the in vitro studies is concerned with the biological properties of the sieve tube proteins. As we have mentioned earlier, the purified proteins showed no ATPase activity and no binding capacity with several nucleotides. The discrepancy between these results and the cytochemical results of Gilder and Cronshaw can hardly be understood at present. The following interpretation may be discussed: (1) cytochemical localization of enzymatic activity included some diffusion artefacts and did not really represent genuine protein activity (Yapa and Spanner, 1974). (2) The cell structures displaying ATPase activity in situ were retained within the sieve tubes and did not occur within the exudate. This would also explain some negative assays for ATPase activity in the phloem exudate of Cucurbita (Eschrich et al, 1971). (3) The in vitro behaviour did not represent the properties of the proteins in the intact cells.

This interpretation should be discussed more precisely. As Gilder and Cronshaw have recently suggested, the enzyme activity might be regulated by changes in the conformation of the P-protein (Gilder and Cronshaw 1973b, 1974). It was shown that the P-protein underwent some characteristic conformation changes during sieve tube differentiation. The onset of ATPase activity was dependent on the dispersal of the tubular protein structures into filaments. Possibly, the enzyme existed in the young sieve elements (or in the companion cells?) in a somehow masked or inactive form and became activated by the conformational changes of the developing protein filaments. Normally these changes of the protein structures may be induced by certain environmental factors. However, alteration of the protein undoubtedly will be caused by external factors too. This means that we have to consider that structure and function of the P-protein change during the process of exudation and/or further manipulation.

Further biochemical investigations now have to prove such a concept. We need more information about conformational changes of the sieve tube proteins with regard to their possible biological activity.

Let us finally point to a central problem in which we should be engaged. Up to now the sieve tube proteins from Cucurbita and Heracleum have been studied biochemically. However, some important results concerning the molecular weights and the isoelectric point of the proteins were not in agreement (see Kleinig, 1974; Walker, 1972; Walker and Thaine, 1971; Yapa and Spanner, 1972). We cannot decide at present whether this discrepancy was due to the different methods employed or whether phloem proteins from different plants show different properties. This problem raises the fundamental question of the universality of the sieve tube protein. Provided that the latter has any essential function in the mechanism of translocation, real differences in its nature should hardly be expected. This consideration may indicate the importance of phloem studies in different plant material under comparable conditions.

References

BEHNKE, H.-D. 1968. Protoplasma 66: 287-310
BEHNKE, H.-D. 1969a. Protoplasma 68: 289-314
BEHNKE, H.-D. 1969b. Protoplasma 68: 377-401
BEHNKE, H.-D. 1971. J. Ultrastr. Res. 36: 493-498
BEHNKE, H.-D. 1972.Botan. Rev. 38: 155-197
BEYENBACH, J., C. WEBER and H. KLEINIG. 1974. Planta: in press
BOUCK, G.B. and J. CRONSHAW. 1965. J. Cell Biol. 25: 79-95
BRAUN, H.J. and J.J. SAUTER. 1964. Planta 60: 543-557
BUVAT, R. 1963. Port. Acta Biol. A7: 249-299
CATESSON, A.-M. et M. LIBERMAN-MAXE. 1974. C.R. Aca. Sci.
   (Paris) 278, Ser. D: 2771-2773
CRONSHAW, J. and K. ESAU. 1967. J. Cell Biol. 34: 801-815

CRONSHAW, J. and K. ESAU. 1968. J. Cell Biol. 38: 25-39
DESHPANDE, B.P. 1974. Ann. Bot. 38: 151-158
DORR, I. 1972. Protoplasma 75: 167-184
DORR, I. and R. KOLLMANN. 1974. Protoplasma: in press
ENGLEMAN, E.M. 1965. Ann. Bot. N.S. 29: 103-118
ESAU, J. 1969. The Phloem. Handbuch der Pflanzenanatomie. Gebr.
    Borntrager, Berlin, Stuttgart
ESAU, K., V.I. CHEADLE and E.B. RISLEY. 1962. Bot. Gaz. 123:
    233-243
ESAU, K. and J. CRONSHAW. 1967. Virology 33: 26-35
ESAU, K. and J. CRONSHAW. 1968. J. Ultrastr. Res. 23: 1-14
ESAU, K. and R.H. GILL. 1970. Protoplasma 69: 373-388
ESAU, K. and R.H. GILL. 1972. J. Ultrastr. Res. 41: 160-175
ESAU, K. and R.H. GILL. 1973. J. Ultrastr. Res. 44: 310-328
ESCHRICH, W., R.F. EVERT and W. HEYSER. 1971. Planta 100:
    208-221
EVERT, R.F., C.H. BORNMAN, V. BUTLER and M.G. GILLILAND. 1973a.
    Protoplasma 76: 1-21
EVERT, R.F., C.H. BORNMAN, V. BUTLER and M.G. GILLILAND. 1973b.
    Protoplasma 76: 23-34
EVERT, R.F., J.D. DAVIS, C.M. TUCKER and F.J. ALFIERI. 1970.
    Planta 95: 281-296
EVERT, R.F. and B.P. DESHPANDE. 1969. Protoplasma 68: 403-432
FRANKE, W.W. 1971. Protoplasma 73: 263-292
GILDER, J. and J. CRONSHAW. 1973a. Planta 110: 189-204
GILDER, J. and J. CRONSHAW. 1973b. J. Ultrastr. Res. 44:
    388-404
GILDER, J. and J. CRONSHAW. 1974. J. Cell Biol. 60: 221-235
JOHNSON, R.P.C. 1973. Nature New Biol. 244: 464-466
KLEINIG, H., I. DORR and R. KOLLMANN. 1971a. Protoplasma 73:
    293-302
KLEINIG, H., I. DORR, C. WEBER and R. KOLLMANN. 1971b. Nature
    New Biol. 229: 152-153
KOLLMANN, R. 1964. Phytomorphology 14: 247-264
KOLLMANN, R. 1968. Funktionelle Morphologie des
    Coniferen-Phloems. "Vortrage aus dem Gesamtgebiet der Botanik",
    Neue Folge Nr. 2, 15-19
KOLLMANN, R. 1973. In: Grundlagen der Cytologie, G.C. Hirsch,
    H. Ruska and P. Sitte (eds.). VEB Gustav Fischer, Jena, s.
    479
KOLLMANN, R. and I. DORR. 1966. Z. Pflanzenphysiol. 55:
    131-141
KOLLMANN, R. and I. DORR. 1969. Ber. Dtsch. Bot. Ges. 82:
    415-425
KOLLMANN, R., I. DORR and H. KLEINIG. 1970. Planta 95: 86-94
KOLLMANN, R. and W. SCHUMACHER. 1962. Planta 58: 366-386
KOLLMANN, R. and W. SCHUMACHER. 1963. Planta 60: 360-389
KOLLMANN, R. and W. SCHUMACHER. 1964. Planta 63: 155-190
KUO, C.F. 1964. Acta Bot. Sinica 12: 100-108
LEHMANN, J. 1973. Planta 111: 187-198

LESTER, H.H. and R.F. EVERT. 1965. Planta 65: 180-185

MAXE, M. 1966. C.R. acad. Sci. (Paris) Ser. D. 262: 2211-2214

NEUBERGER, D.S. and R.F. EVERT. 1974. Amer. J. Bot. 61: 360-374

NORTHCOTE, D.H. and F.B.P. WOODING. 1966. Proc. Roy. Soc. (London) Ser. B, 163: 524-537

PALEVITZ, B.A. and E.H. NEWCOMB. 1970. J. Cell. Biol. 45: 383-398

PALEVITZ, B.A. and E.H. NEWCOMB. 1971. Protoplasma 72: 399-426

PARAMESWARAN, N. 1971. Cytobiologie 3: 70-88

PARTHASARATHY, M.V. 1974a. Protoplasma 79: 59-91

PARTHASARATHY, M.V. 1974b. Protoplasma 79: 93-125

PARTHASARATHY, M.V. and K. MUHLETHALER. 1969. Cytobiologie 1: 17-36

RESCH, A. 1954. Planta 44: 75-98

SABNIS, D.D. and J.W. HART. 1973. Planta 109: 127-133

SABNIS, D.D. and J.W. HART. 1974. Planta 118: 271-281

SRIVASTAVA, L.M. and T.P. O´BRIEN. 1966. Protoplasma 61: 277-293

STEER, M.W. and E.H. NEWCOMB. 1969. J. Cell Sci. 4: 155-169

THOMPSON, R.G. and A.D. THOMPSON. 1973. Can. J. Bot. 51: 933-936

WALKER, T.S. 1972. Biochim. Biophys. Acta 257: 433-444

WALKER, T.S. and R. THAINE. 1971. Ann. Bot. 35: 773-790

WANNER, H. 1953. Ber. D. Schweiz. Bot. Ges. 63: 201-212

WEBER, C., W.W. FRANKE and J. KARTENBECK. 1974. Exp. Cell Res. 87: 79-106

WERGIN, W.P. and E.H. NEWCOMB. 1970. Protoplasma 71: 365-388

WILLIAMSON, R.E. 1972. Planta 106: 149-147

WOODING, F.B.P. 1966. Planta 69: 230-243

WOODING, F.B.P. 1967a. Planta 76: 205-208

WOODING, F.B.P. 1967b. Protoplasma 64: 315-324

YAPA, P.A.J. and D.C. SPANNER. 1972. Planta 106: 369-373

YAPA, P.A.J. and D.C. SPANNER. 1974. Planta 117: 321-328

ZEE, S.-Y. 1969. Aust. J. Bot. 17: 441-456

DISCUSSION

Discussant: H.B. Currier
              Botany Department
              University of California, Davis
              Davis, California

Aronoff: In the analysis of exudate, the TCA precipitates a variety of things. What about other substances that are not precipitated by TCA; for instance lipids, sugars, etc.?

Kleinig: We did not analyze the non-precipitated, low molecular weight substances, as there are many analyses of these substances by Zeigler and others. Phospholipids are generally localized in the membrane, and when they occur free, you find then as myeline structures which are considered to be artefacts. One usually does not see them in sieve tubes.

Gorham: What Dr. Aronoff wants to know, is what is the total composition of the organic components in the exudate? You have the macromolecules, and the phospholipids, but what is the relation of these to the total sucrose and free amino acids, of which there are some, and other such compounds?

Kleinig: We have not done this total recovery.

Eschrich: The macromolecular matter is a minor component of the sieve tube exudate. When you make a cut and collect exudate from a cucurbit plant, at first you get the liquid matter, which runs through by surging, and only small amounts of macromolecules. When you make a second cut, more macromolecules come because they are moving slower than the liquid matter. When you make the third and the fourth cuts, then the amount of liquid coming out is even smaller, but the increase in protein or in macromolecules is much greater. Therefore, what is in the exudate (in the table, up to 30 micrograms per mil were shown) is very difficult to decide quantitatively, as it changes all the time.

Kollmann: With this analysis we only intended to prove the proteinaceous nature of the macromolecular structures seen in the electron microscope. We were not interested in the soluble components.

van Die: In Yucca exudate and in exudates of palms, the total solid concentration is very stable over a long period of time, about 18% on a fresh weight basis, and 0.2 or 0.3% of the volume is protein.

Aronoff: Has anyone studied the turnover of these P-protein components? If you feed labelled sugar, or $^{14}CO_2$ and get out the P-protein, is it hot? Is the P-protein flow a mandatory part of the flow mechanism? It is conceivable that it might be, despite it´s small fraction, if there were continuous synthesis of it at a very high rate.

Mittler: If the thing is being synthesized continuously, because it is moving away from where it is synthesized, then it is presumably going to the natural sinks in the plant, and you would either have to find large aggregations there, or it would have to be changed back into something soluble that could then be synthesized into the proteinaceous components of the growing regions.

Ho: Dr. Kollmann pointed out that he failed to identify protein in the stylet itself. It seems that if you cannot identify protein in the stylet, not in the early stage of the exudation, but in the later stage, then that is an indirect proof that P-protein is not a mobile component of the sieve tube.

Kollmann: Do the aphids have enzymes which alter the protein structures, and is that why we cannot find protein components in the stylet exudate?

Mittler: Some years ago, Dr. Bumstead demonstrated proteinase activity in the homogenates of the alimentary canal of the green peach aphid. Other workers have not been able to demonstrate that; but that means, an aphid, having the proteinase activity, could cope with the ingested protein. I agree that we must find the proteins in the stylet exudate.

MacRobbie: It seems that from the plant point of view, it does not make any sense to transport P-protein. It may only put up the nitrogen content by 50%, but what it does to the transport problems in the sink end, is out of all proportion to that. It is going to have enormous problems getting protein out of the system. If we believe that P-protein has a function, whether in blocking or in the transport mechanism, then equally, it does not make sense to turn over this part and finish with it in the sink end, where the system finishes.

Geiger: Dr. Mittler, could you indicate the relative sizes of the aphids´ stylets compared to the sieve pore, because if we are concerned about plugging a sieve pore with this material, it seems that the aphids stylet might also, in some species, present a problem.

Mittler: It is difficult to generalize as there are big aphids, and little aphids, and this is reflected in the size of the stylets and the stylet food canal, but it is not exactly proportional. In the case of the willow aphid, the maxiliary stylets are close to 1 μm in diameter and, for smaller aphids, they are considerably smaller - that refers to the lumen. The actual diameter of the stylets is somewhat bigger. Also, there is not only an individual stylet, but two stylets that are close to each other, with the food canal in between. You have an additional pair of stylets on the outside, which may be primarily responsible for the insertion of the stylets. Altogether, you have a stylet which looks in section, very much like a floral diagram. The diameter of this entire bundle, which is inserted in, may be 3 - 5 μm.

Swanson: Is there a turnover of the P-protein in a metabolic sense?

Eschrich: We do not have any real data, but we have fed ^{14}C labelled leucine and mixtures of amino acids to cotyledons of cucurbita plants and, after a certain time, even after 24 hours, collected the exudate. We got labelled protein, but the question is, do we have P-protein turnover? Also, what is P-protein? Are we referring to filaments; the old slime, or something else? I believe that P-protein is synthesized only at a certain time in ontogeny of the sieve element, because at that time ribosomes are present. The protein may be stored for a while, and this may be the slime body. In sieve elements that are used, very often longitudinal sieve tubes, in Cucurbita for example, the slime bodies are dispersed very quickly, but in the cortical sieve tubes of Cucurbita, which probably are not used so often, the sieve tubes retain the slime bodies undispersed for a long time - even after the sieve pores are open. Therefore, we cannot decide whether P-protein, in the sense given by Dr. Kollmann, has a turnover or not.

Milburn: It has been suggested tnat the protein cannot possibly move down the system, and it seems that there is good exudate evidence that it can, as well as that palms have been reported to produce liters of sap for long periods of time. We have never really found a satisfactory use for companion cells. I suggest that the companion cells control protein dissolution from the slime body, and the dissolution of slime such that it can disappear rather easily at the sink end of the chain. There are abundant connections between the companion cell and sieve element, and there are some suggestions that a massive surge will actually draw materials, presumably enzymes, from the companion cells into the sieve tubes.

van Die: The aleurone layer of grain seeds produces

amylase that is secreted by these cells into the endosperm. If these cells can secrete protein molecules, why should not transfer cells, or companion cells, synthesize proteins and secrete them into the sieve tubes?

Currier: Meaning then, that the protein is moving and is continually supplied by companion cells? Referring to Dr. MacRobbie's question, how does the sink area handle all this protein?

van Die: There is only very little protein translocated. The protein content is only 0.2 - 0.3%, and perhaps that small quantity can be replenished by cells surrounding the sieve tubes during translocation.

Turgeon: If we supply $^{14}CO_2$ to a leaf, we know that the ^{14}C does not simply go into sugar, but into amino acid and into proteins. The sieve elements in the leaf contain P-protein, and if that protein is being washed away, it would have to be replaced by labelled compounds and the P-protein would be labelled. It is not, according to Dr. Gorham.

Gorham: According to Dr. Mortimer, if you analyze the petiole, while the labelled translocate is still in the phloem and has not proceeded to move radially, you get lots of sugars, and traces of amino acids. Although we were not looking for P-protein, we did not get any gross labelling of protein, or TCA-fraction-labelled.

Eschrich: In Cucurbit and Robinia phloem exudate, there is not only the filaments, but up to 50 different enzymes. We also know that by cutting several enzymes from the companion cell into the sieve element, although some enzymes, for example acid phosphatase and peroxidase, may occur only in the sieve element, and not in the companion cell. I would urge, therefore, not to talk about P-protein as long as it is not clear what P-protein is, in the biochemical sense. So far, P-protein is only an anatomical term, and is nothing else than slime, in the former century.

Aikman: In some species, sieve elements contain endoplasmic reticulum and, in some others, endoplasmic reticulum and P-protein. If these sieve elements are functioning in the same way, would the endoplasmic reticulum also be required to move? If there are similarities and relationships between plasmodesmata and sieve pores, would it also be required that filaments or tubules pass through plasmodesmata?

Weatherley: When an aphid punctures a sieve element, exudation occurs and there is very little protein in this exudate. If the filamentous P-protein is mobile, then it is going to come down and gather in one single sieve element, which has been punctured, and gather at a colossal rate - a flow of one sieve element per second. Surely a continued exudation through the stylet is impossible. I cannot see how these filaments can be draining into one sieve element. It is conclusive that the P-protein is not moving.

Parthasarathy: If one assumes that new P-protein is being formed in differentiating sieve elements, and is being translocated away from that to other places, then in fully expanded leaves, where there is no new differentiation going on, the metaphloem should be completely empty of any filamentous protein. That is not the case.

Mittler: There are a few good examples of stylet exudation, such as in the willow aphid, from which considerable amounts of exudate can be obtained without any apparent blockage, and possibly without any P-protein to block it. However, this is not so readily achieved with other aphids. Therefore, the possibility exists that their stylets are blocked; we cannot generalize. There is the observation that one and the same aphid, feeding on a variety of different host plants, in some cases, contained a completely clear fluid in the esophageous and stomach. Presumably it ingested just such a liquid. In other cases, it contained very dense materials - strands, or "sausages". Perhaps the aphid did not have the ability to digest them. I have heard only one amino acid, leucine, referred to, and I think it should have been lysine, as it was in relation to a basic protein. Could it be, that if we knew more about the relative amino acid composition of these P-proteins as compared to the other amino acids in the sap, that perhaps P-rotein could be a convenient way of condensing and storing certain amino acids, regardless of whether the strands are moving or not?

Fisher: There is well-documented evidence of movement of protein filaments, that is, viruses - for instance, TMV, in the sieve tubes - from the mesophyll to the companion cells, into the sieve elements, down the sieve elements, and into other parts of the plant.

Cronshaw: Yes. If we infect a tobacco leaf with TMV and rub the particles into the surface of the leaf, then the infection spreads, very slowly, from cell to cell. Then all of a sudden, it becomes systemic and the whole plant breaks out with a typical TMV infection. What has happened is that the virus particles have gotten into the sieve elements and have been translocated throughout the whole plant. Therefore, we have good evidence that particles can come, if not from mesophyll cells, at least from companion cells, that are filled with virus particles, down through the phloem of the plant.

Aikman: The comments about virus particle movements are not really relevant. No one is suggesting that the virus particles are anchored. What we are trying to answer, is not whether proteins that are not anchored would move, but whether P-proteins that might be anchored, move.

Currier: There are several possible P-protein functions. One is that the P-protein is anchored and forms a protective holding net, preventing the movement of organelles, the plastids, and the mitochondria on the periphery of the cell. Another one, is that it acts as a sealing mechanism. We know that the first line of defense against a plant bleeding to death when you cut it, is blockage of the sieve pores with P-protein. Another possible function, is it´s involvement in translocation mechanism directly, if for instance, it has an actin-like component.

Christy: If P-protein is involved in the mechanism of translocation, how do you account for translocation in plants that do not have P-protein?

Currier: That is a very good point. Dr. Kollman, would you elucidate whether young sieve elements conduct, because it is common understanding that conduction begins when a sieve tube or sieve element, becomes mature, that is, the pores are open and the organelles have degenerated.

Kollmann: We investigated this point some years ago, and found from 3 lines of evidence: the presence of ^{14}C label; anatomical aspects of albuminous cells, and the presence of aphid stylets in sieve cells next to the cambial zone. We concluded that young sieve elements are able to translocate in the tree Metasequoia, but this does not exclude the older sieve elements.

Lamoureux: It seemed from your autoradiographs, that not only the young sieve elements, but also the cambial cells and the phloem parenchyma cells, had an equal amount of label in them. Are we really looking at the sink here, rather than the channel of translocation?

Kollmann: That is right. It is difficult to localize a single cell with autoradiographical ^{14}C. On the basis of grain count however, we always had the peak of grains in the cells produced just before, and after, the first differentiated fibre band.

Srivastava: Young, undifferentiated parenchyma cells can translocate - they do it all the time in the shoot and root apex. What I would like to know, is whether Dr. Kollmann thinks that translocation occurs exclusively in young, undifferentiated cells, or does it also occur predominantly in the mature cells?

Kollmann: I can only propose a hypothesis.   Perhaps the
active motor for translocation is localized in the youngest cells,
while the older cells are only passively involved.

Mittler: In willow, we could not get aphids to feed or to
produce honeydew in dormant stems; but just before budburst in
spring, there was plenty of feeding and exudation.   The stylets
would exude for many hours - sometimes 3 to 4 days on end. I would
like to ask Dr. Kollmann how long exudation went on in
Metasequoia, where he had established that the exudation was from
young cells.

Kollmann: It is very difficult to get exudate from conifer
sieve cells.   We got exudate for perhaps 1 - 1 1/2 hours after
cutting the stylets. We always followed the pathway of the stylet
by serial sectioning with a cryotome. Both from aphid and ^{14}C
studies, we found that the sieve cells which translocate in the
spring and in autumn, are in an intermediate stage of
differentiation. Some cells have a tonoplast, and in others the
tonoplast is lost.   I must emphasize that not only do the young
cells translocate, the older ones conduct as well. The young cells
can perhaps function under certain conditions in spring and autumn.

Dorr: Cuscuta has marked transfer cells against very young
sieve elements of the host elements which show a tonoplast and a
central vacuole.

Turgeon: Is it known that the aphid - and now I extend this
to Cuscuta - always feed on sieve elements that are active in long
distance transport? In other words, these young cells may not be
involved in long distance transport, but they may still be
efficient at accumulating a certain amount of sugar, which may
account for the autoradiographs, and for the aphids feeding on
them.

Srivastava: In temperate climates, many trees do not seem to
have any functional sieve tubes left in the winter. Yet in the
spring, hormones, as well as food material, have to be mobilized
and translocated, for cambium to become reactivated and to start
dividing. These movements occur over long distances, and although
we do not know the precise velocity at which this transport occurs,
these cells have normal parenchyma-type contents, and are involved
in transport of substances over long distances.   It seems that
among the first cells to become reactivated in the spring, are the
cells that become differentiated as new phloem cells.   If, from
mature wood, you have something like 4, 5, or 6 layers of cells
which are undifferentiated, it is the last cells towards the
periphery that become reactivated first, and apparently start
translocating. Then the cambial cells become reactivated, and then

the cells towards the outside. The cells in the outer phloem, or
bark, may be completely dormant for as much as 4 - 6 weeks after
the activation of these inner cells. Therefore, there is no doubt
that cells with living contents are involved in long distance
transport.

Watson: Dr. Currier, what is the significance of the binding
of P-protein by heavy meromyosin (HMN)?

Currier: We suggest that, if HMN binding is a specific
actin-like property, it would seem that P-protein has a substantial
actin-like component.   The function of P-protein, whether it is
involved in movement, or in other such specialized cellular
activities, such as gel-sol interchanges, or whether it acts
protectively to hold organelles in place, and to block sieve pores
upon wounding, remains a crucial question.

Walsh: Why does P-protein have to have any function at all?
A plant collects and accumulates a number of things that it does
not really need, or does not really have a function for - like a
crystal in a cell.

Currier: There is so much information as to how P-protein
might function, that I would include this as a logical possibility.

van Die: Is there any possibility that P-proteins   are
involved in the maintenance of the living state of the sieve
element, that is in the maintenance of the plasmalemma in a
functional state?   Or is it that the metabolic processes are
regulated via the companion cells?

Kollmann: There is not enough information on this point, but
companion cells do contribute many enzymes to the sieve elements,
certainly after a cut.

Christy: Cytoplasmic streaming is directed, or controlled, by
actin-like proteins (Ed. note: microfilaments).   I wonder if
P-protein could be composed in part, by remnants of this system?

Currier: It is a possibility, if you mean that the immature
sieve element does stream, and in development some of this is
carried over into the mature element.

Phloem Loading and Associated Processes

Donald B. Geiger

Department of Biology, University of Dayton

Dayton, Ohio

A. Phloem loading

1. Definition

Phloem loading is the process by which the major translocate species are selectively and actively taken up by the sieve tubes, especially in the minor veins of source regions.   The entry of other solutes which may readily but passively move into the phloem is not considered to be part of phloem loading.   One of the earliest and possibly the first mention of the phloem loading concept in the literature was in the work by Barrier and Loomis (1).

In defining phloem loading in this manner we are positing that one or more substances selectively enter the phloem by an active uptake process, capable of accumulating these species above their concentration outside the phloem. In this category would be the major translocate sugar or sugars and perhaps other substances such as K^+ and some amino acids. Other phloem-mobile materials, not actively loaded, would not be able to be concentrated in the phloem sap but could be translocated in significant quantity if the concentration in the leaf were reasonably high.   It is not considered necessary or probable that each phloem-mobile substance have a carrier. From the viewpoint of energetics, it seems essential that at least one species be actively loaded to provide the osmoticum for a mass flow mechanism based on water flux down a water potential gradient.   It appears that the osmotic gradient created by active solute uptake constitutes the potential energy which is converted to mass flow by the entry of water. With this introduction we are in a position to investigate the existence of phloem loading.

2. Demonstration of phloem loading

Following the introduction of the Münch flow hypothesis (23) studies on osmotic values in the source path and sink tissues were made by several workers. As early as 1933, Curtis and Scofield found osmotic pressure data which seemed to be inconsistent with the Münch mechanism (3). Studying plants kept in the dark, in which growth was supported by solutes translocated from storage organs, these investigators found rather consistently that the osmotic pressure of the growing tissue as a whole was higher than that of the source region (Table 1). From their data on osmotic concentration at insipient plasmolysis they calculated that the osmotic pressure was 2 to 6 atm higher in the sink tissue than in the source storage tissue in most plants they studied.

Table 1. Freezing point depressions and osmotic pressures of sink and source tissue of various plants growing in darkness. Curtis and Scofield (3).

Experimental Material	Average Freezing Point Depression °C	Osmotic Pressure atm
Onion - Growing Tissue	1.179	14.19
- Storage Tissue	0.918	11.06
Difference	0.262	3.13
Etiolated Bean Seedling		
- Growing Shoot	1.49	17.92
- Cotyledon	1.25	15.04
Difference	0.24	2.88
Bryophyllum - Plantlet	1.0008	12.14
- Parent leaf	0.771	9.29
Difference	0.237	2.85
Squash - Growing Shoot	1.246	14.99
- Cotyledon	1.054	12.69
Difference	0.193	2.30
Potato - Tip of Growing Sprout	2.382	28.60
- Tuber	1.845	22.18
Difference	0.537	6.42

In a study of source tissue in leaves of Robinia, Roeckl (25) found that the osmotic pressure of the leaf mesophyll cells was much lower than that of the sieve sap (Tables 2 and 3). The data of these two studies clearly point out that overall osmotic pressure values for the source and sink tissues do not support the Munch hypothesis.

Table 2. Osmotic pressure values for palisade cells of Robinia pseudoacacia obtained by plasmolysis. Roeckl (25).

Tree Number	Leaf Type	Osmotic Pressure of Palisade Cells	
		M	atm
1	Sun Leaf	0.60	17.8
2	"	0.55	16.2
3	"	0.55	16.2
4	"	0.60	17.8
5	Shade Leaf	0.50	14.3
6	Sun Leaf	0.60	17.8
7	Shade Leaf	0.55	16.2
8	"	0.50	14.3
9	Sun Leaf	0.60	17.8
10	"	0.70	21.5
11	"	0.65	19.1
12	"	0.55	16.2
13	"	0.65	17.8
	Average		17.9

Table 3. Osmotic pressure values of Robinia pseudoacacia leaf tissue obtained by a cryoscopic method and of sieve tube sap measured by refractometry. Roeckl (25).

Experiment Number	Osmotic Pressure of the Leaves atm	Solute Content in Sieve tube Sap	
		%	atm
1	12.76	24.0	29.7
2	13.36	25.0	32.3
3	15.65	27.0	36.5
4	13.60	23.0	26.9
5	22.87	29.3	41.0
6	21.36	28.4	39.8
Average	16.60		34.4

The problem in each case seems to rest in the incorrect designation of the structure acting as the translocation source. Most workers appear to regard the mesophyll cells surrounding the vascular tissue as the source. The studies of Curtis and Scofield (3) and of Roeckl (25) do not rule out the operation of a mass flow mechanism. Rather they point out the need to find a different structure to regard as the Münch translocation source. Freeze substitution studies of the osmotic pressure of the various tissues of the source leaf (11) indicate that the sieve elements and companion cells of the minor veins are more likely to be the translocation source in a mass fow system. It seems quite feasible to redefine the Münch system to be made up of a translocation source region consisting of the sieve elements and companion cells (se-cc complex) of the source leaf, connected to the sink region by the sieve tubes of the translocation path (Fig.1). The mesophyll in which the translocation source is embedded serves as as assimilate source connected to the Münch system by means of active sugar uptake. Operationally or physiologically, it seems that in the source there are two symplastic systems - one, the se-cc complex of the translocation system, embedded in a second, the rest of the plant symplast. In sinks such as young tissues, the se-cc complex may, in fact, be connected to the symplast of the growing region.

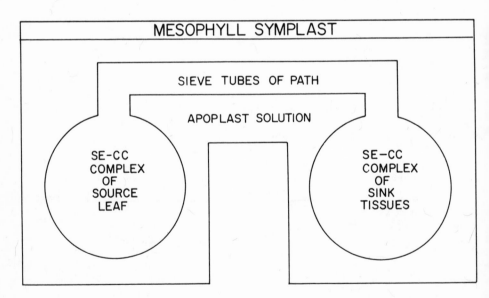

Fig.1. Assignment of structures in a modification of the Münch mass-flow model of translocation.

If this model applies, then the concentration of labelled solute should be much higher in the minor veins of the source leaf than in the surrounding mesophyll. A number of studies have supported this proposal (8, 13, 19, 21). When a leaf photosynthesizes in $^{14}CO_2$ the minor veins show a higher level of ^{14}C than the surrounding mesophyll (Fig.2). Histoautoradiography also supports the conclusion that labelled solutes accumulate to a higher level in the minor vein phloem than in the surrounding mesophyll (10, 32).

Fig.2. Autoradiograph of a sugar beet source leaf supplied with $^{14}CO_2$ during a 5 min pulse. Marker indicates 5 mm.

The considerations presented up to this point suggest that the se-cc complex of the minor veins rather than the source leaf mesophyll be considered as the entity functioning as the origin of mass flow as described in Münch's model.

3. Description of the minor vein phloem in relation to
phloem loading

The specialized structure of the minor vein phloem of most dicots was recognized by Fischer as early as 1884 (5). In a series

of studies, Esau described the structure of the minor veins of the
sugar beet leaf in detail (5,6). Working on the hypothesis that
the phloem of the minor veins constitute the translocaton source
structure of the Münch system, we will look at its structure
quantitatively and in detail.

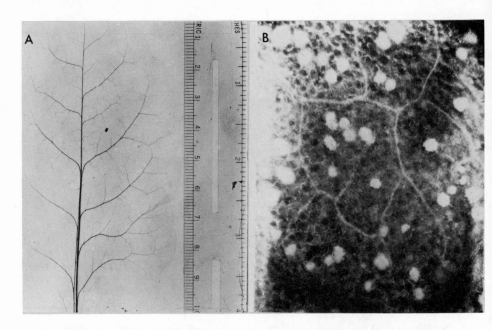

Fig.3. Structural details of a sugar beet leaf with a 10 cm
blade, about 60% of final laminar length. (A) General view. Black
rectangle depicts size of field shown in (B). Detailed view (B)
showing polygon and 5th order of branching. Bright circles are
crystals which are approximately the size of mesophyll cells.
Geiger and Cataldo (10).

In the sugar beet leaf the major veins branch repeatedly,
ending in polygons and short branches of 5th order veins (Fig.3).
In these leaves, the minor veins have an extent of approximately 7
cm/cm^2 blade (Table 4). A length of minor vein equal to 1
mesophyll cell diameter receives translocate from approximately 30
cells situated approximately 2 to 3 cell diameters 65 to 100 μm
from the minor vein (10). Smith and Epstein (28) observed that
uptake of small solute molecules becomes limited when the distance
to a free edge of the leaf tissue piece exceeds 150 μm. Esau (5)
reported that this is the approximate maximum distance between
mesophyll cells and the minor veins in dicot leaves. Although the
aggregate se-cc complex occupies only about 0.6% of the blade
volume, it has a surface of 0.9 cm/cm^2 blade (29).

Table 4. Description of various aspects of structure and function related to phloem loading in a 10 cm sugar beet source leaf. The blade is approximately 60% of the final laminar length. Performance parameters are for photosynthesis in air at 2000 ft c light intensity. Sovonick et al (29).

Parameter	Value
1. Extent of minor veins	70 cm/cm^2 blade
2. Volume se-cc complex in minor veins of 1 cm^2 source leaf	0.15 mm^3/cm^2 blade
3. Proportion of leaf blade volume occupied by se-cc complex	0.6%
4. Surface area of se-cc complex in minor veins of 1 cm^2 source leaf	88 mm^2/cm^2 blade
5. Sucrose concentration in se-cc complex	0.8 M
6. Amount of sucrose present in se-cc complex of minor veins	45 µg/cm^2 blade
7. Proportion of source leaf sucrose in se-cc complex	80%
8. Turnover time for sucrose in se-cc complex of minor veins; tr. rt. = 0.95 µg sucrose/min cm^2 blade	48 min
9. Half time for sucrose in se-cc complex of minor veins; tr. rt. = 0.95 µg sucrose/min cm^2 blade	33 min
10. Sucrose flux through membrane of se-cc complex of minor veins; tr. rate as above	3.2 x 10^{-9} moles/ min cm^2 membrane
11. ATP required to maintain flux of sucrose 1:1 stoichiometry	2.8 n moles ATP/ min cm^2 blade
12. Proportion of photosynthate carbon which would be required to supply ATP (glucose equivalent) at photosynthesis rate of 1.6 µgc/min cm^2	.3% of C fixed
13. Ratio of sucrose required as a source of ATP required in 10 to sucrose translocated	1.4% of transloc. sucrose
14. Water flux into se-cc complex required to produce the sucrose: water ratio observed in the sieve sap	1.5 x 10^{-7} moles/ min cm^2 membrane

Measurement of the osmotic pressure of the cells of the sugar beet source leaf by a freeze-substitution/plasmolysis method confirmed that the osmotic pressure of the se-cc complex is considerably higher than that of the surrounding mesophyll or phloem parenchyma cells (Table 5). The abrupt change in solute concentration between the se-cc complex and all surrounding cells suggests that phloem loading occurs across the plasmalemma of the se-cc complex (Fig.4). the latter appears to function as an integral unit. The extensive network, with a large surface provided by the aggregate se-cc complex, as well as the high osmotic concentration of its contents, support its role as a source of potential energy in a translocation system.

A number of cytological features also support the role of the intermediary or companion cells of the minor veins in active phloem loading. The specialized companion cells of the minor veins contain a large number of mitochondria and ribosomes indicative of high metabolic activity. In some higher plants inwardly directed protuberances form on the cell walls of the intermediary cells in a manner characteristic of transfer cells (16, 17). It appears significant that the time of appearance of the wall projections in pea coincides with the onset of export from the leaf under study (17).

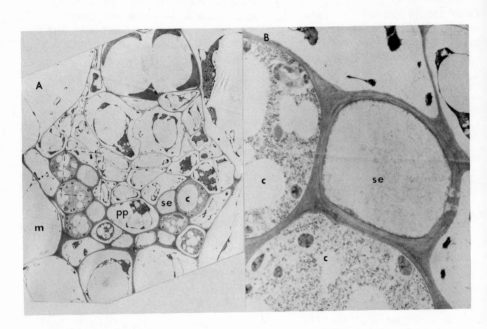

Fig.4. Electron micrograph of source leaf tissue from sugar beet prepared by a freeze substitution-plasmolysis method. (A) General view with detail (B) indicated. c: specialized intermediary or companion cell, se: sieve element, m: mesophyll, pp: phloem parenchyma cell.

Table 5. Osmotic pressure values for various cells of sugar beet source leaves. Geiger et al (11).

Cell Type	Mannitol Concentration Causing Incipient Plasmolysis	Equivalent Osmotic Pressure
	Molar	Atm
Source Leaf Blade		
Sieve Element	1.0-1.1	28-32
Companion Cell	1.0-1.1	28-32
Phloem Parenchyma	0.30	8
Mesophyll	0.50	13
Source Leaf Petiole		
Sieve Element and Companion Cell	1.0-1.1	28-32

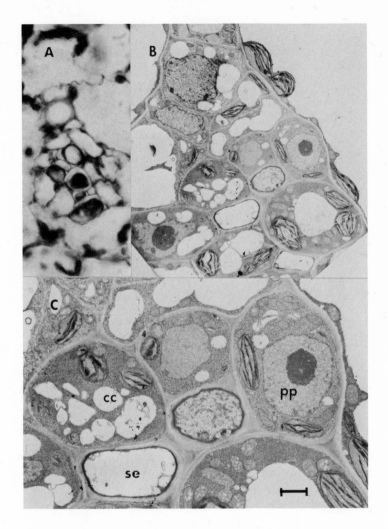

Fig.5. Structure of a 5th order vein from a sugar beet leaf.
(A) Light microscope picture (B) Electron micrograph of a similar
vein (C) Detail of (B) Key as in Figure 4. Marker: 1 μm.

In summary, the results of autoradiographic, structural,
cytological and solute concentration studies indicate that active
uptake of solute occurs at the surface of the se-cc complex of the
minor veins of the phloem. The high concentration of solute in the
se-cc complex in contact with free space solution with its higher
water potential appears to constitute a translocation source region
in the sense posited by Münch.

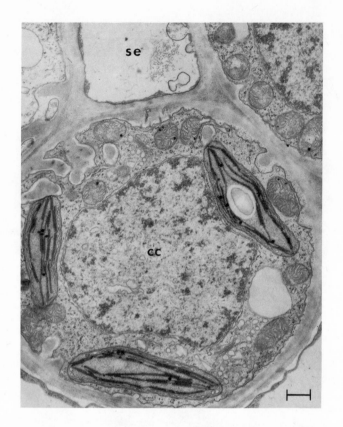

Fig.6. Electron micrograph of transfer cell (CC) and sieve element (SE) from a cross section of a minor vein in <u>Helianthus annuus</u>, L. Arrows indicate wall projections of "A type" transfer cell. Marker: 1 μm. R. Fellows, Department of Botany, University of Illinois.

4. Nature of phloem loading

a. Role of the free space. The abilty of the minor vein phloem to accumulate solute to high concentrations inside the sieve tubes suggests the existence of an active process of phloem loading at the plasmalemma of the se-cc complex. Two mechanisms seem plausible - active transport via plasmodesmata which connect the se-cc complex with surrounding cells or uptake through the plasmalemma by a carrier mechanism. Support for the latter comes from experiments in which exogenous ^{14}C-sucrose was applied to the free space of source leaves. Sucrose at a concentration of 20 mM is loaded from the free space and translocated at approximately the same rate as is photosynthetically fixed ^{14}CO$_2$ (13, 29 and Table 6). In this study, sucrose was presented to the leaf through the abraded upper epidermis while CO_2 at 500 μl/liter was supplied to the lower surface of the leaf held at 2000 ft c illumination. Labelling one, or the other or both forms of applied carbon permits comparison of the rates. Although the findings are consistent with phloem loading from the free space it may be argued that the sucrose is taken up by the mesophyll and loaded from the mesophyll symplast via the plasmodesmata.

Table 6. Comparison of availability of exogenous ^{14}C-sucrose and ^{14}C from photosynthesis for translocation from a source leaf of sugar beet. Sucrose supplied as 20 mM solution applied to abraded upper epidermis. Both carbon dioxide and exogenous sucrose supplied throughout. Geiger et al (13).

Manner in which ^{14}C was supplied	Rate of Translocation of Label from		Total Translocation Rate	Relative Translocation Rate	
	^{14}C-sucrose	[14]CO2		^{14}C-sucrose	[14]CO2
	μgC/min dm^2	μgC/min dm^2	μgC/min dm^2	(% of total)	(% of total)
[14]C-sucrose, then both, then [14]CO2	25	28	53	47	53
	21	27	48	44	56
[14]CO2, then both, then ^{14}C-sucrose	68	9	77	88	12
	24	19	43	56	44
	44	15	59	75	25

Application of isotope trapping appears to offer a means of determining if sucrose passes through the free space during phloem loading. The rationale of the isotope trapping experiments is illustrated in Fig.7.

As sucrose passes through the free space it appears to exchange with a certain amount of the unlabelled 20 mM sucrose added to the free space and ^{14}C-sucrose accumulates in the circulating solution in contact with the upper leaf surface (Fig.8A). When photosynthesis and translocation rates are increased by increasing the light intensity, increased free space sucrose turnover results in increased sucrose trapping. Similarly, when the translocation rate is increased by adding exogenous ATP the rate of sucrose trapping in the free space increases accordingly (Fig.8B). No increase in trapping of glucose occurs when illumination is increased or ATP is added, indicating no change in turnover of free space glucose (Fig.8C,D). When no trapping agent is added there is no continued trapping of label after isotope saturation of sucrose. Increased illumination causes an increase in entry of label into the circulating solution while ATP has no effect on the entry of labelled species (Fig.8E,F).

Sucrose constitutes 80 to 90% of the label in the trapping solution
when sucrose is present while glucose and fructose constitute over
80% of the labelled species when glucose is used as a trapping
agent.   Results of the isotope trapping experiments   clearly
indicate that in sugar beet, sucrose is loaded into the phloem from
the free space.

    b.   Evidence for active loading.   The ability of sugar beet to
load and translocate exogenously applied sucrose provides a means
of determining a number of facts about the loading process.
Application of 4 mM dinitrophenol (DNP) to the free space of the
sugar beet leaf increases carbon dioxide release to approximately
210% of the control and lowers the ATP level of the tissue to
approximately 40% of the control tissue after 3 hours of treatment
(Table 7). When 4 mM DNP is supplied along with ^{14}C-sucrose via a
reverse flap capillary method, translocation is inhibited to an
average of 21% of the control. Supplying 4 mM ATP in place of the
DNP restores translocation to the control rate (Fig.9). DNP at 8
mM appears to produce another type of damage and is not reversible.

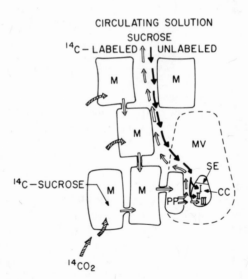

CIRCULATING SOLUTION
SUCROSE
^{14}C – LABELED   UNLABELED

^{14}C – SUCROSE

$^{14}CO_2$

    Fig.7.   Diagram of the sucrose isotope trapping technique.
The added sucrose increases the rate of sucrose translocation (I).
Some labelled sucrose is intercepted upon exiting and is replaced
by unlabelled sucrose (II).   Most of the labelled sucrose which
enters the free space is loaded by the se-cc complex (III).   Solid
arrows, unlabelled sucrose. unfilled arrows, labelled sucrose.
crosshatched arrows, CO_2. Mesophyll cells (M), minor vein (MV),
phloem parenchyma (PP), sieve element (SE) and companion cell (CC).
Geiger et al (13).

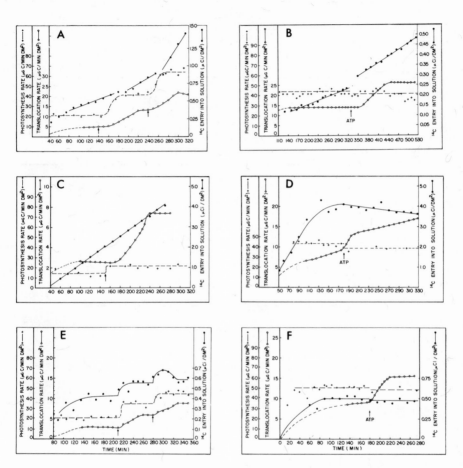

Fig.8.   Time course of translocation rate (O), photosynthesis rate (+), and exit of label into an external solution (●) for leaves treated by stepwise increases in illumination (A, C and E) or addition of exogenous 4 mM ATP (B, D and F). Entry of label into a circulating solution was measured using sucrose (A, B) glucose (C,D) or no trapping agent (E,F). Arrows indicate where treatments were started. Geiger et al (13).

Table 7. Effect of 4 mM DNP and 4 mM ATP supplied to a sugar beet source leaf blade by a reverse-flap capillary method on translocation and other parameters. Sovonick et al (29).

DNP Concentration	Rate of CO Production	ATP Level in Source Leaf Blade	Minimum Translocation Rate +DNP	Maximum Translocation Rate -DNP, +ATP
m M	%	%	%	%
0	100	100	100	183
4.0	209	43	21	97
8.0	(233)[1] 39	36	9	11

[1] Rate after 25 min. Rate gradually declined thereafter to the other reported level.

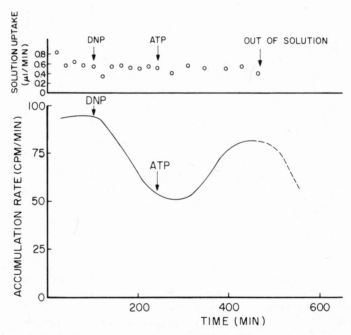

Fig.9. Effect of 4 mM DNP on the rate of translocation of ^{14}C-sucrose supplied to the source leaf via reverse flap method. DNP was replaced with 4 mM as indicated. Decline (dashed line) the result of termination of supplying of sucrose. Sovonick et al (29).

Application of 4 mM ATP generally increases the translocation rate to approximately 180% of the control rate and nearly doubles the ATP level within the tissue (29). This observation supports the data reported by Ullrich (35) and by Kursanov and Brovchenko (22) who found that externally applied ATP promotes translocation from the treated region. The application of 4 mM ATP does not change the rate of photosynthesis significantly but increases the proportion of assimilate which is translocated to a ratio 175% of the pretreatment ratio (13). It appears that externally applied ATP somehow increases phloem loading.

Phloem loading was also studied in the absence of translocation in a leaf disc system. Leaf discs were supplied ^{14}C-sucrose at various concentrations at 25°C and 0°C at 21% O and also at 25°C under a nitrogen atmosphere (29). The Q_{10} for phloem loading was found to be approximately 1.7 over the range of concentrations studied. Anaerobiosis also inhibits phloem loading (Table 8) supporting the active role of respiratory metabolism in this process. The data from application of ATP and inhibitors of energy metabolism indicate that phloem loading is an active process.

Table 8. Values for K_j and J_{max} for sucrose uptake and translocation rates in intact sugar beet leaves and leaf discs under control conditions, cold treatment and anaerobiosis. Sovonick et al (29).

Method of Sugar Presentation	Treatment	Rate Parameter Measured	Kinetic Parameters	
			K_j	J_{max}
			m M	µgC/min dm^2
Abraded Epidermis of Whole Leaf	none	translocation	16	70
Floating Leaf Discs	none	sugar uptake	88	66
	0C	sugar uptake	66	10
	anaerobiosis	sugar uptake	94	23

c. Kinetic parameters. By studying the rate of translocation in whole leaves and by measuring the rate of phloem loading in leaf discs at a series of sucrose concentrations it is possible to arrive at kinetic parameters analogous to those of enzyme studies (24). The rate curves obtained when sucrose from 10 to 400 mM was presented to whole leaves and to leaf discs are shown in Fig.10. The values for rate parameters are given in Table 8. The K_j for translocation is somewhat lower than for sugar uptake by the discs.

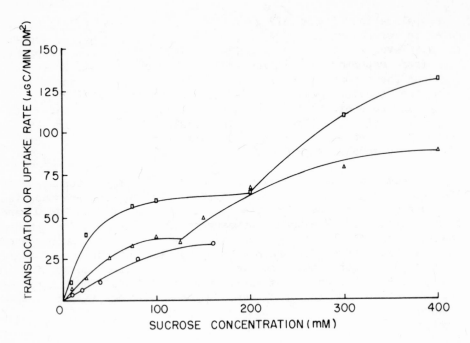

Fig.10. Dependence of sucrose uptake and translocation on the concentration of ^{14}C sucrose supplied to a sugar beet source leaf. Application of sucrose in solution is by reverse flap-capillary method (0), through abraded upper epidermis of whole source leaves (□) or source leaf discs (Δ) of sugar beet. Sovonick et al (29).

Table 9. Results of two experiments in which translocation rate was measured while net photosynthesis rates were kept constant by increasing CO_2 concentration (No. 14) or lowering O_2 concentration (No. 8) while light intensity was varied. Servaites and Geiger (27).

Exp. No.	Measurement No.	CO_2 concn μl/l	Light Intensity ft-c	O_2 concn %	Photosynthesis μgC dm^{-2} min^{-1}	Translocation μgC dm^{-2} min^{-1}	T/P ratio
8	1	600	7200	21	94.9±1.0	24.9±0.1	0.26
	2	600	3700	1	98.9±1.4	23.8±0.4	0.24
14	1	300	7200	21	73.8±2.1	21.3±2.0	0.29
	2	350	3700	21	70.4±1.6	16.3±0.3	0.23
	3	850	2000	21	67.0±2.3	15.3±1.1	0.23

The J_{max} for the first of the two isotherms observed, is larger than the highest rate of translocation observed in the intact sugar beet leaf under high light and high CO_2 (27, 29). It seems that the second isotherm may not be of practical interest under ordinary circumstances. The data from the kinetics experiments indicate the presence of an active carrier responsible for phloem loading of sucrose from the free space of the leaf. Summarizing the results of structural and physiological studies we find that the data support the loading of sucrose present in the free space in the mM concentration range via a carrier system in the plasmalemma of the se-cc complex of the minor veins.

d. Performance and rate data. Based on the reasonable assumption that the plasmalemma of the se-cc complex is the site of active phloem loading, it is possible to examine the feasibility of the proposed process of phloem loading by calculating estimated rates of sucrose and water flux in a sugar beet source leaf. A number of values for the structural features of a 10-cm sugar beet source leaf were obtained for the 7th and 8th leaves, the stage used in a number of physiological studies (10, 29). The data are summarized in Table 4. The value for sucrose flux, 3.2×10^{-9} moles/min cm^2 membrane, is several orders of magnitude above the flux rate reported for passive permeation of the chloroplast membrane by glucose (4) although it seems to be a feasible rate for active transport. Using a 1:1 stoichiometry for sucrose:ATP, it would require 3.2×10^{-9} moles ATP/min cm^2 membrane to drive this sucrose flux. This amount of ATP could be derived by utilizing 1.4% of the sucrose transported. To maintain a sucrose concentration of 0.8 M in the se-cc complex a water flux of 1.5×10^{-7} moles/min cm^2 membrane would be required. Again, this rate is within a range which is feasible for water flux (15). The rates calculated from estimates of solute (11) and membrane area (10, 29) support the operation of an active mechanism for phloem loading at the plasmalemma of the se-cc complex.

e. Relation to photosynthesis. Photosynthesis serves more or less directly as a source of the assimilate which is translocated. In addition, it has been suggested that photophosphorylation may be a source of at least part of the ATP which is used in phloem loading. To investigate this possibility, Servaites and Geiger (27) adjusted net photosynthetic carbon fixation rates to the same level under light intensities of 2000, 3700 and 7200 ft c by adjusting CO_2 concentration. No increase in rate was observed as light intensity was increased under this condition. The rate of translocation was found to increase directly with an increase in the rate of net carbon assimilation whether at low or high light intensity. The higher light intensity, unless accompanied by an increased carbon fixation rate, apparently does not cause a greater amount of sucrose to be produced or translocated. At net photosynthesis rates of from approximately 0 to 260 µg C/min dm^2 the increase in translocation was equal to a constant proportion of the increase in net carbon assimilation rate, averaging approximately 18%. The ratio is nearly identical when the experiments are carried out in an atmophere of approximately 1% oxygen (27). Based on the promotion of translocation by exogenous ATP (13, 29) it was anticipated that increased photophosphorylation or a decrease in photorespiration might produce higher translocation rates for a given net photosynthesis rate but this was not observed. The data suggest the existence of regulatory process which controls the proportion of net photosynthate which is translocated. In addition, there appears to be a constant amount of material translocated independent of light intensity (zero intercept of Fig.11).

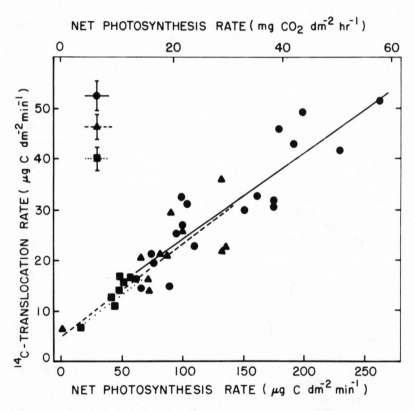

Fig.11. Relationship between net photosynthesis rate and translocation rate for 39 measurements in 19 plants. Source leaf in 21% O_2 at 2000 (dotted lines, squares), 3700 (broken lines, triangles) or 7200 ft c (solid line, circles). Photosynthesis rate was adjusted by controlling the CO_2 concentration. Servaites and Geiger (27).

As discussed earlier, data from studies with exogenous sucrose showed the existence of saturation kinetics for the carrier system of phloem loading (29). Experiments in which the rate of net photosynthesis was increased by increasing CO_2 level, decreasing O_2 and increasing light intensity to 7200 ft c failed to show saturation of loading. Any saturation observed appeared to be the result of saturation of photosynthesis (27). Apparently the rate of formation of assimilate rather than phloem loading is limiting translocation in sugar beet under the conditions used in this study.

B. Role of phloem loading in translocation

There appears to be two principle functions served by phloem loading, selectivity of the major translocate species and generation of the water potential difference which produces mass flow of the translocation stream. These two points will be discussed in more detail in the sections which follow.

1. Selection of major translocate species

Phloem exudate studies with trees (38) and with Yucca inflorescences that have been supplied exogenous sugars (30) point up the fact that the composition of sieve tube exudate is limited in the number of major species present. The spectrum of sugars present and the proportions of each is considerably different from the surrounding tissue. At least part of the selectivity may arise from the selectivity of the carrier in the plasmalemma of the se-cc complex. Trip et al (31) observed that only the non-reducing sugars verbascose, stachyose, raffinose, sucrose and mannitol were translocated when a variety of reducing and non-reducing sugars were supplied by reverse-flap technique to the leaves of white ash and lilac plants. The long times of application make interpretation difficult. Tammes et al (30) supplied a number of labelled compounds to the xylem of detached tops of Yucca inflorescences. Some sugars were entirely converted to sucrose, while when other compounds such as sorbitol and K glycerate are presented, they were found along with a small amount of sucrose in the phloem exudate. Others were exuded as such and not converted to sucrose (Table 10). These studies suggest selectivity of certain compounds but do not directly address the question of selective phloem loading. Preliminary data support the selection of translocate species by specific carriers during phloem loading. However, it is premature to discuss these findings in detail until further experiments are performed.

Table 10.   Summary   of   compounds   applied   to   the   base   of detached   Yucca inflorescence stalks and the compounds recovered in the bleeding sap from the upper end of the   stalks.   Tammes,   Vonk and van Die (30).

Supplied in external solution	Labelled compounds present in the exudate	Remarks
^{14}C(U)-sucrose	sucrose	asymmetric labelling of the hexose moieties ^{14}C-gluc./^{14}C-fruct. = approximately 3:1
idem + excess of ^{14}C-fructose	sucrose	asymmetric labelling as above.
idem + excess of ^{14}C-glucose	sucrose	asymmetric labelling as above
^{14}C(U)-maltose	sucrose	^{14}C-maltose not detectable on chromatograms of the exudate sugars.
^{14}C(U)-lactose	sucrose raffinose(?)	raffinose "identified" by its R-sucrose only; ^{14}C-galactose not detectable in the exudate.
^{14}C(U)D-glucose	sucrose	asymmetric labelling of the hexose moieties; ^{14}C-gluc./^{14}C-fruc. = ≫1.
^{14}C(U)D-fructose	sucrose	asymmetric labelling of the hexose moieties; ^{14}C-gluc./^{14}C-fruc. = approximately 1:7.
^{14}C(U)D-galactose	sucrose	galactose not detectable on exudate chromatograms.
^{14}C(U)L-sorbose	sorbose	
^{14}C(U)D-sorbitol	sorbitol sucrose	sorbitol was main constituent.
UDP-^{14}C(U)D-glucose	sucrose glucose	sucrose was main constituent
^{14}C-fructose-1,6-di-phosphate(NH -salt)	sucrose	
^{14}C-fructose-6-phosphate (NH -salt)	sucrose	
^{14}C-glyceric acid (K-salt)	glyceric acid sucrose glucose	glyceric acid was the main labelled constituent, followed in decreasing by sucrose and glucose.
^{14}C-glycollic acid (K-salt)	glycollic acid unkown substances	glycollic acid was the main constituent beside considerable amounts of an unknown spot. Traces of other unkown spots were also present in the exudate.
^{14}C-glycerol	glycerol	An unknown, distinct spot was also present on chromatograms
^{14}C(U)L-glutamine	glutamine	

2. Source of the driving force for mass flow
The feasibility of phloem loading as a source of the force
which generates mass flow depends on a demonstration of several key
conditions.   First it must be shown that phloem loading is
generally capable of producing sufficient potential energy to
continue to provide a driving force along the entire length of the
translocation path.   Secondly, the translocation path must be shown
to be sufficiently free of obstructions to allow passage of the
translocation stream without an excessive expenditure of energy.
Finally, if phloem loading is the major cause of the movement of
translocate, the onset of mass flow out of a developing leaf should
coincide closely with the attainment of the necessary water
potential gradient.

a. Magnitude of the potential energy source. To serve as the
driving force for mass flow, phloem loading must be capable of
developing sufficient potential energy to move the translocate
along the length of the path. In order for the system consisting
of the se-cc complex of the source, the path and the sink to
function as a Münch system, it appears that two key conditions must
be met. First, the potential energy source, in this case, the
osmotic pressure of the solutes in the se-cc complex of the source
region in contact with a solution with high water potental, must be
capable of generating the observed hydrostatic pressures.   From a
feasibility standpoint it is somewhat academic to discuss whether
the pressures are cause or effect of mass flow (36).   In fact, the
water potential difference across the membranes is the cause of
mass flow and a pressure gradient will follow as long as there is
resistance to flow.   Furthermore, for flow to occur, the osmotic
pressure in the source region must be large enough so that the
water potential difference between the source and path and their
surroundings exceeds the water potential difference at the sink
end.

In a study of red oak phloem Hammel found both turgor and
osmotic pressure gradients (Table 11).   The gradients reported were
generally of the order of 0.3 to 0.4 bars/m, with the upper end
higher as required for mass flow (18). As noted earlier (Tables 2,
3, 5) the osmotic pressure of the mesophyll is generally too low to
produce the pressures observed in the sieve tubes.   It appears that
the driving force for the translocation stream is the free energy
of the water of the free space solution moving into the se-cc
complex solution where the water potential has been lowered by
solute accumulation.   The osmotic pressure of the contents of the
se-cc complex of the source region of sugar beet is 28 to 32 bars.
If the xylem water and presumably also the free space water is
under a tension of -3 to -5 bars, a hydrostatic pressure of 23 to
29 bars can be generated by the influx of water from the apoplast.

Table 11. Turgor pressure and osmotic pressure values in phloem of the trunk of red oak at upper (6.3+0.3 m) and lower (1.5+0.3) levels. Hammel (18).

Sample	Conditions	Level	Average Osmotic Pressure	Gradient	Average Turgor Pressure	Gradient
			atm	atm/m	atm	atm/m
1	Overcast after rain 17 - 18°	U	24.0		17.5	
		L	22.7		16.2	
		difference	1.3	0.26	1.3	0.26
2	Bright sun 12 - 13°	U	23.7		13.2	
		L	21.9		11.7	
		difference	1.8	0.36	1.5	0.30
3	Clear sky 7 - 7.5°	U	24.2		15.9	
		L	22.7		13.8	
		difference	1.5	0.30	2.1	0.42
4	Clear with breeze 13 - 14°	U	24.3		17.4	
		L	22.5		13.0	
		difference	1.8	0.36	4.4	0.88
5	Hazy sun, breeze 11 - 14°	U	24.9		15.2	
		L	22.8		13.3	
		difference	2.1	0.42	1.9	0.38
6	Hazy, sunny 14 - 16°	U	24.4		14.3	
		L	22.3		14.5	
		difference	2.1	0.42	-0.2	-0.04
7	Cloudy, misty 14 - 15°	U	23.4		17.2	
		L	22.3		15.4	
		difference	1.1	0.22	1.8	0.36
8	Cloudy,foggy,rain 15 - 16°	U	23.6		17.1	
		L	22.1		16.0	
		difference	1.5	0.30	1.1	0.22
9	Sunny,hazy, broken clouds 11 - 13°	U	22.9		14.0	
		L	21.2		13.0	
		difference	1.7	0.34	1.0	0.20
		Average		0.33		0.33

Christy and Ferrier (2) have developed a steady state solution of two similar mathematical models to describe the various performace parameters of a sugar beet translocation system. Based on this work Tyree (34) computed that translocation can continue along a path of 318 cm in the model before the turgor pressure in the sieve tubes reached zero. With a linear decrease in turgor along the path, the drop is osmotic pressure expected between the source and a sink 30 cm away would be approximately 10% of the original osmotic pressure value. Geiger et al (13) did not find a detectable difference between the osmotic pressure in se-cc complexes of the source and those of the path 30 cm away. The estimated sensitivity of the freeze substitution/plasmolysis method as originally applied is approximately a 10% difference. However, the data of their study do indicate that the resistance to flow is not so large as to cause a large lateral influx of water along the path. Stated another way, only a small part of the original osmotic pressure is expended (through dilution by lateral influx of water) to provide a turgor pressure gradient along the 30 cm sugar beet translocation path. Recent evidence for the open condition of intact sieve element lumens and pores is consistent with this physiological data (9, 14 and Fig.12).

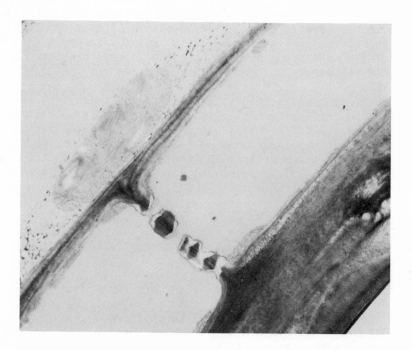

Fig.12. Electron micrograph of a sieve element of Phaseolus vulgaris prepared by freeze substitution. Note open lumen and pores and position of organelles along the wall. Giaquinta and Geiger (14).

A subsequent study with a refined freeze substitution/plasmolysis method indicates that there is an osmotic pressure difference of 5 to 7 bars between the se-cc complexes of the source and those in a sink leaf on the same plant some 30 cm distant (8). The osmotic pressure studies in sugar beet demonstrate that the potential energy in the se-cc complexes of the source leaf as a result of phloem loading are more than sufficient to generate the motive force for mass flow in the sugar beet system (8, 13). Further, the difference in osmotic pressure between the source and sink ends of the Munch system is sufficient to produce flow in the direction of source to sink (8).

In trees, gradients of turgor pressure in the xylem and phloem are larger. Even with the leaf free space water under a tension of -15 to -20 bars (26), the water would possess sufficient free energy to produce a turgor pressure of 30 to 35 bars. The turgor pressure drop in the phloem can be made up in part by lateral influx of water along the path, an aspect stressed recently by Eschrich et al (7). Perhaps it is significant that as the osmotic pressure of the contents of the sieve elements decreases in going from the crown to the roots, the free energy of the xylem water increases. The gradient appears to be approximately plus 0.1 bar/m change in xylem water potential compared to minus 0.2 to 0.4 bar/m change in the osmotic pressure of the sieve element contents. At the base of the tree where the phloem solutes are most dilute, the water of the free space solution has the lowest tension, and the highest free energy.

One of the recent reports in series of studies by Humphreys and Garrard offers an example of a mass flow proportional to an osmotic pressure difference (20). Although the exact mechanism is not clear, it appears that mass flow from the cut phloem of the corn scutellum is proportional to the difference in the osmotic pressure of the bathing medium and the level of accumulated sucrose in the cells of the scutellum (Fig.13).

A recent report by Wu and Thrower (37) of the ability of the feeding of aphids to induce flow into a mature leaf, specifically into the infested areas offers further demonstration of the operation of a pressure flow system.

a. Correlation between the onset of export and the osmotic pressure of the contents of the se-cc complex. As young sink leaves mature they gradually cease to import and start to export sugar (Figs. 14 and 15). If phloem loading is responsible for generating the potential energy which drives translocation there should be evidence of a close causal link between loading and the onset of export. Examination of the events which precede the beginning of translocation out of the leaf reveals that the phloem of the minor veins is mature, the lumens and plates open and loading is started before export can be detected (8). The onset of loading and export sweeps basipetally down the developing leaf. Table 12 is a summary of the values of osmotic pressure for the se-cc cells of a mature source leaf and of a developing leaf which contains non-loading (sink), transition, and loading (source) regions. It can be seen that the osmotic pressure is highest in the phloem of all orders of veins in the region which is exporting (tip), is slightly lower in the intermediary region which is just beginning to export and is lowest in the importing (basal) region.

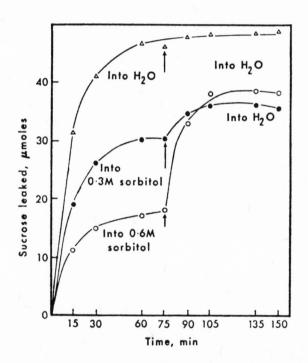

Fig.13. Release of sucrose from the cytoplasmic synthesis compartments of maize scutellum slices. The tissue was first incubated in 0.9 M fructose for 3 hr at 30°C, during which sucrose was synthesized. At time zero the slices were transferred to the media as identified. A further transfer to water occurred at 75 min. Humphreys and Garrard (20).

The middle region which has an osmotic pressure approximately equal
to that of the mature source leaf se–cc complex is no longer able
to import from source regions (8).

The data support the concept that import stops when phloem
loading raises the osmotic pressure to a point in the se–cc
complexes that the source is no longer able to develop sufficient
potential energy by loading to generate flow into the region.
Further, it indicates that export begins when the osmotic potential
is sufficient to develop a hydrostatic pressure gradient large
enough to overcome loss of energy by frictional resistance along
the translocation path.

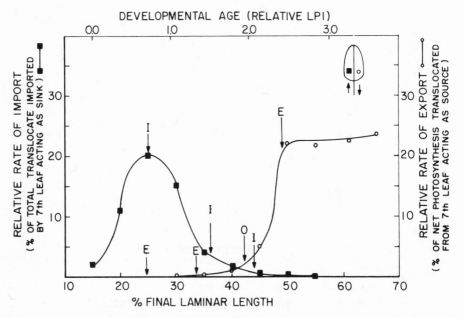

Fig. 14. Relative rate of import (■) and export (O) by the
developing 7th leaf of sugar beet. Arrows indicate stages of
leaves in autoradiographs of importing (I) and exporting (E) leaves
shown in Fig. 1. Fellows and Geiger (8).

Fig.15. Autoradiographs of developing 7th leaves of sugar beet. A to C: label imported from a mature 4th and 5th leaf supplied with $^{14}CO_2$. D to F: label fixed from $^{14}CO_2$ by these leaves. Marer 1 cm. A,D: 25%FLL B: 36%FLL C: 44%FLL e: 34%FLL F: 49%FLL. Fellows and Geiger (8).

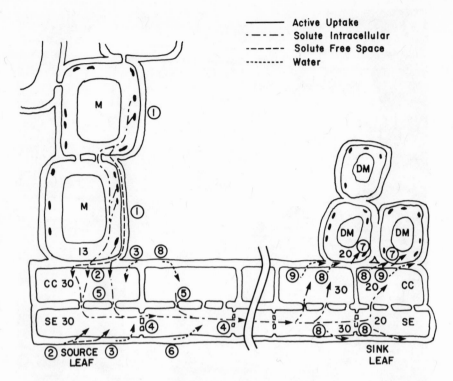

Fig. 16. Model of movement of sugar from chloroplast and of phloem loading in sugar beet. Circled numbers refer to steps in the translocation process. Other numbers in cells refer to osmotic pressure in bars.

Table 12. Osmotic pressure values for estimated 50% plasmolysis of cells (companion cell = cc; sieve element = se) from the tip (loading), middle (intermediate), and base (non-loading) from the 42% FLL leaf and source leaf from the same plant. Osmotic pressure expressed in atm. Fellows and Geiger (8).

Vein Order	Osmotic Pressure Values							
	42% FLL Leaf						Source Leaf	
	Tip Loading		Middle Intermediate		Base Non-Loading			
	cc	se	cc	se	cc	se	cc	se
	atm	atm	atm	atm	atm	atm	atm	atm
I	--	--	24.2	24.5	19.0	19.0	23.9	24.5
II	25.5	32.5	25.5	25.8	19.0	19.0	23.9	24.5
III	25.8	32.5	23.9	25.8	19.0	19.0	23.2	24.5
IV	25.8	32.5	25.2	27.2	19.0	23.2	24.5	27.8
V	25.8	32.5	22.0	25.8	21.0	23.2	23.6	27.8

C. Model of phloem loading and associated processes

From the data just presented it is possible to develop a model, built on a number of working hypotheses, for translocation in sugar beet (Fig. 16). The steps and supporting arguments can be outlined as follows:

1. Sucrose in the chloroplasts moves via the symplast to the vicinity of the minor vein phloem where it enters the free space. Symplastic transport is feasible for small solute molecules (33). Sucrose passes through the free space prior to phloem loading (13).

2. Sucrose is actively loaded from the free space through the plasmalemma of the se-cc complex source regions. Solute content of the source region se-cc complex is high (8,11). There is an abrupt change in solute concentration inside the se-cc complex of the source region (11). Phloem loading is dependent on energy metabolism (29). Performance parameters for phloem loading via se-cc complex plasmalemma are in feasible range (10,29).

3. Active phloem loading builds up potental energy for driving mass flow, stored in the form of high solute concentration. Water entering from free space moves assimilate solution into the sieve elements and down the sieve tubes. Solute content of se-cc complex of source leaf is sufficient to generate pressures of the observed size (8,11). Models based on this hypothesis demonstrate feasibility (2,24). Mass flow proportional to osmotic pressure difference can be demonstrated (20,37).

4. As energy is expended overcoming frictional forces, turgor pressure in the sieve tubes decreases and water enters, lessening potential energy of the sieve sap through dilution and reduction of the osmotic pressure. Osmotic and hydrostatic gradients have been reported along oak phloem (18) and in other trees (38).

5. In sink regions, sucrose and water exit from the se-cc complex into the sink symplast, providing a region with a lower osmotic pressure at the sink end of the phloem. Osmotic pressure in the sink se-cc complex is lower than the path or sources (8, 11).

This model should offer a number of aspects which can be tested to determine how well it describes the mechanism of translocation. It is more a point of departure than a last word on the elusive subject of translocation mechanism.

References

1. BARRIER,G.E. and W.E. LOOMIS. 1957. Plant Physiol. 32: 225-231
2. CHRISTY, A.L. and J.M. FERRIER. 1973. Plant Physiol. 52: 531-538
3. CURTIS, O.F. and H.T. SCOFIELD. 1933. Amer. J. Bot. 20: 502-513
4. EDELMAN, J., A.I. SCHOOLAR and W.B. BONNER. 1971. J. Exp. Bot. 22: 534-545
5. ESAU, K. 1967. Proc. Amer. Phil. Soc. 111: 219-233
6. ESAU, K. 1972. New Phytol. 71: 161-168
7. ESCHRICH, W., R.E. EVERT and J.H. YOUNG. 1972. Planta 107: 279-300
8. FELLOWS, R.J. and D.B. GEIGER. 1974. Plant Physiol.
9. FISHER, D.B. 1971. Plant Physiol. 47: 41 Abstract 243
10. GEIGER, D.B. and D.A. CATALDO. 1969. Plant Physiol. 44: 45-54
11. GEIGER, D.B., R.T. GIAQUINTA, S.A. SOVONICK and R.J. FELLOWS. 1973. Plant Physiol. 52: 585-589
12. GEIGER, D.B. 1974. In: Encyclopedia of Plant Physiology. Phloem Transport. Eds. M.H. Zimmermann and J.A. Milburn. Springer Verlag. Berlin.
13. GEIGER, D.B., S.A. SOVONICK, T.L. SHOCK and R.J. FELLOWS. 1974. Plant Physiol.
14. GIAQUINTA, R.T. and D.B. GEIGER. 1973. Plant Physiol. 51: 372-377
15. GIESE, A.C. 1973. Cell Physiology. Saunders Co. Philadelphia. p. 301.
16. GUNNING, B.E.S., J.S. PATE and L.G. BRIARTY. 1968. J. Cell Biol. 37: C7-C12
17. GUNNING, B.E.S. and J.S. PATE. 1969. Protoplasma 68: 107-133
18. HAMMEL, H.T. 1968. Plant Physiol. 43: 1042-1048
19. HO, L.C. and D.C. MORTIMER. 1971. Can. J. Bot. 49: 1769-1775
20. HUMPHREYS, T.E. and L.A. GARRARD. 1971. Phytochemistry 10: 2891-2904
21. KOCHER, H. and O.A. LEONARD. 1971. Plant Physiol. 47: 212-216
22. KURSANOV, A.L. and M.I. BROVCHENKO. 1961. Soviet Plant Physiol. 8: 211-217
23. MUNCH, E. 1930. Die Stoffbewegungen in der Pflanze. Fischer. Jena
24. NOBEL, P.S. 1974. Introduction to Biophysical Plant Physiology. W.H. Freemand and Co., San Francisco
25. ROECKL, B. 1949. Planta 36: 530-550
26. SCHOLANDER, R.F., H.T. HAMMEL, E.D. BRADSTREET and E.A. HEMMINGSEN. 1965. Science 148: 339-346
27. SERVAITES, J.C. and D.B. GEIGER. 1974. Plant Physiol.

28. SMITH, R.C. and E. EPSTEIN. 1964. Plant Physiol. 39: 338-241
29. SOVONICK, S.A., D.B. GEIGER and R.J. FELLOWS. 1974. Plant Physiol.
30. TAMMES, M.L., C.R. VONK and J. VAN DIE. 1973. Acta Bot. Neerl. 22: 233-237
31. TRIP, P., C.D. NELSON and G. KROTKOV. 1965. Plant Physiol. 40: 740-747
32. TRIP, P. 1969. Plant Physiol. 44: 717-725
33. TYREE, M.T. 1970. J. Theor. Biol. 26: 181-214
34. TYREE, M.T. 1974. Plant Physiol.
35. ULLRICH, W. 1962. Planta 57: 713-717
36. WEATHERLEY, P.E. 1973. Planta 110: 183-187
37. WU, A. and L.B. THROWER. 1973. Plant and Cell Physiol. 14: 1225-1228
38. ZIMMERMANN, M.H. and C.L. BROWN. 1971. Trees: Structure and Function. Springer Verlag. New York

DISCUSSION

Discussants: Susan Sovonick Dunford
                Maria Moors Cabot Foundation for Botanical Research
                Harvard University
                Petersham, Mass. 01366

                Robert J. Fellows
                Botany Department
                University of Illinois
                Urbana, Illinois 61801

Dunford: Drs. Sovonick Dunford and Fellows were invited to present their paper on "Phloem Loading". The paper follows.

Phloem Loading

Susan Sovonick Dunford and Robert S. Fellows

Although the phenomenon of phloem loading has been a widely accepted fact for some time, it is clear that the details of the process require further study.   In the presentation on phloem loading, several areas either need clarification or suggest further lines of experimentation.

1. Nature of phloem loading: The role of free space.   Not all research workers in sugar transport agree that the apoplastic route is of predominant importance.   Symplastic transport of organic solutes has been shown to occur in various tissues and organs. For example, Arisz (1) demonstrated symplastic movement in Vallisneria leaves for sugars, amino acids, and other organic solutes, basing his conclusions partly on the fact that he measured no leakage or exchange of solutes accumulated in the leaves. Using a system which presumably includes a loading step, Humphreys (9) and Humphreys and Garrard (10) observed that no leakage occurred into an external solution from maize scutella until the root-shoot axis was removed, indicating that the apoplast was not the leakage site. One of the most direct studies in this regard is the recent paper by Cataldo (4), who isolated mesophyll cells and minor veins from tobacco leaf discs and compared the abilty of the isolates and of the discs to accumulate various sugars.   The author concluded that in tobacco leaves the apoplastic route is less likely than the symplastic one for two reasons:

a. The minor veins showed a reduced ability to accumulate sucrose compared to discs of leaf tissue, as shown by measurements of V_{max}.   If the apoplastic route were the predominant one, "isolated bundles should have had a V_{max} for exogenous sugar at

283

least approximating the rates measured for discs," since the minor
veins would be the major site of uptake in that case.

    b. Competitive studies show that the presence of sucrose
above 6.6 mM increased the rate of glucose uptake. One explanation
for ths observation states that excess glucose is transferred to a
special metabolic compartment for subsequent transport to the minor
veins when sucrose is in high concentration. Cataldo suggested
that the transport compartment may in fact be the endoplasmic
reticulum and the plasmodesmata of the short distance pathway.
Finally, the work of Milburn (12) (that the ability of a Ricinus
plant to exude sap from an incision depended on the ability of the
organ being studied to mobilze and secrete solutes. The lateral
hydraulic conductivity of the stem was low when such exudation did
not occur, and increased only when solute secretion began. Thus,
solute secretion and water conductivity appeared to be linked, and
both were postulated to occur from storage cells across lateral
sieve areas. Whether this model includes only companion cells or
the total mesophyll as storage areas was not stated.

    Such contradictory evidence with regard to sugar movement need
not negate the data presented by Geiger. However, it does point
out that free space transport in loading systems may not be
universal. The measurement of free space turn-over rates is the
most direct means of detecting apoplastic movement and should be
applied to other experimental systems, for example, those of
Cataldo and Milburn. In addition, such a study should be extended
to include C4 plants, in which leakage does not seem to be part of
the loading process (15).

    The question of water movement in the free space must also be
resolved. Boyer (2) has recently demonstrated in sunflower that
the mesophyll protoplasts of the leaf can represent a high
resistance to water flow. A considerable portion of the water
movement through the leaf must therefore bypass them and travel in
the free space. While the accumulating action of an active carrier
in the membranes of the vascular tissue may serve to steepen the
concentration gradient of solutes, the rapid flow of water of
transpiration may have a tendency to sweep these solutes away. To
date, however, no deposits of sugar or other translocatable
materials have been reported near stomatal cavities.

    More detail must also be paid to the other specialized cells
adjacent to, and included in, the vascular bundle. As Geiger has
mentioned, there are a number of different types of transfer cells
(13), all believed involved in some sort of transport. The
ancillary role, if any, of these adjacent phloem and xylem
parenchyma as well as the mesophyll bundle sheath should be
considered when investigating phloem loading. This becomes more
compelling in the C4 plants, as witnessed by the controversy over
the function of the bundle sheath in photosynthesis (11).

2. Nature of phloem loading: evidence for active loading. Humphreys (9) has demonstrated that sucrose leadage from excised maize scutella is a mass flow, dependent on an osmotic pressure difference between the tissue and the external solution (described by Geiger on p.251-79). It should be stated that Humphreys also concluded that the flow extends from the mesophyll cells and that there is no active loading process. The evidence included a low Q_{10} for leakage, contained in the phloem alone. Since the majority of evidence points clearly to an active loading step in most cases, data derived from this experimental system should be interpreted with care. The question remains whether there are any mass flow systems that do not require an active loading step. The need for data from $C4$ plants is evident in this regard also.

3. Selection of major translocate species. The crucial question here seems to be the degree of specificity of the loading steps. If for example we present stachyose to a species that does not normally produce it, can the sugar be loaded? The data presented here indicate that it cannot. A further question in this regard is whether a plant can be forced to load a sugar that is normally produced in the leaves, but is not translocated. Hendrix (8) has concluded that the loading system in squash will load any non-reducing sugar that is supplied to it and that the phloem loading step is not the point at which a specific sugar is selected for transport. However, sucrose was the only sugar tested and squash may normally transport traces of sucrose anyway (16, 17). These results seem far from conclusive. Earlier work by Trip et al. with lilac and white ash (16) lends support to the hypothesis of Geiger. Supplying excess amounts of non-transported sugars to leaves did not lead to their translocation. A significant problem in all these studies is that sugars in transit and those absorbed by surrounding tissues cannot be distinguished by the techniques used. Sugar interconversions are quite likely in these storage cells.

4. Control of loading. While export may be stimulated by concentration gradients near the sieve elements, the actual mechanism, or mechanisms, controlling the vectorization of the loading process is yet to be determined. Enzyme and isozyme patterns change during the maturation of many plant tissues (14). Gaylor and Glaziou (6) have reported the existence of two forms of invertase in internodal storage tissue of sugar cane, one of which, an outer compartment acid invertase, disappears when the tissue matures. Webb and Gorham (17) stated that in squash, end products of photosynthesis, i.e. sucrose and stachyose, depend on the maturity of the leaf and that transport out of the leaf appeared to depend on the amount of stachyose present. Recently, Dickson (5) observed that in developing cottonwood leaves, sugars translocated

into the leaf entered different metabolic pools than those produced within the leaf by photosynthesis. If the postulated carrier is specific for a single substrate or group of substrates, the question remains how soon in the ontogenetic development of the leaf this specificity appears. Thus, it may be that in various species, the transition from sink to source may be triggered by the production of a specific enzyme, by the appearance of the transport sugar(s), or by the attainment of a certain sugar concentration within the vascular tissue.

Structural observations, conducted on the maturing membranes of the sieve element and companion cells, could be useful in this regard. Briarty (3) has recently reported the presence of 11 nm particles forming hexagonal arrays on the walls of root nodule transfer cells of Trifolium ripens. The appearance and organization of these particles may be related to some enzymatic or physical process such as phloem loading.

It would appear that a key role in the control of phloem loading in a mature leaf could be played by the carrier enzyme postulated to be present in the plasmalemma of the vascular tissue. Increased amounts of photosynthate in the free space of sugar beet would lead to increased loading into the companion cells and sieve elements. However, the rate of loading must also be controlled in some way by the activity of sinks (7). Milburn 12) has suggested that the loading process can be triggered by pressure differences in the phloem. When Ricinus plants were caused to exude sap by incision, the sap concentration increased when solutes were available for secretion, presumably due to increased loading. According to the author, if loading were controlled by sap concentration, then the sap concentration should be maintained at a maximum in intact sieve tubes, and dilution should accompany exudation. The interaction between control of loading by supply (solute concentration in the free space) and control by demand (either sap concentration or osmotic pressure) needs to be further clarified.

Finally, and perhaps prematurely, the exact nature of the phloem loading step needs to be defined. General areas of inquiry include: is there actually a physical entity functioning as a carrier or could binding to specific sites (followed by pinocytosis?) be the crucial event? How is the carrier actually powered? Studies on energetics have almost exclusively tested only the effect of ATP. Could some other high energy species or even a proton pump be the primary energy source under normal conditions? The final answers to such questions may have to await further conceptual and technical developments in the fields of membranes and transport processes.

References

1. ARISZ, W.H. 1969. Acta Bot. Neerl. 18: 14-38
2. BOYER, J.S. 1974. Planta 117: 187-207
3. BRIARTY, L.G. 1973. Planta 113: 373-377
4. CATALDO, D.A. 1974. Plant Physiol. 53: 912-917
5. DICKSON, R.E. Plant Physiol. (suppl.) 53: 44
6. GAYLOR, K.R. and K.T. GLASZIOU. 1972. Physiol. Plant 27: 25-31
7. HABESHAW, D. 1973. Planta (Berl.) 110: 213-226
8. HENDRIX, J.E. Plant Physiol. 52: 688-689
9. HUMPHREYS, T.E. Phytochemistry 11: 1311-1320
10. HUMPHREYS, T.E. and L.A. GARRARD. 1971. Phytochemistry 10: 2891-2904
11. LAETSCH, W.M. 1974. Ann. Rev. Plant Physiol. 25: 27-52
12. MILBURN, J.A. 1974. Planta (in press)
13. PATE, J.S. and B.E.S. GUNNING. 1969. Protoplasma 68: 135-156
14. SCANDALIOS, J.G. 1974. Ann. Rev. Plant Physiol. 25: 225-258
15. SCHOOLAR, A.I. and J. EDELMAN. 1971. J. Exp. Bot. 22: 809-817
16. TRIP, P., C.D. NELSON and G. KROTKOV. 1965. Plant Physiol. 40: 740-747
17. WEBB, J.A. and R.P. GORHAM. 1964. 39: 663-672

Dunford: Dr. Fellows and I have reviewed Dr. Geiger's paper and we believe there are 3 major areas that are open for considerable discussion: apoplastic vs. symplastic transport, the specificity of the loading systems, and control systems – whether the carrier controls loading in the majority of cases, what the role of the sinks might be, whether this occurs through pressure differences or concentration differences. I suggest that we begin with the apoplastic-symplastic question.

Willenbrink: I have three minor questions about techniques: what was the concentration of ATP used, what was the pH, and what was your basis for our assumed 1 to 1 stoichiometry of the ATP requirement?

Geiger: We supplied 4 mM ATP, resulting roughly in about 50% elevation of the tissue concentration (much of the added ATP was, of course, broken down). The pH was 6.5 in phosphate buffer. As for the ATP requirement, we posited only one active transport step, i.e., across the plasmalemma into the sieve element-companion cell complex, and we made the very simple assumption of 1 to 1 for lack of any better information at present.

Tyree: Were you able to estimate the concentration of sucrose in the apoplast in the loading region?

Geiger: Based on the solution without a trapping agent, the measurements would indicate a low conentration - on the order of 20 micromolar at equilibrium.   The 20 millimolar concentration that was used in certain experiments was based on what was needed to produce a translocation rate similar to that with CO_2.

Tyree: This 20 millimolar figure, then, is the concentration you needed based on Michaelis-Menton kinetics to get sufficient loading?

Geiger: That is correct. One other thing we did was to infiltrate the leaf and then centrifuge out the free-space solution. Its sugar concentration was about 20 millimolar.

Tyree: In your graph showing the concentrations required for incipient plasmolysis in the different cells of a leaf undergoing transition from import to export, it appeared that the companion cells were consistently lower than the sieve elements. Was this difference significant?

Geiger: Probably not, so it was much easier to detect incipient plasmolysis in the companion cells.

Tyree: I am puzzled by one additional point. My recollection from the earlier manuscript which you kindly supplied me about a year ago, is that you stated then that the osmotic pressure of the mesophyll cells in the sink region was much lower than that of the sieve element-companion cell complex, but I thought from today's lecture you had changed your story. Is that right?

Geiger: The data presented today were based on the recent studies by Dr.   Fellows which were much more extensive. Data in the earlier manuscript were perhaps in error because of faulty sampling.   In a developing leaf at the proper stage one can distinguish three regions: a region which is both loading and exporting, a region which is loading but not exporting, and a region which is importing only. It is specifically in the region which truely importing that one finds this close similarity in osmotic pressures between the sieve elements and the mesophyll cells.

Tyree: Then it appears that you are suggesting that as a leaf changes from importing to exporting the sieve element-companion cell complex goes from a state in which it is a part of the symplast continuum, physiologically speaking, with the rest of the leaf, but then later separates itself. Do you have any speculations on how this might come about?

Geiger: If we take the sink tissue and put it into a sucrose solution we get the similar dual isotherms, similar kinetics, but lower maximum velocities. So what we might have is a pump-leak system, where it can take up sugar as a sink. It is symplastically connected . If the symplast seals off at some boundary, then what is pumped would remain within, rather than being able to leak throughout.

Weatherley: One of the problems which besets the choice of paths from the mesophyll to the sieve tubes is the fact that transpiration gives a sort of countercurrent, and hence the diffusion from mesophyll cells to the sieve tubes would be against a stream. Do you think it is possible to detect this effect by comparing zero and high transpiration rates? In the latter case the sugar would have to build up to a higher concentration to get to the companion cells.

Geiger: In my experimental set-up, one side of the leaf is in contact with a solution, so I don't think the transpiration rate would be very high. It is an interesting question, however, and something that should be tried.

Weatherley: You showed us that there was no saturation of the translocation rate with increasing photosynthesis. How do you reconcile this with the commonly observed fact that there is an accumulation of dry weight during the day in the leaves, presumably because the translocation does not equal the phosynthetic rate?

Geiger: In sugar beet a rather small percentage of the sucrose seems to be retained in the leaf normally, but starch or glycogen-type polysaccharides are retained to a large degree. Amino acids and hexoses also accumulate, so there is some sort of control leading to metabolic diversion. The addition of ATP appears to bring about the scavenging of more sucrose than was apparently made normally, so the ATP effect would fit along with the idea of control by channelling of matabolic pathways rather than by compartmentation.

Milburn: I would like clarification on one or two points in connection with the way these experiments were conducted. As I understand it, Beta vulgaris is a good exuder if one cuts the appropriate tissues, so the first point is, does abrasion of the leaf cause exudation to occur from anywhere in the cellular regions? Normally we don't think of cellular regions exuding because the amounts of exudate are so small as not to be useful for analysis. Nevertheless, it does seem to me that if symplastic transport were to work in any way in a manner analogous to sieve tube transport in the rest of the plant, but on a smaller scale, this kind of phenomenon might occur. Now if it were to occur, then one might get the escape of sap from wounded symplastic connections which would be rather like leaky but semi-plugged sieve plates. In this case one couldn't distinguish between true apoplastic solution and symplastic escaping solution.

In your graphs it was shown that changing the light into new regimes increased the uptake of materials and the labelling. It strikes me that if the pressure were to increase as a result, this would increase the leakage. One futher point bearing on all this is how you imagine the pressure gradients to go in the system. One can imagine in the very simplest form the pressure gradient decreasing in some fashion more or less from the chloroplast right down to somewhere near the sink. Clearly you imagine a gradient with a step in it.   Do you have in mind a decreasing pressure gradient within the mesophyll, followed by a step, followed by a pressure gradient down the sieve tube complex?

Geiger: First, with respect to your question on leakage, we carried out plasmoysis tests to see if the symplast is fairly open. All of the cells with the exception of the epidermal cells and an occasional palisade cell were able to be plasmolyzed within a half-hour period in the mannitol solution. As far as the pressure steps are concerned, I would anticipate cyclosis in the mesophyll cells, diffusion across the plasmodesma down a very steep gradient, and so to a point near the phloem, at which point I would perhaps think of facilitated diffusion. Then, of course, I would think in terms of a large increase in turgor pressure in the sieve element-companion cell complex of the minor veins of the source region. I would anticipate in sugar beet a rather small decline in pressure along the path but in the sink I would tend to think of this pressure being expended in terms of movement into the symplast. There remains the question of where the water goes.

Milburn: I would agree with this description. We've got concentration data from Ricinus recently published in Planta that indicate just this kind of pattern - a rather low concentration drop off in the phloem but a steep rise and fall in the appropriate directions at source and sink. Getting rid of the water is not such a problem since the xylem system would apply to both sink and source and would be under tension and water can travel very fast in that.

Swanson: Referring back to your isotope trapping system in the mode in which no trapping agent was used, I have one question: because of the extremely high ratio of the escape volume of the trapping system relative to the volume of the apoplast, then would not this large volume itself act as a trapping agent? I don't see how such a system could come to an equilibrium level within the time periods you used.

Geiger: I believe there was some sort of steady-state loss to the mesophyll cells on the way out. Perhaps someone who is a little more used to thinking in these dynamic terms could correct me but I picture the sucrose as being trapped on the way out at a certain point.

Reinhold: I want to refer to the ATP experiment. It seems to me there are two possibilities here as to where ATP is acting. One possibility is that the ATP enhances the pump, and the other is that ATP increases the loss from the mesophyll, i.e., a toxicity effect owing to a high concentration of ATP. Now if it were the former case, wouldn't one expect less ^{14}C being trapped because of that increase in the pumping capacity, whereas in the latter case one would expect more ^{14}C to appear outside? As in fact you did observe more ^{14}C outside, is it not less likely that the ATP enhanced the pumping capacity?

Geiger: That is a good point, and I've spent quite a few hours thinking about this problem. I can't say I have a very good answer for it. As you implied, the data tend to support better the leakage effect. If you recall, I kind of indicated this when I said that I thought ATP and DNP were working on two different systems, i.e., that ATP was not just reversing the DNP effect.

Dunford: If ATP increased leakage in the mesophyll cells, it would also probably increase leakage in the vascular tissue, and it would be difficult then to explain how we got an increased translocation to the sink. Also I would add that the trapping experiment would, I think, measure more how fast the sucrose is moving to the free space, rather than the level, except when no trapping agent is used. In fact, in the experiments where no trap was used, you got a slightly lower sucrose concentration in the free space. That doesn't indicate that the turn-over rate is lower, in fact the turn-over rate appeared to be higher.

Tyree: I would like to reply to Dr. Milburn's comment on what the transpiration stream might do to the distribution of sucrose in the apoplast. Briggs and Slatyer and I have all made calculations (my own are unpublished) on the pile-up of solutes, any solutes, in the apoplast at the evaporating surface during high transpiration. Although the calculations are of necessity crude, because we do not have quantitative data on the distribution of cell walls, it can be shown that the pile-up is negligible - not enough to be worrisome. What I am worried about is the very high concentration of sucrose required, according to Dr. Geiger's data - some 20 millimolar, which seems to me very high.

Aronoff: I would like to ask two questions. The first concerns the role of ATP again. If the mechanism of sugar transport into the sieve tube complex is that of facilitated diffusion, one needs no energy for this process, consequently there would have to be a completely different role for ATP. What would you consider this role to be? Somewhat related is the second question: have you, in your feeding experiments, put in some substance like glucose and then sampled the sucrose to see whether it is asymmetrically labelled?

Geiger: One of the problems with ATP studies is that ATP is a lot of things. For example, it can be a membrane modifier. With respect to your second question, we are in the process, particularly in studies being carried out by Ms. Fonde, of beginning experiments along the lines you suggested.

Dunford: Was your question, Dr. Aronoff, whether or not there is facilitated diffusion into the vascular tissue? I believe the point that was made was that facilitated diffusion might carry the sugar out of the mesophyll or parenchyma cells but that there was an active step into the vascular tissue.

Aronoff: Yes, my question was, where the was energy required. If it were connected in some way with the transformation of the sugar, which was indicated by Kursanov long ago, this might be suggested by the point at which some precursor was in fact transformed into sugar - at least it might serve in an indirect way to indicate where the pump might be. I also simply wanted to clarify that "facilitated transport" needs no energy.

Ho: Soviet investigators working on phloem loading also suggest that the apoplast may be involved. They also suggest that sucrose is first hydrolyzed and then resynthesized from the hexoses during loading. Do you have any views, Dr. Geiger, on how this activity may take place during active loading?

Geiger: One of the difficulties with the Russian studies, to the extent I am familiar with this work, is that their data are all based on leaf discs which have been soaked for extended periods of time. As you know, this induces invertase activity in many tissues. I am not certain that this criticism applies to all of their studies- in some cases their methods are not clear on this point.

Crisp: The laboratory I am associated with is concerned with the molecular design of symplastic foreign molecules. In that sense, we are deeply involved, of course, in the apoplast-symplast question. To get at this question, we use isolated cells and protoplasts. I am pleased to find that Dr. Geiger's and our results agree within 6%, in terms of uptake per square millimeter per minute.

Cataldo: Referring back to the Russian studies on leaching, they usually bathe the leaf discs in water, which makes the leaf tissues very fragile and they often rupture. The analyses of the leachates in these cases show ratios of sucrose to glucose to be the same as that found in the cells.

Moorby: In your exchange experiments, Dr. Geiger, how long did it take to detect label in the bathing medium? Dr. Troughton and I have been doing ^{11}C experiments and, surprisingly, it takes 3 to 5 min. before we can detect any export from the fed region - a fairly long time considering it takes only a second or less for ^{11}C to get into the chloroplast. Another point, on the matter of interconversion, we may be forgetting that when label gets into a leaf, some of it may go straight through and out into the sieve tubes, other of the labelled atoms may go into starch or other storage products which are in fact turning over. In connection with this we have ^{11}C data showing that if you put in a 1-minute burst of $^{11}CO_2$, the actual half-width of the pulse leaving the leaf is 45 minutes. There is an enormous spread here, probably due to interconversion.

Geiger: Regarding your queston on timing, I was able to detect trapping in 5 to 10 min., but I doubt that I would be able to establish the fine kinetics that I think you would need, to talk about that 5 min. delay.

Milburn: I wish to come back briefly to apoplastic movement in relating to the transpiration stream. I haven't examined the calculations cited by Dr. Tyree - clearly I must do this - but in the meantime, it appears to me that the evidence against the idea that the flow of solutes with the transpiration stream is negligible is considerable. Admittedly the evidence is mainly qualitative rather than quantitative. One can very easily show with suitable non-toxic dyes that they are wafted to the very end of the transpiration stream. Secondly there are a number of experiments which have been carried out on the transport of soluble gold which is a very sizable "solute" and which can be wafted along also by the transpiration stream, accumulating at the site of evaporation. But if we are talking about sugars, then the situation can be quite different because sucrose can be taken up by the leaf cells and therefore sucrose would have to run the guantlet before it could be carried to the end of the transpiration stream and dumped. I can't imagine, though, that the system does not operate with solutes in which there is no uptake system in the mesophyll.

Tyree: The amount of accumulation depends on the ratio of water flux to the diffusion coefficient, and gold particles have such a low diffusion coefficient that there is little mystery why they accumulate at the surface.

Eschrich: In your slide showing a cross-section of a bundle, how do you explain that the sieve elements were so perfectly fixed, considering the fact that surging had undoubtedly occurred?

Geiger: In general, when we look at bundles in a leaf blade, we find less surging than we do in the path because we cut simultaneously from two sides and the part in the center does not move very far. Also I think we get sealing at the cut edges very fast, as shown by the fact we can plasmolyze the sieve elements.

Fisher: I am concerned with the high osmotic pressure reported by Dr Geiger for the sieve element-companion cell complex, equivalent to around 35% sucrose, and we know that in the exudate it is only about 8%. This is a large discrepancy. As far as unloading is concerned, one idea that appeals to me is that this is not an active process but is a passive leakage, in which case the reflection coefficient of the membranes might be quite low, requiring a high concentration of osmoticum to plasmolyze the sieve elements in the sink. I think that the concentration data based on plasmolytic methods must be accepted with considerable skepticism.

Spanner: My comment is apropos to a remark by Dr. Geiger that there is little surging in the exceedingly minute sieve tubes of the minor veins, nevertheless, published micrographs show the sieve plate pores nicely plugged with P-protein.

Fellows: I would like to comment here because I did much of this work by the method of freeze-substitution. When I observed the minor vein of sugar beet leaf, particularly in the sink leaf, mature veins with developed sieve elements would show some veins open, some plugged, but a predominance of the open veins.

Willenbrink: We must remember that within the mesophyll the last steps of sucrose synthesis are localized in the cytoplasm and not in the chloroplast, and there we must include an active transport process between the outer membrane of the chloroplast and the sieve tube itself. That's one of the obstacles against the speculation of pure symplastic movement from the chloroplast to the sieve tube.

Turgeon: I think that is a very important question and it brings up the whole question of what plasmodesmata can do and cannot do. Certainly the plasmodesma is not just a simple hole. Perhaps with further studies we will find that we can get actual active transport across the plasmodesma.

MacRobbie: I think we are getting away a bit from what Dr. Geiger was actually proposing. What he was proposing was a leak into the apoplast very close to the phloem and not over the whole mesophyll. Now that seems to me to remove the difficulties of water flow, because he has put us in a situation where we would all agree the water flow is just into the phloem and not outwards.

Another point concerns C-4 plants where it seems to me we must
talk about apoplastic transport. We believe there are very good
symplastic   connections   between   the   bundle   sheath   cells   and
mesophyll cells in C-4 plants, and if these connections are so good
that all the carbon traffic can go on for the acids, then I think
we would be in serious difficulties if we were to allow sucrose
transport back out into the mesophyll.

Crisp: With respect to the movement of the acids, I can only
tell you that in the isolated protoplast studies, acids were found
to have free mobility across the plasmalemma - so it is possible
for them to leave the plasmalemma, move across the apoplast, and be
taken up again.   There seems to be no barrier.

Aronoff: Can you amplify, Dr.   Crisp, as to whether that
includes the phosphates or not?

Crisp: In   the   course   of   looking   at   all   of   the
organophosphates alkylated in different positions on the phosphates
we've looked at those which are acids. All of the monophosphoric
acids are certainly mobile, and by that, I mean, mobile back and
forth across the plasmalemma.

Schmitt: As to the question of apoplast-symplast in the
translocation path of the algae, I believe we have good reason to
believe we are dealing with a true symplastic system, as we have
good evidence that everything photoassimilated is translocated.
The translocate actually represents the photoassimilate.

Physiological and Structural Ontogeny

of the Source Leaf

Robert Turgeon
Rockefeller University
Department of Botany
New York, N.Y. 10021

and

John A. Webb
Department of Biology
Carleton University
Ottawa, Ontario K1S 5B6

A leaf is transformed during ontogenesis from a heterotrophic, to an autotrophic, organ. Initial dependence on phloem imported nutrients is lost as increasing photosynthetic capacity enables the lamina to satisfy its own carbon requirements, and to export excess photosynthate. This transition is interesting from both a practical and a theoretical viewpoint. The leaf becomes an asset to the carbon economy of the plant only when it is autotrophic and has begun to export. Theoretically, the study of conditions which contribute to the start of assimilate flow should provide insight into the mechanisms of vein loading and long-distance transport.

A. The Import-Export Transition

Early studies on the transport of ^{14}C-labelled compounds to and from developing dicotyledonous leaves demonstrated that, for a short period, at approximately 50% expansion, the leaf is capable of simultaneous import and export (Jones et al. 1959: Thrower, 1962: Webb and Gorham, 1964). It was first proposed by Jones and Eagles (1962) that bidirectional transport within the leaf results from localized growth. Expansion and maturation of dicotyledonous leaves proceeds in the basipetal direction (Esau, 1965). Jones and Eagles (1962) suggested that during the transitional phase the leaf tip exports photosynthate while the more immature leaf base continues to import. Our initial experiments were designed to test this hypothesis.

All experiments were carried out on leaf 5 of the Early
Prolific Straight-neck squash, Cucurbita pepo L. var melopepo f.
torticollis Bailey. The leaf plastochron index (LPI) (Erickson and
Michelini, 1957) was used to measure leaf age (Turgeon and Webb,
1973). In preliminary experiments it was determined that the LPI
is linearly related to time, one plastochron being equal to
approximately 2 days.

Progressive basipetal cessation of import is readily
demonstrated by the technique of whole-leaf autoradiography. We
labelled the mature third leaf of C. pepo with a 5 min. pulse of
$^{14}CO_2$ and, following a 2 hr. period, to allow distribution of ^{14}C
compounds throughout the plant, severed leaf 5. The lamina of leaf
5 was quickly frozen between two glass plates, pre-cooled at 38°C
and applied to x-ray film for 2 or 3 days at -38°C. The
autoradiographs from four such experiments are illustrated in
Fig.1.

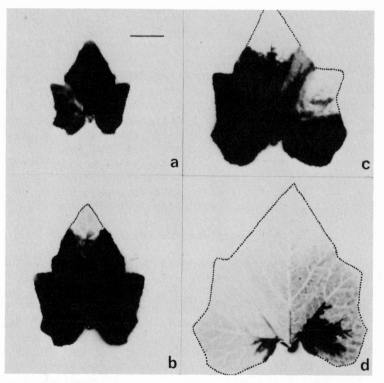

Fig.1a-d. Autoradiographs of lamina 5 of four Cucurbita pepo
plants which imported ^{14}C from leaf 3 for 2 hr. Leaf ages were:
(a) LPI 0.2. (b) LPI 0.4. (c) LPI 0.8. (d) LPI 1.1. Scale = 2 cm.

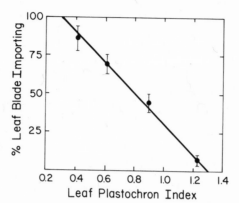

Fig. 2. Development of leaf 5 expressed as the percentage of the total lamina length still importing ^{14}C. Vertical bars equal twice the standard error of the mean.

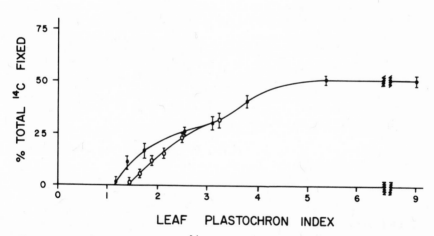

Fig. 3. The amount of ^{14}C translocated from the lamina of leaf 5 in 2 hrs. expressed as a percentage of the total ^{14}C fixed. Either the entire lamina (●) or just the extreme base of the lamina (O) was allowed to fix $^{14}CO_2$.

Progressive basipetal loss of import capacity is linearly related to LPI and therefore to time (Fig.2). The leaf tip stops importing at LPI 0.3 (10% leaf expansion) while the basal region stops at LPI 1.3 (45% of leaf expansion). Basipetal loss of import capacity continues from the leaf base through the length of the petiole beginning at the junction of the lamina and petiole at LPI 1.3 and reaching the petiole-stem junction at LPI 2.0 (Turgeon and Webb, 1973).

While these experiments clearly demonstrate that import capacity is lost in the basipetal direction they do not determine whether cessation of import is immediately followed by the beginning of assimilate export. To examine this question, export from a localized region of the lamina was measured. All but the extreme base of lamina 5 was excised and the remaining basal tissue was exposed to $^{14}CO_2$ for 5 min. Two hours later the ^{14}C in the rest of the plant, including the petiole of leaf 5, was counted. The lamina base begins to export when import stops at LPI 1.3 (Fig.3). This suggests that, throughout the lamina, cessation of import is immediately followed by assimilate export. However, when intact leaves are exposed to $^{14}CO_2$, export from the lamina does not begin until LPI 1.1 (Fig.3). Since the leaf tip stops importing at LPI 0.3 this implies either that cessation of import is not immediately followed by export at the tip or that initial export by the tip is not noticed because the material exported is retrieved by the still-importing basal region. The latter hypothesis was confirmed after exposing only the extreme tip of the intact leaf to $^{14}CO_2$ and detecting transport to the base by whole leaf autoradiography (Fig.4). Transport from the distal to the proximal regions of the lamina could be detected as early as LPI 0.7. When crushed ice was applied to the lamina below the area exposed to $^{14}CO_2$, transport to the base was inhibited. This response of intralaminar transport to temperature change is indicative of phloem transport (Webb, 1970).

Intra-laminar transport provides nutrients for the importing region of a developing lamina. In similar circumstances the ^{14}C-labelled compounds initially exported by the lamina are retrieved primarily by the developing petiole (Fig.5) which also continues to import from other leaves until LPI 2.0. As petiole 5 decreases its capacity to import translocate from other leaves during the period LPI 1.3-2.0 the percentage of ^{14}C exported through the petiole from lamina 5 to the rest of the plant steadily increases.

B. Carbon balance and sugar synthesis

The transition from the heterotrophic to the autotrophic state is the direct result of increasing photosynthetic potential and

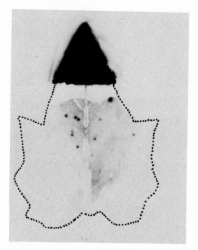

Fig. 4. Autoradiograph of lamina 5. Although the leaf blade was too young to export (LPI 0.7) ^{14}C was present in the midrib and lamina outside the fed tip-region 2 hrs. after $^{14}CO_2$ was administered.

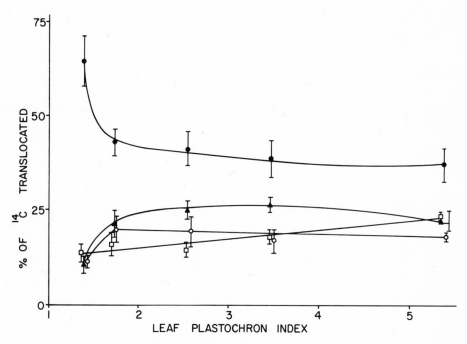

Fig. 5. The amount of ^{14}C translocated in 2 hrs. from lamina 5 to: petiole 5 (●), above leaf 5 (▲), between leaf 5 and roots (□), and roots (O).

decreasing requirements for organic nutrients to sustain a rapid rate of leaf growth. Export begins when the lamina achieves a net positive carbon balance. This relationship can be summarized for a 24 hr. period by the following equation:

$$\text{Net photosynthesis} = \text{Respiration} + \text{Growth} \pm \text{Translocation}$$

Where net photosynthesis is the net weight of carbon assimilated during the light period, respiration is the weight of carbon respired during the night period, growth is the increase in weight of carbon within the lamina in 24 hrs. and translocation is the weight of carbon either imported (-) or exported (+) during the same period.

We measured rates of photosynthesis, respiration and growth in growing leaves and calculated the contribution to the carbon balance of lamina 5 during the period from unfolding to maturity. For these studies plants were grown in a controlled environment chamber. Light was supplied by incandescent lamps at an intensity of 110 microeinsteins m^{-2} sec^{-1} for 18 hrs. each day. The temperature in the cabinet was $26^{\circ}C$. Under these conditions the duration of one plastochron was 1.62 days. Photosynthesis and respiration were measured by infrared gas analysis. Growth was measured by dry weight increase and assuming the dry weight to be 45% carbon (Stout, 1961). The quantity of carbon translocated was calculated by difference from the above equation.

In preliminary experiments it was shown that the rate of photosynthesis remains approximately constant throughout the light period under the growth conditions employed. Night respiration in mature leaves is at first equal to the rate of dark respiration during the day but declines rapidly to a lower, constant level until the beginning of the next light period. In immature importing leaves, rates of respiration during the day and night are approximately equal and are higher, per unit surface area, than in mature leaves (Fig.6). Net photosynthesis becomes measureable at LPI 0.1 and increases to a maximum rate at approximately LPI 2 before slowly declining.

Carbon balance calculations are presented graphically in Fig.7. Until approximately LPI 0.3, when leaf tip stops importing, the amount of carbon translocated into the lamina is equal to that incorporated as dry weight. After LPI 0.3 an increasing percentage of the carbon required for growth and night respiration is derived from photosynthesis until at LPI 1.1 import stops and export begins.

The lamina undergoes the transition from import to export during the period of declining growth rate. This is especially clear when carbon balance calculations are expressed in relative or

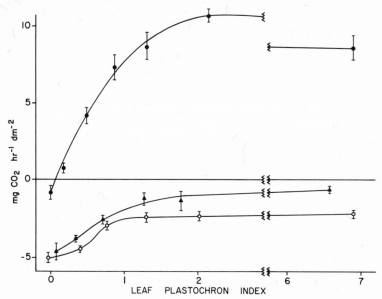

Fig. 6. Net photosynthesis (•), night respiration (▲), and dark daytime respiration (O) of lamina 5 plotted against leaf age.

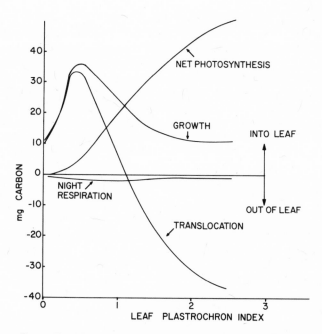

Fig. 7. Carbon balance of lamina 5 expressed as the instantaneous rate at which carbon is lost or gained (dw/dt) over a 24 hr period. Rates for net photosynthesis and night respiration were calculated for 18 hr and 6 hr periods respectively.

percentage terms (Fig.8). The rate of relative growth decreases
rapidly from LPI 0.3 to LPI 1.3 and it is during this period that
import is gradually terminated.

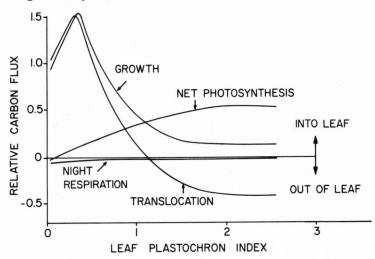

Fig. 8. Relative carbon balance (dw/dt)x(1/w) of lamina 5.
A value of 1.0 on the ordinate indicates a 100 % increase in weight
of carbon in 24 hrs.

      The preceeding calculations are derived from measurements made
on the whole lamina. However, the import-export transition is
gradual and therefore it is necessary to consider local carbon
balance. Since the lamina acquires photosynthetic potential in the
basipetal direction (Larson et al, 1972) and growth also proceeds
basipetally (Avery, 1933) there is reason to believe that the above
equation may be applied, to a first approximation, to local regions
of the lamina and even to single mesophyll cells. If this
assumption is correct the achievement of the autotrophic state by
single cells occurs following the period of intensive relative
growth. There is an important advantage to this pattern of
development. Early, intensive, growth is sustained by imported
nutrients. Therefore, during this critical phase food reserves of
the plant act as a nutrient buffer. At night, and during the day
when photosynthesis is restricted by sub-optimal conditions, an
adequate supply of nutrients is maintained. It has already been
seen that the rate of night respiration is equal to the rate of
dark respiration during the day in importing leaves but is only
one-third the rate of dark daytime respiration in mature leaves
(Fig.6). It is during the import-export transition (LPI 0.3-1.3)
that the difference between day and night respiration first becomes
apparent.

It is interesting and perhaps relevant to this discussion that the early period of intense relative growth in many animal species is also supported by imported nutrients. Calculations of relative growth rates from the primary data of Byerly (1930) and Streeter (1920) show that hatching of the chick and the birth of the human infant occur only after the initial rapid rate of relative growth has, in each case, declined to a low, slowly decreasing value.

C. Sugar synthesis

Webb and Gorham (1964) have indicated that the transport sugars stachyose and raffinose are probably first synthesized at approximately the same time as export from the leaf commences.   We re-examine this correlation using the LPI as a more accurate

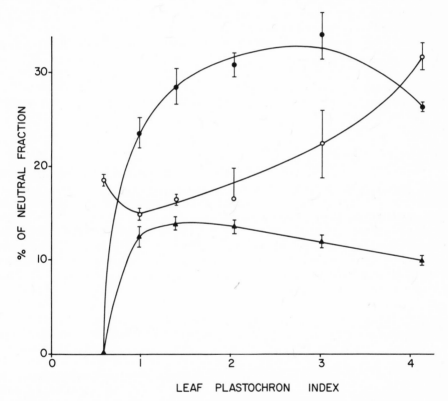

LEAF PLASTOCHRON INDEX

Fig. 9. ^{14}C in the translocation sugars: sucrose (O), stachyose (●) and raffinose (▲) expressed as a percentage of the total ^{14}C in the neutral fraction. Lamina 5 was exposed to $^{14}CO_2$ for 5 min. and extracted with 80% alcohol. Extracts were passed through cation and anion exchange resins and the sugars in the neutral fraction separated by paper chromatography using N-propanol:ethyl acetate:water (7:1:2) as solvent.

measure of leaf age. ^{14}C-sucrose is synthesized from $^{14}CO_2$ by very
young leaves soon after unfolding (Fig.9). However, ^{14}C-stachyose
and ^{14}C-raffinose are not detectably synthesized under similar
conditions until LPI 0.6. This is 0.3 plastochrons after the leaf
tip stops importing. It should be recalled that intra-laminar
transport from the distal to the proximal region of the leaf is
first detectable at LPI 0.7.

These results suggest that stachyose and raffinose are
specific transport sugars in this species, specific in the sense
that their primary function is in translocation. This idea is
supported by the fact that stachyose and raffinose are slowly
metabolized in mature leaves while in the importing leaf they are
readily hydrolyzed (Webb, 1971). Arnold (1969) suggested that
sucrose and sucrose derivatives are preferred transport sugars
because they require hydrolysis before use and are therefore
protected derivatives of glucose. Stachyose and raffinose require
in addition an α-galactopyranosidase for complete hydrolysis and
this provides an added metabolic constraint. Perhaps these
metabolic constraints are of greatest value in the process of vein
loading rather than during long distance transport. Suppression of
α-galactopyranosidase activity enables stachyose and raffinose to
be transported from the mesophyll to the minor veins through what
would otherwise be an enzymatically hostile environment.

D.   Ultrastructure of the minor veins

Before export can begin mesophyll cells must presumably have
access to a functional sieve tube conduit. We undertook a
microscopic examination of minor vein maturation during the
import-export transition to determine the degree of correlation
between initial export and the structural development of sieve
tubes surrounding the areole (Turgeon et al: in press). Prior to
this study we examined the ultrastructure of minor veins in mature
leaves. Fischer (1885) studied the minor veins of mature leaves
with the light microscope.

Minor veins of C. pepo are arranged in the reticulate pattern
typical of dicotyledonous leaves. Terminal vascular bundles end

---

Figs. 10-13: Fig. 10. Transsection of minor vein nearing
maturity. Cell types, from top to bottom: adaxial companion cell
(cc), adaxial sieve element (se), parenchyma (P), tracheids (T),
abaxial sieve elements (SE), intermediary cells (I), bundle sheath
cells (BS). One tracheid has deposited secondary wall material but
is still filled with cytoplasm. Clusters of plasmodesmata traverse
the common walls between intermediary cells and bundle sheath cells
(arrowheads). Scale = 2.5μ. Fig. 11. Paradermal section of an
immature adaxial sieve element. Fibrillar (arrow) and tubular

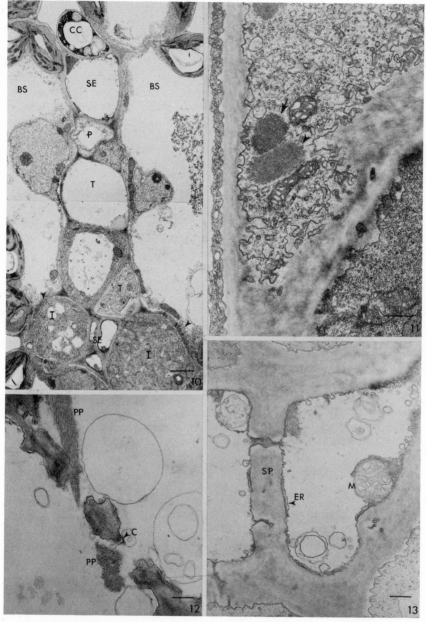

(arrowhead) P-protein bodies are present in the cytoplasm. Scale = 0.5μ. Fig. 12. Paradermal section of sieve plate between two mature adaxial sieve elements. Callose at C. Tubular P-protein at PP. Scale = 0.25μ. Fig. 13. Paradermal section of a sieve plate (SP) between two mature abaxial sieve elements. The sieve plate pore is filled with P-protein. Profiles of endoplasmic reticulum (ER) line the walls and sieve plate. Mitochondrion at M. Scale = 0.5μ.

blindly within many areoles. Cell arrangement within the minor
vein is highly regular (Fig.10) and this greatly facilitates the
examination of veins in paradermal section. The minor veins are
bicollateral. Adaxial (upper) phloem consists, in transection, of
a single companion cell and directly below it a single sieve
element of approximately equal diameter. A parenchyma cell is
situated between the adaxial sieve tube and the xylem which, in
seventh order veins, is composed of one file of tracheary elements.
The abaxial phloem, which lies directly below the xylem, consists
of one to three sieve tubes and associated companion cells. The
companion cells are large and dwarf the diminutive sieve tubes and
termed them Übergangszellen (intermediary cells).

Sieve elements in the minor veins are structurally similar,
except for their reduced size, to those in the phloem of the shoot.
Sieve plates separate the sieve tube members and are callose-lined
(Figs.12 and 13). The sieve elements are enucleate and lack
tonoplast and ribosomes. P-protein occurs in both tubular (19-19.5
nm diameter) and fibrillar forms. The ultrastructure of sieve
elements in the adaxial phloem resembles that of the
extrafascicular phloem of the stem. Aggregates of fibrillar and
tubular P-protein remain undispersed at maturity and the sieve
plate pores, which are wide and easily observed in the light
microscope, are usually unobstructed (Fig.12). In the abaxial
phloem, fibrillar but not tubular P-protein is dispersed at
maturity. The pores of the sieve plates in abaxial sieve elements
are narrow and visible only in the electron microscope. They are
typically plugged with P-protein in our preparations (Fig.13).
Both adaxial (Fig.10) and abaxial (Fig.13) sieve elements are
connected to their respective companion cells by plasmodesmata
which are branched on the companion cell side.

Intermediary cells are structurally distinct from the typical
companion cells of the larger veins and the stem. They are much
larger than associated sieve elements. The cytoplasm of
intermediary cells contains numerous vacuoles and is densely packed
with ribosomes (Fig.10). Mitochondria are primarily located close
to the plasmalemma suggesting that they supply energy for a
membrane mediated process.

The common walls between intermediary cells and bundle sheath
cells are traversed by clusters of plasmodesmata (Fig.10). Over
300 plasmodesmata have been counted in a single cluster. Fischer
(1885) noted that intermediary cells are restricted to those veins
which are embedded directly in the mesophyll (fourth order and
smaller). In the larger of these veins intermediary cells abut
bundle sheath cells and, as in the smallest veins, are connected to
them by characteristic clusters of plasmodesmata. The presence of
intermediary cells probably defines those veins capable of
appreciable vein loading.

Evidence of extensive symplastic connection between companion cells and the bundle sheath contrasts sharply with a report that plasmodesmata do not join these cell types in species where the walls of the companion cell are greatly ingrown forming the so-called transfer cells, (Gunning et al, 1968). Wall ingrowths apparently facilitate cell to cell transport across the apoplast by increasing the surface area of the plasmalemma. In $\underline{C}$. $\underline{pepo}$, where intermediary cells are not transfer cells, the symplastic connection between intermediary cells and the bundle sheath is extreme. It appears therefore that two routes for the movement of solute across the bundle sheath-phloem parenchyma boundary are possible depending on the species. In $\underline{C}$. $\underline{pepo}$ there is an uninterrupted symplastic pathway from the mesophyll cells to sieve elements. Whether or not transport takes place along an entirely symplastic route in this species remains an open question and one that cannot be answered by structural evidence alone.

While the abaxial phloem is apparently specialized for vein loading the role of the adaxial phloem in mature minor veins is unclear. Assimilate flow could probably take place in these elements since the contents of the cells often show the characteristic signs of damage to conducting phloem elements due to pressure release upon cutting. Perhaps the adaxial phloem retrieves a limited amount of assimilate from the free space as does the path phloem (Bieleski, 1966), thereby acting as a supplementary exporter.

Minor vein development

We use the technique of whole leaf autoradiography to correlate the structural maturation of the minor veins with the beginning of assimilate flow. The mature third leaf was exposed to 14-Carbon dioxide for 5 min. Two hours later leaf 5 was excised and two sections of the lamina, one from the distal and one from the proximal region, were removed and processed for microscopic examination. The remainder of the lamina was immediately frozen and autoradiographed as described above. This procedure made it possible to refer to the autoradiograph and determine in which direction assimilates were moving at the time the tissue was fixed. Glutaraldehyde (6%) and osmic acid were used as fixatives and the tissue was embedded in epoxy resin. Serial sections were cut to 1μ thick, placed on glass slides, and stained with methylene blue. When required thin sections were obtained and examined in the electron microscope. The vascular elements of selected areoles were studied in detail by thorough examination of serial sections in the light microscope. Schematic diagrams of single areoles were reconstructed at the levels of the adaxial sieve elements, the xylem and the abaxial sieve elements. Representative diagrams are drawn in Fig.14 along with the autoradiographs of the leaves from which the tissue originated. Variability is low and the data

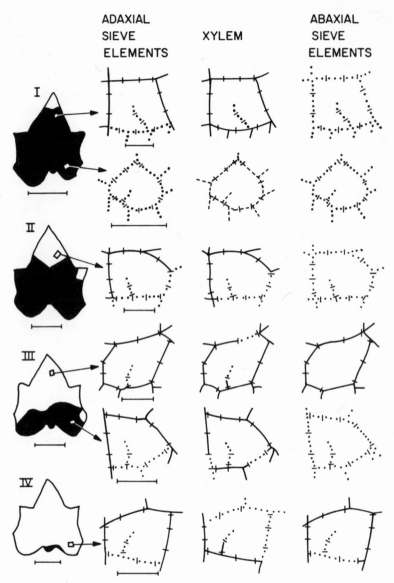

Fig. 14. Schematic outline drawings of developing areoles in paradermal view. Autoradiographs from which the samples were taken are drawn at the left. Arrows point from the position in the leaf where the sample originated to a series of drawings of a representative areole from that sample. The areoles are drawn at three levels: adaxial sieve elements, xylem, and abaxial sieve elements. Lines normal to the longitudinal axes of the cells indicate the position of end walls. Scales beneath autoradiographs = 4 cm, beneath areoles = 100μ. Fig.15 is a light micrograph of the areole from the proximal position of leaf III at the level of the adaxial sieve elements.

obtained from one or two areoles gives a good indication of the state of maturity in minor veins throughout the sample. Where the abaxial phloem contains two or three sieve elements, these elements mature together.

Prior to the transition from import to export, cell division stops. At this point elements of the vascular system are already formed. Maturation of both adaxial and abaxial sieve elements proceeds continuously from the largest toward the smallest veins. Sieve elements in the terminal veinlet are last to mature. Xylem maturation also proceeds, in general, from the largest toward the smallest veins but is not always continuous (e.g. leaf III proximal sample).

The relationship between structural maturation and initial export is best examined in the proximal region of the lamina where the import-export transition is uncomplicated by intra-laminar transport. Approximately 0.8 plastochrons before tissue at the leaf base stops importing (leaf I) the vascular tissue in this region is quite immature. A few of the tracheary elements have begun to lay down secondary walls. The adaxial and abaxial sieve elements appear undifferentiated in the light microscope although P-protein is evident in electron microscope views (Fig.11). Shortly before the leaf base of leaf III begins to export the adaxial phloem matures to the extent that sieve elements with open sieve plate pores surround most of the areole (Fig.15). Similarly many mature tracheids are present in the xylem. However, in the abaxial phloem all sieve elements are filled with cytoplasm. The presence of mature adaxial sieve elements surrounding the areole suggests that during this period the adaxial phloem is the route of solute import. This recalls the experiments of Peterson and Currier (1969) who found that fluorescent dye travels toward the leaf blade in the internal (adaxial) phloem when applied laterally to bidirectionally-transporting petioles of Ecballium (Curcurbitaceae).

The appearance of structurally mature sieve elements in the abaxial phloem coincides with initial assimilate export (leaf IV). In the light microscope these sieve elements appear empty of cytoplasm. When viewed in the electron microscope some, but not

---

Legend: Adaxial sieve elements - with open sieve plate pores (——), with closed sieve plate pores (....). Xylem - tracheids with secondary walls, no cytoplasm (——). tracheids with secondary walls, containing cytoplasm (....). tracheid precursors without secondary walls (----). Abaxial sieve elements - with no cytoplasm visible in light microscope (——). filled with cytoplasm (....).

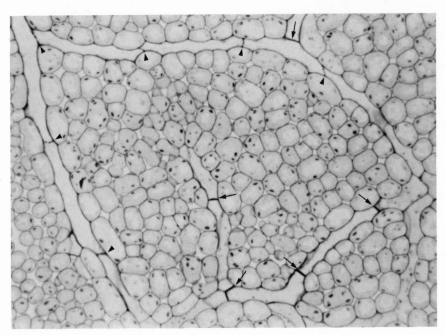

Fig. 15. Paradermal section of developing areole in the
plane of the adaxial sieve elements. Light micrograph. Pores in
sieve plates between most of the sieve elements are open (arrow
heads) but toward the terminal ending they are closed (arrows).
This areole is drawn schematically in Fig. 14 (proximal position
leaf III).

all, of the sieve plate pores of these cells have been perforated.
In certain areoles it was difficult to find sieve plates with open
pores in sixth or seventh order veins. It should be recalled
however that veins of the fourth and fifth order contain
intermediary cells and are probably capable of loading assimilate
from the limited number of mesophyll cells to which they have
access. When export begins the sieve plate pores of fourth and
fifth order veins are open. It seems likely that a limited amount
of photoassimilate is loaded into the abaxial sieve elements of the
larger veins at the beginning of the export period. As the minor
veins mature a larger number of phloem elements begin to load and
export.

The same developmental pattern is apparent in the distal
region of the lamina except that maturation of the abaxial sieve
elements does not occur as soon as import stops (leaf II). Here it
should be recalled that there is a brief period of approximately
0.3 plastochrons between the end of the import period and the first
detectable intra-laminar transport originating at the leaf tip.

It could be inferred from these studies that export is initiated by the structural maturation of the minor veins. This was the conclusion Isebrands and Larson (1973) reached from a study of Populus deltoides leaves. We feel, however, that this is an oversimplification. Any number of developmental advances toward the initiation of export, such as the synthesis of transport-specific sugars, could also be construed as initiating steps. It is more likely that progress toward the fixation of excess carbon is paralleled by the orchestrated maturation of interrelated structural and physiological systems involved in its removal. There is no reason, at present, to suspect that any one particular part of this developmental system is of paramount importance in the sense that it performs a specific regulatory role.

References

ARNOLD, W.N. 1969. J. Theoret. Biol. 21: 13-20

AVERY, G.S., Jr. 1933. Amer. J. Bot. 20: 565-592

BIELESKI, R.L. 1966. Plant Physiol. 41: 447-454

BYERLY, T.C. 1930. Jour. Morph. and Physiol. 50: 341-359

ERICKSON, R.O., and F.J. MICHELINE, 1957. Amer. J. Bot. 44: 297-305

ESAU, K. 1965. Plant Anatomy 2nd Ed., John Wiley and Sons, New York.

FISCHER, A. 1885. Ber. Verh. Kon. Sachs. Ges. Wiss. Liepzig, Math-Phys. Cl.37: 245-290

GUNNING, B.E.S., J.S. PATE and L.G. BRIARTY, 1968. J. Cell Biol. 37: C7-C12

ISEBRANDS, J.G. and P.R. LARSON, 1973. Amer. J. Bot. 60: 199-208

JONES, H. and J.E. EAGLES, 1962. Annals Bot. 26: 505-510

JONES, H., R.V. MARTIN and H.K. PORTER. 1959. Annals Bot. 23: 493-508

LARSON, P.R., J.G. ISEBRANDS and R.E. DICKSON. 1972. Planta 107: 301-314

PETERSON, C.A. and H.B. CURRIER. 1969. Physiol. Planta. 22: 1238-1250

STOUT, P.R. 1961. Calif. Fert. Conf. Proc. 9: 21-23

STREETER, G.L. 1920. Contrib. Embryol. 11: 143-170

THROWER, S.K. 1962. Aust. J. Biol. Sci. 15: 629-649

TURGEON, R. and J.A. WEBB. 1973. Planta 113: 179-191

TURGEON, R., J.A. WEBB and R.F. EVERT. Protoplasma: in press.

WEBB, J.A. 1970. Can. J. Bot. 48: 935-942

WEBB, J.A. 1971. Can. J. Bot. 49: 717-733

WEBB, J.A. and P.R. GORHAM. 1964. Plant Physiol. 39: 663-672

Discussant: C.A. Swanson
              Botany Department
              Ohio State University
              Columbus, Ohio 43210

Eschrich: Cucurbit leaves are very good starch formers.   At what developmental stage is leaf starch formed in the chloroplasts? Does it disappear during the night?

Turgeon: Starch is formed much earlier than I expected. Translocation goes on during the night from the mature leaves into the immature leaves and other regions of the plant. I would assume that unless there is an extremely large pool of sucrose and other transport sugars, starch is probably broken down.

Srivastava: Would you explain xylem development a bit more?

Turgeon: Xylem does not mature continuously like phloem. There does not seem to be a significant correlation between the loss of import capacity or the onset of export and xylem maturation.   In the last fully mature tracheid, I noticed a large amount of material that has been pushed up, presumably, from disintegrating tracheids below it.

Kollmann: How did you differentiate between mature and immature sieve elements in the minor veins? Do the outermost sieve elements, which you called immature, remain immature or do they get mature by opening and enlargement of the sieve pores during further differentiation of the leaf?

Turgeon: In the fully mature leaf, the maturation of the veins goes to completion and every single sieve element in the minor veins is structurally mature, that is, it has open pores, very little cytoplasm, no tonoplast, and shows surging. I would hesitate to make the assumption that the leaf will not export while the sieve pores are closed, however.   Export is a developmental process.   You expect tissues to mature when they are going to be used, so it's difficult to put that interpretation on it.

Kollmann: I think the outermost sieve elements, which you call immature, are really mature, but have very narrow pores which can not be identified with a light microscope.

Turgeon: They showed no further structure of the sieve element pores.   There were plasmodesmata between the sieve elements, which had P-protein bodies that were beginning to hydrolyze, and there were callose platelets on the pores that had

315

not yet been perforated. I don't mean to suggest that immature sieve elements can not translocate. If such sieve elements were given sugar, for example, and loaded, they might translocate. There is no way of determining in a developmental sequence like this whether immature sieve elements can or can not translocate. All one can say is that structurally mature ones can. A mature leaf has only mature sieve elements, one above and two below the xylem, in the vein endings.

Aronoff: Is it possible that export develops because of developing photosynthetic capacity and that structural maturity is only incidental? Are the colourless parts of variegated leaves fed intralaminarly or do they continue to import?

Turgeon: We are dealing with a developmental sequence and it is not valid to think of it only in terms of limiting factors. These things naturally coincide. Dr. Geiger suggests that in the sugar beet leaf everything comes to maturity but export does not begin, even after loading begins, until a certain osmotic pressure is built up. That's not entirely contrary to my view. It is, perhaps, more regulatory than developmental. If you put a leaf in the dark, the osmotic potential would eventually disappear, but the sieve elements wouldn't de-differentiate. I think of that as a regulatory process, and all the rest as developmental processes.

Geiger: Since this is an evolved system that has been successful, you'd expect a lot of correlation between preparatory events to occur. As judged by autoradiography, exogenous sugars supplied to maturing leaves do not show phloem loading until a certain stage is reached. The top, then the middle, then the bottom show the minor veins standing out as accumulating. Thus, you can supply sugar to immature veins and it will not load and be exported. it is not limiting.

Aikman: When an immature leaf is importing, through what is it importing? Immature sieve elements? You say that immature sieve elements are connected by plasmodesmata and do not translocate - yet plasmodesmata in any other situation would be assumed to be involved with transport.

Turgeon: The vein of _Cucurbita_ is bi-collateral. The correlation between export and structural maturation of the abaxial sieve elements is quite good. The adaxial sieve elements were mature before that, and it is possible, although there is no direct evidence, that the adaxial sieve elements were importing. Peterson and Currier have data which indicates that in Ichbellium in the Cucurbitaceae, the internal phloem imports while the outer phloem exports. In sugar beet there is no adaxial and abaxial phloem so, perhaps, that is why Geiger and Fellows find the finest fifth order vein mature during the importing phase - maturation takes place earlier to allow importing to occur.

Swanson: This seems to connect three generalizations: (1) acropetal maturation of major veins; (2) basipetal maturation of minor veins; and (3) maturation of the adaxial and abaxial axis.

Schmitz: In Cucurbita, translocation towards the base of the plant occurs mainly in the outer, external or abaxial phloem and towards the apex in the adaxial phloem. This has also been shown very clearly for the Solanaceae. Unidirectional differentiation may also be true in the leaf.

Turgeon: Autoradiographs of Cucurbita leaves show both adaxial and abaxial phloem are involved in export. Whether adaxial phloem imports and abaxial phloem exports needs proof. Once the leaf begins to export, I imagine that adaxial phloem is a supplementary path for export.

Moorby: In cucumber, the start of export seems to correspond pretty closely, as judged from your plastochron area measurement, with cessation of cell division in the leaf. Moreover, the export of ions, certainly phosphate, seems to start about the time that cell expansion ceases.

Turgeon: (Pointing to a slide) In this tissue which is still importing, maturation of the adaxial sieve elements is starting but the abaxials are immature.   Cell division has ceased and the development of intracellular spaces has begun even earlier - quite some time before export begins.

Lamoureaux: Was steam or ice used to girdle the bi-collateral bundles of Cucurbita? What was the zero point for your plastochron index, first leaf cell divisions in the apical meristem or when the leaf becomes visible?

Turgeon: It was steam. Erickson and Michellini originally used leaf cell divisions in the apex to define a plastochron. The index was evolved to avoid cutting sections.   An index leaf is described and called zero plastochrons old, and when the next leaf reaches the same size as the index leaf, this is one plastochron. There is a mathematical way to get between integral plastochrons.

Fellows: It should be remembered that veins mature at different times in a very small leaf - the largest order first and then the smallest orders - and there is a sort of meeting back together.   Third- or second order branchings in a small leaf are much closer together in a small leaf than in a large leaf and are more mature than the smaller order veins. These are the veins that are actually involved in importing whereas the onset of export involves the smaller order veins. Bi-directional movement probably occurs in the higher order veins rather than the smaller veins.

Turgeon: In young, importing leaves, intercellular spaces are small. There is a lot of cell contact. Solutes probably don't have to move to the end of a mature vein to be unloaded and move around the lamina fairly easily. In fairly mature leaves there may be 25% air space.

Swanson: There are experimental conditions when we cannot use the leaf plastochron index (LPI) and must express development as percent of final expansion, instead. What is the comparability of these two scales?

Turgeon: If you're not too concerned with precision, there's not a great difference, especially during the logarithmic growth phase. The nice thing about the LPI is that you don't have to follow it in the leaf you are studying and can have a continuous index through to maturity. It is difficult to get meaningful expansion statistics when a leaf is almost mature.

Dunford: I got the important point that abaxial phloem maturity is correlated with onset of export. Could there be loading without export? Stachyose and raffinose don't appear until export begins. Is this because they are actively hydrolyzed in very young leaves?

Turgeon: I found intralaminar transport to the base at about 0.4-0.5 plastochrons in a leaf that had stopped importing at the tip at 0.3 plastochrons. I'm uncertain whether this represents a true lag or inability to detect transport because there is not much of it. Transport sugars are not synthesized or don't accumulate until 0.4-0.5 plastochrons. Again, there seems to be a slight lag, which may or may not be significant, of 0.2 plastochrons (0.4 days or 9-10 hours). It could very well be that ^{14}C-stachyose and ^{14}C-raffinose could not be detected in young leaves because they were hydrolyzed as fast as they were made. There could be loading without export.

Swanson: Dickson and colleagues at Rhinelander, Wisconsin, assert that there is no intralaminar transport. Any comment?

Turgeon: Fellows and Geiger are reporting in Plant Physiology that they could not find intralaminar transport such as I have found in a cucurbit leaf. Dickson's paper is unsatisfactory. They studied transport over an 8 plastochron span in one plastochron intervals and I suspect they missed it altogether since they supplied $^{14}CO_2$ to the leaf and checked for differences in gradients. Maybe intralaminar transport occurs only in leaves that have the proper vascular connections. We did another interesting experiment. By cutting away all the basal tissue, but leaving the tip and midrib intact, we were able to induce precocious transport out of these leaves that was statistically significant.

Ho: There are errors involved in the carbon economy approach of Terry and Mortimer for measuring rates of carbon translocate. These can be overcome by using freeze drying instead of oven drying, a wet combustion method for determining organic carbon, and recognizing that the carbon: dry weight ratio is not a constant.

Swanson: During the early stage a leaf is completely heterotrophic, photosynthesis just balances the accrued respiratory losses of day and night. Dry weight increase in such a leaf is due to import translocation in a net sense. When Stella Thrower darkened young leaves, why did they not become better or remain equally good as sinks? Instead, they diminished as sinks and abscissed.

Turgeon: I've confirmed her experiment. The darkened leaf didn't grow, because expansion is highly dependent on light. Not only that, it was yellow and, eventually, died. This implies more photomorphogenetic reactions.

MacRobbie: What happens if you darken a leaf that has just begun to export. Is it still able to import or has it lost the capacity?

Turgeon: It is very difficult to make an exporting leaf into an importing leaf. I think certain hormone treatments might cause the leaf to import again but I don't know. This preventive mechanism is important from an ecological standpoint. Once a leaf is mature and begins to be shaded by leaves above, it would otherwise become a natural sink and draw food away from the growing regions.

Swanson: Dickson has shown that export from one mature leaf to another is extremely small and principally by way of the xylem.

Gorham: Dr. Webb, who opened up export and import studies with cucurbit leaves, found no import by mature leaves. As development proceeded, the next, immature leaf that was formed would import from the various mature source leaves.

Swanson: There was, as I recall, slight import detected, by way of radial movement from phloem to xylem, then up the xylem to leaves or various ages. It took quite a while and required very sensitive detection methods.

Turgeon: It doesn't take any more than 2 hrs. Autoradiographs take a long time to show the small amount of labelled compounds that are imported via the transpiration stream.

Gorham: In answer to Dr. MacRobbie's question, a fully mature leaf is not an effective sink, even if you darken it. When measuring intralaminar transport, the leaf is still attached to the plant. Unlabelled translocate must be coming from external sources while labelled translocate is coming from the tip. That is one reason why you detected so little labelled translocate, I would suggest. Did you try a detached leaf?

Turgeon: No. I have a philosophical prejudice against working with detached leaves.

Aronoff: I'm disturbed by Dr. Gorham's remark. Does it imply that nothing of importance comes from the roots to the leaves?

Turgeon: No. We were only considering phloem transport. Certainly, substances come from the root but mainly in the transpiration stream.

Aronoff: What fraction of the import is coming from the root to a new leaf and an old leaf (which must import things from the root throughout its life)?

Gorham: This depends on elapsed time. It takes a matter of hours for translocate to circulate from a source leaf to the roots, become elaborated primarily into amines, and for these to be transported back up in the transpiration stream to both young and old leaves.

Turgeon: Right. In two hours, the material imported in the transpiration stream by leaves may not have come from roots. Maybe it will have come from lateral displacement within its own petiole, node, or stem.

MacRobbie: What sort of change has occurred in a leaf that was able to unload the phloem and, suddenly, is not? Does it tell us anything about connections between the phloem and the other cells in the leaf?

Geiger: The change seems to be in the high osmotic pressure that develops in the conducting cells once the leaf gets to a certain stage of maturity. When importing is going on, osmotic pressure is quite low - about equal to that in the mesophyll. I have no data on the effect of darkening.

Reinhold: The change is really a secondary effect. Translocate is following growth and expansion of the sink leaf which will be hormone-controlled - but not by one simple hormone. Under some conditions, kinin induces a darkened leaf to import.

Turgeon: A darkened leaf must metabolize and, eventually, exhaust its own reserves, even without expansion.

Gorham: Leaves must be darkened quite awhile to bring reserve levels down. It takes all night to export the reserves accumulated during the day. Eight or nine hours of darkness are required before the rate of translocation from a mature sugar beet leaf falls off significantly.

Cataldo: Darkened sinks, such as cereal florets, cease to grow and to import. My idea is that in darkened tissues there is a drop in the normally high rate of nitrate reduction needed for growth. Since there is very little transport of reduced nitrogen in young, growing plants, there won't be growth, no matter how much carbon is available.

Weatherley: Dr. Mittler has autoradiographs which suggest that aphids cause import of ^{14}C into mature leaves.

Mittler: (Using slides, Dr. Mittler described an experiment using a mature kidney bean leaf attached to the plant. A 2 min. label of $^{14}CO_2$ was applied to the middle leaflet. During the following 8 hrs., 25-30 aphids were caged in a planchet on the under side of one of the lateral leaflets. The aphids were removed and the ^{14}C in them and the honeydew determined. The leaf was freeze-dried and autoradiographed. There was ^{14}C in the aphids, in the honeydew, and in the veins of the most lateral leaflets but none in the control lateral leaflet. The aphids converted a source into a sink.)

Aronoff: Does wounding of any leaf cause the same effect or is it unique to aphids? Would squeezing, irritating or anything mechanical normally unacceptable to a leaf produce the same result?

Mittler: This was not tried. There is no question that the aphids created an appreciable sink and were superimposing their own will on the system. It would represent a considerable proportion of the available assimilate, but an abnormal situation which the plant has learned to cope with to a certain extent. Too many aphids, however, and the plant succumbs.

Turgeon: Fungal infections will apparently do the same thing.

Aikman: Are aphids selective with respect to exporting vs importing phloem?

Mittler: It might be possible to find out with a cucurbit leaf. The consensus would be that aphids feed near sinks, competing with natural sinks since they feed, in general, near growing parts. Young leaves and senescent leaves are colonized and aphids grow quite well. They may insert their stylets in the actively dividing cambial layer, rather than the adjacent undifferentiated sieve tubes of willow or Metasequoia, it being a tissue the plant needs to supply with nutrients from a distance up or down or laterally.

Gorham: Since stylets pierce and cause exudation, aren't aphids opening the system to atmospheric pressure and diverting normal flow as a result? Since aphids are capable of sucking, are they, perhaps, actually creating a suction force which accounts for some, if not all, of the diversion of translocate into what was, initially, a source leaf? I would think that was the most plausible explanation.

Mittler: It would be interesting to see whether cut stylets could cause the same diversion. Contrary to earlier views, aphids can suck since they feed quite well on an artificial diet without any pressure. Their sucking ability, I would imagine, enhances their uptake and therefore the drain which they impose on the plant.

Weatherley: Notwithstanding what Drs. Mittler and Gorham have just said, the fact is that reduction of turgor has caused an import into that leaflet that was photosynthesizing, and, therefore, mainly exporting. The aphids caused a movement against the stream. Here, we've got a case of bi-directional movement, in different sieve tubes, of course, helped a bit by any suction possibility the insects may have. It is a reduction in turgor that we're suggesting causes an ingress into that leaflet. Perhaps the opposite is true. Normally the sieve tubes can't import into mature leaves because turgor pressures are too high - as Dr. Geiger implied.

Heyser: Maybe our results will help to answer this question. (Using slides, Dr. Heyser described a series of experiments using 8 replicate strips of maize leaves of uniform age exposed in a special, compartmented chamber, to different external conditions of light and CO_2 concentration at the basal end, in the middle, and at the apical end. Following a variety of pre-treatments to inhibit photosynthesis, sections were labelled with $^{14}CO_2$ for twenty minutes and examined by autoradiography after 2 hrs. to see where the labelled photosynthate had been translocated. Regardless of morphological base or apex, labelled assimilates went to the end or section which was unable to do photosynthesis. Storage leaf starch had to be degraded by a 48 hr. dark pre-treatment to obtain the most reproducible results.)

Swanson: That was a very interesting contribution to intralaminar translocation. Does it apply only to monocotyledons or can it be demonstrated in dicotyledons?

Heyser: Mostly to parallel-veined monoctoyledons. In dicots you always have veins going to the side and, therefore, not convenient.

Schmitz: May I present further evidence that it is possible to induce a sink by darkening? (Using slides, Dr. Schmitz described ^{14}C translocation in phylloids of Laminaria marginata. Translocation occurred only to the base in light, but 1-2% or 15% to the apex after 12 hr. or 48 hr. predarkening, respectively.)

Willenbrink: In 1956 or 1957, Bachofen induced translocation from illuminated Phaseolus vulgaris fruits to predarkened leaves.

van Die: Miss Hartt reported that darkened portions of sugar cane leaves imported sugars from illuminated parts and we also found it in Yucca leaves. Perhaps it has something to do with monocotyledons.

Heyser: We were unable to confirm Miss Hartt's experiments using cut leaves on so-called phototranslocation - perhaps because of the starch content of the leaves.

Swanson: (Using a slide, Dr. Swanson compared data for 4 species on the development of export and import capacity as a function of per cent of final leaf length. Sugar beet leaf, when it is about 15% of final size, is importing at about 10% of maximum, whereas a squash leaf is importing at or near its maximum at the same stage of development. Sugar beet leaf has reached its full export capacity at 60% of its final size, whereas a soybean leaf doesn't reach full export capacity until near final size and a squash leaf until after it has reached final size.)

Troughton: Dr. Heyser, have you done the same experiments on intact maize leaves? Our results, to date, on shading different parts of intact maize leaves are in conflict with yours. It only induces extremely low import along the length of the leaf.

Heyser: We haven't done such experiments. It's not possible to handle every external condition if the leaf is attached to the plant.

Moorby: How efficient was the translocation system?   The speed Karl April's student found was about two orders down from what we're getting with intact maize.

Heyser: We did some experiments feeding labelled sugar to the 25 cm strips of maize leaf through the troughs at the ends.   The sugar in the sieve tubes reached the middle of the strip after 20 min., from which we could calculate the speed. The experiment was done in the dark.

Swanson: Dr. Moorby has observed speeds of about 4 cm/min., as I recall.

Moorby: We have measured speeds of 2-4 cm/min. in plants that have been in the light for a long time and 0.5 cm/min. in plants that have been in the dark for two days. Speeds doubled soon after we put the lights back on, built up over a period, then dropped again when the plants were darkened.

Aronoff: If you darkened the middle part towards the base and fed $^{14}CO_2$ at one end and ^3H_2O at the other end would the sugars cross?

Heyser: We did a similar experiment. We put 0.75 M labelled sucrose at one end and 0.75 M sucrose with a little bit of label at the other end. The heavy label reached the middle after 45 min. and stayed there for 24 hrs., moving only 1 cm further along.

Mittler: (Using slides of work done in collaboration with Dr. Leonard, University of California, Davis, Dr. Mittler outlined aphid experiments based on the maize leaf strip method described by Dr. Heyser. Aphids caused ^{14}C-assimilate to move either basally or apically, depending on where they were caged. Likewise, aphids were able to counteract the pull of dilute sucrose or mannose solutions applied at one end and water at the other on the movement of ^{14}C-assimilate from the middle.)

Quebedeaux: Dr. Turgeon, what are some of the changes that occur later on in development after a leaf has been a source?

Turgeon: Does it become a sink later on? I think not. Photosynthesis begins to decline almost as soon as it reaches a peak and, eventually, reaches zero. The leaf then senesces and dies without becoming an importer again. Hormonal treatments, aphids, a fungus infection or something that creates an artificial sink can, apparently, pull material in again. There seems to be a mechanism to prevent old or shaded leaves from becoming a liability.

Walsh: There would be approximately 100 vascular bundles in maize strips that are 1 cm x 25 cm. Did you notice whether shorter strips had shown movement?

Heyser: No. We always used 25 cm strips. But bigger bundles translocate further than smaller ones.

Fellows: Dr. Turgeon, if we can have a sink and a source within the same developing leaf or create them in a mature leaf, what type of translocation mechanism would you favour for export from a leaf once it has become a source?

Turgeon: While there is symplastic connection all the way from the chloroplast to the sieve element, I have no evidence for symplastic transport. It may be apoplastic, at least part of the way. There are no differences in the structure of sieve elements that would lead me to believe that there would be any difference in the translocation mechanism in the leaf and the rest of the plant.

Gorham: Dr. Trip and I used the Biddulph flap-feeding technique followed by freeze-drying and autoradiography to examine the loading of ^{3}H-sucrose by the minor veins and neighbouring intermediary (ubergangen) cells of a squash leaf. Pronounced loading of the intermediary cells was observed but no loading of the sieve elements between them. However, back along the vein, beyond the region of the labelled intermediary cells, there were labelled sieve elements. There appeared to be longitudinal transport between the intermediary cells as a first step and then phloem loading further along. I think this is of some interest.

Swanson: I agree. Dr. Geiger has comments along this line, and I would ask him to point out possible artefacts he has encountered in autoradiography.

Geiger: In the freeze-drying of sieve tubes, it seems that the sugars have a hard time staying in the middle (of the lumen), they seem to get swept out, creating the appearance that there isn't anything there. With the degree of resolution that is possible, I've abandoned freeze-drying for this reason. This is a possible explanation for Dr. Gorham's observation. I don't know if that's the whole answer.

In trying to make gray strips for quantitative autoradiography, we dried some leaves, soaked them in sucrose, rinsed, heat-dried, and autoradiographed them. To our surprise, the minor veins showed up very nicely.

Heat drying of leaves at temperatures higher or lower than 56°C can create artefacts that look like vein loading. This may just be absorption onto colloids in the tissues. We're using freeze-drying when it's feasible.

Aronoff: How then can you locate sucrose autoradiographically in a reliable manner?

van Die: By freeze-drying whole leaves at -80 or -100°C before you autoradiograph them.

Turgeon: That technique may be suitable for histoautoradiography but not for whole leaves. I just freeze them and carry out the whole autoradiographic procedure at -40°C, without drying, and no loss of resolution.

Eschrich: There should be a control, exposed for 2 days, to make sure there is no blackening of, say, the veins, caused by chemical reactions.

Turgeon: Correct. Fortunately, this is not necessary with cucurbit leaves. Chemical reactions can be prevented by interposing a layer of cellophane between tissue and film, but this, of course, decreases resolution.

Translocation Kinetics of Photosynthates

Donald B. Fisher

Department of Botany, University of Georgia

Athens, Georgia 30602

It is hard to imagine what working on translocation must have been like before radioactive tracers were available. So much of the information we have, particularly about physiological aspects, is based on tracer methodology that it would be difficult even to list it. Nevertheless, one of the most interesting - and important - promises of the tracer approach to translocation has not been fulfilled. Even though the first experiments with radioctive carbon dioxide were run more than twenty years ago, there is no general agreement on what translocation kinetics have to say about the translocation mechanism. It hasn't been possible to reach a consensus even where such different possibilities as "active diffusion" and plug flow are concerned. Nevertheless, I personally feel that this confusion is much greater than it needs to be, and that the fundamental reasons for the observed kinetic patterns can be identified with a fair degree of confidence.

The sort of experiment I will be mainly concerned with is one in which a leaf is labelled with radioactive CO_2 and the distribution kinetics of radioactive photosynthate are followed by some means or other. One approach is simply to cut the translocation pathway into segments and plot the amount of activity versus distance from the fed leaf. This approach was first used in 1952 by Vernon and Aronoff (30) in their studies of translocation in soybean, and this kind of a plot is referred to as a "translocation profile". Vernon and Aronoff found that there was a roughly exponential decrease in activity with increasing distance

1. The unpublished work reported herein was conducted jointly by the author and A. Lawrence Christy, Thomas L. Housley or William H. Outlaw, as indicated.
2. The author's work has been supported by National Science Foundation Grants GB14719 and GB33903.

from the fed leaf. This sort of distribution has been observed frequently since then, not only with ^{14}C but with other tracers. However, I will be concerned mainly with those experiments in which a leaf is labelled with radioactive CO_2. An explanation for the kinetic behaviour of labelled sugars need not apply to other compounds.

Since the logarithmic distribution of tracer is so frequently found, it has been very tempting, on the part of some workers at least, to believe that it has some special significance with regard to the way in which sugars move in sieve tubes, particularly because it suggests a diffusion-like mechanism of transport. This was all the more agreeable, since the idea of "active diffusion" as a transport mechanism had its supporters, and supporting data, even before radioactive tracers came into use. Nevertheless, I feel that the exponential shape of the translocation profile has been a red herring, and has nothing to do with the way in which sugars are transported in sieve tubes.

The factors which might affect the kinetics of labelled photosynthates in the path fall into three categories. The first is simply the loss of tracer from the translocation stream. Certainly the labelling of compounds like CO_2 and starch must be taken to indicate a loss of tracer from the sieve tubes, and these compounds do, in fact, appear during the movement of labelled photosynthate through a stem or petiole. The second category is the possibility that there might be a range of translocation velocities, including movement in opposite directions. In considering the implication of velocity gradients, one must obviously distinguish between different velocities in different sieve tubes and different velocities in the same tube. Certainly the conclusions one would draw about the transport mechanism would be very different in the two cases. As far as velocity differences in the same tube are concerned, there is certainly one effect that can be eliminated immediately. If solution flow does occur in sieve tubes, the effects of any accompanying velocity profile would be undetectable. Radial diffusion is comparatively so rapid in tubes of this diameter that all of the tracer would move at the same average velocity. This is apparent from a comparison of the diffusion coefficient to the sieve tube diameter, and has been treated quantitatively by Taylor (28).

Finally, there is the question of the rate at which labelled photosynthate enters the translocation pathway from the source. Vernon and Aronoff recognized that at least two important factors contribute to this rate. First of all, since different leaf areas are different distances from the petiole, they will not be making equal contributions of ^{14}C to the translocation stream as it leaves the leaf. I've called this the effect of "kinetic size", and have shown by modelling that the effect lasts for about the same time it takes for tracer to reach the petiole from the farthest part of the leaf (10). Second, there is the problem of how quickly the labelled sugars move from their sites of synthesis into the translocation stream. Vernon and Aronoff described this as "the rate of diffusion of photosynthate to the sieve tubes," but "diffusion" is not a good description, since it implies a particular mechanism of movement and excludes other effects like compartmentation. Furthermore, as far as the kinetics in the translocation stream are concerned it doesn't make any difference how tracer gets to the sieve tubes. A more useful way of stating the question with respect to translocation kinetics is simply: "What are the kinetics in the source pool for translocation?" It is this kinetic behaviour which will determine the rate at which tracer enters the translocation stream. As it turns out, these last two factors, kinetic size and source pool kinetics, which together determine the rate at which radioactive sugars enter the translocation pathway, are mainly responsible for the kinetics along the pathway.

For the moment, however, I'd like to address only one question, that is: "Do all sugar molecules in the translocation stream move with the same velocity?" Or is there a range of velocities? The reason for stating the problem in this way is that it gives us something fairly simple to look for in interpreting kinetic studies of translocation. Consider for a moment what happens when you drive your car; if all goes well, all of its parts move with the same velocity and it has the same appearance when you reach your destination as when you started. If some parts of the car move with respect to its other parts - that is, if different parts move at different velocities - it will take on a very different shape from when you started. In fact, the two ways of looking at the problem are identical; an object will change in shape if and only if its different parts move (have different velocities) with respect to each other. Asking the question, "Do all sugar molecules move with the same velocity during translocation?" is tantamount to asking "Does the translocation profile change in shape as it moves along the pathway?" One question is precisely the same as the other.

Before looking at experimental data, let me illustrate more plainly the points I'm trying to make. In the first place, the determination of profile shape can be approached experimentally from several directions. Since each approach gives somewhat different kinds of data we have to know how those results can be compared. In the second place, the analogy between the translocation profile and a moving car is not an entirely valid one.

Fig. 1 illustrates some of these points. For purposes of illustration, tracer is assumed to enter the pipe at a linearly increasing rate, followed by a decrease in rate which is also linear but with only one-third the slope. Consider for the moment only the first part of the pipe, where the velocity is the same throughout. These kinetics are shown in the first curve in the

lower graph. Obviously the same kinetic curve would result if we
held a Geiger tube at this point. In fact, since we are assuming
that all of the tracer molecules move with the same velocity, one
would observe the same results at any point along the pipe. The
only difference would be a transport lag equivalent to the distance
divided by the velocity. With this sort of movement, there is also
a simple relationship between these kinetics and the distribution
of tracer along the pipe. This is shown in the upper graph. The
stippling in the pipe under the graph is meant to indicate the
tracer distribution represented in the graph. This distribution is
simply a mirror image of the curve which describes the rate of
tracer entry into the pipe. The curve describing tracer
distribution will be stretched or compressed relative to the
kinetic curve, depending on whether the velocity is greater or less
than 1 cm min^{-1}.

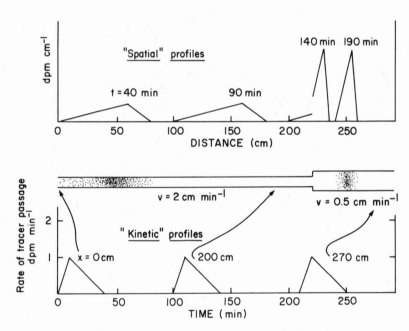

Fig. 1. The relationship between the kinetics observed at
some   point along the path (the "kinetic" profile) and the
distribution of tracer along the path (the "spatial" profile)
during the plug flow.

Because both of these curves are, in a sense, "translocation
profile", one with respect to distance and the other with respect
to time, I've tried to distinguish between them by referring to one
as a "spatial translocation profile" and to the other one as a

"kinetic translocation profile". (These terms are synonomous with Canny's "translocation-distance profile" and "translocation-time profile" (2).) Obviously, the theoretical relationship between the two will depend on one's interpretation of translocation kinetics, but I think that it is worthwhile to think in terms of both kinds of kinetics. I hope the utility of the terms will become more apparent in the discussion that follows.

One of the reasons for introducing the idea of a kinetic profile is illustrated by the profile behaviour in the wider part of the pipe. If we are dealing with a bulk flow mechanism, there ought to be changes in velocity where there is a change in the total cross-sectional area of the stream. We would also find a change in shape of the spatial translocation profile in the sense that, if the area increased, it would be narrower and taller. In this case, since there is a hypothetical four-fold increase in the cross-sectional area, the height increases by four-fold and the length is cut by one-fourth. Not that the total amount of tracer in both profiles (the area under the curve) is the same, as it should be. In the analogy with a moving car, just the opposite would occur: its spatial profile would be the same no matter how fast it was moving, but if we plotted the height of the car versus time as it passed some particular point (that is, its kinetic profile), its maximum height would always be the same, but its

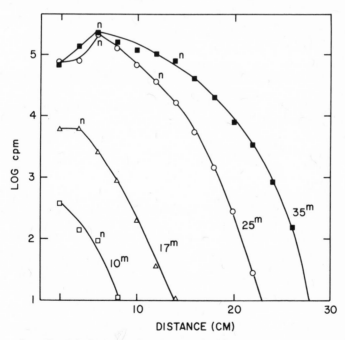

Fig. 2. Spatial translocation profiles in soybean stems at various times during steady state labelling of the oldest trifoliate leaf with $^{14}CO_2$ (from Fisher (8)).

length (measured in time) would depend on its velocity. The reason, of course, is that a car's shape is defined in spatial terms; it has a fixed height and fixed distance between its parts. A translocation profile, on the other hand, is defined in kinetic terms. Its appearance in space is very much dependent on its velocity even if we're talking about plug flow. When I ask the question, then, "Does the translocation profile change in shape?", I am referring to the kinetic profile, although it will also refer to the spatial profile if the velocity does not change with distance. The problem I am addressing is not whether there are velocity changes along the path but whether there is a velocity gradiant across the path. Wardlaw (31) has, in fact, demonstrated that sudden velocity change occurs in wheat in passing from the sheath to the stem, and it probably also occurs in soybean in passing from the petiole to the stem.

There is one other way of approaching translocation kinetics which should be mentioned. This was introduced by Geiger and Swanson (15) in their studies on translocation kinetics in sugar beet, and Larry Christy and I have used a modified version of it to follow translocation kinetics along morning glory stems. If one follows the kinetics of total tracer in a sink, the rate at which it enters the sink may be determined from the slope. Geiger and Swanson's plants were trimmed so there was only one sink leaf, which they monitored with a Geiger tube. In our morning glory experiments, we monitored sink leaves at various points along the stem. Consequently, we were tapping only a small fraction of the passing activity.

Figures 2-4 provide some selected illustrations of kinetic behaviour by translocation profiles. Figures 2 and 3, although appearing at first glance to be quite different, actually demonstrate equivalent kinetics in soybean plants, but results from quite different approaches. The profiles in Fig. 2 are spatial profiles which were found at various times after steady-state labelling single leaves with $^{14}CO_2$ (8). The curves in Fig. 3 are taken from the work of Moorby, Ebert and Evans (22), and represent various positions along the stem of a plant which had been pulse labelled with $^{11}CO_2$. To understand the kinetic similarity between Figures 2 and 3, one must realize that the response observed in a pulse labelling experiment will be proportional to the first derivative of that seen in a steady-state labelling experiment (12). In a steady state labelling experiment the curves expected on the basis of the data in Fig. 3 would be exponential at first, changing to a linear form later. (This was, in fact, demonstrated by Moorby, Ebert and Evans; conversely, a linear spatial profile was demonstrated by Fisher 25 min after pulse labelling.) This shape would be the mirror image of the data, represented in Fig. 2, which show exponential shapes near the front, but become more linear near the leaf in longer experiments. However, the main

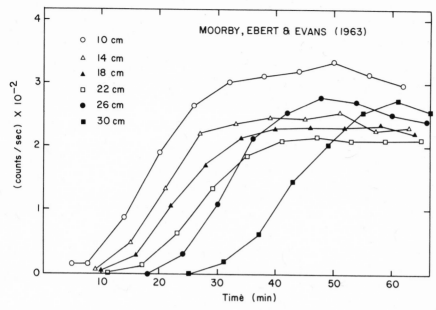

Fig. 3. Kinetic translocation profiles at various distances along a soybean stem after pulse labelling with $^{11}CO_2$. (Redrawn from Moorby et al (22).)

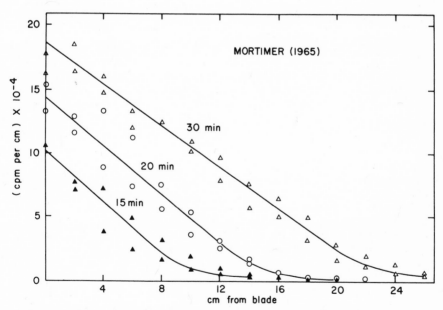

Fig. 4. Spatial translocation profiles in sugar beet petioles at various times after pulse labelling the leaf with $^{14}CO_2$. (Redrawn from Mortimer (23).)

point to be taken from Figures 2 and 3 is simply that no systematic changes in shape occur during the movement of the profile along the translocation pathway.

A similar conclusion may be drawn from Mortimer's observations (23) on the behaviour of translocation profiles in sugar beet petioles after pulse labelling (Fig. 4). Other reports clearly demonstrate similar behaviour (15, 20, 21, 24, 31, 33) and none seriously conflict.

The foregoing data justify my contention, at least to a reasonable approximation, that the range of translocation velocities cannot be very wide. But more precise data do, in fact demonstrate the occurrance of a range of velocities. More and Troughton's (20) recent observations on kinetic profiles in corn leaves after pulse labelling with $^{11}CO_2$ clearly showed profile spreading. Qureshi and Spanner (24) found a slow but consistent decrease in the slope of spatial profiles moving along Saxifrage stolons after pulse labelling with $^{14}CO_2$. They attributed it to the ocurrence of different velocities in parallel sieve tubes.

In work done recently by Larry Christy and me, we, too, observed limited, but definite, spreading of the translocation profile. Our work was done with morning glory vines, which we chose for two reasons. First, one of the principal difficulties in understanding translocation lies in explaining how it can proceed over such long distances. Second, many of the measurements which must be made to describe accurately the characteristics of translocation are made more readily with long paths. Where translocation kinetics are concerned, the longer path allows a longer time to observe the kinetics during transport, including the visualization of the entire spatial translocation profile. One of the problems with the short paths that have been used (most are only about 30 cm long) is that only part of the spatial translocation profile can be visualized at a time.

The vines we worked with ranged from about 2.5-5 meters long. During growth the lower mature leaves were continually removed, leaving about five or six mature leaves near the apex. Before the experiment was started the growing tip and all mature leaves except one were removed. Thin end-window Geiger tubes were used to monitor expanding leaves on side shoots at various points along the stem. As mentioned earlier, the rate of ^{14}C arrival in the sink leaf was taken to indicate the ^{14}C content of the translocation stream at that point. The data are expressed as the per cent of the final amount in the leaf which arrived per minute after the source leaf was pulse labelled with $^{14}CO_2$. By expressing it in this manner the data are normalized to a form which allows a theoretically sound basis for comparing data between different sinks. Some typical results from such an experiment are shown in

Fig. 5.   Except for a slow widening, the shapes of the kinetic
profile   seen   at   successive   nodes   were   quite similar.   The
translocation velocity in these experiments, calculated from the
time   at   which tracer first appeared, was usually about 0.7-0.8 cm
min^{-1}.   There was no correlation of velocity with distance along
the   stem.   In some experiments it was somewhat faster in the upper
part of the stem, but in others it was somewhat slower or about the
same along the entire stem.

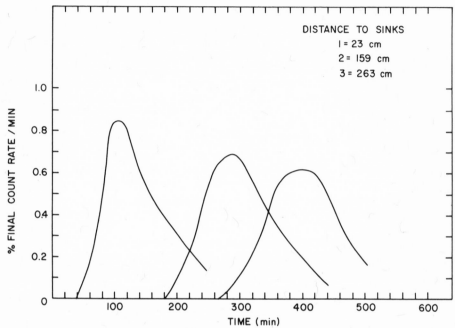

DISTANCE TO SINKS
1 = 23 cm
2 = 159 cm
3 = 263 cm

Fig. 5.   Kinetics of ^{14}C-arrival in sink leaves on axillary
shoots along a morning glory stem.   The arrival rate was calculated
from   the   kinetics   of   total ^{14}C in   the   sink leaf after pulse
labelling a mature leaf with $^{14}CO_2$.   The arrival rate is presumed
to be a measure of the kinetic translocation profile, or part of
it, at that node.

A progressive widening in the shape of the kinetic profile was
a consistent feature of all experiments   (see Fig.12 for a more
extensive summary).   We have attempted a quantitative assessment of
profile   spreading   by   assuming   that   there are perhaps different
rates of movement in different   sieve   tubes.   This   is   a   simple
proposal   to   model, and we have done so by assuming that there are
40 translocation streams moving at slightly different velocities
and   summing   their   contribution   to   the translocation profile at
several points along the stem.   Some kinetic curves based on this
approach   are   illustrated in Fig. 6.   These calculations are based
on an average velocity of 0.65 cm min^{-1} and   the   assumption   that

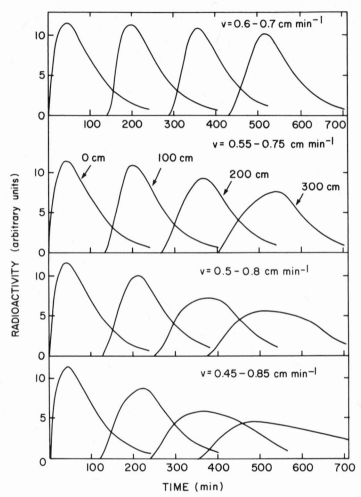

Fig. 6. Computer-based predictions of profile spreading in
morning glory stems,   using various velocity ranges and a uniform
velocity distribution.   The kinetics of the source are indicated by
the first curve, and the succeeding profiles were calculated by
summing the contributions of 40 separate streams, each moving with
its own velocity.

there is a linear velocity distribution over the velocity range. Judging from our data, which now include about 20 experiments, we would compare the profile behaviour in morning glory vines with that in the last two velocity ranges (0.5-0.8 and 0.45-0.85 cm min^{-1}), depending on the experiment. In a few experiments, the range has probably been greater.

In attempting to account for the presence of a velocity distribution, certainly one of the obvious possibilities is that there might be a dependence of velocity on the size of the sieve tube. The range of sieve tube sizes involved in translocation was determined from microautoradiographs, an example of which is illustrated in Fig. 9. These observations demonstrated that a wide range of sieve tube sizes participated in translocation, and that each size range contributed about the same fraction to the total area of the translocation stream, although there was a somewhat higher proportion in the middle range. We have made fairly complete measurements of sieve tube dimensions, and on the basis of the calculated resistance to a pressure flow mechanism, it seems that the size variation might be sufficient to cause the observed range of velocities.

We also tried to verify the predicted relationship between the spatial translocation profile and the kinetic translocation profile. In one such experiment, kinetic profiles were followed at two nodes along the stem. A third sink leaf was monitored near the base, and when activity appeared there, the experiment was terminated by cutting the stem into 2 cm segments for extraction. If the predicted relationship between spatial and kinetic profiles is accurate (see Fig.1), we should have been able to predict the position of the spatial profile along the stem. For example, since the kinetic profile showed a peak 75 minutes after detection and the translocation velocity was 0.73 cm min^{-1} without rapid spreading, the peak in the stem should have been about 55 cm upstream from the last sink, since the experiment ended just as ^{14}C appeared there. As Fig. 7 illustrates, the predicted relationship was not quite accurate. The general shape of the profile was similar to that predicted, except for the "tail" of activity at the front. As a result, the predicted position of the peak was about 30 cm ahead of its actual position. This was also the approximate position predicted from the kinetics at the second sink. We have run two other experiments like this and they, too, predicted spatial profiles which were more advanced than the actual profile. The kinetics obtained from the sink leaves obviously did not reflect those of the main profile. Instead, we apparently were seeing a minor component which moved mostly to the sink leaves at a somewhat higher velocity than most of the ^{14}C. Because of the similarity in shapes of the spatial and kinetic profiles, we feel that the conclusions we have drawn from the kinetic profiles are valid ones, but the situation was more complicated than we had hoped for.

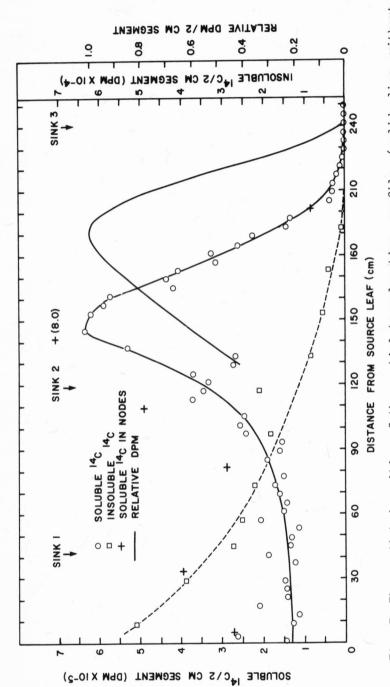

Fig. 7. The predicted position of the spatial translocation profile (solid line without points; predicted from the kinetic profiles observed at Sinks 1 and 2) compared with its actual position (solid line with circles; obtained by extraction when activity was initially detected at Sink No. 3). The kinetic profiles predicted a somewhat more advanced position of the spatial profile (by about 30 cm) than was actually found. Most of the soluble ^{14}C (about 90% in the peak region) was in the form of sucrose. Note that the scale for insoluble activity differs from that for the soluble activity by ten-fold.

One of the principal difficulties in interpreting translocation profiles, at least as they are determined by counting whole segments of the path, is whether all of the activity is in the translocation stream. The discussion so far hasn't been particularly concerned with that problem or with the indentity of the compounds involved. To a degree, this has been deliberate, because I feel that the basic features of translocation kinetics are fairly straightforward and I wanted to demonstrate how far one could go with them before complicating the picture with what I regard as second order considerations. But inevitably one must consider the question of where this activity is before kinetic interpretations can be convincing, particularly since some workers have claimed that it is this loss which is mainly responsible for the profile kinetics. Furthermore, the question of loss is important because it could lead to an osmotic gradient along the pathway, although conclusions to be drawn here also depend on what happens to the velocity and cross-sectional area of the translocation stream.

As a conservative estimate of what chemical compounds are in the translocation stream, only non-reducing sugars can be included. However, even several hours after labelling with $^{14}CO_2$ most of the activity along the translocation stream is typically in the translocated compound (e.g. Fig.7). But this doesn't give an accurate estimate of leakage because the labelled compound which accumulates in greatest abundance outside the translocation stream is the translocated sugar itself, at least for periods of up to several hours. This has been clearly demonstrated for sugar beet petioles by Geiger et al (17) and for soybean petioles (8). In the case of sugar beet, sucrose in the translocation stream accounted for less than 10% of the total ^{14}C-sucrose in the petiole after 8 hours of steady state labelling. At first glance, this might seem to demonstrate that the kinetics of loss would indeed have an important effect on translocation kinetics. Nevertheless, this is not true. To the extent that translocation kinetics are concerned with movement along the translocation stream, there was no information available on translocation after 90 minutes because there was no further change in specific activity in the translocation stream. To put it somewhat differently, the informational content of the translocation stream reached zero because the tracer in it reached a state of maximum entropy (25). If any information on translocation in the petiole were to be obtained it would have to be taken in the first 90 minutes; after that the only available information would relate to rates of leakage. At 90 minutes, however, two-thirds of the sucrose was still in the translocation stream; much higher fractions would be in the translocation sream at times closer to zero. During the first 50 minutes or so, when most of the information on movement in the translocation stream is available, it would be quite reasonable to ignore the effects of loss.

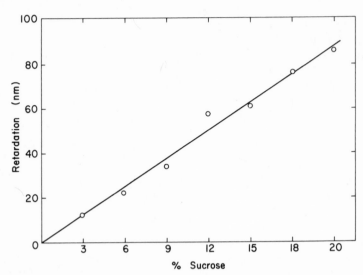

Fig. 8.   The   phase   retardation   caused   by   various concentrations   of   sucrose   embedded   in   methacrylate.   Sucrose solutions   were   infiltrated   into   pith   blocks,   quick-fozen, freeze-substituted   in   acetone   and embedded in methacrylate (11). Measurements were made in 4μ sections at 546 nm.

Table 1.   "Leakage ratios" for sugar   movement   through   sieve tubes (= amount lost per cm/amount passing through).

Leakage ratio(%)	Plant	Author
2.7	Sugar beet (petiole)	Geiger et al (17)
∿1	Soybean (petiole)	Fisher (10)
0.8	Soybean (stem)	Evans et al (6)
0.8	Sugar beet (petiole)	Mortimer (23)
<1	Wheat (stem)	Carr & Wardlaw (1)
0.6	Squash (petiole)	Webb & Gorham (32)
0.3	Morning glory (stem)	Christy & Fisher

It is worth noting that the above problem is the reason for Spanner and Prebble's failure to extract velocity from their experiments with ^{137}Cs movement through Nymphoides petioles (26). Virtually all of the observed kinetic changes took place after changes in the specific activity within the translocation stream had almost ceased. In spite of its appearance of completeness, their data were not taken over a time interval when the information for a velocity measurement was available. It related instead to leakage rates.

In general, the parameter which is of greatest importance in describing the loss of tracer from the translocation stream is the amount lost per unit length in comparison to the amount passing through that same length (Table 1). This sort of data is available in several forms. The simplest and most direct way to calculate the ratio is from a pulse labelling experiment. After several hours one can simply look at the amount of ^{14}C in some section of the translocation pathway and compare this amount to the amount of ^{14}C which passed through it. In the case of morning glory, for example, all one has to do is integrate the spatial translocation profile and compare this value with the amount of ^{14}C in a stem segment close to the leaf. On this basis, the "leakage ratio" in morning glory is only 0.3% per cm. The same approach can be taken with Mortimer's data for sugar beet petioles (23), Carr and Wardlaw's data for wheat stem (1), and Webb and Gorham's data for squash petiole (32). With soybean, the figure is about 0.8-1% per cm, based on quite different kinetic data obtained by me (10) and by Evans, Ebert and Moorby (6). The leakiest sieve tubes so far examined (at least for which there are reasonably clear data) have been in the sugar beet petioles used by Geiger et al, where the ratio was 2.7% per cm. My argument against a significant role of leakage in determining profile shape therefore applies even more strongly to these other systems.

The question of the histological localization of tracer is, of course, answered most directly by autoradiographic techniques. High resolution autoradiographs, in particular, demonstrate the almost exclusive localization of translocated ^{14}C in the sieve tubes (13, 14; Fig.9).

One of the disconcerting aspects about translocation kinetics is that different authors can look at the same kinetic data and come to diametrically opposing explanations for it. One model for translocation kinetics which is certainly very different from the viewpoint just presented is that of Canny and Phillips (5), which presumes the presence of oppositely moving transcellular strands in the sieve tube lumen. This model has so many serious problems that it probably could be eliminated on several grounds from consideration as a possible mechanism of phloem transport. But it supposedly receives its strongest support from studies on

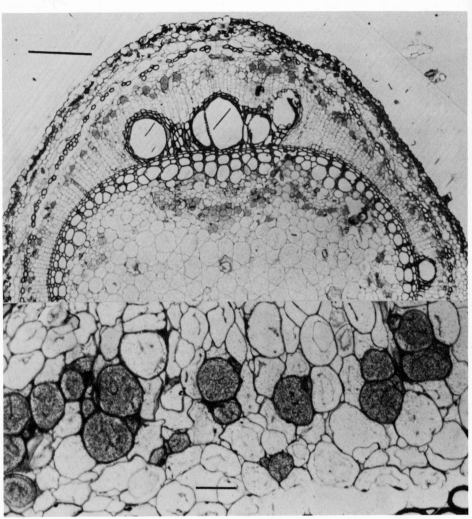

Fig. 9. Microautoradiographs of 2 μ epon sections from a morning glory stem which had been quick-frozen, freeze substituted in acetone, embedded in epon and dipped in Ilford L4 emulsion (13). A single remaining mature leaf had been steady state labelled with $^{14}CO_2$ for 6 hours. The sections were stained with methyl violet. A. Low magnification: the bar indicates 200 μ .   B.   Higher magnification: the bar indicates 20 μ.

translocation kinetics, so it seems particularly appropriate to examine its success in that context. Under certain conditions, the Canny and Phillips model predicts that the movement of tracer from a leaf should follow diffusion-like kinetics. These kinetics supposedly are describable by the error function, which Canny has systematically fitted to many translocation profiles (4). There are a few he did not fit, for reasons that will become obvious.

Probably the best data to consider are my own for soybean (Fig.2), since Canny had described it as offering the most convincing available support for his model. This is because he was able to fit successive kinetic curves with the same value for the apparent diffusion coefficient. In most cases he had to assume progressive changes, sometimes by two orders of magnitude, in the apparent diffusion coefficient, supposedly a constant. With this data for soybean he did not. But a condition which must apply if the error function is to be a valid description of the kinetics is that the level of tracer at the start of the path must remain constant, or nearly so. For example, fitting of an error profile to the curve observed at 10 min implies that the level of ^{14}C in the petiole remained constant during the entire 10 min of the experiment. Likewise, fitting the curve observed at 25 min with an error curve implies that the level of ^{14}C in the petiole remained constant during the entire 25 min. Obviously this is a gross distortion of the data; from 10 min to 25 min the level of activity in the petiole increased by almost 2 orders of magnitude.

Another serious problem with their model, and with any model that predicts diffusion-like kinetics, is that it would be impossible to propagate a pulse of radioactivity along the translocation stream. This certainly does occur in morning glory, and has been clearly demonstrated in many other species, including soybean (Fig.10), wheat (31), sugar beet (15,16,23), corn (20), castor bean (18), palm (29), squash (32), and willow (2). In the latter report, Canny demonstrated not only the movement of a pulse, but the strong resemblance between the spatial and kinetic profiles. That observation can be explained only if movement approximated plug flow with minimal leakage.

The Canny-Phillips equations do predict kinetics which are qualitatively similar to experimental behaviour, but only when changes in the source kinetics are taken into consideration. However, the equations then describe wave movement, not diffusion, and the solution to the equations arises by ignoring reverse movement of tracer in the oppositely moving strands. This amounts to plug flow with slow leakage, the kinetics of which are hardly peculiar to a bidirectional streaming mechanism.

The kind of kinetics that have been observed is consistent with any mechanism which requires unidirectional bulk flow. This

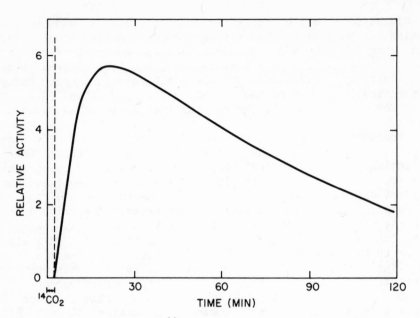

Fig. 10.   Kinetics of ^{14}C efflux from a pulse labelled soybean leaf, as followed by monitoring the petiole   with   a   Geiger   tube. (Redrawn from (10)).   Although this is not the most reliable method of determining efflux kinetics, it agrees in this case with the profile kinetics and with subsequent   determinatons   based   on   the arrival kinetics of a source leaf.

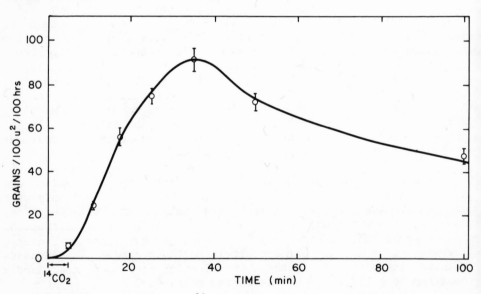

Fig. 11.   Kinetics of ^{14}C in the companion cells in the   minor veins   of   a pulse labelled soybean leaf, as   determined   by quantitative autoradiography. Note the   similarity   between   these kinetics and the efflux kinetics shown in Fig. 10.

would include osmotically generated pressure flow, peristaltic flow or electro-osmotic flow. For various reasons which extend beyond kinetic data, I'm inclined to think in terms of osmotically-generated pressure flow. But there still has to be a reasonable explanation provided for the observed spreading of translocation profiles. If it occurs because of different rates of movement in different sieve tubes, that's understandable. But if there is a detectable velocity gradient in individual sieve tubes, that's much more difficult to explain.

The following material digresses somewhat from translocation kinetics, although it is aimed in large part at obtaining kinetic data. This includes some recent developments in the autoradiography of sugars, but I particularly want to emphasize at the outset that the methodology is not limited just to autoradiography. In fact, to narrow one's perceptions that far would result in overlooking other applications which are just as useful, or even more so.

In 1967, Stirling and Kinter (27) reported a vastly improved procedure for the autoradiographic localization of sugars by embedding freeze-dried tissue in epon. Since very little sugar was lost from this material when it was exposed to water, sections could be cut onto a water surface, dried onto a slide and autoradiographed by dipping the slide into liquid emulsion. In 1970, Fritz and Eschrich (14) published an adaptation of Stirling and Kinter's procedures for plant tissue, with excellent results.

At the same time I, too, was trying to apply Stirling and Kinter's procedures to plant tissue. However, I found, at least with the procedures that I used, that freeze-drying resulted in a fairly serious artifact in that the water-soluble compounds in each cell underwent drastic shrinkage (11). This sort of artifact almost certainly explains the very patchy distribution of silver grains in almost all high resolution autoradiographs of sieve tubes, with the exception of those by Fritz and Eschrich.

I had also been trying to use freeze-substitution as an alternative to freeze-drying, primarily because it is a much easier procedure. As it turned out, freeze-substitution with acetone or propylene oxide not only allows dehydration without loss of water soluble compounds, but with sufficient care the problem of shrinkage can be avoided, since it is due to minute amounts of water in the air and in solvents. This is taken care of simply by thoroughly drying all solvents and resin monomers over molecular sieves before they contact the tissue. In addition, it is necessary to carry out all manipulations of the tissue in a dry box.

Depending on the purpose of the experiment, the tissue is embedded in either epon or methacrylate. For autoradiography we use epon, but if we want to see the water soluble compounds by phase microscopy, or if we want to extract them later, we use methacrylate.

If non-pigmented compounds are embedded in epon, there's no way of seeing them because their refractive index too closely matches that of the epon. But if methacrylate is used, the refractive indices are different, and sugars can easily be seen by their phase retardation. In fact, since phase retardation can be measured by interference microscopy, there is a reasonable possibility that sugar concentration could be measured in individual sieve elements. If one constructs a standard curve of phase retardation versus sugar concentration in sections of standard thickness, (Fig.8), one might use the retardation of the sieve tube contents to determine the sugar concentration. I've made some efforts at this, obviously, but the numbers for both the standard curve and for the sieve tube contents jump around more than I'd like. Judging from the phase retardation of the sieve tube contents in pumpkin, their sugar concentration would be about 18%. Methacrylate embedded material has been very useful for other purposes, too, but they are more readily described in the context of their application (see below).

Tom Housley and I have spent considerable effort at developing quantitative procedures for the autoradiography of sugars. The results so far have not been entirely satisfactory although, with some caution, I think they're usable. The problem lies in the variable retention of sucrose in the sections when they contact water. With material which is simply embedded in Spurr's epon mixture we have seen retention which varied from 25 to 75%, although it is usually closer to 40%. The reason for any retention at all is not clear, but it is probably not simply physical entrapment. Aqueous stains penetrate the sections fairly readily, and the sucrose which is retained is apparently chemically unreative (13). The retention might be due to reaction of some of the hydroxyl groups with epoxide groups of the resin, but that is difficult to demonstrate.

Aside from the problem of variable retention, we have worked out satisfactory procedures for the quantitative autoradiography of ^{14}C. We prepare our own stripping film by dipping parlodion covered slides in emulsion, giving a uniform emulsion layer about 0.2 μ thick. After drying, the film is stripped into a water surface and a slide with sections of measured thicknesses was coated with the film. Detection efficiency was determined from ^{14}C-methacrylate sections of known activity.

If the interpretation I have made so far of profile kinetics is at all correct, then there is a simple corollary to that interpretation: if there's not much change in the shape of the profile after it leaves the leaf, then that shape - whatever it is - must be determined by factors operating within the leaf. As I mentioned earlier, there are basically two factors which must be considered in trying to account for the rate of tracer efflux from a leaf: the kinetic size of the leaf, and the source pool kinetics. There may be others, but these two must be considered. We have tried to evaluate their role in morning glory and soybean by following the kinetics of ^{14}C in the source pool by quantitative autoradiography and, if necessary, adjusting that data to account for the effect of kinetic size.

With the relatively small soybean leaf, the velocity is too high for the kinetic size to be any more than 2-4 min (10). Consequently, the source pool kinetics should be sufficient to explain the efflux kinetics. That this explanation does in fact seem to be sufficient can be seen by comparing Figures 10 and 11.

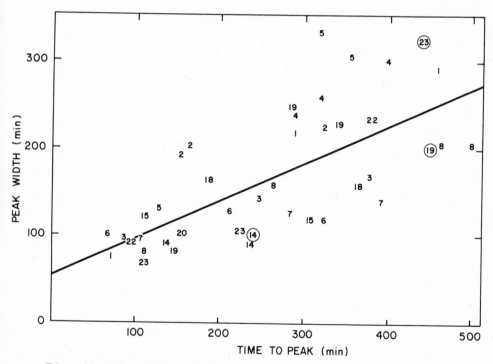

Fig. 12. Spreading of the kinetic translocation profile in morning glory stems. Numbers indicate individual experiments. The regression line has slope of 0.41 min min^{-1}. Data for some (spatial) profiles are shown by the circled numbers. They have been "converted" to kinetic profiles by correcting for velocity (see Fig. I in (12)).

Both sets of data are taken from pulse labelling experiments, and both show a roughly linear increase for about 30 min, followed by a slower decline. This behaviour is quite different from that of the sucrose specific activity, which reached a maximum in about 5-7 min and then showed an exponential decay (9).

The case of the morning glory leaf is more complicated than soybean since it's fairly large (about 15 cm long) and the velocity is probably less than 1 cm min^{-1}. It's kinetic size, then, is probably about 15-20 minutes, and this effect on the efflux kinetics will have to be considered along with the role of source pool kinetics. The efflux kinetics from the leaf were somewhat uncertain because of the profile spreading, but a reasonable judgement may be made from Fig. 12. This is a graph of the peak width, measured at half its maximum value, plotted against the time at which the maximum value occurred. The slope gives an indication of how rapidly the peak was spreading. Although one is tempted to extrapolate the curve back to the ordinate, this procedure, of course, isn't justified because even at the petiole of the fed leaf the peak in activity wouldn't come until about 35 minutes, at which time it would have a width of about 70 minutes.

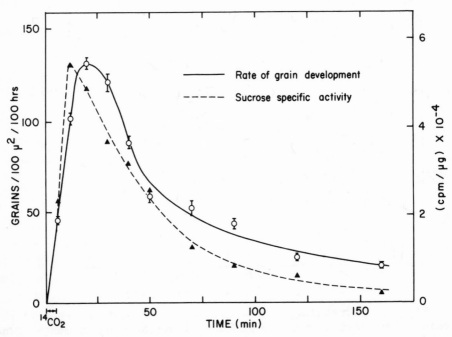

Fig. 13. Kinetics of ^{14}C in the companion cells in the minor veins of a pulse labelled morning glory leaf, as determined by quantitative microautoradiography.

The data for companion cell kinetics, as determined by quantitative autoradiography, are shown in Fig. 13, along with the kinetics of sucrose specific activity in the same leaf. In morning glory, the kinetics for these two were more similar than in soybean, although labelling in the companion cells still seemed to lag behind the sucrose specific activity. The maximum activity came at about 25 minutes. When the effect on tracer efflux of kinetic size was included (12) the predicted efflux kinetics were as shown in Fig. 14.

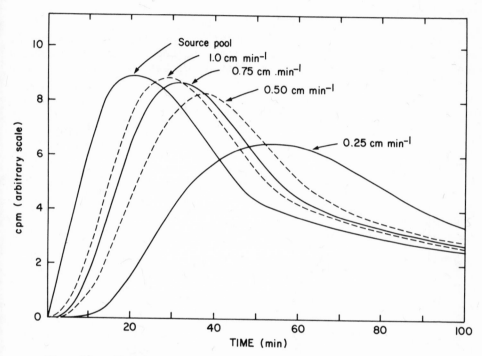

Fig. 14. The influence of kinetic size on the predicted efflux kinetics from a pulse labelled morning glory leaf, assuming the source pool kinetics in Fig. 13 and using the numerical approach described in (12).

With a velocity of 0.75 cm min^{-1}, which is representative of the velocities found in these experiments, there is a peak at about 31 minutes and a peak width of about 45 minutes. These values are about 4 and 25 minutes less, respectively, than we were looking for. At the moment we can't explain this large a difference, particularly for the peak width. One possibility may be the presence of a velocity gradient. In any case, it seems apparent that the effect of these two factors, source pool kinetics and kinetic size, which in any case must be included in trying to account for efflux kinetics, come not too far by themselves from actually explaining the observed kinetics.

A note of caution is in order, however, since the amount of ^{14}C sucrose lost from the sections is unpredictable. Nevertheless, I think that the smoothness of the data, for both soybean and morning glory, encourages some faith that this source of variability was not a serious source of error in these experiments. The experiments have been duplicated but the results are not yet available.

The foregoing data, particularly for soybean, obviously suggest that occurrence of significant sucrose compartmentation in leaves. Since there is very little direct evidence on this

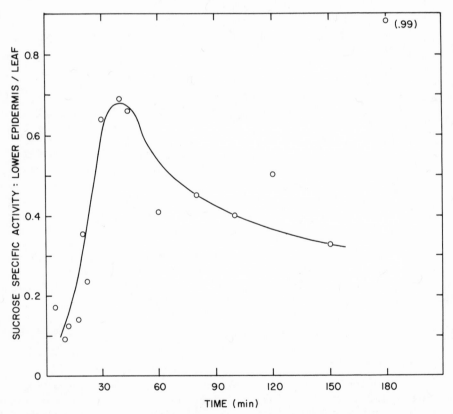

Fig. 15. Ratio of sucrose specific activity in the lower epidermis to that in the whole leaf, followed by peeling the lower epidermis from pulse labelled leaves at various times after labelling.

problem, ьill Outlaw and I examined the kinetics of
^{14}C-photosynthates in the tissues of <u>Vicia</u> <u>faba</u> leaves. We found a
fairly complex pattern of sucrose compartmentation and transport.

The reason for using <u>Vicia</u> was that we could get some data on
compartmentation fairly quickly just by peeling the lower epidermis
at various times after pulse labelling and comparing sucrose
specific activity in the epidermis to that in the leaf as a whole.
The composite results of several such experiments are shown in
Fig. 15. Since each point had to come from a separate leaflet
there was considerable scatter in the absolute values, but the
ratio of epidermal to whole leaf specific activity followed a
smooth kinetic curve. The data clearly demonstrate kinetically
distinct compartments, but it is difficult to suggest a reason for
the particular behaviour which was found.

To make a more detailed analysis of photosynthate movement in
the leaf we embedded freeze-substituted tissue from pulse labelled
leaves in methacrylate and sectioned the blocks paradermally. This
gave us sections which included only one tissue type, except for
the veins, which also included spongy parenchyma. The volume of
each tissue sample was determined by including a fluor in the
methacrylate. After the counting data had been obtained the
sections were dissolved in toluene and the amount of fluorescence
was determined. The relationship between fluorescence and plastic
weight was determined for each block by measuring the fluorescence
from sections which had been weighted on a quartz fiber fishpole
balance.

The results from a pulse labelling experiment are shown in
Fig. 16, and are expressed as the total activity in each tissue per
square millimeter of leaf surface. All samples came from the same
leaf. Although we followed the kinetics of both insoluble and
soluble ^{14}C, there was very little insoluble activity and it showed
virtually no change with time. In the mesophyll, the palisade
parenchyma initially contained more ^{14}C than the spongy parenchyma
and there was a continual decline after the labelling period.
Neither of these claims is necessarily convincing here, but it is a
fair summary of several experiments. In the spongy parenchyma
there was typically an increase in activity, and then a decline.
The data for veins was quite erratic, which is to be expected
because they are so narrow and occur at different levels in the
spongy parenchyma. It is simply impossible to get reproducible
samples of vascular tissue by this procedure. However, the basic
features of ^{14}C in the veins are illustrated here. Even though
their volume was much less than the spongy or palisade parenchyma,
they contained as much or more total ^{14}C. Secondly, the amount of
^{14}C was high soon after labelling and showed a general decline with

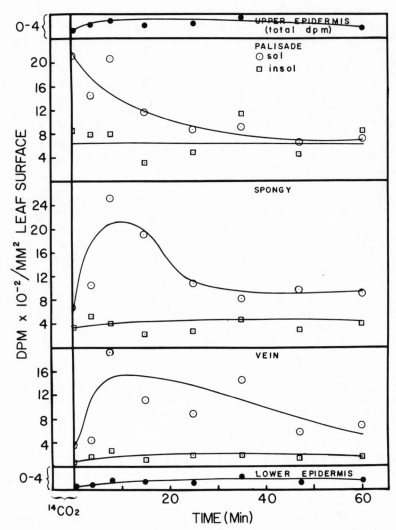

Fig. 16. The kinetics of ^{14}C in various histological samples from a pulse labelled <u>Vicia faba</u> leaflet after pulse labelling. The samples were obtained by taking paradermal sections from methacrylate-embedded, freeze-substituted leaf punches.

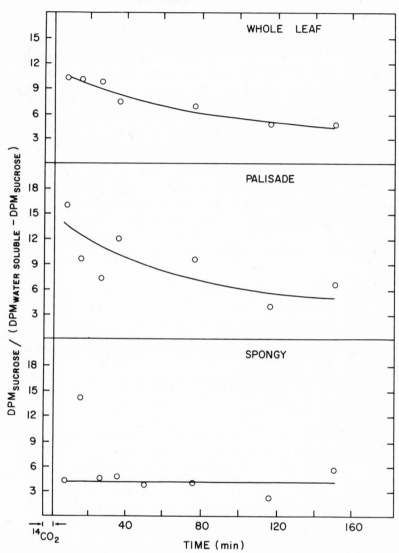

Fig. 17.   Relative changes in sucrose and non-sucrose ^{14}C in the spongy parenchyma, palisade parenchyma and whole leaf after pulse labelling with $^{14}CO_2$. Mesophyll samples were obtained as for Fig.   16.

time. Other than this general description we do not feel that much else can be said about the kinetics of total ^{14}C in the veins, particularly with regard to when it reaches a maximum. There is just too much variability in this sample.

     To get an idea of what compounds were involved in these changes we chromatographed extract from the palisade parenchyma, spongy parenchyma and whole leaf tissue. These results are shown in Fig. 17, which illustrates the kinetic changes in water-soluble compounds expressed as the ratio of sucrose to non-sucrose ^{14}C. This ratio was quite high, particularly early in the experiment, when it was about 10:1 in the whole leaf extract. In the palisade parenchyma it was even higher starting out at about 15:1 and declining to about 5:1. Since the total amount of ^{14}C in the palisade parenchyma also dropped to about 1/3 of its initial level over the same time interval, essentially all of this loss must have been in the form of sucrose. There was no apparent change in the proportion of ^{14}C-sucrose in the spongy parenchyma; likewise there was no change in total ^{14}C. We infer from this data that sucrose is the compound actually involved in the intercellular transport of ^{14}C.

     Obviously there is abundant evidence in the data presented so far for the compartmentation of sucrose in Vicia faba leaves. But data on the compartmentation is difficult to interpret without some knowledge of specific activities, and these values are difficult to get when such small quantitites are involved. Nevertheless, it is possible to assay the amounts of sucrose found in these samples of ulta-microtome sections by an enzyme cycling procedure developed by Lowry (19). The version which we used had a range of usefulness from 10^{-12} to 10^{-11} moles.

     Fig. 18 illustrates the kinetics of sucrose specific activity in the palisade parenchyma, spongy parenchyma and veins of a Vicia faba leaflet after pulse labelling. The difference in specific activity between the samples was quite striking and consistent. The sucrose specific activity in the spongy parenchyma was the lowest of the three throughout the entire experiment. The highest specific activities occurred in the sections containing veins, even though they presumably included some lower specific activity sucrose from the spongy parenchyma. Values for the palisade parenchyma were generally intermediate between the veins and spongy parenchyma. Since sucrose reached a higher specific activity in the veins than in the mesophyll tissues where it was actually synthesized, the recently synthesized sucrose must have been transported preferentially to the veins without mixing completely with sucrose already present in the photosynthetic cells. The most obvious basis for this sort of compartmentation is the vacuole,

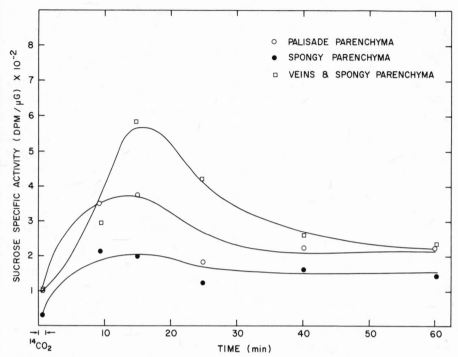

Fig. 18. The kinetics of sucrose specific activity in various histological samples after pulse labelling a $\underline{Vicia}$ $\underline{faba}$ lealet with $^{14}CO_2$. Sucrose was assayed by A.L. Christy, using a sensitive enzyme cycling procedure developed by Lowry (19).

which would represent the relatively non-mobile pool of previously formed sucrose, and the cytoplasm, which would represent the more mobile component. However, the data do not indicate whether intercellular movement actually occurred in the cytoplasm; it could just as easily occur in the cell walls.

The picture of sucrose compartmentation which emerges from our work with $\underline{Vicia}$ $\underline{faba}$ is fairly complex. There is compartmentation at the intracellular level into mobile and non-mobile sucrose pools, and there is compartmentation which arises from differences between different tissues. We refer to the latter as "histological compartmentaton" to clearly distinguish it from the more generally recognized intracellular level of compartmentation. In all, we infer 9 separate sucrose compartments from the Vicia faba data. Two are in the spongy parenchyma and two are in the palisade parenchyma. We have to guess at what the situation is in the upper and lower epidermis, but it seems reasonable to suppose that there are mobile and non-mobile pools there, too. In fact it becomes difficult to explain the generally low specific activity in the lower epidermis if that assumption is not made. This adds four more compartments. In the veins, we assume that there is only one compartment because most of the sucrose there seems to be concentrated in the transfer cells, and is in equilibrium with translocated sucrose. We have attempted a formal compartmental analysis of sucrose compartmentation in Vicia faba leaves; some aspects of it are discussed in an accompanying paper in this volume (12). Nevertheless, the model is necessarily incomplete; we simply don´t have enough data to pin down all the numbers that are needed.

As far as explaining the kinetics of ^{14}C in the source pool, the general significance of our observations on leaf sucrose compartmentation are not yet clear. In the sense that an explanation might require the existence of kinetically distinct sucrose pools in the leaf, the data clearly provide evidence for the fact of compartmentation. But even with the evidence for compartmentation that we found in Vicia faba, the kinetics of total ^{14}C in the veins were still similar to those for whole leaf sucrose, except for perhaps a slight lag. There also appears to be a general similarity between whole leaf sucrose kinetics and minor vein kinetics in morning glory leaves and apparently in sugar beet leaves (15). Only in soybean leaves, so far, does there seem to be a particularly striking difference between whole leaf sucrose kinetics and minor vein kinetics. As suggested earlier (10) this difference may be related to the presence in soybean leaves of a unique layer of mesophyll cells, the "paraveinal mesophyll", which extends between the spongy and palisade parenchyma at the level of the veins (7).

Before concluding, let me take a few steps away from the trees and point out the main features of the forest, at least as I see them. As a first-order approximation, the kinetics of photosynthate translocation along the path can be described by the simplest possible interpretation; plug flow without loss of tracer from the translocation stream. The most straighforward support for this interpretation comes from the limited change in prcfile shape during movement, from the strong resemblance between the kinetic and spatial profiles, and from the almost exclusive localization of ^{14}C in sieve tubes while the most rapid kinetic changes are occurring, as demonstrated by autoradiography. For a more accurate description of tracer kinetics, it is necessary to include, as a second-order effect, the presence of a velocity gradient - since the profile does spread slowly - and loss of tracer from the translocation stream. The actual shape of the translocation profile is determined not by events occurring along the path but by the rate at which tracer leaves the source leaf. This rate, in turn, is determined primarily by the kinetic size of the leaf and by the rate at which ^{14}C-sugars are secreted into the translocation stream from companion cells in the minor veins. The source pool kinetics may differ from those of sugar in the leaf as a whole, but the difference can be attributed reasonably to compartmentation.

References

1. CARR, G.J., and I.F. WARDLAW. 1965. Aust. J. Biol. Sci. 18: 711
2. CANNY, M.J. 1961. Ann. Bot. (NS) 25: 152
3. CANNY, M.J. 1962. Ann. Bot. (NS) 26: 181
4. CANNY, M.J. 1973. Phloem Translocation, Cambridge University Press, London
5. CANNY, M.J. and O.M. PHILLIPS. 1963. Ann. Bot. (NS) 27: 379
6. EVANS, N.T.S., M. EBERT and J. MOORBY. 1963. J. Exp. Bot. 14: 221
7. FISHER, D.B. 1967. Bot. Gaz. 128: 215
8. FISHER, D.B. 1970. Plant Physiol. 45: 107
9. FISHER, D.B. 1970. Plant Physiol. 45: 114
10. FISHER, D.B. 1970. Plant Physiol. 45: 119
11. FISHER, D.B. 1972. Plant Physiol. 49: 161
12. FISHER, D.B. (This volume, p. 513)
13. FISHER, D.B. and T.L. HOUSLEY. 1972. Plant Physiol. 49: 166
14. FRITZ, E. and W. ESCHRICH. 1970. Planta 92: 267
15. GEIGER, D.R. and C.A. SWANSON. 1965. Plant Physiol. 40: 685
16. GEIGER, D.R. and C.A. SWANSON. 1965. Plant Physiol. 40: 942
17. GIEGER, D.R., M.A. SAUNDERS and D.A. CATALDO. 1969. Plant Physiol. 44: 1657
18. HALL, S.M., D.A. BAKER and J.A. MILBURN. 1971. Planta 100: 200
19. LOWRY, O.H. and J.V. PASSONNEAU. 1972. A Flexible System of Enzymatic Analysis. Academic Press, New York
20. MORE, R.D. and J.H. TROUGHTON. 1973. Photosynthetica 7: 271
21. MOORBY, J.M., in M.J. Canny. 1973. Phloem Translocation, Cambridge University Press, London
22. MOORBY, J., M. EBERT and N.T.S. EVANS. 1963. J. Exp. Bot. 14: 210
23. MORTIMER, D.C. 1965. Can. J. Bot. 43: 269
24. QURESHRI, F.A. and D.C. SPANNER. 1973. Planta 110: 145
25. SHEPPARD, C.W. 1962. Basic Principles of the Tracer Method, J. Wiley & Sons, Inc., New York
26. SPANNER, D.C. and J.N. PREBBLE. 1962. J. Exp. Bot. 13: 294

27. STIRLING, C.E. and W.B. KINTER. 1967. J. Cell Biol.   35: 585

28. TAYLOR, G.I. 1953. Proc. Roy. Soc. A219: 186

29. VAN DIE, J. and P.M.L. TAMMES.   1964.   Acta Bot.   Neerl. 13: 84

30. VERNON, L.P. and S. ARONOFF.   1952. Arch.   Biochem. Biophys. 36: 383

31. WARDLAW, I.F. 1965. Aust. J. Biol. Sci. 18: 269

32. WEBB, J.A. and P.R. GORHAM. 1964. Plant Physiol. 39: 663

33. ZIMMERMANN, M.H. 1969. Planta 84: 272

DISCUSSION

Discussant: W.H. Outlaw
               Biochemistry Department
               Michigan State University
               East Lansing, Michigan 48823

Outlaw: Drs. Troughton and Moorby have some interesting
kinetic data to present.

Troughton: (Using a film and slides, Dr. Troughton described
facilities assembled at the Institute of Nuclear Sciences, Lower
Hutt, New Zealand, for using a 3.5 MEV accelerator or cyclotron to
produce ^{11}C with sufficient activity for translocation experiments
with maize. ^{11}C is a positron emitter, which, on annihilation,
emits gamma rays with a half-life of 20.5 min. that can be
monitored with sodium iodide crystal detectors. The whole plant
was enclosed in a plant chamber in which temperature, humidity, and
light were very accurately controlled. The transport of ^{11}C was
monitored along a leaf by detectors which were set under it at 10
cm intervals. Photosynthesis and transpiration were monitored
simultaneously with ^{11}C movement. With a PDP8 and a PDP11 computer
and a high speed plotter, data corrected for decay provided graphs
of the radioactive material as it passed along the leaf.   A 2-cm
wide strip of maize leaf was fed $^{11}CO_2$ for 2 min. and movement of
the resultant pulse of labelled assimilate either upstream or
downstream from the feeding position was followed. Counts moving
upstream were less than 5% no matter whether the leaf and the whole
plant were pretreated in the dark for 48 hrs. or whether it was
shaded upstream, shaded downstream, or the feeding chamber was
shaded. 95% of the carbon moved basipetally. Virtually all the
^{11}C fed was gone from the fed area in 1 hr. The velocities and
shapes of the moving impulses were not constant but depended on the
treatments. Using three detectors located 10, 20 and 30 cm from
the labelling area, measurements were made every 2 hrs. throughout
the day on the same leaf (possible because of the short half-life),
first thing in the morning or 2 hrs. later. There was a very
slight increase in half-width of the pulse from the 1st to the 2nd
counter, but it was negligible compared to the increase in
half-width of the feeding pulse, which was 1 min., to the
half-width of 40 min. at the first detector, which represents
loading into and along the phloem. Sometimes the half-width at the
first counter was narrower than at the third, wider than at the
other counters. By shading the fed area, but not the pathway, the
shape of the pulse could be changed dramatically, from a half-width
of 40 min. to 120 min. Speeds for one leaf were identical at the
same time on 3 successive days. Shading the upstream or downstream
portions did not affect the speed very much. By shading the area
between the feeding chamber and the first counter an enhanced speed

of 5.5 cm min.[-1] over a 20 cm distance was observed. Speeds between counters varied. In the morning, speed between counters 2 and 3 was faster than between counters 1 and 2. Speeds of 10 cm min.[-1] were recorded on 3 occasions. When the light was turned off there was an exponential decay in the speed of translocation with time. The half-time for this decay varied from 10 hrs. to 2 hrs., depending on whether the pretreatment light period was 40 hrs. or 6 hrs., respectively. Thus, there is a light effect which builds up in the light. When speed was plotted as a function of light intensity, an approximately linear relation was obtained for the range of intensities used.)

   Outlaw: (Using slides, Dr. Outlaw described the experiments on compartmentation of sucrose in leaves which Dr. Fisher had mentioned. He pointed out that histoautoradiographs published by various people have shown that veins preferentially accumulate [14] C following a pulse of $^{14}CO_2$. Fisher showed that in a steady state the specific activity of sucrose in a leaflet never reached that of the $^{14}CO_2$ with which it was supplied. The first experiments were designed to establish that compartmentation occurs between the lower epidermis and the remainder of the leaf of Vicia faba. The specific activity of the whole leaflet peaked between 10 and 20 min. after the start of pulse labelling. Then it declined rapidly, followed by a period in which decline was less rapid. The specific activity of sucrose in the lower epidermis peaked some 20 min. after the peak for the whole leaf. Similar experiments for the upper epidermis were not conclusive because of greater contamination by palisade parenchyma. They indicated, however, that the upper epidermis was in exchange with a pool of much higher specific activity than the lower epidermis. To study the kinetics of photosynthate movement in the mesophyll tissues leaflets were pulse-labelled, frozen, freeze-substituted in propylene oxide for 2 wks. at -80°C, and embedded in methacrylate containing 1% BBOT as a fluor. The tissue was sectioned paradermally and any extra plastic or cell types were trimmed away. By relating section weight to fluorescence it as possible to calculate the volume of the tissue. This was done after first extracting the water-soluble compounds and counting them, then dissolving the plastic in toluene and measuring the fluorescence. In a typical experiment, ^{14}C was initially high in the palisade parenchyma and then declined during the course of an experiment. Insoluble activity showed little change. Activity increased in the spongy parenchyma, then decreased. Activity increased slowly in each epidermis. Data for the veins were calculated by assaying ^{14}C in the spongy parenchyma plus veins and then correcting for the amount of spongy parenchyma that was present. While the results were erratic, the veins which accounted for roughly 10% of the tissue volume, had 30% of the activity. In the palisade, 95% of the ^{14}C was in sucrose; in the spongy layer, 75-80% was in sucrose. Dr. Outlaw described a simulation model that he had developed in collaboration with Drs.

Christy and Fisher, to express the most probable ^{14}C-sucrose kinetics in the Vicia faba leaflet. The model has separate compartments for the lower epidermis, palisade parenchyma, spongy parenchyma and veins, and upper epidermis. Histological pool sizes were determined. They found that it was necessary to divide the spongy parenchyma into two layers in order to simulate the kinetics and get appreciable amounts of activity into the veinal sections. Rate constants were adjusted to fit empirical data wth some restricting assumptions (e.g., the rate constant to the upper epidermis was set equal to the rate constant to the lower epidermis and turnover times for all non-mobile components were set equal). The data predicted by the model were in good agreement with the empirical data, with activity peaking in the veins about 18 min. after pulse labelling. The important finding was that changes in pool sizes didn't alter the kinetics appreciably, and this was interpretated as indicating how a leaf functions. Rapid increases in activity in the cytoplasm of the palisade parenchyma were dependent on the pool sizes assigned to the non-mobile compartment (which, they speculate, was the vacuole). Dr. Outlaw concluded by pointing out that the model was presented in terms of total activity, whereas that which Dr. Fisher had presented was in terms of specific activity. He had multiplied specific activities by pool sizes to get total activities.)

Johnson: Could you explain the sucrose assay that was used?

Outlaw: Dr. Fisher mentioned that Dr. Christy was assaying sucrose at the 10^{-12} mole level. He hydrolyzed it to glucose, used hexokinase and produced NADPH in stoichiometric amounts. NADP was destroyed by heating to 60° in pH 12. Using excess α-ketoglutarate and ammonia cycling in one direction and glucose-6-phosphate cycling in the other direction and allowing it to cycle around 6000 times, the amount of gluconate formed was related in a catalytic manner to the amount of reduced pyridine nucleotide produced which was measured fluorimetrically. The strength of the method lies in the 6000-fold multiplication factor. H. Lowry and J.V. Passonneau have published the method in a book by Academic Press entitled "A flexible system of enzymatic analysis" (1972).

Swanson: I'd like to congratulate the Australian and New Zealand plant physiologists. They now hold two records. Except for Nelson's so-called fast translocation, which is difficult to pick up on a predictable basis, phloem transport at 600 cm hr.$^{-1}$ is the highest I know of. Recently, xylem transport at 29,000 cm hr.$^{-1}$ has been reported.

Geiger: Dr. Fisher, would you comment on the localization of labelled sugar in sieve tubes in the path and in companion cells in the source? I think your slides actually indicated it was in both sieve tubes and companion cells in the path and in the source.

Fisher: Yes, in the microautoradiographs of morning glory and soybean phloem you find ^{14}C localized in the companion cells as well as in the sieve elements. The rate of exchange between the two is at least moderately rapid. In 20 min. you can see varying ratios between the two cell types. We're undecided whether the concentration in companion cells may be the same, greater or less than in sieve tubes. Gage and Aronoff, using <u>Cucurbita</u> petioles found label in companion cells but not in sieve tubes. I'm sure that this is the reason.

Geiger: Dr. Fisher, in terms of the Poiseuille profile that is often assumed, do you conceive of a very easy wall slippage as the main reason for plug flow, or do you have any ideas on how to get plug flow?

Fisher: The reason the flow appears to be plug flow is that there is rapid radial diffusion and this is essentially what Taylor has shown in his mathematical treatment of viscous flow in small capillaries.

It would be very difficult to imagine anything but the normal parabolic velocity profile in that the walls are highly hydrated, with water moving back and forth very easily, so it's not very reasonable to imagine that there is not a hydrated layer along the plasmalemma; and once this has been admitted, velocity gradients between the wall and the central part of the lumen have to be postulated. If pressure flow is accepted, it's very difficult to argue against the velocity profile in sieve tubes.

Eschrich: In the first microautoradiograph of a cross section of <u>Convolvulus</u> stem, you showed that companion cells and sieve tubes were equally labelled and later you showed a slide in which the companion cells were much more labelled than the sieve elements. (Using a slide, Dr. Eschrich showed an autoradiograph of a longitudinal section of <u>Vicia faba</u> stem in which two companion cells associated with a sieve element were labelled, one much more than the other. He stressed the importance of the observation that, even with fluorescein, companion cells are labelled, some more than others. He thought it important with regard to distribution of assimilates and also for pools, as described by Dr. Outlaw, and suggested the possibility that import and export laterally (radially) is regulated by companion cells.)

Moorby: Returning to flow gradients across the sieve tube, we calculated what this was likely to be, and it's of the order of seconds, in the arrival of a front as it is going through, assuming laminar flow all the way across at average sizes. Once pores are considered, there will be increased turbulence and things of that sort and it will disappear. I think you'd be very lucky to find it experimentally.

Dr. Outlaw, in doing some compartmental analysis using whole leaves and total ^{14}C we have to use non-linear kinetics - in other words the exchange constants between the storage pools and the labile pool is pool-size dependent. Have you observed this?

Outlaw: No, we haven't. We haven't done any really sophisticated kinetic analysis and I don't feel we should extend anything that we've done any further without more experimental evidence. I think we simulated the kinetics relatively well. The empirical data and the simulation seemed to be relatively close. On the changing pool sizes or rate constants, I really can't say how much they vary. We found, however, from leaflet to leaflet, there were different pool sizes. The amount of sucrose in a palisade cell may vary, but the ratio of sucrose in a palisade cell compared to the sucrose in a spongy cell from the same leaf was relatively constant.

Willenbrink: Dr. Fisher, in your experiments with soybean I calculated that there were different velocities with time in the translocation profiles. How do you explain these data? Another question - are the shapes of the curves obtained for morning glory not influenced by the removal of the apex?

Fisher: I would not try to make too much of the variation in velocity along the length of the morning glory stem or the soybean stems. Dr. Moorby may wish to comment since they made a more detailed analysis. As to the second question, we haven't run an experiment with the apex present. We were afraid that we might get too much movement toward the tip if we left it on, and what was going toward the base wouldn't resemble normal transport. We had vines much longer than 5.5 m. We had one 26 m long which had to be cut into segments to get it out of the growth chamber. With longer plants we got lower velocities. There almost seems to be a critical length to the plant. With reduced velocities the profiles begin spreading in an erratic fashion. You can't make much sense of the data, because, during the length of time of the experiment, the size of the sink leaves is changing, too.

Currier: Dr. Fisher, I think you said: "The distribution of sucrose in the sieve tube is unknown." Don't we assume that it completely fills the lumen?

Fisher: I was referring to the assumptions by some that there are transcellular strands of several microns diameter in the sieve tube lumen and supposedly there are different concentrations in these strands. I don't want to start with the assumption of uniform distribution of sucrose in the sieve elements. I wanted to be able to start from somewhere else and compare my observations with the sieve tube.

Currier: It seems to me that sucrose should completely fill the lumen or it should be completely dissolved in the mictoplasm. It's distribution shouldn't be disturbed too much by any P-protein or strand-like material that ought to be present.

Fisher: I certainly assume that sucrose is evenly distributed in sieve elements and I think our data shows that. It's a solution. It's not adsorbed on anything, either.

Lamoureaux: In _Vicia_ _faba_ do small veins in the leaves have bundle sheath extensions ?

Outlaw: I don't know.

Lamoureaux: Wiley, some yours ago, proposed the idea of bundle sheath extensions to the upper and sometimes lower epidermis. When these are present in numbers, there is some experimental evidence that the epidermis becomes a main channel of conduction between the palisade parenchyma and the vein. Palisade parenchyma cells are frequently not in lateral contact. Solutes could move up to the epidermis and down these extensions.

Outlaw: In _Vicia_ _faba_ I didn't see any extensions that would go up and down. The veins were confined to a layer of spongy parenchyma which was usually separated by cells that are somewhat intermediate between spongy and palisade.

Lamoureaux: There are some old data using dyes which show that xylem movement may go this way and, I think, there are tracer data showing the movement may go back into phloem through these extensions. If present, one might expect a fairly high concentration of labelled sugar in the upper epidermis if this is a pathway between the producing cells and the veins.

Outlaw: I'd completely discount that possibility. In 6 experiments, I never saw a surprising concentration of radioactivity in the upper epidermis. I only saw an amount that indicated it had moved there at a very slow rate.

Swanson: The concentration that you reported in the sieve tubes was 18% ? I noticed that Dr. Geiger reported that it took a mannitol concentration of 0.32 M to plasmolyze. This would be around 35%. There seems to be difference by a factor of 2.

Fisher: I doubt that the sucrose concentration in sugar beet sieve tubes is 35%. It would be closer to 8.5%, and, offhand, I would regard that figure of 18% a bit on the high side. The method of measurement isn't too accurate. Maybe a variation will work better.

Geiger: The data of Recko indicate osmotic potentials for sieve sap that are higher than those I've reported from plasmolysis experiments. There is a non-linearity with sucrose in terms of osmotic pressure and concentration. From composition data I estimate about 20% of the solute to be electrolytes plus some amino acids. On the basis of osmotic pressure data, I would estimate sucrose concentration at 25-27%, which may be high.

Dr. Fisher, would you comment on the differences in the position of the pulse profile with the lateral sink versus along the stem? Could the stem exert an average pull of roots plus lateral sinks. The lateral sink was, if you wish, the pull of developing leaves. Under these conditions, perhaps, the roots, either by being further away or somewhat less effective as a sink, would move the profile less.

Fisher: I don't know how to interpret the apparently more rapid movement of sucrose to the lateral leaf in comparison to the roots. I think it is fairly obvious that there is a component that is moving more rapidly than the main peak, but, as I recall, the lateral leaf was only aout 15 cm from the roots. It almost looks as though the main profile moves down and then, just before it reaches the sink, part of it starts to move faster.

Christy: In morning glory, the sinks at the bottom of the stem seem to become sources faster than sinks at the top of the stem. Maybe they were limited in their supply of assimilates and didn't expand as much as sinks at the top of the stem. Maybe they weren't getting as good a supply of the material coming down.

Eschrich: Earlier, we agreed, I think, that an old leaf will not serve as a sink unless treated with air minus CO_2 or by darkening. I do not understand how these Convolvulus leaves serve as sinks. Perhaps developing buds are sinks in the lower part of the stem. Dr. Christy, you say that an old leaf is a greater source than a young leaf - it would contribute sugar to the stem and dilute the activity. But you should consider leaf traces which may be very long in this plant, because the internodes are long. You can hardly expect that a complete set of sieve tubes is going through to the roots without any connection to other parts.

Outlaw: There seems to be a bit of confusion about the experimental protocol.

Fisher: I wouldn't envision the sieve tubes remaining separate all the way from the source to the roots. One of the perplexities is that the 30 cm discrepancy between the predicted and actual spatial profiles is longer than a single internode - which is between 15 and 20 cm. I assume that there are inner connections between all of the sieve elements. The autoradiograph that I showed came from the first internode below the fed leaf. All the sieve elements were labelled, no matter where we looked on the plant. So, I think all sieve elements are interconnected. I don´t understand why there is faster movement.

Eschrich: A profile is a measurement of the amount of activity, not the velocity. In Convolvulus, the internodes near the top are greater than at the base and the leaf traces are going down over several internodes. There could be recycling, just at the spot where you monitor. If you have recycling going down and up and down then you could have more activity in the cross section than you have in the parts above and below.

Fisher: I´m aware of the possibility of circulation back and forth between adjacent nodes. The kinetic data says it doesn´t happen, because that would spread the profile out very quickly. I´m glad to learn that leaf traces may run through more than one node. We plan to get some answers to the question of whether different velocities occur in different sieve tubes by autoradiography.

Geiger: I heard Dr. Eschrich say something which may be part of the reason for the confusion. Those Convolvulus leaves are immature, small sink leaves, as I understand it, not sources with the bud as the sink.

Some of the work with trimming leaves off sugar beet indicates that quite different patterns can be set up very quickly. I envision, at least in part, a sort of dedication of a file of sieve tubes to some particular sink. Not that it was growing that way initially - maybe the rapidly growing sink leaves that you have just trimmed are stimulated to a lot of growth and constitute a stronger sink, perhaps, than the root.

Willenbrink: Dr. Troughton, how do you explain the increase in velocity with the amount of light?

Troughton: We haven´t got supporting data to show what the explanation is for the increase in speed with light. We only know that the possibility that it is due to photophosphorylation isn´t tenable because, if we darkened the leaf there is not necessarily an immediate drop in speed. The speed can be maintained for some time depending on the pretreatment in light. We would, perhaps, envision the probability that substances, such as starch, are built up in the light and that these contribute to sucrose in the dark, and therefore to the size of the sucrose pool, which in some way determines the mass of material moved under speed.

The Effect of Externally Applied Factors on the

Translocation of Sugars in the Phloem

Leonora Reinhold

The Hebrew University of Jerusalem

Jerusalem, Israel

In this paper I shall select certain factors which are suspected of affecting phloem translocation and shall examine what we know about the possible mechanism of their action.

The influence of light, the influence of K-deficiency, and the interaction between these two factors. It has been noted in several laboratories that translocation seems to proceed faster in the light than in the dark (e.g. 28, 18, 14, 53).

In many of the reported cases the leaf sugar level at the start of the translocation period was very low, as in Thrower's (67) experiment with leaves which had been in darkness for 15 hrs., or in Hartt's (28) sugar cane leaves collected in the early morning. A considerable difference in leaf sugar level will therefore have existed between darkened and light-treated plants during the translocation period. If the driving force for translocation is closely related to overall leaf sugar level, then this would seem to offer a ready explanation of the effect. But in an investigation we carried out some years ago (53), we made an effort to minimize this difference in sugar level between illuminated and darkened plants. Our translocation period was short - 15 or 30 min. - and followed immediately on a period of 30 hrs. in bright light. The sugar levels do not change more than about 10 percent during this brief translocaton period in the dark. Nevertheless, we saw a clear light effect on translocation. Geiger and Batey (14), using a more elegant method of continuous monitoring, have also observed a rapid effect of darkening on translocation in plants that have been in the light for several hours (see Fig.1). They state that within 5 min. a decline in translocation rate may be detected. After 60 min. the rate is 50 percent of that in the light. Although their paper shows that the curve for sucrose concentration also falls, the fall in translocation rate is far steeper and flattens out at a relatively lower level.

What might the mechanism of this light effect be? One obvious possibility is that photosynthesis is at the root of it in some manner yet to be clarified. Another possibility is that it is an entirely unrelated light effect - for example, a possible light effect on the permeability of membranes to sugar. Hartt (28)

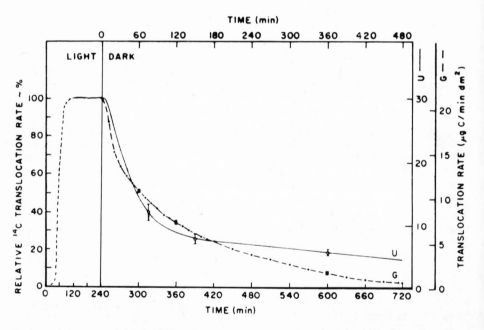

Figure 1. Time-course of ^{14}C translocation rate. The average rate in light = 100%. Lower abcissa, time 0 = beginning of steady state labelling (approximately 0930 EST.). Upper abscissa, time 0 = beginning of dark period. Curve U is the average of computer curve-fit data from 5 ungirdled plants; curve G is the average from 3 girdled plants. Standard deviations for several points are given (from Geiger and Batey, 14).

believes it to be independent of CO_2 assimilation. This belief is based on a comparison of the 2 curves relating translocation and photosynthesis respectively, to light density. The translocation curve was obtained by feeding [14]CO_2 to the apical section of detached sugar cane leaves for 5 min. and allowing translocation to proceed for the subsequent 24 hrs. at various light intensities. The results are expressed as the percentage of the radioactivity reaching the basal portion. The curve shows that translocation was apparently saturated at about 100 to 500 f.c. The curve for photosynthesis on the other hand, goes on rising over the whole range (up to 3,300 f.c.).

Our own experiments (53) to test the possible involvement of photosynthesis in our light effect stand in sharp contrast to this result. We have, for example, examined whether the light stimulation was related to CO_2 fixation by placing half our plants in CO_2-free air. In plants translocating in the light lack of CO_2 inhibited [14]C_2 transport, an effect which was highly significant statistically. In the dark, however, depriving the plants of CO_2 had no effect. The translocation performance of plants both with and without CO_2 in the dark was approximately equal to that of plants in the light without CO_2. Similarly, in further experiments we observed that translocation in the light (but not in the dark) was susceptible of inhibition by the photosynthetic inhibitor 3(3,4-dichlorophenyl)-1,1-dimethylurea (DCMU) - a clear, statistically significant effect. Again the translocation performance of both control and treated plants in the dark was approximately equal to that in DCMU-treated plants in the light.

The results of our experiments with CO_2-free air and with DCMU thus strongly suggest that simultaneous photosynthesis (not past photosynthesis) is involved in the light effect on translocation. How, then, can we explain the apparently conflicting evidence of Hartt's (28) curves against light intensity? On re-examination of Hartt's results I suggest that a likely explanation is as follows: Hartt's translocation period was very long - 24 hrs. - after which time 50 to 60 percent of the counts had been transported to the base of the leaf at saturating light intensities. In another experiment in the same paper, translocation was measured over 6 hrs., after which 70 percent of the counts had been transported basipetally. This strongly suggests that the curve relating amount transported to time flattens out after about 6 hrs. In other words, virtually all that can be transported has been transported long before the end of the 24 hrs. period. Thus the 24 hrs. experiment on the effect of light intensity may tell us nothing at all about the relative rates of transport at the various light intensities - it only gives us the total amount eventually translocated after the system has run to completion. At the lower light intensities the translocation rates may have been much lower than at the higher intensities.

But why in that case, it may be asked, was any effect of light visible in this experiment at all? I suggest that less ^{14}C was detected in the basal part of the blades in darkened leaves because it was respired; and that irradiation during the translocation period, even at very low intensity, may have a "sparing" effect on the previously assimilated ^{14}C, i.e. it allows sufficient ^{12}C assimilation to spare some of the ^{14}C from being respired. There is strong support for this suggestion in another interesting paper of Hartt´s (29), which reports the effect of different wavelengths on translocation. From her Table III, where absolute figures are given instead of her more usual "percentage translocation", it can be calculated that the total amount of ^{14}C detectable in a darkened blade is substantially less than that in an illuminated blade.

It may also be noted from this paper of Hartt´s (29), that the results of experiments with light of various colours would be compatible with chlorophyll being the photoreceptor pigment involved.

If, then, on consideration of all the evidence, we accept the view that light exerts its effect on translocation via concomitant photosynthesis, what product of photosynthesis is involved? It might be photosynthetic ATP. This has been suggested by both Bianchetti (4) and Hartt (28). One likely point at which ATP might intervene is in the supply of energy for vein-loading, known to proceed against the chemical potential gradient (39). I do not mean to suggest by this that passage through the membrane involves phosphorylation of the sugars. Indeed, a very recent investigation of ours (23) led us to the contrary conclusion. We studied uptake of a nonmetabolizable sugar analogue by Ricinus cotyledons, organs adapted to take up sugars from the endosperm and to transfer them to the developing embryo axis. Our observations, including the kinetics of the rise in specific activity of internal sugar and sugar-phosphate pools respectively after isotopic sugar supply, suggested that phosphorylation is not a necessary step in the entry process, a conclusion also recently arrived at by investigators of various fungal and mammalian cells (see 23).

On the other hand, the photosynthetic product involved in promoting translocation may not be ATP. It may be nascent sugar - by which I mean newly synthesized sugar. The pool of storage sugar in the leaf may well be spatially separated from the compartment containing the sugar newly synthesized by the chloroplast. In the light the sugar translocated may be principally drawn from this nascent sucrose compartment. When the light is switched off the concentration in this compartment will fall sharply, even though the overall sugar concentration may not change appreciably.

This view conflicts, however, with an important paper by Geiger and Swanson (18) in which evidence is presented that all of the sucrose in the source leaf is about equally available for translocation, and hence the sucrose pool is not divisible into a transport pool and a storage pool. These authors set up a mathematical model on the assumption that newly-synthesized sucrose mixes immediately and uniformly with the total pool of indigenous sucrose, and the agreement between their experimental data and their theoretical curves is very impressive. If these authors are right, then the suggestion that newly-synthesized sugar is translocated in preference to stored sugar must fall away. Fisher, (13), however, has concluded from kinetic studies that there is compartmentation in the Soya leaf, though he indentifies the compartmentation with different cell types, not with different intracellular sucrose pools; and Shiroya et al. (60) suggested a tentative scheme to explain some puzzling results with pine seedlings according to which stored sugar is translocated only when photosynthetic sugar is not available.

I want to postpone further discussion of the light effect for a while and turn to the question of the effect of potassium deficiency on translocation, since the two effects seem to be related. It is well known that sugars accumulate in the leaves of K-deficient plants, and this has given rise to the suggestion that sugar translocation is depressed in these plants. But of course this is not a necessary inference - the sugar accumulation might well be the result of a disturbance in the equilibrium between sugars and their metabolic products. Direct investigation into the effect of lack of potassium on sugar transport has only been undertaken relatively recently. Anisimov (3) in Russia and Hartt (31) in Hawaii reported a decrease in translocation of labelled photosynthate in K-deficient plants and we ourselves (2) fed ^{14}C-sucrose (via leaf vein flaps) to bean seedlings which had had their cotyledons removed and were growing in culture solutions, and saw a clear, statistically significant depression of ^{14}C translocation under conditions of mild K deficiency - well before any visible symptoms, e.g. chlorosis, were apparent. We found that addition of KNO_3 to the culture solution as little as 3 hrs. before the experiment brought a statistically significant improvement in ^{14}C translocation. A particularly interesting observation was the dependence of the K effect on light - no effect of deficiency was detectible on ^{14}C translocation in the dark. There was a statistically significant interaction between presence of potassium and light.

You may have noticed that I have carefully referred to an effect on the translocation of [14]C - I have not spoken of an effect on translocation accumulation in K-deficient plants which I mentioned earlier. Before transport labelled sugars might get diluted to different extents in control and deficient plants because of the differing sizes of the endogenous sugar pools. Decreased transport of ^{14}C, as also reported by Hartt (31),

Table 1. Effect of the supply of different quantities of exogenous sugar on ^{14}C translocation in control and K-deficient bean plants. Translocation period 2 hrs. Leaf denotes treated leaf; Int.+leaf = trifoliate leaf and 2nd internode; the figures in cols. 10 and 11 based on direct analyses of total sucrose and radioactivity. Column 5 gives ^{14}C translocated out of the treated leaf as percent of total ^{14}C in the plant. Column 12: as calculated from specific activity of sucrose in petiole. (From Amir and Reinhold (2)).

Dose	Specific activity mCi/mmole	Treatment	^{14}C translocated cpm	Translocation %	^{14}C, % of ^{14}C translocated		Sucrose detected, µmol		cpm per µmol sucrose		Sucrose translocated, µmol
					Petiole	Int.+ Leaf	Leaf	Petiole	Leaf	Petiole	
1 µCi	10.4	-K	35,500	4.8	30.0	9.7	10.2	2.2	67,700	4,900	7.25
		Control .	104,000	17.3	9.3	23.4	7.5	0.9	66,700	8,450	12.30
2 µCi	0.05	-K	47,100	9.7	17.0	14.0	13.2	1.1	33,600	11,700	4.02
		Control .	14,700	5.6	26.3	16.1	8.8	1.3	36,200	3,700	3.98
		SE	7,000	1.2	1.6	3.7	1.4	0.2	4,800	1,400	

therefore, does not in itself establish an effect on amount of sugar transported. The observed differences in ^{14}C movement might merely be due to the difference in specific activity of the transported sugar.

We tested this possibility by supplying sucrose at 2 widely differing specific activities. Our reasoning was that, if we diluted the ^{14}C-sucrose considerably with "cold" carrier sucrose while it was still outside the cells, then any further dilution inside the cells whould be relatively less important. In this experiment we also analyzed the sugar content of various plant parts in an attempt to estimate specific activity directly. Column 5 in Table 1 shows that the percentage of ^{14}C-sucrose translocated from the leaf was significantly affected by K deficiency when the specific activity was high - but not when it was low, i.e. when a large amount of carrier sugar had been added. Similarly the significant effect on upwards translocation disappeared when carrier sugar had been added (Column 7). So, on the face of it, the results so far hint that there is no real effect of K deficiency on translocation - the effects observed may be due to differing degrees of dilution by endogenous pools. In order to examine the situation quantitatively we were obliged to compare the absolute amount of sugar (labelled + endogenous) translocated in each case, i.e. we had to know how many molecules of sugar each labelled molecule represented. We could not base our estimate of specific activity on the analysis of the treated leaf, because almost certainly there is no complete mixing of the endogenous and exogenous sugar here. But most of the sugar in the petiole may be "in transit", as suggested by Geiger (XI Int. Congress Bot.) and Fisher (12), so the specific activity of petiolar sucrose should give us an upper limit for dilution by endogenous sugar. Column 12 in the Table gives the absolute amount of sucrose translocated on the basis of this maximum correction. The effect of K deficiency is still apparent when the labelled sugar supplied is of high specific activity. We may therefore conclude that there is indeed a real effect of this deficiency on sugar translocations.

But we were confronted with the baffling disappearance of the K effect when exogenous sugar was added in relatively large amounts. Since the effect of K deficiency had been found to interact with light, and since we knew that the light effect was related to photosynthesis, we asked ourselves if the disappearance could be related to a depression of photosynthesis consequent on the application of sugar to the leaf. So we performed an experiment where we measured both CO_2 fixation and sucrose translocation using a double labelling technique. We observed that adding exogenous sugar did indeed depress CO_2 fixation - but only in the control plants, not in the K-deficient. This interaction between the effects of added sugar and K deficiency was statistically significant. On the other hand there was no significant effect of mild K deficiency as such on CO_2 fixation - the latter was depressed by deficiency only in plants fed high specific activity sugar.

I want now to propose a scheme which would account for all our results.

The statistically-significant interaction between K deficiency and amount of exogenous sugar added indicates that both these factors were acting at the same point. One may reasonably suggest that the effect of added sguar in CO_2 fixation was a mass action phenomenon - some steps in the chain of reactions between photosynthesis and veinloading were driven backwards. The fact that no depression of CO_2 fixation by added sugar was detectable in K-deficient plants shows that K is necessary for the backward reactions as well as for the forward reactions leading to translocation.

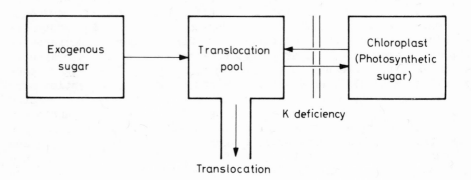

Our scheme (see above) proposes that exogenous sugar mixes with photosynthetic sugar prior to translocation in what we term the "translocation pool". There is an important role for potassium in the chain of reactions between photosynthesis and the entry of photosynthetic sugar into this pool. The chain is reversible. The scheme thus accounts for the fact that no effect of K deficiency is seen when photosynthetic sugar is not being translocated, for instance in plants in the dark, or in plants fed relatively large amounts of exogenous sugar. In the latter case we suggest that mass action effects are keeping photosynthetic sugar out of the translocation pool.

By contrast, a clear effect of K deficiency is seen when the sugar being translocated is almost entirely photosynthetic, i.e. when the concentration of the exogenous sugar added is very low and its specific activity high. In this situation mild K deficiency also depresses CO_2 fixation, because sugar accumulates at the K block and by mass action or feed-back inhibits photosynthesis. K deficiency did not directly inhibit photosynthesis in our experiments, as appears to happen when deficiency is more severe. Hartt (31) has also found that translocation is more sensitive than photosynthesis to lack of K.

To return now to the consideration of the effect of light on translocation. We have just been led to the conclusion that a major role for K in translocation is in influencing the level of photosynthetic sugar in the "translocation pool". I suggest that light, too, exerts its effect via the sugar level in this pool, either by provision of newly synthesized sucrose along some sort of express way between chloroplast and vein; or, if Geiger and Swanson (18) are correct in deducing that nascent sugar mixes immediately with the total leaf sugar, then because light facilitates the transfer of photosynthate to the pool, a function possibly connected with photosynthetic ATP supply.

A paper (54) has recently appeared which would suggest that light affects the movement of sugar, not on its way to the sieve tubes, but when it is already inside them. These workers investigated ^{14}C movement along very long stolons connecting young Saxifrage plants to their mother plant. The stolons, which are green, were passed through a polythene tube about 30 cm long, either black or translucent. A moist airstream was passed through the sleeve for 18 hrs. under continuous light or continuous dark for the translucent and black sleeves respectively, after which labelled CO_2 was supplied to the parent plant. There was an obvious difference in the profiles of radioactivity obtained. One question which suggests itself is whether part of the difference may not be explained on the basis of consumption of ^{14}C by starving cells along the route in the black sleeves - they have been unable to photosynthesize during the long pre-treatment. The airstream would have swept away the $^{14}CO_2$ respired. But, if this is not the case, then in the presence of light, sugar transfer along the stolon has been speeded. I shall be referring again to this interesting system later in my paper.

I. Water stress

Theoretically, a mass flow mechanism would not necessarily be expected to be slowed by water stress. Provided that loading of sugars into the sieve tubes continued at the same rate, the reduction in volume of the assimilate stream consequent on the altered water potential of the surrounding tissues would be balanced by the increase in its concentration. Weatherley, Peel and Hill (77) irrigated willow stem segments and bark strips with mannitol solutions and indeed found that the decline in rate of exudation was almost balanced by the rise in concentration of the exudate. Nevertheless, a number of investigators working with intact plants have observed a reduction in carbohydrate translocation under conditions of moderate or severe water stress (e.g. 79, 58, 51, 52, 30, 73, 74). The fact of an effect is thus widely acknowledged; but one of the several questions which remains wide open to argument concerns the stage of the translocation process affected.

From our own work (51) we have concluded that, apart from any possible effects on phloem loading, it is extemely likely that stress retards movement within the phloem itself. One of our reasons for this conclusion is that a clear effect of stress is visible on ^{14}C distribution through the plants (at various periods after isotope supply to a leaf) whether absolute values for radioactivity are considered, or whether the results are expressed as a percentage of the total ^{14}C recovered in all fractions from mid-vein to root. The latter basis only takes into account ^{14}C that is within, or that has passed through, the phloem and therefore minimizes differences between treatments due to differing rates of uptake and vein-loading. Our results strongly suggest that the ^{14}C within the phloem has moved more rapidly towards the roots in the control plants.

Wardlaw (73, 74), by contrast, has concluded that, though movement of assimilates out of the leaves of wheat plants is delayed under water stress, translocation within the conducting vessels is not affected. He reports an experiment where the flag leaf blades of stressed plants were allowed to assimilate $^{14}CO_2$ for 10 min. Individual plants were harvested every few minutes and divided into segments for analysis. He followed the change in radioactivity with time in the upper and lower sections of the leaf sheath respectively. From the interval that elapsed before half the maximum value was achieved at these 2 points in the leaf sheath, he calculated that the velocity of transport through the sheath was in the order of 40 cm/h. This was in the same order as observed in other experiments with non-wilted plants. In these experiments with wheat, the flag leaves were exporting to the grain, the development of which is resistant to water stress. In another species, Lolium temulentum, Wardlaw (70) did detect decreased velocity of movement under stress, but he attributed this to reduction in sink activity. This may certainly be part of the explanation of retarded movement along sieve tubes in stressed plants, but I doubt whether it is the whole explanation. Wardlaw´s method of determining velocity in the wheat experiment could only be expected to show up very large changes. It seems highly likely that several factors, apart from increase viscosity of the phloem sap, would affect flow. These include a decrease in L_p, the coefficient of hydraulic conductivity of the membranes. Christy and Ferrier (7) have recently stressed the potential importance of L_p to a pressure flow mechanism. Increasing osmotic concentration on the inner or outer face of the membrane decreases L_p (10, 60). The effective radius of the transport channel may also decrease due to the effect of dehydration.

Wardlaw (69) believes that the main effect of stress is to reduce the rate of movement of assimilates into, but not through, the conducting tissue. Now we also believe that movement into the conducting tissue must eventually be reduced - but secondarily, as a result of the high concentration of sugar piling up in the veins. Wardlaw, on the other hand, thinks of the effect on vein-loading as primary, possibly relating to reduction in "the ability of the leaf to utilize available energy", for instance due to uncoupling of phosphorylation and respiration. We are unwilling to go along with him in this. Several of our results suggest to us that loading has been efficient in a stressed leaf. We have observed a very strong temporary acceleration of transport as compared with fully irrigated controls when wilted bean plants are irrigated (51). This response seems to be almost immediate, and occurs well before the leaves recover their turgidity as assessed by relative turgidity measurements. It seems reasonable to suggest that the veins were heavily loaded in these stressed leaves, and that water passed rapidly straight from the xylem to the phloem on irrigation, generating a stronger driving force than existed in the control.

We have also observed that when a drop of ^{14}C-sucrose solution is applied to the primary leaf, and then after various translocation periods the small application site is removed, far higher amounts of ^{14}C are detected in the leaves of stressed plants than in controls. Experiments including radioautography, double labelling of the water and the sucrose, and girdling have shown that phloem transport through the leaf tissue is the process involved (52). Crafts and Crisp (9) have pointed out that the volume of the droplet of sucrose solution applied, though small (0.01 ml), would account for the total volume of the protophloem sieve tubes in several centimeters of leaf, vein, and petiole. A reasonable explanation for the striking enhanced ^{14}C movement through the stressed leaf would therefore seem to be the same as that offered for the recovery effect - the veins are heavily loaded, and the local supply of a small volume of water allows the development of a more powerful driving force than in the control.

An alternative explanation is worth serious consideration but must be rejected. This is that sugars had accumulated, not in the veins but in the mesophyll or apoplast of the stressed leaves. Loading was therefore faster than in the controls as soon as the collecting cells were relieved from stress. The reason for rejecting this possibility is that when the labelled sugar was supplied it would in this case have had to compete for the loading mechanism with a large number of endogenous sugar molecules; and the result of this keener competition would be a reduction in labelled sugar movement as compared with control plants, not a large increase as actually observed. This objection does not apply to the first explanation offered, i.e. accumulation of sugar in the veins, because once in the veins, there need be no competition between sugar molecules - movement is by mass flow or some form of diffusion.

Hall and Milburn (25), studying the dependence of phloem exudation in Ricinus on the water supply to the tissues, have also concluded that sugar loading occurs even under conditions of considerable water stress.

If, as we have maintained, sugar transfer from mesophyll to vein can indeed continue efficiently in stressed plants, a legitimate question would be: why has a decrease in mass transfer been observed? Why has the higher concentration not compensated for the reduction in the volume moving? I suggest an answer based on the assumption that the "active" sugar accumulation mechanism is located in phloem parenchyma cells, and that the sugar then passes into the sieve tubes along plasmodesmata. Geiger et al (16) and Ziegler (85) have stressed the remarkably high incidence of plasmodesmata between companion cells and sieve tubes. Water stress may well interfere with the passage of the sugar through the plasmodesmata. Dehydration may for instance, bring about a decrease in the diameter of the transport channel. This slowing down of the flow of sugar between collecting cell and sieve tube may result in a situation where the sugar concentration in the latter is considerably below that in the former; and thus will depress mass transfer.

II. "Hormone-directed" transport

A considerable literature has accumulated demonstrating that treatment of various plant organs with hormones leads to changes in the amounts of metabolites transported to these organs (see Moorby 1968). Wareing and his group, for instance, have noted that if IAA and other growth regulators are applied to the stump of decapitated seedlings, transport of assimilates into the stump is substantially enhanced (see 5 for earlier references; 49, 50). Weaver and his colleagues (see 59 for references) have applied kinins to fruit clusters, shoot tips, leaves or roots of grape vines and have in each case observed increased movement of labelled assimilates into the treated part. Similarly Harris, Jeffcoat and Garrod (27) noted that the application of gibberellin raised the amount of labelled assimilate transported to a developing bud.

The movement of various solutes within single leaves from untreated areas towards a kinin treated "mobilizing centre" has also received considerable attention (44, 22, 45).

A vital question here from the point of view of phloem physiologists is, has the translocation system been directly affected by hormones in any of these systems? Or have the hormones only been fulfilling their well-known roles as stimulators of protein synthesis, cell division, cell elongation and other consumption processes in the sink areas? Shindy, Kliewer and Weaver (59) in fact consider that enhancement of sink activity is a reasonable explanation of their results.

But here I would like to focus on a problem. Enhanced mass transfer as observed in the hormone experiments must mean that the rate of vein-loading has been raised. How does increased sink activity bring about a rise in loading rate? One can visualize that, as a result of speedier unloading by an improved sink, the translocation system would tend towards a new steady state in which the osmotic gradient down the phloem was steeper, and this would in turn lead to increased velocity of transport. But higher velocity would not account for the observed effect on mass transfer unless loading rate were increased. So by what mechanism does increased sink activity bring about a very substantial higher rate of loading at a point geographically remote from the sink? Is it related to a drop in sugar concentration in the translocation stream, assuming this occurs? The possible dependence of loading rate on internal concentration is a factor which has so far not attracted the close attention of the model makers. Christy and Ferrier (7) for instance put r, loading rate, constant, though internal concentration was changing. It has been suggested in the past (e.g. 77, 64) that sugar in the sieve tubes is in a state of dynamic equilibrium with that in surrounding cells. Obviously, then, a drop in internal concentration would increase net flux inwards (i.e. loading) by decreasing efflux. However, I want to point out that control of loading by means of the dynamic equilibrium between influx and efflux would be a very insensitive control mechanism. Efflux from loaded phloem cells in the source area is likely to be extremely small in comparison with influx. This can be deduced from radioautographic studies, and also from wash-out experiments carried out with conducting bundles excised from leaves (37). If we are considering a process which is the difference between a large influx and a small efflux, then even considerable changes in the magnitude of the efflux will have only a relatively small effect on the net process.

And now to analyze this situation more quantitatively. Ever since the classic demonstration by Mason and Maskell (39) that sugar passes from mesophyll to vein against the concentration gradient, loading has been recognized as an active transport process. In common with other carrier-mediated transport mechanisms it is likely to conform with Michaelis-Menten kinetics, at any rate over a limited concentration range (80,48). Net influx, v, may therefore be expressed as

$$v = V_{max} \left[\frac{S_1}{S_1 + K_1} - \frac{S_2}{S_2 + K_2} \right] \qquad (1)$$

where V_{max} = maximum rate, S_1 and S_2 external and internal concentration respectively, and K_1 and K_2 the Michaelis constants on the outer and the inner sides of the membrane respectively. (I am here assuming that efflux as well as influx is a carrier-mediated process, because many workers seem to be tending towards this view. If, however, it is a diffusional leak it can be written as AS_2, where A is a constant, and this will not alter in principle the rest of my argument).

K_1 will be smaller than K_2, according to widely held views on the mechanism of uphill transport. An estimate of K_1/K_2 can be obtained from the maximum accumulation ratio, since substrate will move uphill until a steady state is reached when v = 0 and consequently

$$\frac{S_1}{S_2} = \frac{K_1}{K_2} \qquad (2)$$

(For simplicity I am not introducing the complication of possible additional asymmetry between the velocity constants for carrier-substrate fluxes in the inward and outward directions.) Bacteria accumulate sugars up to several hundred fold, and it is highly likely that specialized higher plant cells do the same. We (57) have observed a ratio of 75/1 in carrot for a non-metabolizable glucose analogue, and washing experiments (37) suggest a far higher ratio for conducting cells. Let us assume this ratio is 100/1, and also take the following values from the literature: S_1 (mesophyll) = 33 mM; S_2 (conducting bundles of 3rd and 4th orders of branching) 200 mM (from (6); I have taken the figures for total sugars (August) since all sugars are possibly taken up by bundles and converted at some point to sucrose. In the case of S_2 I have taken into consideration that approx. 1/3 the volume of the bundle is xylem, (15)). K_1 = 10 mM (based on the average K_m for uptake of sucrose, glucose, and fructose respectively by conducting bundles (68, 62). When these values are substituted in Eq.1, the loading rate (v) is found to be 0.73 x V_{max}.

Now let us suppose that, due to increased sink activity following hormone treatment, S_2 falls as much as 50 per cent. The new loading rate as calculated from Eq.1 is 0.81 x V_{max}. It has increased by only 14 per cent. This modest increase must be compared with data in the literature indicating hormone-induced

increases   in mass transfer of several hundred per cent as measured
in translocation periods of 30 min. (32) or several hours (59,
50).

"Dynamic equilibrium" would therefore not appear to be a
sufficient explanation of the control of net vein loading exerted
by increased sink activity. Is there some more sensitive form of
feedback   control   exerted   by   internal   concentration,   such as a
loading mechanism highly sensitive to "transinhibition" ?   (i.e.
inhibition by intracellular sugar - see 35). Or is the signal
which controls loading, flow rate through the sieve tubes, rather
than internal concentration ?

While some workers, as I noted, consider that increased sink
activity could account for their results with hormone treatments
this is not the view of Wareing and his group.   They report that
IAA had no effect on the "sink activities" they measured, e.g.
respiration, growth rate, ^{14}C incorporation into protein, within
the short term of the experiments (49 and Parick and Wareing in
preparation). Their most persuasive place of evidence is their
finding that more than 90 per cent of the ^{14}C transported to the
treated stump was present in alcohol-soluble form and that sucrose
formed at least as large a proportion of the alcohol-soluble
fraction in treated as in control. These results indicate that the
pool of free translocated ^{14}C-sucrose was very much higher in the
hormone treated case, since more than 5 times as much radioactivity
was detected here. (Incidentally, these results also indicate that
sink activity was higher here, though the authors´ measured
parameters did not detect it, since the amount of non-sucrose [14]C
is also much higher in the auxin-treated case). The possibility
that the large pool of translocated free ^{14}C-sucrose in the
hormone-treated stump reflects a different kind of sink activity -
accumulation in some interior compartment of the cells - is
weakened by the fact that the authors failed to detect any effect
of IAA treatment on sugar uptake by stem tissues incubated in
solutions. This evidence thus does suggest an effect of IAA on
translocation other than sink stimulation. But in interpreting it
further one must bear in mind (see 47) that a decapitated internode
is a declining system, and that an important part of the role of
hormones here may lie in maintaining the integrity of the tissue in
general and the phloem in particular.

The possibility that the hormones act, not at the site of
application, but at some locus to which they have been transported
has been considered by some workers. Hew, Nelson and Krotkov (32)
observed enhanced downwards transport away from the treated site
and concluded from studies with ^{14}C-IAA that the primary site of
its action was in the stem. Mullins (46) has pointed out that in
these experiments the mobilized ^{14}C-sucrose may have moved
downwards first before transfer to an ascending pathway due to the

pattern of anastomosis of the vascular bundles. His own experiments (47) with triiodobenzoic acid at concentrations which inhibited polar IAA transport led him to conclude that the descent of auxin into the stem was not involved in the effect on translocation. Lepp and Peel (38) presented some data obtained with willow bark strips which, they suggested, indicated that hormones may enhance loading of sugars into sieve elements. This finding recalls Vernon and Aronoff´s (72) observation that the application of small amounts of 2,4-D to Soya leaves resulted in increased translocation, and would seem to merit deeper investigation.

The conclusion so far must be that we are still awaiting crucial evidence that hormones act directly on any stage of the phloem translocation process.

III. Anaerobic conditions and metabolic inhibitors

Controversy is still exceedingly brisk in this area. Mason and Phillis (40) found that translocation was resistant to lack of O_2, but succeeded in demonstrating an effect under extremely drastic conditions. Vernon and Aronoff (72), Willenbrink (81) and Ullrich (69) have all failed to detect an effect on sieve tube transport, and recently Sij and Swanson (61) demonstrated that depriving a 15 cm zone of squash petiole of O_2 only brings about a transient decline in translocation, full recovery occurring within about an hour. Against these results must be set the recent report of Qureshi and Spanner (54) that translocation of tracers is depressed if 20 - 30 cm stretches of stolon connecting mother and daughter Saxifrage plants are submitted to anaerobic conditions.

Reported results with metabolic inhibitors are also conflicting and this is in part due to the technical difficulties involved. Since sugar loading proceeds against the chemical potential gradient (39) it must be coupled to an energy yielding process. Investigators whose aim it is to examine the effect of metabolic inhibitors on movement within the sieve tubes must therefore take adequate precautions to ensure that their results are not due to the inhibitor reaching the loading sites. Since this was not sufficiently taken into account in the earlier work (e.g. 34) the conclusions are unreliable.

Although CN is very highly mobile in the plant (see 33), there are nevertheless, indications that sieve tube transport may have been more sensitive to CN than was loading under the conditions of several investigations (69, 82, 83, 33, 56). If the basis of this sensitivity is inhibited respiration and consequently a lowered ATP level, then one would expect that uncouplers of oxidative phosphorylation such as 2,4-dinitrophenyl (DNP) would also inhibit sieve tube transport. Yet it is by no means clear that this is the case. We ourselves have reported (26) that translocation is stable to DNP when care is taken that the inhibitor does not interfere with loading. Willenbrink (83) has also reported the failure of DNP to inhibit sieve tube transport in his experiments.

Qureshi and Spanner's (55) results once more stand in contrast, and it is of interest to consider why their Saxifrage stolons differ from the other experimental tissues investigated. The authors themselves suggest (54) that the length of the inhibited axis and the magnitude of the transpiration stream may each play a part in the response to anoxia. But their treated length of stolon (20 - 30 cm) is not so very different from the 15 cm length of petiole deprived of O_2 by Sij and Swanson (61). Moreover the latter authors have calculated the extent to which the transpiration stream and the assimulate stream could be supplying dissolved O_2 to the deprived phloem and have found it to be equal to only about 0.6 per cent of the potential aerobic demand. One might speculate that, even if the Munch hypothesis is the correct one, mass flow along these extremely long, thin stolons is in need of "boosting" along the route, i.e. a certain amount of sugar and water must flow into the stream as it passes along the stolon. Since the latter is very thin and probably contains neglgible sugar reserves, the "boosting" will depend on concomitant or recent photosynthesis carried out in its chlorenchyma, as well as on the metabolic energy necessary for phloem loading and for maintaining adequate permeability to water. This would explain the apparent requirement for light and for aerobic respiration as well as the inhibition by DNP - it has been shown in our own and other laboratories that not only membrane sugar transport (see 11) but water permeability (84, 29, 21) is sensitive to DNP.

The puzzling fact that susceptibility to CN can apparently be demonstrated in systems where sensitivity to anoxia cannot has recently been discussed by Sij and Swanson (61). Among other alternatives they mention the possibility, first raised by Mason and Phillis (40) that a cyanide-sensitive peroxidase-peroxide transport system maintains oxygenation of the phloem. It seems to me that if this were the case these systems should show DNP sensitivity, since it is implicit in this idea that oxidative phosphorylation is proceeding normally even in O_2-deprived tissue. DNP sensitivity is not shown; and moreover it is reported (70) that though application of ATP can stimulate fluorescein transport in the phloem it cannot counter CN inhibition of this transport. These results suggest that inhibition by CN is not connected with lowered ATP supply. It seems more likely it is a specific effect on a CN-sensitive enzyme concerned in keeping the pathway open - possibly by countering a sealing mechanism such as callose synthesis, since the latter is promoted by CN treatment (71).

IV. Temperature

Once more conflicting evidence abounds. I cannot enter into a detailed discussion of this evidence (for a useful review see 9). Here   again   it is very important to distinguish between a temperature effect on the "active" vein-loading step, and that on transport of sugar within the sieve tubes; and this point underlies part of the conflict.   Nevertheless, where cooling has been strictly confined to the translocation path itself, remarkable resistance to the treatment has been observed in some investigations while in others translocation has been substantially reduced.   It would probably be fair to sum up the evidence (cf. 19) by saying that cooling to temperatures near $0^{\circ}C$ depresses translocation in non-hardy plants which are also liable to other types of chilling injury. On the other hand, cooling has little effect on translocation if the plant is cold-hardy.   Thus Weatherley and Watson ((76) and in press) observed little or no falling off in translocation performance when willow stems were cooled to $0^{\circ}C$, even when lengths up to 60 cm long had been pre-cooled for 18 hrs. before the experiment to exhaust reserve supplies of ATP, or similar compounds which might conceivably provide energy to drive transport. Respiration of the bark was reduced by 95 per cent. Wardlaw (75) could detect no effect on the velocity of transport in the leaf blade of <u>Lolium</u> <u>temulentum</u> when he cooled a 2-cm length to $0^{\circ}C$.

This resistance of the translocation system to temperatures which almost halt respiration accords with the lack of sensitivity to anoxia and to DNP discussed earlier. The question may, however, be asked - if a CN-sensitive enzyme has a role in keeping the pathway open as was earlier suggested, why is the temperature effect on its activity not manifest in translocation experiments? A possible answer is suggested by the observation (65) that the sealing process which tends to halt phloem exudation in <u>Yucca</u> is strongly retarded by low temperature. If, as was proposed in the last section, the function of the enzyme is to counter a sealing process, inhibition of the latter by low temperature would obviate the need for the counter measure.

Localized high temperature treatments (approx. 30° - 50°) depress translocation (see e.g. 18, 42).   There is evidence (42,41) that this effect may to a large degree be accounted for by reversible callose formation.

V. The "adaptation" of the translocation system to cold

Swanson and Geiger (63) have shown that, in sugar beet, cooling the translocation path from $25^{\circ}C$ to approximately $1.5^{\circ}C$ initially brings about a sharp drop in mass transfer rate, but that recovery starts immediately and the former translocation rate is regained within 90 min. Several explanations for this effect have been thought of and disposed of. The possibility that increased loading rate at the source might compensate for low velocity by increased solute concentration, was abandoned when pulse labelling experiments showed that recovery occurred mainly because of a restoration of velocity (17). A temporary closure of the stomata, with a consequent reduction in photosynthesis, was excluded by the work of Coulson et al (8). The latter investigation also showed that recovery of translocation rate was not accompanying recovery of metabolic processes. The latest suggestion (19) is that the fall is due to the increased viscosity of the sieve sap, and recovery probably occurs by virtue of an increased pressure gradient. But the earler data (63, 8) suggest a larger effect than can be accounted for by the Q_{10} for viscosity, which is only approximately 1.33 (19) and I wish to suggest a possible alternative explanation.

We have recently made a study (1) of the effect of various forms of shock on membrane transport in plant cells. We observed that the various shock treatments, for example osmotic shock, were far more effective if they were combined with cold shock i.e. if the cells were suddenly shifted from 25°C to 2°C. Cold shock not only greatly reinforced the effect of osmotic shock but also produced an effect of its own on membrane transport. Provided the shock was not too severe, recovery took place in about 15 - 30 min. Now in these experiments recovery was occurring at room temperature, but I want to suggest some ways in which recovery could be expected even if the cells are maintained at the lower temperature. The effects of cold shock have also been observed not only in bacterial membrane but in model membrane liposomes, and have tentatively been related to lipid phase transitions in the membrane (see 24). Upon rapid cooling, transition of certain lipids constitutents from the liquid crystalline to the gel state is postulated to cause discontinuities in the packing of the membrane. it is also postulated (36) that the association between a membrane protein and a liquid crystalline lipid is more favourable than that between a protein and a lipid in the gel state. When a lipid associated with a protein gels, it will be exchanged for a liquid crystalline molecular species. Therefore, if only some of the membrane lipid constituents have changed phase as a result of a sudden temperature shift, the membrane might recover its integrity by gradual rearrangement of the molecules so that their packing once more resulted in a continuous shell, and by re-association of membrane proteins with liquid crystalline lipid species.

In other words, I suggest that the transient inhibition of translocation observed by Swanson and Geiger and their colleagues may be an example of "cold shock" to membranes consequent on the sudden change in temperature in these experiments. Discontinuities in the packing of the phloem membrane, may for instance, have caused temporary loss of sugar from the transport channel.

Acknowledgements

It is a pleasure to acknowledge my indebtedness to Professor Weatherley and his colleagues in Aberdeen, and to Professor van Die and his colleagues in Utrecht, for very helpful discussions and hospitality while I was preparing this paper.

References

1. AMAR, L. and REINHOLD, L. 1973. Plant Physiol. 51: 620-625
2. AMIR, S. and REINHOLD, L. 1971. Physiol. Plant 24: 226-231
3. ANISIMOV, A.A. 1959. Fiziol. Rast. 6: 138-143
4. BIANCHETTI, R. 1963. Nuovo Giorn. Bot. Ital. 70: 329-337
5. BOWEN, M.R. and P.F. WAREING. 1971. Planta (Berl.) 99: 120-132
6. BROVCHENKO, M. 1965. Fiziol. Rast. 12: 270-279
7. CHRISTY, A.L. and J.M. FERRIER. 1973. Plant Physiol. 52: 531-538
8. COULSON, C.L., A.L. CHRISTY, D.A. CATALDO and C.A. SWANSON. 1972. Plant Physiol. 49: 919-923
9. CRAFTS, A.S. and C.E. CRISP. 1971. Phloem Transport in Plants. W.H. Freeman and Co., San Francisco
10. DAINTY, J. and B.Z. GINZBURG. 1964. Biochim. Biophys. Acta 79: 102-111.
11. FINKELMAN, I and L. REINHOLD. 1963. Israel. J. Bot. 12: 97-105
12. FISHER, D.B. 1970a. Plant Physiol. 45: 107-113
13. FISHER, D.B. 1970b. Plant Physiol. 45: 114-118
14. GEIGER, D.R. and J.W. BATEY. 1967. Plant Physiol. 42: 1743-1749
15. GEIGER, D.R. and D.A. CATALDO. 1969. Plant Physiol. 44: 45-54
16. GEIGER, D.R., R.T. GIAQUITA, S.A. SOVONICK and R.J. FELLOWS. 1973. Plant Physiol. 52: 585-589
17. GEIGER, D.R. and S.A. SOVONICK. 1970. Plant Physiol. 46: 847-849
18. GEIGER, D.R. and C.A. SWANSON. 1965. Plant Physiol. 40: 942-947
19. GIAQUITA, R.T. and D.R. GEIGER. 1973. Plant Physiol. 51: 372-377
20. GINSBURG, H. and B.Z. GINZBURG. 1970. J. Exp. Bot. 21: 580-592
21. GLINKA, Z. and L. REINHOLD. 1972. Plant Physiol. 49: 602-606
22. GUNNING, B.E.S and W.K. BARKLEY. 1963. Nature 199: 262-265
23. GUY, M. and L. REINHOLD. 1974. Physiol. Plant. 31: 4-10
24. HAEST, C.W.M., J. DE GIER, G.A. VAN ES, A.J. VERKLEIJ and L.L.M. VAN DEENEN. 1972. Biochim. Biophys. Acta 288:
25. HALL, S. and J.A. MILBURN. 1973. Planta (Berl.) 109: 1-10

26. HAREL,S. and L.REINHOLD. 1966. Physiol. Plant. 19: 634-643
27. HARRIS, G.P., B. JEFFCOAT and GARROD. 1969. Nature 223: 1071
28. HARTT, C.E. 1965. Plant Physiol. 40: 718-724
29. HARTT, C.E. 1966. Plant Physiol. 41: 369-372
30. HARTT, C.E. 1967. Plant Physiol. 42: 338-346
31. HARTT, C.E. 1969. Plant Physiol. 44: 1461-1469
32. HEW, C.S., C.D. NELSON and G. KROTKOV. 1967. Amer. J. Bot. 54: 252-256
33. HO, L.C. and D.C. MORTIMER. 1971. Can. J. Bot. 49: 1769-1775
34. KENDALL, W.A. 1955. Plant Physiol. 30: 347-350
35. KOTYK, A. and L. RIHOVA. 1972. Biochim. Biophys. Acta 288: 380-389
36. de KRUIJFF, B., P.W.M. VAN DIJCK, R.W. GOLDBACH, R.A. DEMEL and L.L.M. VAN DEENEN. 1973. Biochim. Biophys. Acta 330: 269
37. KURSANOV, A.L. and M.I. BROVECHENKO. 1970. Can. J. Bot. 48: 1243-1250
38. LEPP, N.W. and A.J. PEEL. 1970. Planta (Berl.) 90: 230-235
39. MASON, T.G. and E.J. MASKELL. 1928. Ann. Bot. 42: 571-636
40. MASON, T.G. and E. PHILLIS. 1936. Ann. Bot. 50: 455-499
41. McNAIRN, R.B. 1972. Plant Physiol. 50: 366-370
42. McNAIRN, R.B. and H.B. CURRIER. 1968. Planta (Berl.) 82: 369-380
43. MOORBY, J. 1968. "The Transport of Plant Hormones" (ed. Y. Vardar) North Holland Publishing Co., Amsterdam
44. MOTHES, K. and L. ENGELBRECHT. 1961. Phytochemistry 1: 59-62
45. MULLER, K. and A.C. LEOPOLD. 1966. Planta (Berl.) 68: 186-205
46. MULLINS, M.G. 1970a. Ann. Bot. 34: 889-96
47. MULLINS, M.G. 1970b. Ann. Bot. 34: 897-909
48. NISSEN, P. 1974. Ann. Rev. Plant Physiol. (in press)
49. PATRICK, J.W. and P.F. WAREING. 1972. "Plant Growth Substances 1970" (ed. D.J. Carr) Springer Verslag Berlin pp. 695-700
50. PATRICK, J.W. and P.F. WAREING. 1974. J. Expt. Bot. (in press)
51. PLAUT, Z. and L. REINHOLD. 1965. Austral. J. Biol. Sci. 18: 1143-1155
52. PLAUT, Z. and L. REINHOLD. 1967. Austral. J. Biol. Sci. 20: 297-307
53. PLAUT, Z. and L. REINHOLD. 1969. Austral. J. Biol. Sci. 22: 1105-1111
54. QURESHI, F.A. and D.C. SPANNER. 1973a. Planta (Berl.) 110: 131-144

55. QURESHI, F.A. and D.C. SPANNER. 1973b. Planta (Berl.) 111: 1-12

56. QURESHI, F.A. and D.C. SPANNER. 1973c. J. Exp. Bot. 24: 751-762

57. REINHOLD, L. and Z. ESHHAR. 1968. Plant Physiol. 43: 1023-1030

58. ROBERTS, B.R. 1964. "The Formation of Wood in Forest Trees" (ed. M. Zimmermann) Academic Press, Inc., New York

59. SHINDY, W.W., W.M. KLIEWER and R.J. WEAVER. 1973. Plant Physiol. 51: 345-349

60. SHIROYA, T., G.R. LISTER, V. SLANKIS, G. KROTKOV and C.D. NELSON. 1962. Can. J. Bot. 40: 1125-1135

61. SIJ, J.W. and C.A. SWANSON. 1973. Plant Physiol. 51: 368-371

62. SOKOLOVA, S.V. 1972. Fiziol. Rast. 19: 1282-1291

63. SWANSON, C.A. and D.R. GEIGER. 1967. Plant Physiol. 42: 751-756

64. TAMMES, P.M.L., C.R. VONK and J. VAN DIE. 1967. Acta Bot. Neerl. 16: 244-246

65. TAMMES, P.M.L., C.R. VONK and J. VAN DIE. 1969. Acta Bot Neerl. 18: 224-229

66. TAZAWA. M., and N. KAMIYA. 1965. Ann. Report of Biological Works, Osaka University 13: 123-157

67. THROWER, S.L. 1962. Austral. J. Biol. Sci. 15: 629-649

68. TURKINA, M.V. and S.V. SOKOLOVA. 1972. Fiziol. Rast. 19: 912-919

69. ULLRICH, W. 1961. Planta (Berl.) 57: 402-429

70. ULLRICH, W. 1962. Planta (Berl.) 57: 713-717

71. ULLRICH, W. 1963. Planta (Berl.) 59: 387-390

72. VERNON, L.P. and S. ARONOFF. 1952. Arch. Biochem. Biophys. 36: 383-398

73. WARDLAW, I.F. 1967. Austral. J. Biol. Sci. 20: 25-39

74. WARDLAW, I.F. 1969. Austral. J. Biol. Sci. 22: 1-16

75. WARDLAW, I.F. 1972. Planta (Berl.) 104:18-34

76. WEATHERLEY, P.E. and B.T. WATSON. 1969. Ann. Bot. 33: 845-853

77. WEATHERLEY, P.E., A.J. PEEL and G.P. HILL. 1959. J. Exp. Bot. 10: 1-16

78. WEBB, J.A. 1967. Plant Physiol. 42: 881-885

79. WIEBE, H.H. and S.E. WIHRHEIM. 1962. "Radioisotopes in Soil-Plant Nutrition Studies" (Int. Atomic Energy Agency, Vienna)

80. WILBRANDT, W. and T. ROSENBERG. 1961. Pharmacol. Rev. 13: 109-183

81. WILLENBRINK, J. 1957. Planta (Berl.) 48: 269-342

82. WILLENBRINK, J. 1966. Z. Pflanzenphysiol. 55: 119-130

83. WILLENBRINK, J. 1968. "Symposium Stofftransport" (Deutsch. Bot. Ges.) Fischer Verlag, Stuttgart

84. WOOLEY, J.T. 1965. Plant Physiol. 40: 711-717

85. ZIEGLER, H. 1974. Proc. Julich Workshop on Membrane Transport. Springer-Verlag (Berlin) in press.

DISCUSSION

Discussant: A.L. Christy
Agricultural Research Dept.
Monsanto Company
800 N. Lindbergh Blvd
St. Louis, Missouri 63166

Christy: (Using slides, Dr. Christy described experiments on the relation between steady-state and unsteady-state photosynthesis and translocation velocity. With pulse-labelling of a sugar beet leaf under steady-state conditions of photosynthesis, there was a strong correlation between translocation velocity (av.=0.9 cm min^{-1}), sucrose concentration in the leaf, and photosynthetic rate. Under unsteady-state conditions, pre-treating the leaves with above- or below-normal CO_2 concentrations for different periods of time prior to pulse labelling, translocation velocity was not correlated with photosynthetic rate or sucrose concentraton in the leaf.

Dr. Christy also described a mathematical model which he and Dr. Ferrier used to examine the effects of water stress on translocation. The model had a source region, a path region, and a sink region. The water potential surrounding the source was varied from 0 to -10 atm., the sink was maintained at 0 atm. and the path at a gradient between the values in the source and the sink. Holding the translocation mass transfer rate constant, when the water potential gradient between source and sink was increased from 0 to 10 atm., the sucrose concentration and osmotic pressure in the path increased from 10.6 to 18.3% and 16.1 to 97.1 atm. min^{-1}, respectively. The hydrostatic pressure gradient decreased from 7.9 to 6.2 atm. m^{-1} and velocity decreased from 1.17 to 0.66 cm min^{-1}. This indicates that translocaton can function under considerable water stress. When a cold block followed by recovery was simulated, translocation velocity recovered 100% if the unloading rate in the sink was constant. The viscosity of a 12% sucrose solution would double in resistance from 26° to 0°. In the model system, it was necessary to increase the resistance to 50 times normal to obtain the experimentally-observed inhibition of translocation. This indicated that cold inhibition is caused by other changes besides increased viscosity. Dr. Christy suggested that these might be physical blockage of sieve plates, as proposed by Giaquinta and Geiger, with less blockage occurring in chill-resistant plants and increased hydrostatic pressure tending to promote recovery.)

Currier: Dr. Reinhold, what do you mean by phloem membrane and osmotic shock to the phloem membrane?

389

Reinhold: I mean the sieve tube plasmalemma. By osmotic shock I mean the phenomenon described by bacteriologists. The loss of a small fraction (3%) of protein into the solution when cells are transferred from solutions of high to low osmotic concentration, which is correlated with a considerable loss in membrane transport ability. We have been unable to show in vitro binding of solutes to the protein, however. Over the 2 cm of stem that are rapidly cooled, "cracking" of the membrane, as occurs with artificial membrane liposomes, would cause sugar to leak out and continue until regeneration of the entire membrane has occurred.

Giaquinta: What species did you use for the osmotic shock experiments?

Reinhold: Phaseolus; bean.

Giaquinta: We've studied both chilling-resistant (sugar beet) and chilling-sensitive (bean) species. We propose that the site of low temperature-induced damage in the bean is, indeed, the plasma membrane, or the endoplasmic reticulum membrane, on the basis of a liquid phase transition in membranes which have a high dgree of saturated fatty acids. Perhaps osmotic shock wouldn't occur in resistant species which have a more fluid membrane and a higher degree of unsaturated fatty acids.

Reinhold: Perhaps, in a chilling-sensitive species, a large number of the lipid components in the membranes will undergo an irreversible phase transition. With model liposome membranes, osmotic shock is a temporary thing. The lost proteins can go back into the membrane if supplied in concentrated form, or, given more time, they appear to be resynthesized in the cells and the membrane fully recovers.

Giaquinta: Did you observe recovery, with or without rewarming?

Reinhold: Recovery was observed at room temperature. The Netherlands' biophysicists would expect recovery at low temperature simply because the lipids that are associated with the membrane proteins are in the most favourable association with a liquid crystalline species which would exchange for a gelled one and, by repacking of molecules after a sudden shift of temperature, membrane integrity would be restored - provided only a small percentage of the lipids have undergone phase transition.

Giaquinta: I agree. But, I'm not sure that there would be a rapid efflux or change in permeability in a chilling-resistant species like a sugar beet. The introduction of spin labels into the lipid matrix has not been able to show a phase transition upon chilling.

Christy: Are you suggesting that the lateral membrane of the sieve tube is becoming more permeable?

Giaquinta: We didn´t localize it to the plasma membrane.   It could also be involved in the endoplasmic reticulum because we saw structural disruption during low temperature treatment to chilling sensitive   species.   This would block sieve pores, increase resistance, and require increased pressure to overcome it in a manner such as you´ve shown in your model.

Ferrier: Cataldo and I and others have found that there is a marked decrease in the exchange of tritiated water between the sieve tube and the surroundings - it seemed that there was a decrease in permeability.

Troughton: It is necessary to distinguish between effects on metabolism and on membranes.   Unless you have done spin label experiments on a variety of sensitive and non-sensitive plants to show phase transition in particular membranes, it is very dangerous to speculate, that, in fact, chilling has an effect on the membrane.   In virtually all our studies we´ve not yet been able to show that there is a phase change in chilling sensitive plants.

Aikman: Fensom, Moorby, Williams, Dale and I have recently been Using ^{11}C to see if there was, indeed, loss of sucrose from the translocation stream in the region of chilling.   Preliminary analysis indicates none.   There could have been some increase in viscous resistance.

Watson: The effect is more complicated than viscosity changes alone.   In a 20 cm pathway, doubling the resistance to flow in a 2 cm length would cause, perhaps, a 10% extra resistance, whereas Swanson and Geiger found a temporary fall in translocation of as much as 90%.

Ferrier: Dr. Watson has a good point.

Hall: Dr.   Reinhold,   Bidwell   showed   recently   that indoleacetic acid (IAA) applied to a lateral bud of a decapitated bean plant was transported very rapidly into the adjacent leaf and there stimulated photosynthesis and export. Wareing believes that exogenous IAA may affect sucrose movement to the treated area but is not, itself, translocated in the phloem. We have found by mass spectrometry that IAA is present in both xylem and phloem exudate at about 5 mg l^{-1} and, therefore, is freely available to move throughout the whole plant. If it does have a direct effect on translocation one would expect that it would be administered along the whole pathway.

Reinhold: I agree. We ought to look more carefully at the effect of IAA on loading or on some other step in the process. With Wareing´s decapitated seedling, this can´t be the explanation because Mullens has treated these seedlngs with TIBA at concentrations that inhibit IAA transport and yet this hasn´t affected the IAA mobilizing effect on assimilate.

Hall: 0.1% TIBA has rather drastic effects on bean plants if you leave it on for several hours or days. First, the buds drop off, then the leaves, then the whole plant dies, so the experiment has not much value.

Cronshaw: In a study we did on tension wood formation we found that by putting a ring of TIBA around woody seedlings (e.g. Acer, Ulmus), we always got a swelling above the ring which we think is caused by stoppage of transport and accumulation of assimilates above the ring. The trees live after that and appear to be perfectly healthy. We have shown by cytochemical localization that there is ATPase activity in the plasma membrane of the sieve elements and companion cells. Could some of the effects that have been described, be mediated by ATP? Could we have some structural change in the membrane that would alter the ATPase activity in the membranes during osmotic shock situations? Could the effect of light be through ATP produced by photophosphorylation - having nothing to do with sucrose, but, perhaps, on levels of ATP?

Reinhold: While some believe that uncoupling oxidative phosphorylation depresses translocation, Dr. Willenbrink and I have not been able to find any effect of uncouplers on translocation (provided great care is taken to prevent them from reaching the loading sites). That is one reason why we don´t think that ATP can be very intimately connected with the translocation process. Another reason is that Ulrich has found that cyanide inhibition of fluorescein transport cannot be countered by ATP, even though there is some sort of ATP effect on fluorescein transport in the phloem. He thinks it has to do with loading and not with translocation along the sieve tubes.

Christy: If photophosphorylation is to drive translocation, it will require a very sophisticated system to get ATP from the chloroplast through 2 or 3 cells to the sieve tube without being hydrolyzed. Moreover, there is no agreement as to whether or not the outer membrane of the chloroplast is permeable to ATP. Are there any other comments on this?

Geiger: This may seem unrelated to the discussion, but one of the problems is to localize these pools. The sugar beet leaf seems to store much of the material that is not translocated as hexose and as starch, or a glycogen type of polysaccharide, rather than as sucrose. Why there isn't a great build-up of sucrose in sugar beet leaves under our situation may mean that the non-translocate pool is relatively small, and, in fact, sucrose builds up at a rather slow rate. One place to put the translocatable sucrose pool would be in the minor veins - in the sieve elements and companion cells. At the meetings at Calgary, last year, it was reported that rather small potassium deficiencies caused the retention of sucrose in isolated chloroplasts. I know there is disagreement that sucrose is even made in the chloroplast. Sucrose retention by chloroplasts would fit very nicely into the diagram which showed that potassium would, in fact, block the supply. Exogenous sucrose, however, might well be able to get to the sieve elements by another way and would not be similarly affected.

Outlaw: Elaborating on what Dr. Geiger has just said, Vicia faba leaves have about 5 times more sucrose per unit area than sugar beet leaves.

Swanson: Others have repeated Dr. Bidwell's work on IAA vs. photosynthesis but have been unable to confirm it. Any conclusions should be reserved for the moment. With the ^{11}C-technique that Dr. Troughton is using, many problems can be handled very elegantly. Dr. Troughton, have you investigated Hartt's so-called phototranslocation effect for corn leaves? This is the effect of light of low intensity - about 50 ft. c - on the basipetal portion of a sugar cane leaf preventing normal basipetal translocation of labelled sugars from a zone higher up?

Troughton: We've done light response curves on the speed of translocation only. In a C_4 plant like maize, the ^{11}C-labelled pulse that we look at, probably represents the mass of materials being moved at the time of labelling. Irrespective of light intensity, the bulk of the labelled material moves out of the fed area in, say, 90 min. Movement is directional and always towards the base, regardless of whether the leaf is in high light or in the dark. The amount that moves to the tip of the leaf is always very small - less than 2-3%. Now, this is with plants kept under a normal diurnal cycle. If, however, we keep the plant under abnormal conditions, e.g. 40 hr in the dark to de-starch it, then loss of material from the fed area is reduced quite dramatically; but, even under those conditions, there is no enhanced movement towards the tip. In our experiments, movement is highly directional, generally towards the base of the leaf.

Ho: We measured the mass transfer of carbon, using ^{14}C, in tomato during 12-hr light, 12-hr dark cycles. When the leaf was about 50% of full size, export was higher in the light period than in the dark, but when the leaf was nearly to full size and after, the export was higher in the dark than in the light. Does anyone have similar observations?

Eschrich: Translocation dependent on photosynthesis, rather
than phototranslocation, is surely a fact. What about potatoes
sprouting in the cellar where no light is available for
photosynthesis? Or reactivation in spring? There must be a phloem
loading process that is independent of light which depends on other
energy sources. I would suggest that Dr. Heyser show us some of
his results with sugar loading of maize leaves which will help to
define the independence of transport from light or photosynthesis.

Heyser: (Using slides, Dr. Heyser reported on results of
experiments with strips of maize leaf given a pulse label with
$^{14}CO_2$ in a band in the middle, the zones basipetal or aeropetal to
the fed area exposed to various treatments, and the direction of
translocation of the labelled assimilate observed. Using
pre-darkened leaves with both sides under normal air conditions in
light, translocation was always basipetal. If both sides were in
atmospheres minus CO_2, translocation went about equally to both
sides. That is, normal basipetal translocation did not occur. Dr.
Heyser then described the directonal effects of applying various
sugar solutions to the maize leaves. If he supplied sucrose or
other sugar solutions to one part of the leaf, for example,
^{14}C-translocate tended to move away from the point of application.
He didn´t observe the strong mannose effect that Dr. Mitter
described. Assimilates moved towards, rather than away from, the
site of application of inositol, arabinose,
ethylenediaminetetraacetic acid (EDTA) or propylene glycol. With
low concentrations of mannose applied at the apex, translocation
was basipetal, but at higher molarities, it suddenly changed and
went to the apex - towards the appled sugar solution - presumably
in response to plasmolyzing concentrations.)

Christy: A 0.012 M sucrose solution would be sufficient to
draw water out of a sieve tube. If sieve tubes contain 25% sucrose
- that´s approaching 1 M - it would tend to pull water in.

Milburn: If cells are initially under full turgor, and they
are elastic, and you change to 0.2 M mannitol (it doesn´t matter if
its only a minor change) you will get an elastic turgor change,
causing a change in the balance, but not a plasmolytic change,
that´s necessary to get the flux.

Tyree: I agree with Dr. Milburn. All that is needed is to
change the water potential difference between the inside (of the
phloem) and the out, and there will be a water flow for quite some
time - even well beyond the adjusting of volumes.

Weatherley: Mr. Pakinathan, a research student of mine,
subjected sunflower and cotton plants to a period of light, then
followed the loss of dry weight and of sucrose content per unit

area of leaf during darkness. He found that the decline in rate of translocation and in sucrose content were strongly correlated. Dr. Reinhold referred to some experiments by Weatherley and Peel in which the xylem in segments of willow was infiltrated with various concentrations of mannitol and the rate and concentration of phloem exudate from aphid stylets was determined. From this data, the mannitol secretion rate into the sieve tubes was calculated. With increasing osmotic pressures in the xylem, the rate of exudation declined, but the concentration rose. The product of the two - the secretion rate - very nearly balanced. In Peel's recent book there is a graph showing that the rate of secretion of sugar vs concentration of sap is a fairly steep straight line.

Dr. Christy, in your model, when you cooled the translocating stem and increased the resistance, wouldn't there be a build-up of concentration above the cooled zone, hence an increase in pressure and perhaps a greater recovery of rate? Would not this increase in concentration in the sieve tubes slightly reduce secretion or transfer into them and, in turn, reduce the specific mass transfer rate? Does the model show this?

Christy: The model doesn't have any feedback to change the loading rate. Loading rate is constant. It's very difficult to say how loading interacts with the concentration of sucrose in the sieve tube.

Weatherley: It's clearly what we need to know, because, at the end of the day, this is what is going to control the rate of translocation - the rate of loading and specific mass transfer.

Tyree: The modelling would be easy if Dr. Weatherley or Dr. Reinhold could tell us what Michaelis-Menton kinetics to use for loading.

Reinhold: You can safely start with Michaelis-Menton kinetics because all carrier processes so far studied approximate to it reasonably well.

Moorby: In sprouting potato tubers you get bigger transfer in the dark than in the light. You can even get new tubers in the dark and go on like this for 4 generations. There is a big feedback from the sink to the source. Dr. Troughton and I found with tomatoes that by cutting off the fruit you can, in fact, cut down the speed of ^{11}C translocation very rapidly, and cut down the rate of translocation within a matter of hours, while there is no effect on the rate of photosynthesis for 3 days. There are some big, baffling effects here.

Geiger: Some of Dr. Sovonic Dunford's data with discs may suggest a possible explanation. When exogenous sucrose is applied to a leaf that can translocate, V_{max} is considerably lower when sucrose is not removed by translocation than when it is. This is an indication that there may be a feedback mechanism as Dr. Reinhold suggested.

Willenbrink: (Using a slide, Dr. Willenbrink reported on the effects of 6 metabolic inhibitors that affect membrane properties of plant mitochondria on long distance translocation of ^{14}C-labelled assimilates in the petiole of Pelargonium zonale. Because of their insolubility in water, the inhibitors were dissolved in 1% ethanol and 1% aqueous dimethylsulfoxide (both of which had been tested beforehand and had no effect on translocation). The inhibitor solutions were applied to the exposed central bundle of the petiole 15 hrs prior to ^{14}C-labelling. Cytochalasin B (10-100 μg/ml) had no effect. The uncoupling agent CCCP (10^{-7} - 2.10^{-6} M) caused ^{14}C-labelled substances to accumulate above the treated zone. Antimycin A (10 μg/ml), atractyloside (5.10^{-5} M), ouabain (strophanthin) (10^{-2} M), and valinomycin (5.10^{-7} M) caused a 50% inhibition compared to controls, in the rate of ^{14}C-labelled substances which reached and passed the treated zone during translocation times of 4-to-6 hours. None of these inhibitors caused ^{14}C-labelled substances to accumulate above the treated zone, however, valinomycin renders mitochondrial and other membranes sensitive to an uncontrolled uptake of potassium. The effective concentration of valinomycin was only 20-100 times that for mitochondrial membranes. An intimate linkage of translocation with a functional compartmentation of potassium, either with pumps at the sieve plates or along the sieve tube-companion cell units which control the concentration gradient of a mass flow, was suggested. Dr. Willenbrink stressed the need for his results to be confirmed by simultaneous cytology and autoradiography of the inhibited phloem.)

Giaquinta: Dr. Willenbrink alluded to the difficulty of deciding whether these metabolic inhibitors are affecting metabolic processes which are necessary for the driving force of translocation or whether they inhibit translocation through some side effects. There are similarities between the effects of inhibitors and low temperature. for example, the 90 minutes required for translocation to recover after removal of the inhibitors or of low temperature. In bean, we believe low temperature damage was not metabolic, but caused by physical blockage of the sieve plate pores. Like Dr. Ho, we've found that low concentrations of potassium cyanide (as used for metabolic inhibitors) are not effective in inhibiting translocation. At very high concentrations, we get inhibition and the ultrastructure, by freeze substitution, shows blockage of the sieve plates as we see at low temperature. Your concentrations could give rise to conflicting data; there could be two modes of interaction of these metabolic inhibitors.

Willenbrink: We always used an atmosphere of cyanide around the exposed central bundle of the petiole. Therefore, it could be that a lower concentration in solution would have been effective. I am not convinced that the effect can only be explained by a massive blockage of the sieve plates.

Christy: One possible effect that Dr. Giaquinta was referring to was on membrane permeability - either through metabolism or through a direct effect of the inhibitors on membranes.

Willenbrink: I would agree. Most of the effects of valinomycin are intimately linked with membranes. Perhaps the mitochondria were directly disturbed in supplying energy to keep the whole sieve tube system open for flow.

Christy: Have you tried to plasmolyze sieve tubes that have been treated with metabolic inhibitors?

Willenbrink: No yet. We tested plasmolysis in earlier experiments with fluorescein and with inhibition by cyanide.

Reinhold: Dr. Willenbrink, in the very long, 15 hr pre-treatment such as you used, might not some of the inhibitor reach the loading sites? What effects did a shorter pretreatment have?

Willenbrink: We've done only 3 or 4 experiments pretreating the phloem with inhibitors for 2 hr prior to $^{14}CO_2$ fixation. Inhibition occurred at concentrations nearly 10 times higher than the concentrations reported earlier. There are several processes mixed together, such as the time required for the inhibitor to penetrate to the conducting tissue. A shorter time with higher concentrations might be better.

Milburn: There has been a good deal of attention paid to the action of the source and rather little to the action of the sink, although the interaction of the two seems to be generally accepted. One of the most interesting methods of controlling the loading process is that induced by exudation - by either an aphid or by a razor blade cut. The rate at which loading must occur to sustain exudation must be very considerable. A castor been plant that is 0.5 m tall with a stem about as thick as a finger will produce, from a small razor blade cut, about 1.0 ml of concentrated sap per hour for many hours. Now, this is an increase in the total volume of sap that would otherwise be in the plant but for the cut. Furthermore, a very fascinating aspect is that by means of this razor blade cut what is normally the sink, i.e. the roots and lower stem, is converted into a source so that it will bleed sap and keep on doing this for considerable periods of time. These two effects suggest that the source/sink relation is a somewhat symmetrical one that can reverse quite quickly. A further point is that on making the razor blade cut the first effect is undoubtedly turgor release - a massive drop of pressure inside the opened sieve tubes. Immediately after this there must be a very rapid mobilization, which indicates that loading from the various organs, be they roots or leaves, has been "switched on" at a remarkably rapid rate. It is difficult to imagine how this could be induced, but I would suggest pressure since the concentration changes are relatively minor and you can easily send a pressure change through a long pipe at great speeds. But we run into problems when we come to the mesophyll in the leaf or the cell in a sink like a root. If there is no pressure transport in these cells, what could the mechanism be?

Reinhold: It is quite possible, and even probable, that turgor may control loading. At the Membrane Transport Conference at Ulich it was stated that active transport of potassium (in Valonia) is greatly enhanced if the tension of the plasmalemma is decreased. If might well be that the carrier mechanism for loading is very sensitive to the degree of tension in the membrane.

Dainty: I agree. It's becoming clear that active transport is sensitive to turgor. It is arguable that it is tension in the plasmalemma that is responsible, but all osmoregulating plants may well act through a turgor-sensitive active transport mechanism. We seem to be looking for very simple explanations but the loading process is far from simple. Dr. Reinhold used a rather simplified example, using Michaelis-Menton kinetics; but, in fact, the membrane process involved in loading may well depend not only on sucrose concentration, but also on potassium ion concentration, because there may be co-transport with potassium. It may well depend on hydrogen-ion pumping, again because of co-transport, counter-transport, or some complicated arrangement between sucrose, potassium ions and hydrogen ions. There must be an energy supply such as ATP. Turgor may be a factor, as just discussed, and membrane potential may enter in because the carrier would be charged if there is a co-transport system. It is not being very realistic to look for simple, single explanations. We know very little at the present time.

Lobban: Returning to the question of feedback, what happens to hormone relations in a plant that has more than one shoot, such as a grass or a strawberry with stolons? Is there any kind of inhibition of hormone producton in the secondary shoots by the main shoot and how might transport be affected if the apex of the main shoot is cut off, thereby removing the source of hormone?

Reinhold: I can't comment. I don't know the answer.

Christy: Using inhibitors of photosynthesis, would it be expected that translocation would go on during the night or stop altogether? We have some preliminary indications with herbicides like triazines or ureas which affect photosynthesis, that translocation stops altogether.

Willenbrink: Many inhibitors are not specific for photosynthesis. CCCP inhibits photophosphorylation and oligomycin inhibits electron transport. It would be difficult to sort out relations between inhibition of photosynthesis and the translocation process.

Moorby: I don't know about triazines and sugar transport. However, I did some experiments on the effects of substituted triazines on the transport of cesium and they were very similar to the effects of darkening. In addition to reducing the export of cesium from the leaf, triazines altered the proportions that went up and down. This is another effect on source/sink relations.

Price: I wish to introduce a note of caution about the interpretation of experiments involving the application of exogenous sugar solutions. On a leaf surface it dries and there is some concentration, some penetration, and some movement, of the solute throughout the leaf to the vascular tissues. With a stressed plant, actual mass movements of the solution into the leaf occur. My evidence is somewhat indirect and mainly concerned with reverse movements of herbicide compounds that normally move in the xylem but, under conditions of severe water stress, when the roots are short of water, they move in the phloem down towards the roots.

Reinhold: We've examined this question of droplet application very thoroughly and done double labelling experiments with tritiated water and ^{14}C-sucrose and found quite opposite effects of stress on water movement and sucrose movement. Girdling effectively stopped sucrose movement in stressed plants. Had there been a mass flow through the xylem capillaries you wouldn't have expected girdling to stop it.

Fellows: Dr. Reinhold, when doing the experiments with water stressed vs control plants, was photosynthesis measured at the same time? If photosynthesis was low prior to labelling there would be a lower amount of synthate available for export and the addition of the labelled material would alter the balance already there.

Reinhold: That's a very valid criticism. At that time, we weren't so specific activity conscious as we became later on. However, I don't think that is the explanation of our results. The degree of water stress we had in those plants was not high. In separate experiments run, unfortunately, a year or two later, we found only a slight effect of such mild water stress on photosynthesis.

Bidirectional Transport

Walter Eschrich

Forstbotanisches Institut der Universität Göttingen

34 Göttingen-Weende, Busgenweg 2, West Germany

I. Brief historical survey

The basic problem which prompted experimental work on bidirectional transport, was to elucidate the mechanism of assimilate translocation. Until recently, the goal of investigations on bidirectional transport was to determine whether simultaneous movement of solutes in opposite directions can take place in a single sieve tube. Such experiments were designed either by analyzing sieve tubes between two sources, or by analyzing bidirectional spreading from a single source.

Curtis (1920a,b) was one of the first investigators who suggested that carbohydrates may be transferred longitudinally through the phloem in both basipetal and acropetal directions. This was at a time when the transpiration stream of water was named the "ascending", and the stream of assimilates was called "descending" sap stream.

Later, Mason, Maskell and Phillis (1936), working on the transport of carbohydrates and nitrogenous compounds in cotton plants, came to similar conclusions. Eventually, the concept of source-to-sink movement of assimilates was advanced by Mason and Phillis (1937), and that of a simply descending sap stream became obsolete. Nevertheless, descending sap streams still wet a few pages of some of our present-day textbooks.

It was at that time that Schumacher (1933) introduced potassium fluorescein as a tracer for the sap movement in sieve tubes. Although fluorescein may be discernible only when adsorbed by P-protein in sieve tubes, its movement must be governed by a mechanism similar to that by which assimilates are moved. Schumacher was able to observe fluorescing sieve tubes in the stem above and below the node where the source leaf was attached. Up until now, potassium fluorescein is the only tracer of assimilate movement in sieve tubes which can be seen in vivo (Fig.1).

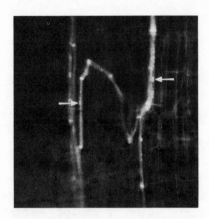

Figure 1. Sieve tubes and companion cells of Impatiens holstii transporting a fluorescent dye solution. Arrows pointing to labelled nuclei of companion cells. 150x. (from Eschrich, 1953).

When radioactive tracers became available as tools for biological research, short-term experiments with two different isotopes became possible. Chen (1951) demonstrated the simultaneous movement of ^{32}P and ^{14}C in opposite directions in the phloem.

Biddulph and Cory (1960), utilizing the same isotopes, were able to differentiate between ^{14}C- and ^{32}P-radiation in autoradiographs of bean plants. Their results provided the first evidence that bidirectional movement could occur in a single vascular bundle. With Biddulph and Cory's paper, a series of investigations were initiated which, besides their original purpose of elucidating the problem of bidirectional transport, introduced new techniques for the study of sieve-tube translocation.

Biddulph and Cory (1960) applied $^{14}CO_2$ to the lowest ternate leaf of a bean plant (Phaseolus vulgaris) and $^{32}PO_4$ to the next higher leaf. After 30 min the bark was peeled off, frozen and dried. The cambial surface of the flattened bark was exposed to X-ray film either directly to record ^{32}P- and ^{14}C-radiation or with an interface of aluminum to reduce almost completely the ^{14}C-radiation. In addition, autoradiographs were prepared after the decay (10 half lives) of ^{32}P. The autoradiographs showed that the acropetal transport of the ^{14}C-label and the basipetal transport of the ^{32}P-label took place in separate bundles of the stem. However, movement of both tracers against each other also occurred in the same bundle, but obviously only in the most recently formed sieve tubes of the secondary phloem.

Trip and Gorham (1968) administered glucose-6-^{3}H solution to a mature leaf of a <u>Cucurbita</u> seedling, and after 160 min a half-grown leaf of the same plant was allowed to assimilate [14]CO_2. Twenty min later the plant was frozen and parts of it were processed for liquid scintillation counting. The results confirmed that tritium label had entered the petiole of the [14]CO_2-fed leaf and ^{14}C-label had passed the advancing tritium label. From that region freeze-dried petiole tissue was infiltrated with paraffin and 10 μm sections were exposed to Kodak NTP-10 plates. Part of the emulsion had been coated previously with gelatine to screen out ^{3}H-radiation. Unfortunately, microautoradiographs obtained by this procedure seldom yield immaculate preparations. So, Trip and Gorham's microautoradiographs scarcely give convincing evidence for a bidirectional transport in the same sieve tube. However, this experiment shows clearly that the tracers moved in opposite directions in the same phloem bundle.

Another technique has been used by Eschrich (1967) and by Ho and Peel (1969). When a single aphid is placed between two sources of differing tracers, its honeydew should contain both tracers when it is feeding in a sieve tube where bidirectional movement occurs.

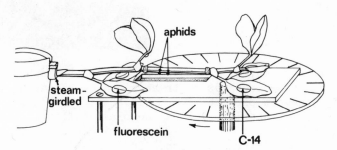

Figure 2. Set up for the demonstration of bidirectional transport by the time course of doubly-labelled honeydew production of single aphids. (from Eschrich, 1967)

In one case (Eschrich, 1967), <u>Vicia faba</u> plants with 4 primary leaves were investigated, and individuals of the aphid species <u>Acyrthosiphon pisum</u> were applied to the internode below the 3rd leaf (Fig.2). The honeydew was collected on a turntable rotating once in 24 h. The leaves of <u>Vicia faba</u> are arranged in two orthostichies, leaf 3 positioned over leaf 1. The latter was supplied with K-fluorescein, while leaf 3 was fed with ^{14}C-labelled urea or bicarbonate. In preliminary experiments it was shown by fluorescence microscopy and microautoradiography that both tracers were present in the same phloem bundles of the stem part between the two source leaves. In 12 experiments, 68 aphids produced rows of honeydew drops which were recorded by photographing the "honeydew-chronogram" under UV-light, as well as by exposing it to X-ray film (Fig.3). The results showed that 42% of the rows of droplets were doubly-labelled from the first drop. 7% first labelled singly, later became doubly-labelled. 21% were only fluorescein-labelled, 28% were only ^{14}C-labelled, and 2% were unlabelled.

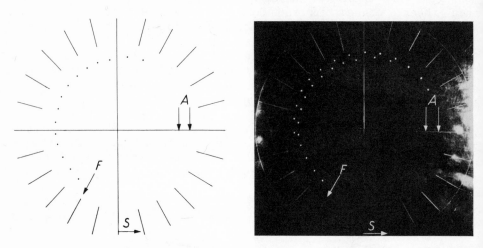

Figure 3. Honeydew chronograms (24h) of two aphids feeding in sieve tubes between sources of tracers. A, applicaton of aphids. F, end of honeydew production. S, start of experiment. The inner row of honeydew droplets shows double-labelling with ^{14}C-urea (autoradiograph left) and K-fluorescein (photograph taken under UV-light, right). (from Eschrich, 1967)

Similar results were obtained by Ho and Peel (1969). Shoots of willow (Salx viminalis), 15 to 30 cm long, were supplied on both ends with two different tracers, either ^{14}C-sucrose and ^{3}H-glucose or Na_2[14]CO_3 and [32]PO_4. Single aphids of the species Tuberolachnus salignus were positioned between the sources of tracers and the honeydew of each was collected on cellophane on a turntable. The single drops of honeydew were cut out and assayed by liquid scintillation counting. In most cases doubly-labelled honeydew was produced during a period of 24 to 48 hours.

A third group of investigations has been concerned with bidirectional spreading of a tracer from a single source. Such experiments were carried out by Peterson and Currier (1969) and by Fritz (1973). In both cases <u>Vicia</u> <u>faba</u> was utilized as the experimental plant.

Peterson and Currier (1969) supplied a buffered solution of K-fluorescein from a trough around the stem of the plant and investigated cross sections made by hand under the fluorescence microscope. Of 130 bundles from 13 treated plants only 6 bundles were found in which the dye was located in both apical and basal sections.

Fritz (1973) concentrated his studies on internodes of <u>Vicia</u> <u>faba</u>, which showed in gross autoradiographs a bidirectional spreading of the ^{14}C-labelled tracer. Very low concentrations of phenylalanine but with high specific activity were applied outside of one bundle of the stem. Using an autoradiographic technique which allows the exact localization of soluble radioactivity in any sieve tube of a bundle, he got the following results. From 26 treated plants 12 showed bidirectional spreading. 5 of them had labelled sieve tubes in the same region of the phloem above and below the site of application of the label. In the remaining 7 plants, the acropetally and basipetally transporting sieve tubes were located in different regions of a cross-sectional area.

II. Anatomical aspect

The results reported so far allow a wide variety of interpretations. Those of Peterson and Currier show that both acropetal and basipetal transport are restricted to different bundles. The same interpretations can be applied to Biddulph and Cory's results, although some sieve tubes, probably the youngest of the secondary phloem, showed bidirectional transport. The results of Trip and Gorham and those of Ho and Peel have been interpreted by the authors as bidirectional transport in single sieve tubes. The same interpretation could be applied to my own results obtained with aphids and to some experiments in the paper of Fritz. However, before drawing conclusions on a possible mechanism, we should spend some time on thoughts on the anatomy of the phloem.

Long-distance translocation of assimilates is restricted to plants equipped with phloem or tissues equivalent to those occurring in mosses and red brown algae. Among these plants there apparently are no species or plant organs with a single sieve tube. Sieve tubes or equivalent rows of cells are branched and interconnected forming a system which traverses the whole plant and has numerous blind endings. Thus, one incorrectly speaks of a "single sieve tube" but means a "single part of the sieve tube system." Keeping this fact in mind, any result in bidirectional transport in single sieve tubes can be interpreted by the following schema (Fig.4).

^{14}C-assimilates

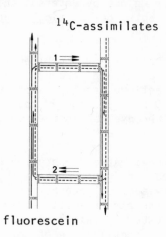

fluorescein

Figure 4. Homodromous loop path of two tracers moving in separate sieve tubes in opposite directions feasible by anastomoses or lateral sieve plates (according to Eschrich, 1967).

If two different tracers are moving in opposite directions, but in different sieve tubes, exchange must be possible when lateral sieve plates or anastomoses are present. This exchange was termed a "homodromous loop path" (Eschrich, 1967). It would allow the mixing of both tracers in one sieve tube without opposing directions of flow.

III. Vicia faba, a paradigm

During my thesis work (Eschrich, 1953) I tried to find out whether wounded sieve tubes, which are formed after severing a bundle, are able to merge with the old sieve tubes by forming a sieve plate.   This was never the case; instead, the newly-formed wound sieve tubes were continous with newly-formed longitudinal sieve tubes which run along the old bundle, and, it is unknown how far those may extend in either direction.   This shows that regeneration does not occur in mature sieve tubes. In addition, it suggests that probably more than one sieve tube system can exist in a single plant and that such sieve tube systems are of different origins.

Vicia faba, the main object of studies on bidirectional transport, may serve as a paradigm. When analyzing the pattern of venation in a Vicia faba stem by serial cross sections (Fig.5), it can be seen that from the 5 prominent bundles of the petiole only 3 can be followed backwards into the stem. There is one median leaf trace, which immediately enters the ring of stem bundles, and two

lateral leaf traces, which run isolated inside the wings of the
stem over a distance of two internodes before they branch and merge
with the stem bundles. From cross sections at any level of the
young plant we can estimate that such leaf traces branch repeatedly
(Fig.6) and at every second node, and may be recognized as far as 8
to 10 internodes below the node where the leaf is attached. The
size of the leaf traces diminishes continuously in the downward
direction. In the first and second internodes below the leaf, the
associated traces are represented by primary phloem and primary
xylem, including both protophloem and protoxylem. In the 7th and
8th internodes below the leaf the branches of the leaf traces are
represented only by metaphloem. Xylem is lacking at that level and
will not be formed until secondary growth is initiated there.

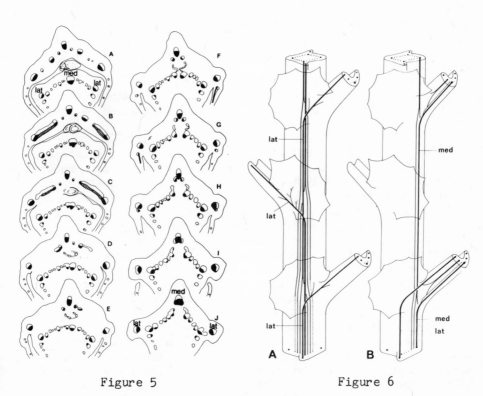

Figure 5   Figure 6

Figure 5. Serial cross sections of <u>Vicia</u> <u>faba</u> stem from above
(A) to below (J) a node. Merging and splitting leaf traces lead to
the diagram in Fig.6. Black parts of bundles represent xylem (from
Fritz, 1973).

Figure 6. Course of lateral (A) and median (B) leaf traces in
the stem of <u>Vicia</u> <u>faba</u> (from Fritz, 1973).

Translating the anatomical pattern into developmental terms, it can be stated that the leaf traces of a given leaf are initiated 8 to 10 leaf plastochrons prior to the formation of the primordium of that leaf. Accordingly, the procambium appears in the stem before it does in the leaf and it differentiates acropetally (Esau, 1965b). Within the growing primordium acropetal differentiation, in general, proceeds until formation of the leaf blade is initiated. Concomitantly with the elongation of the primordium, protophloem elements differentiate acropetally. It is not known how far back in the stem leaf traces are established with protophloem. With increasing size of the leaf and elongation of the internodes, formation of protophloem sieve tubes is discontinued and metaphloem elements differentiate. Metaphloem develops bidirectionally, acropetally and basipetally, beginning at the level of the stem where the leaf is inserted (Esau, 1965a). Before the leaf reaches maturity, it passes through a developmental stage involving overall basipetal maturation of the leaf blade. The upper limit of this stage has been characterized for <u>Populus deltoides</u> leaves by Larson et al (1972): leaf plastochron index 4, the pattern of smallest veins can be recognized and the mesophyll exhibits intercellular spaces.

After considering the anatomical and developmental aspects, we are now able to draw conclusions about the regulation of assimilate transport.

The shoot tip and all growing leaves import assimilates which are provided by exporting leaves. According to the pattern of leaf traces, exported assimilates move basipetally. Acropetal recycling results when the leaf traces of the exporting leaf come into contact with leaf traces of a younger, still importing leaf. Recycling from basipetal to acropetal transport can occur as far as 10 internodes below the exporting leaf, provided leaf traces of importing (growing) leaves are available at that level of the stem. Thus, the shoot tip and growing leaves draw assimilates from 8 recently matured leaves and in part from partly mature leaves, which are exporting and importing simultaneously.

The latter deserve special consideration because their leaf traces show simultaneous basipetal and acropetal transport. Such bidirectional transport within a single bundle is restricted to sieve tubes of the metaphloem (Fritz, 1973). Apparently, the first-formed sieve tubes of the metaphloem are concerned primarily with acropetal transport. Since differentiation of the sieve elements progresses basipetally in the leaf traces and acropetally in the leaf, imported assimilates first move against the gradient of sieve tube differentiation. After entering the leaf, imported assimilates move with the differentiation gradient. Exported assimilates, in turn, have first to move against, and then, after entering the stem, with the gradient of sieve element differentiation. Thus, there seems to be no relation between direction of sieve element differentiation and direction of transport.

The number of metaphloem sieve tubes transporting in the acropetal direction is relatively few (1 to 3 in the pictures published by Fritz, 1973). It is not known whether those sieve tubes can switch over to basipetal transport. Microautoradiographs of leaf traces of a mature (exporting) leaf, which previously was exposed to $^{14}CO_2$, show many labelled sieve tubes of the metaphloem concerned with basipetal transport (Fig.7). However, few of them appear unlabelled. It is not known whether the unlabelled sieve tubes represent those which initially transported in an acropetal direction.

We can assume that bidirectional transport in general takes place in separate bundles, a process which has been termed recycling. This is the case when assimilates exported basipetally from a mature leaf meet leaf traces of a growing organ. Therein they move acropetally.

One might argue that assimilates exported from eldest leaves will not find contact with leaf traces from growing leaves, and have to move continuously in basipetal direction. In this case it might be expected that cambial activity has provided a new set of sieve tubes which probably are continuous with younger leaf traces in the upper part of the stem. However, secondary phloem has an origin other than the primary metaphloem and both may constitute separate sieve tube systems. This problem deserves a careful anatomical investigation. The same question is whether increments of secondary phloem following each other, year after year, in perennial dicotyledons are interconnected or not.

Besides the phenomenon of recycling in separate bundles, simultaneous bidirectional transport in a single bundle seems to be restricted to photosynthesizing organs of a certain developmental stage. This stage encompasses the period from initiation of assimilate export to the cessation of assimilate import as a requirement for growth.

No data are available which unequivocally show a simultaneous bidirectional transport in a single sieve tube, i.e. a part of the sieve tube system composed of one row of sieve elements.

There remain some other questions to be answered. Does the phloem of the roots show bidirectional transport, or are all assimilates arriving from the shoot moving towards the root tip? If bundles of perennial monocotyledons with only primary phloem would sooner or later switch over to basipetal transport, how would the aerial organs of such plants draw their supply of nitrogen and phosphorus? What is the situation in storage organs, when import

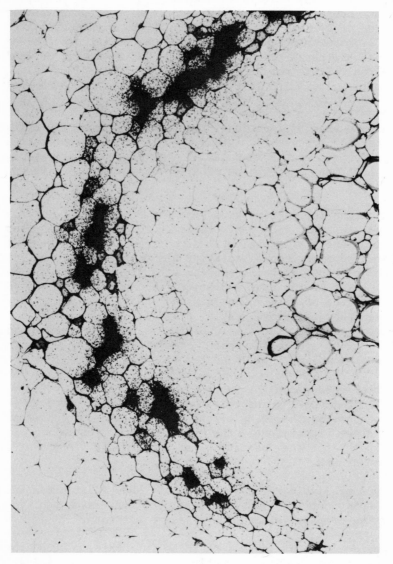

Figure 7. Microautoradiograph of a cross section of *Vicia faba* stem. Only sieve tubes and companion cells of the metaphloem are labelled. One mature leaf above was treated with $^{14}CO_2$. Some sieve tubes appear less or unlabelled (from Eschrich and Fritz, 1972).

switches over to export? Are there different sets of sieve tubes
involved ?

IV. Mechanism of bidirectional transport

The principle of apical growth results in plant individuals
which have undifferentiated, mature and senescent leaves at the
same time. When all tissues of the stem follow the same gradient
of differentiation, assimilates provided by a mature leaf have to
pass senescent parts of the sieve tube system when moving to the
roots.

This situation occurs perhaps in protostelic cryptograms.
However, the rootless gametophytes of the moss Polytrichum show
only acropetal transport (Eschrich a. Steiner, 1967) (Fig.8).

Higher plants have surmounted this problem by evolution of the
vascular cambium and, eventually, by the formation of leaf traces
with basipetal differentiation. By saving procambium for a period
of 10 or more leaf plastochrons, direct connections of leaf traces
belonging to leaves of different age become possible at any level
of the stem (Fig.9). The optimal use and distribution of
assimilates is provided by this system. In addition, basipetal
differentiation suits especially the supply of the roots with
photosynthetic assimilates.

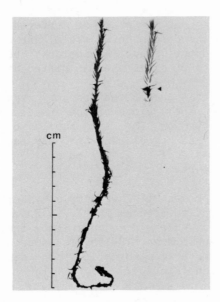

Figure 8. Gametophyte of <u>Polytrichum commune</u> (left) and its
autoradiograph (right) after labelling a single leaf (arrow head)
with ^{14}C-sucrose. Only acropetal transport can be recorded
(according to Eschrich and Steiner, 1967).

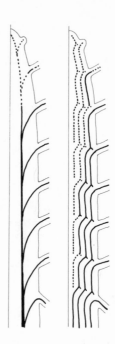

Figure 9. Pattern of leaf-trace differentiaion in a protostelic plant (left) and an eustelic plant (right). Solid lines, leaf traces of mature leaves. Stippled lines, leaf traces of immature leaves.

One might argue that the state of differentiation of a phloem bundle influences the direction of assimilate transport. However, as I already pointed out, recycling includes both transport with and against gradients of sieve element differentiation.

Bidirectional transport seems to be regulated entirely by source and sink. The joints of leaf traces merely seem to open successively new sources for the supply of growing organs. Therefore, it can be concluded that any part of the sieve tube system as soon and as long as it is mature can be utilized for bidirectional transport.

This has been proven with two simple systems: (1) The silks of female corn inflorescences are 15 to 20 cm long. Each stigma contains 2 bundles which are separated from each other by parenchymatic tissue. When ^{14}C-labelled sucrose is applied to one or the other end, the tracer moves in both directions, and with velocities of 60 cm/h (Fig.10).

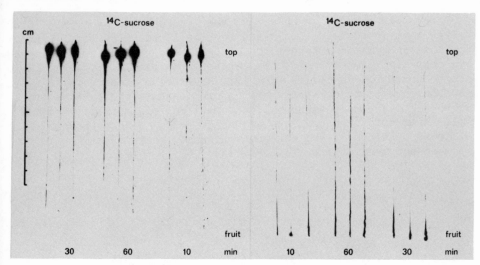

Figure 10.   Autoradiographs of silks of young cobs of <u>Zea</u> <u>mays</u>.   Movement of ^{14}C-labelled sucrose basipetally to the fruit (left part of Fig.10) and acropetally from the fruit to the apex of the silk (right part of Fig.10).   Equal transport velocities in either direction.

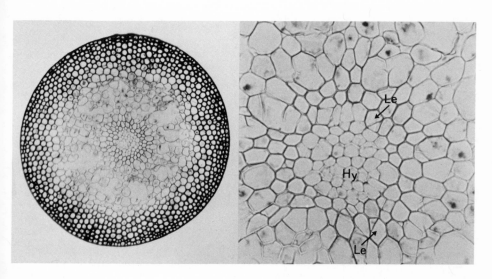

Figure 11.   Cross sections of the mature seta (left side) of <u>Polytrichum</u> <u>formosum</u>.   Diameter: 300 m.   Central strand magnified on right side shows hydroids (Hy) surrounded by groups of leptoids (Le).

(2) The second system is the sporophyte of the moss
Polytrichum commune. Its seta becomes 12 to 15 cm long. Inside
the sclerenchymatic metaderm the central bundle with leptoides and
hydroids is loosely connected to the outer mantle (Fig.11). This
hadrocentric bundle begins in the haustorium and ends in the
capsule. Autoradiographs have shown that a radioactive tracer
moves simultaneously acropetally as well as basipetally (Fig.12).
Velocities of transport are around 50 cm/h. Microautoradiographs
of cross sections of the seta show highest grain density in the
ring-shaped area of leptoids (Fig.13).

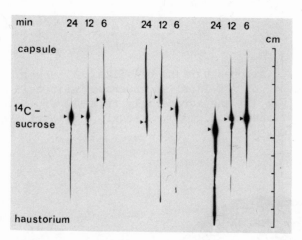

Figure 12. Autoradiographs of isolated cores of the seta of
Polytrichum commune. ^{14}C-sucrose, applied to the middle of the
strands (arrow heads), moved acropetally as well as basipetally.

Since bidirectional spreading also occurs in the isolated
seta, where haustorium and capsule are cut off, the influence of
assimilate consuming sinks can be ruled out. P-protein is lacking
in leptoids. Any mechanism based on participation of P-protein is
not applicable. In addition, leptoids have no sieve pores
(Paolillo, 1967; Hébant, 1970; Eschrich and Steiner, 1968;
Stevenson, 1974), their protoplasts are connected by plasmodesmata.
Therefore, it can be guessed that the Münch-type pressure flow
would meet extreme frictional resistance in the plasmodesmata. A
Münch-type flow with no longitudinal pressure gradient seems more
suitable to generate a transport velocity of 50 cm/h. Such a
mechanism, which has been termed volume-flow (Eschrich et al, 1972;
Young et al, 1973) is applicable not only to sieve tubes but also
to systems of endoplasmatic reticulum which are continuous with the
plasmodesmata.

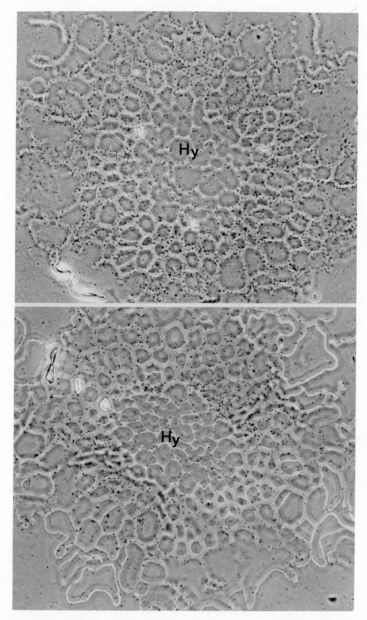

Figure 13. Microautoradiographs of the central core of the seta of <u>Polytrichum</u> <u>commune</u>. Freeze-dried and resin-embedded tissue of two different experiments were exposed to Kodak stripping film. ^{14}C-label caused scattered silver grains most prominent in the ring of leptoids (courtesy of Dr. E. Fritz, Gottingen).

References

BIDDULPH, O. and CORY, R.   1960. Plant Physiol. 35: 689-695
CHEN, S.L.   1951. Amer. J. Bot. 88: 203-211
CURTIS, O.F.   1920a. Amer. J. Bot. 7: 101-134
CURTIS, O.F.   1920b. Amer. J. Bot. 7: 286-295
ESAU, K.   1965a. Plant Anatomy, 2nd Ed. John Wiley & Sons, New
     York
ESAU, K.   1965b.   Vascular differentiation in plants.   Holt,
     Rinehart and Winston, New York
ESCHRICH, W.   1953. Planta 44: 37-74
ESCHRICH, W.   1967. Planta 73: 37-49
ESCHRICH, W.   1972. Planta 107: 279-300
ESCHRICH, W. and FRITZ, E.   Microautoradiography and electron
     probe analysis: their application to plant physiology, ed. U.
     Luttge. pp. 99-122
ESCHRICH, W. and STEINER, M.   1967. Planta 74: 330-349
ESCHRICH, W. and STEINER, M.   1968. Planta 82: 321-336
FRITZ, E.   1973. Planta (Berl.) 112: 169-179
HEBANT, C.   1970. Phytomorphology 20: 390-410
HO, L.C. and PEEL, A.J. 1969. Ann. Bot. 33: 833-844
LARSON, P.R., ISEBRANDS, J.G. and DICKSON, R.E.   1972.   Planta
     107: 301-314
MASON, T.G., MASKELL, E.G. and PHILLIS, E. 1936. Ann. Bot. 50:
     23-58
MASON, T.G. and PHILLIS, E. 1937. Bot. Rev. 3: 47-71
PAOLILLO, D.J. and REIGHARD, J.A.   1967.   The Bryologist 70:
     61-69
PETERSON, C.A. and CURRIER, H.B.   1969.   Physiol.   plant 22:
     1238-1250
SCHUMACHER, W.   1933. Jahrb. wiss. Bot. 77: 685-732
STEVENSON, D.W.   1974. Amer. J. Bot. 61: 414-421
TRIP, P. and GORHAM, P.R. 1968. Plant Physiol. 43: 877-882
YOUNG, J.H., EVERT, R.F. and ESCHRICH, W.   1973.   Planta 113:
     355-366

DISCUSSION

Discussant: H.B. Currier
              Botany Department
              University of California, Davis
              Davis, California

     Currier: I have no particular or serious reservations about
any of the views expressed by Dr. Eschrich. We have always told
our students that if bidirectional transport is proven without any
doubt, then the mass flow theory is "out the window", that is, mass
flow would be impossible. And I think this statement still holds
if we´re thinking of a single sieve tube or a single sieve element
series in a sieve tube system. I think it is quite clear from Dr.
Eschrich´s remarks that the term bidirectional transport has been
redefined over the sense used in previous years and in previous
publications. Now the definition is much broader.   It would
include two-way transport in different bundles at any particular
time. It would include two-way transport occurring simultaneously
in two different sieve tubes in the same bundle. It would also
include movement toward the apex at one time and then reversing at
another time.   These examples are logical and proven, but the
conclusion that a single sieve tube, or a single part of the sieve
tube system, can transport simultaneously two or more species of
solutes in opposite directions, seems unproven as yet.   This new
definition I think is better - we do then have bidirectional
transport under the conditions and limitations that I´ve talked
about. The main point is that regulation is by source and sink and
the source and sink relationship can change with time. I was very
much interested in the presentation by Dr. Turgeon where he
referred to the physiological maturity of a leaf in conjunction
with the morphological maturity of a leaf. It is evident that, as
a leaf develops, there is a point, to which both Dr. Turgeon and
Dr. Eschrich referred, where portions of the leaf might be
conducting downward toward the base and other portions of the leaf
would continue to behave as sinks.   This, then could explain
bidirectional movement in different sieve tubes of the same or
different bundles.

     I was happy to see the emphasis on plant anatomy in Dr.
Eschrich´s presentation. Dr. Eschrich is thoroughly familiar with
this discipline and we can all be envious of his competency in this
area. I was also happy that Dr. Eschrich included a brief history
of the fluorescein method. Dr. Schumacher is represented at this
conference by several of his former students.   His work has
initiated much subsequent studies at many other universities. At
Davis, Ms. Peterson is applying the fluorescein method to studies
on Vicia faba, the long-suffering plant of translocation. Although

417

she has found bidirectional transport in different bundles, she has
not found instances of bidirectional transport in the same bundle
in this species. However, in the squirting cucumber plant, she has
discovered bidirectional transport in a single bundle. The outer
phloem in these bundles seems to be conducting downward and the
inner phloem upward. She found this relationship only when she
allowed a considerable time to elapse. Her interpretation is that
there is cycling in the sense of the diagram that Dr. Eschrich put
on the board. (Ed. note: a homodromous loop)

I think we need more work on the anatomy where cycling can
occur, where the anastomoses are. In our work with the grape
phloem, we have occasionally seen a sieve tube connecting one group
of phloem elements with another group of phloem elements, moving in
the tangential direction. We definitely must know more about
anatomy.

Srivastava: I would like to ask this of Dr. Eschrich:
bidirectionality in different bundles or in different sieve tubes
of the same bundle seems to have been shown generally in dicots.
What about gymnosperms, for instance, where there are no clear-cut
sieve tubes? Are there any reports of bidirectional transport in
these cases?

Eschrich: So far as I know, there are no studies of this kind
based on gymnosperms. What we know about gymnosperms is mainly
about secondary phloem. I think there are only a few papers
concerning the primary phloem. As I pointed out, it is not known
whether there is a connection between the increments of secondary
phloem in the radial direction.

Tyree: With respect to bidirectional transport, I would like
to include one additional type which I think is important. This is
the case where sucrose might enter a branch sieve tube and then
from that point of branching go in opposite directions. So we have
in this case bidirectional transport in the same sieve tube at the
same time, but in different regions of that sieve tube. A reverse
situation could occur if you have an aphid stylet - in this case
you would have bidirectional transport to the stylet.

I would like to ask a question about homodromous loops. I
understand in principle how one can envisage some mechanism by
which one can get looping, but I don't understand how this can
occur with the Münch pressure-flow mechanism. The pressure drop
where the loop is completed would be negative. (The discussion was
clarified by reference to diagrams drawn on the blackboard.)

Eshrich: One can postulate variable or changing pressure
differences, if there are any pressure differences at all. I think
parting in the assimilate streams can also occur between two sieve

tubes which have only lateral sieve plates, and we have seen many pictures where young sieve elements may have connections where sieve plates are divided into two sieve parts.

Currier: Another kind of cycling that can occur is between the phloem and xylem. We accept the fact that such circulation occurs, as has been well shown by the work of Dr. Biddulph at Pullman and Dr. Crafts at Davis - down the phloem, across to the xylem, up the xylem, and across to the phloem. With sufficient time such a cycle can be completed. This would tend to mix up the tracers and then you might find two tracers in the same sieve element under these conditions.

Walsh: I am referring back to your "out the window" comment, Dr. Currier, if we should ever find bidirectional movement in the same sieve tube. If there is one common denominator as far as sieve tube structure is concerned, it is that they contain a plasmalemma. This is a semi-permeable tube. What if we have a sieve tube through which there are many smaller semi-permeable tubes running the entire length? Perhaps such a structure could account for bidirectional movement within a single sieve tube. There is a good deal of endoplasmic reticulum within the sieve element, at least in some species. Tubular ER has been shown, as well as cysternal ER in mature sieve elements.

Currier: My own picture of a mature sieve element does not agree with your suggestion. I do not picture membranes blocking the lumen of the sieve tubes. The ER is essentially peripheral and I don't think anyone has demonstrated any permeability characteristics of the P-proteins.

Walsch: It is true that the ER is peripheral but if ones uses a bit of imagination, perhaps we can picture this ER as being continuous peripherally through the sieve tube. It is just a possibility, but I think we should give more thought to this.

Milburn: I would like to oppose this idea of extending the term bidirectional flow in the manner proposed. It seems to me quite out of court. It is well known that if we feed in the middle of the sieve tube, e.g. ^{14}C-sucrose, we are putting an artificial source in the system. It is very "Münchian" in principle to have movement away from this source in two different directions. It is a source that we apply in sufficiently large dose to overcome the effects of the natural source. So the autoradiographs shown by Dr. Eschrich do not surprise me in the least. If I can turn the argument around, one can do exactly the opposite effect, which is to make some kind of wound in the system which gives you an artificial sink. Under these conditions, either an exudation cut or an aphid stylet, flow would occur towards that point in both directions. But this is not bidirectional flow in my opinion. It should be distinguished from it very carefully, because this is an example of opposing flow and is not to be confused with bidirectional flow which would be a serious problem for the Munch hypothesis.

This brings me to another objection to Dr. Eschrich´s paper. It was assumed that one can use an aphid as a passive monitor of what is going on in the sieve tubes, but there is good evidence against this. We cannot use aphids as passive monitoring systems for they are themselves natural sinks. If we think of the somewhat complicated homodromous loop patterns which are adduced to explain mixing of opposing channels, I would suggest that, given the pressure fluctuations that must be going on to give complete mixing and given all of the complexities of this anatomical situation, it seems an unnecessarily complicated hypothesis. It is far simpler to say that the aphids simply allow the escape of sap, forming a natural sink which draws from two opposite directions. We have seen excellent evidence of this from the work of Mittler that this kind of thing can occur.

Currier: Admittedly the new definition of bidirectional transport is less restrictive than the earlier one, and my "out the window" remark was obviously in reference to the earlier definition. I am a little more content with the newer definition, but it is certainly a debatable point.

Swanson: I would like Dr. Eschrich to clarify a matter for me. Referring to the experiments of Tripp and Gorham in which they deduced evidence for the simultaneous bidirectional transport of ^{3}H-sucrose and ^{14}C-sucrose in the same sieve tube, I have attempted to explain this paradox in terms of a sort of homodromous loop between phloem and xylem, the ^{3}H-sucrose moving radially from the xylem into the sieve tubes of the ^{14}C-sucrose labelled leaf. I thought, however, that you presented a distinctly different possible explanation in your paper, and I would appreciate it if you would elaborate a bit more on your explanation.

Eschrich: I hope Drs. Gorham and Tripp will forgive me if I criticize their technique because I have used it myself. In cutting radial sections, you will never cut a single sieve tube twice. Autoradiographs in their paper show different sections, so you can´t decide whether activity coming from one section is activity in the same sieve tube which is shown in the next picture. If you make a tangential section, then it would be possible but not with a radial section because there are thousands of radial sections possible but only one tangential.

Gorham: I would agree with Dr. Eschrich´s over-all interpretations. As I recall, however, the sections were tangential and not radial. In my own interpretation we do not

involve   homodromous   loops.   I   think   that   is   excessively
complicated. We need simply imagine two contiguous sieve tubes,
back to back. In the process of freeze-drying, carried out as
carefully as we could, it was observed that the cytoplasm
containing the labelled sucrose tended to be displaced. We have
only to allow for the possibility that we would occasionally have a
back-to-back, almost a wall-to-wall, displacement of the cytoplasm
in sieve elements of the contiguous sieve tubes. Then when we cut
10-micron sections we could have two different sieve elements back
to back, ^{14}C-sucrose in one and ^{3}H-sucrose in the other. I think
this is the most probable explanation of our data. What we proved,
if we proved anything, was the up and down flow in adjacent
"pipes". (Diagram was drawn to clarify this discussion.)

Fisher: I would like to comment on the possibility of
recycling between internodes in our morning glory experiments.   In
the experiments I was commenting on, I was chiefly interested in
kinetics, and our data therefore came from situations where one
could hope he was able to relate whatever observations you made to
what was happening in individual sieve elements.   In choosing
experimental material for this sort of work you want to avoid, of
course, recirculation between nodes. It has been demonstrated that
you can get recirculation between nodes, but to get at the sort of
questions involving relationships between kinetics and mechanism,
one certainly must choose situations where you can avoid that as
much as possible.   I would say our data in morning glory would
indicate there is virtually no recyclng between nodes.   It would
cause spreading much more rapidly. I would say, however, that we
could not eliminate the possibility of some significant recylcing
in the first internode below the fed leaf. This may be a possible
explanation for the discrepancy we observed between the pulse width
coming out and kinetic size effects.

I would also like to comment on the autoradiographs Dr.
Eschrich showed of moss seta because it seemed to me that the
silver grains visible were undeniably associated with the walls and
not with the lumens. Can we assume here that we are really talking
about the sort of movement that we think we are talking about?

Currier: As far as cycling is concerned, time is very
important and my impression is that initially it is hard to
demonstrate, but if sufficient time is provided, it can be shown to
occur.

Fisher (responding chiefly to questions raised by Dr.
Eschrich): The sink leaves we used were on side shoots at the
nodes. Each of these lateral shoots had, of course, expanding
leaves on it.   The data that we presented represented the net
accumulation of ^{14}C with time in one of these sink leaves after a
mature leaf at the apex was pulse-labelled, and the curve which

described this was S-shaped. No activity was measured for a substantial period of time, representing the time required for the front to reach that position, followed by a rapidly increasing rate which tailed off, so you had a skewed S-shaped curve. The rate of appearance of ^{14}C in the sink leaf, was calculated from the first derivative of that curve, so the curve we have on the board is a derived curve from the total ^{14}C in the leaf.

Weatherley: May I say that I agree entirely with what Dr. Milburn said in regards to the term "bidirectional transport". It seems to me most unfortunate that there should be a change of meaning in this term, unless we now believe bidirectional transport in the proper sense is out. The literature has a definite meaning for bidirectional flow, meaning within one sieve tube. We cannot lightly change the meaning of such a term. It is going to be confusing and I would suggest that we try to come up with another term.

In this connection I would like to ask Dr. Eschrich about the model which he and his colleagues have recently put forward in Planta, in which there was sugar supplied at both ends of a length of permeable tubing, each colored with a different dye, one blue and the other yellow, and these moved up to the middle as you would expect on the basis of ordinary pressure flow, but then the mysterious thing was that they crossed over, each going down the opposite arm and in the end the whole system was colored uniformly green. This does seem a surprising result.

Eschrich: It is quite difficult with this model system to have high sugar concentrations at both ends and a completely filled tube. Turgor pressure becomes very high, and you probably missed the point in our text where we noted that this shuttle movement after the meeting of the two fronts may have been caused by leakage at one or the other end.

About the terminology matter, bidirectional transport or movement is a very old term, and in the literature has been applied to several things including all the interpretations that have been given in Crafts. Therefore, I do not think we do an injustice to use this expression in the sense I have indicated. At least in my view, this definition is appropriate, but I will not insist on it.

Milburn: I'd like to ask for clarification on one point. I alluded earlier to feeding with ^{14}C-sucrose in the middle of a bundle and the fact that this might constitute an artificial source in the sense that one had elevated the local availability of sucrose to a high level. When one does pulse-labelling this is also a danger, it seems to me. I would like to know what precautions are used to ensure that unlabelled CO_2 is given before and after the pulse of labelled CO_2 so that there won't be any shift in the source, because this might give the slightly distorted curves of the type sketched on the board.

Fisher: Our procedure is to have the leaf in a chamber with gas circulating through it and then inject the labelled CO_2 into a closed circulating system, during which time there would be, of course, a varying CO_2 concentration. Our approach is simply to keep the pulse as short as possible so that there won't be much effect on the sugar level in the leaf. At the end of the pulse application, we open the chamber to the original conditions, so I don't think we are disturbing the steady state much although admittedly we are to a degree.

Moorby: Concerning the ^{14}C-profiles reported by Dr. Fisher, they are surprisingly similar to what we have found with ^{14}C, using either soybean where we monitored the stem or maize where we monitored just the leaf. Here we did have an estimate of the spread of the pulse and this was very, very minor indeed, and we thought we could account for it by the increasing sieve-tube area, etc. The other point I'd like to raise is the matter of having a sink supplied by transport converging from different directions. A good example of this is in the developing monocot leaf, where the intercalary meristem acts as a sink for its own leaf, that is, for the older parts of its lamina, and for other older leaves. I am pretty sure in this case, although the actual data escapes me for the present, that assimilates from the older parts of the lamina may also go through the intercalary meristem. In other words, you have a sink which is actually being transported through. There is very little recirculation - the pulse simply moves straight through.

Eschrich: (The discussion at this point moved to a consideration of results obtained with excised corn leaves as models for simple translocation studies. Continuous reference was made to drawings and slides, and hence cannot be readily summarized.)

van Die: I fear we neglect too much the ability of parenchymatic cells to transport (referring to results obtained by Arens et al. on transport in Vallisneria leaves around cut midrib sections and to basipetal transport of ^{14}C in exuding stalks of Yucca when supplied with labelled $^{14}CO_2$ at an intermediate position on the stalk. A facilitated transport rate approximately 50-fold greater than that which could be accounted for by simple diffusion was observed).

Geiger: What is the velocity that one might expect if you had a situation involving cyclosis in the parenchyma cells and diffusion through the plasmalemma?

Swanson: What velocity did you observe through the parenchyma bridges? Are we talking in terms of centimeters per minute or centimeters per hours?

van Die: Several centimeters per hour - three to four.

Currier: In the bundles?

van Die: In both the bundles and the parenchyma.

Dainty: With reference to this diffusion problem, you cannot ascribe a single velocity to a diffusion front because the velocity of diffusion is inversely proportional to the distance diffused. So without knowing the distance diffused, it is almost meaningless. But just taking the figure of 50 at its face value, as a ball-park figure, it's about right, because the ratio of the lumen of a parenchyma cell to the width of the cell wall is of that order of magnitude, but it is really relatively meaningless to talk about this because of the way that velocity of diffusion depends inversely on the distance so that without knowing the distance we do not know what we are talking about.

van Die: With regard to the velocity of diffusion, I have plotted the log of the activity against distance and according to the slope of this line the coefficient of diffusion was, if you consider the translocation a kind of facilitated diffusion, about 50 times the slope you could expect if it were pure diffusion of a small molecule in water.

Dainty: If the log activity against time gives a straight line then it is not a diffusion process.

Tyree: A straight diffusion process from a fixed source gives an error function profile, as in the Canny-type models.

van Die: I also made calculations according to the error function and got similar values of about 50 times. If you have diffusion for a relatively long period, a number of days, the error curve will be almost a straight line.

Aronoff: If I remember correctly, if diffusion occurs not from a point source but from a circle, the solution is not an error function but a zero-order Bessel function.

Tyree: Yes, that is correct, but it is a question of how small a block you need to have a point.

Moorby: I'd like to get back to bidirectional transport. The thing that I have never been able to understand in terms of bidirectional transport are some experiments done, I think, in the early 60's by Jones and Eagle where they used variegated leaves with white margins. It was found that ^{14}C fixed in the chlorenchymous center was exported from the leaf but was not transported into the white margins. Since the leaves were certainly transpiring, any re-cycling of the labelled assimilates in the xylem was virtually zero. Yet if ^{14}C was supplied to a lower leaf, labelled assimilates were transported to the white margin of the leaf above. In other words, the white part of an exporting leaf was dependent on another leaf for its carbohydrates. So there must be bidirectional transport in the petiole. I am not familiar with the petiole anatomy of this leaf but it seems unlikely to me that there would be bundles concerned only with supplying the white portions of the leaf.

Eschrich: Was the labelled carbon supplied as CO_2 or as sugar? If labelled sugar was supplied to an abraded surface of a lower leaf, then surely some label could enter the xylem stream and ultimately reach the white parts of leaves above. We must also know the type of albinism – certain apparently albino regions do contain some chlorophyll. This must be clarified before we can discuss this problem further.

The Use of Phloem Exudates from Several Representatives

of the Agavaceae and Palmae in the Study of

Translocation of Assimilates

J. van Die

Botanisch laboratorium van de Rijksuniversiteit

Lange Nieuwstraat 106, Utrecht, The Netherlands

I. Introduction

The bleeding phenomenon that can be observed after the severing of the apical part of the inflorescence of Yucca flaccida HAW. (Agavaceae) very much resembles the bleeding from similarly treated inflorescences of several palm species (e.g. Gibbs, 1911; Tammes, 1933; van Die, 1968), and probably does not differ from the well known bleeding from Agave americana and related species (van Die and Tammes, 1974). In palms and Yucca the bleeding sap can be regarded as phloem exudate (Tammes, 1933; van Die and Tammes, 1966). Typical for the phloem exudates of these Agavaceae and Palmae are the large amounts in which they are produced (several hundreds of millilitres to many hundreds of litres per plant) and the fact that they probably can be obtained from plants in the productive phase only.

Although outside the group of these monocotyledons, sieve tube or phloem bleeding has often experimentally been provoked (e.g. Hartig, 1858; Munch, 1930 Ziegler, 1956 Zimmermann, 1960), there are only a few species known which can deliver more than a few ml of exudate. Fraxinus ornus and some related species cultivated on Sicily (Huber, 1953) and Fraxinus americana (Zimmermann, 1960, 1969) are well known examples. Apart from these exudates resulting from man-made wounds, the sieve tube saps normally utilized by Aphidae and Coccidae can be experimentally obtained by severing these insects from their stylets (Kennedy and Mittler, 1953; Mittler, 1957, 1958).

427

This paper consists of two parts. Part I will be mainly concerned with experiments carried out with Yucca flaccida, one of the best known plant species as far as phloem bleeding, with reference to assimilate translocation, is concerned. Part II will relate the findings and conclusions of the first part with data on assimilate translocation in bleeding and in intact palm trees and discuss the question of whether the developed views are of general significance for our knowledge on assimilate translocation through sieve tubes.

A. Arguments in favour of the assumed phloem origin of the exudates from Yucca flaccida and several palm species

The available evidence for the phloem origin of the exudates can be summed up as follows: (1) The bleeding sap exudes from the phloem part of the severed vascular bundles of the Yucca inflorescence (van Die and Tammes, 1966). (2) The bleeding takes place during intense transpiration by the leaves; the existence of a negative pressure within the xylem vessel system of a bleeding specimen of Yucca flaccida can be demonstrated (Tammes and van Die, 1964). Even under conditions of high water stress (plant cut off from its rhizomes and without any supply of water) bleeding may continue for many days, but with a reduced rate (Tammes, unpublished results). (3) In all the bleeding palms and Agavaceae, exudaton is only continuous if at least three times a day a thin slice is removed from the bleeding end of the inflorescence. This slice may be as thin as a fraction of a mm. Longitudinal sections of the wound area coloured with resorcin blue (according to Eschrich, 1956) shows a blue colour in the phloem elements at the wound surface (van Die and Tammes, 1966). This indicates that the formation of callose near the wound surface is the probable cause of the cessation of bleeding. (4) At about $20^{\circ}C$ exudation stops within 7-8 hours. At $0^{\circ}C$ complete sealing of the wound takes place in 24-36 hours (Tammes et al, 1969). One may conclude that a chemical process is responsible for the sealing of the sieve pores. (5) The exudate from Yucca inflorescences contains an enzyme which catalyzes the synthesis of callose-like material from UDPG and a primer (Eschrich et al, 1972). (6) Thirty measurements gave 464 m as the average length of a sieve element in an inflorescence stalk of Yucca (Tammes et al, 1971). It means that during bleeding, it is mainly the sieve plates of the sieve tube elements actually cut that are sealed with callose. (7) The compositions of the exudate from palms and the bleeding Agavaceae are very similar. No fundamental differences can be demonstrated for (a) the dry matter content, (b) the sucrose content, (c) the virtual absence of reducing sugars, (d) the high potassium content, (e) the very low

Table 1.   Some parameters used in calculations on flow through sieve tubes of <u>Yucca flaccida</u> HAW.   (van Die and Tammes,   1966;   Ie et al, 1966; Tammes et al, 1971).

Cross section of a stalk at the bleeding site	154 mm^2
Cross section of the vascular bundles	22 mm^2
Sieve tube area (phloem area minus companion cells)	2.5 mm^2
Average exudation rate during 29 h	1.1 ml h^{-1}
Rate of translocation through the sieve tubes	44 cm h^{-1}
Organic matter which exuded during 29 h	5.0 g
g dry matter translocated per h and cm^2 of phloem	5.7 g
g dry matter translocated per h and cm^2 of sieve tube	6.8 g
Average diameter of the sieve tube	21-22 μm
Average length of the sieve tube member	464 μm
<u>in vivo</u> thickness of the sieve tube	0.4 μm
Surface area of a sieve plate (oblique position)	942 m
Surface area of the average pore	0.21 μm^2
Total surface area of the pores per plate	370 μm^2
Number of filaments (not always present) per pore	10 - 20
Estimated distance between the filaments	100 nm
Pressure drop per m of sieve tube required to cause a flow velocity of 44 cm h^{-1}	
(a) if all the pores are open	0.35 atm
(b) if each pore is traversed by 10-20 filaments	5.3-5.6 atm

calcium content, (f) a pH around 8.0 (Tammes and van Die, 1964; van Die, 1968). (8) A very similar composition has been found for phloem exudates from dicotyledons obtained either by cuts in the secondary phloem, or by means of aphids (e.g. Crafts and Crisp, 1971). (9) The sucrose exuding from <u>Yucca flaccida</u> is largely derived from the current products of photosynthesis, as have been demonstrated with the aid of $^{14}CO_2$ (van Die and Tammes, 1964); van Die et al, 1973). (10) A solution flow in the sieve tubes of <u>Yucca flaccida</u> is theoretically possible (Tammes et al, 1971) see also Table 1.

From the foregoing we have concluded that the bleeding sap may be regarded as a phloem exudate in the monocotyledons investigated. Anatomically, the sieve tubes are the only cells present in the phloem that are capable of conducting sucrose in a longitudinal direction with a mass transfer value of up to 5.7 g dry weight per cm of phloem per hour (van Die and Tammes, 1966; see also Table 1). The bleeding sap consequently will be a sieve tube exudate. However, since the sieve tubes remain functionally intact during bleeding, the exudate apparently does not sweep out the entire cell content. The best explanation, therefore, that may be offered for the bleeding phenomenon is that the exudate represents the aqueous phase of the sieve tube lumen and does not include the parietal layer which is a more static, structural, plasmatic phase.

<u>Yucca</u> inflorescences may bleed for about 3 weeks. During this period a few hundred ml of exudate can be collected, containing up to 70 g of dry matter, or about twice the dry weight of a fully developed inflorescence (van Die, 1968). <u>Arenga</u> inflorescences may bleed for about 6 weeks, producing in this period about 220 l of exudate with about 25 kg of dry matter, 27 times the dry weight of the inflorescence stump from which the fluid has exuded (Tammes, 1933). <u>Corypha</u> palms even produce considerably larger amounts of exudate and sucrose (Gibbs, 1911). During the whole bleeding period the composition of the exudates remains approximately constant (Tammes and van Die, 1964) or shows a characteristic maximum in concentration after about 5 weeks of bleeding (Gibbs, 1911; Tammes, 1933). Therefore, not only is the aqueous phase of the sieve tube system rapidly replenished during the bleeding, but the mechanism which replaces the exuded solutes is not inactivated by the severe cut made. The system apparently has the capacity to deliver the many exudate compounds in their respective amounts with a rate of mass transfer similar to (and in palms even much greater than) that found for normal assimilate translocation through sieve tubes.

B. The origin of the sucrose in the exudate

Two sources of carbohydrates should be considered in connection with the origin of the exudate sucrose: (1) stored carbohydrates present in the rhizome or in aerial stem parts ; (2) current photosynthates present as such, or in a more or less transitory state in the leaves.

The literature data available to a relation between the origin of the exudate sucrose and the way of flowering of the plant species concerned. In palms with a hapaxanthic way of flowering, and therefore with a terminal inflorescence (e.g. Metroxylon rumphii MART., and Corypha elata ROXB.), starch accumulates in the trunk during the vegetative phase of its life, which ranges between about 10 and about 40 years. With the onset of the reproductive phase the starch becomes mobilized (Gibbs, 1911) and exudate can be collected. The exudate sucrose is apparently formed at the expense of the starch in the trunk since this starch disappears during exudation (Gibbs, 1911 ; Tammes, 1933) and the amounts of both are of the same order of magnitude (van Die, 1974). The starch in the trunk will also be the main source of carbohydrates for the production of fruits. Zimmermann (1973) has reported the production of 600 kg dry fruits by a single Corypha elata specimen. Fruit and exudate production are apparently alternatives, both depending on the same carbohydrate resources available during their formation (van Die, 1974; see Fig.1).

In palms with a pleonanthic way of flowering, and therefore with lateral inflorescences, we notice that the cocos palm (Cocos nucifera L.),for example, does not store starch in its stem (Reyne, 1948). The fruits are formed at the expense of the current products of photosynthesis. Possibly a similar situation exists for the date palm (Phoenix dactylifera L.) and the oil palm (Elaeis guineensis JACQ.).

Yucca flaccida has a hapaxanthic way of flowering. A shoot with about 25 leaves forms a terminal inflorescence and then dies. The plant, however, possesses a large fleshy rhizome by which new shoots are formed regularly, and which stay in a vegetative phase for a few years before forming an inforescence. We could demonstrate that the photosynthates of the leaves of the flowering shoot are all, or at least for a large part, translocated to the site of bleeding (van Die and Tammes, 1964; van Die et al, 1973). Probably the photosynthates of the non-flowering shoots supply the carbohydrates for storage in, and growth of, the rhizome (Fig.1).

Following the supply of $^{14}CO_2$ to a single leaf of a bleeding Yucca shoot, the first detectable activity exuded from the cut end

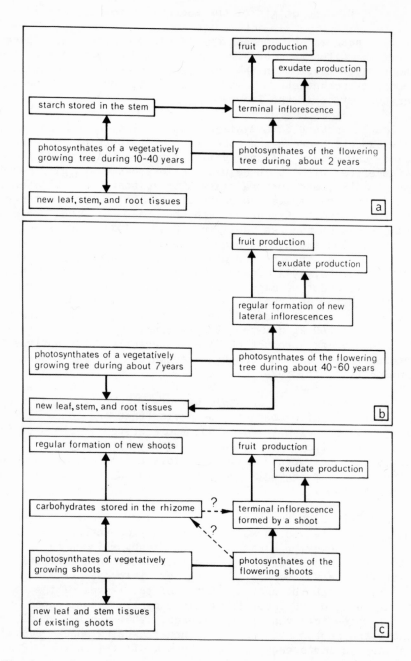

Fig.1.   Distribution,   storage   and   utilization   of
photosynthates during the vegetative and reproductive phases of
growth in (a) a hapaxanthic palm species (e.g. Corypha elata
ROXB.), (b) a pleonanthic palm species (e.g. Cocos nucifera L.)
and (c) Yucca flaccida HAW.

of an inflorescence occurs after about 50 minutes (Fig.2). The
specific activity of the exudate sucrose rapidly increased with
time and reached its maximum after about 7 hrs. Just as in earlier
experiments (van Die and Tammes, 1964), the exudation of the
labelled sucrose continued for more than one week (van Die et al,
1973), indicating the existence of a pool of ^{14}C-photosynthates
somewhere in the bleeding plant. From ths pool the exudate sucrose
was apparently continuously withdrawn. We could demonstrate that
this pool was localized in the treated leaf since removal of the
leaf resulted in a rapid drop in specific activity of the exudate
sucrose (van die and Tammes, 1974). This indicates that outside
the labelled leaf and outside the aqueous phase of the sieve tubes
only traces of labelled compounds were available for translocation.
From the slope of the declining part of the curve in Fig.2 we
calculated a turnover rate of the ^{14}C-carbohydrate pool in the
treated leaf of about 21% per day.

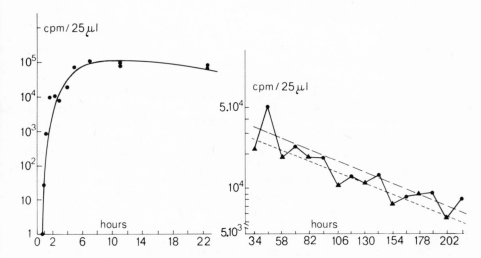

Fig.2. The ^{14}C-content of phloem exudate from the wounded top
of the inflorescence stalk during a 215 hr. bleeding period
following pulse-labelling of a single leaf. In the figure on the
right the solid circles represent aliquots of exudate taken at
07:00-07:30 hrs.; the triangles represent aliquots taken at 19:00
hrs. (van Die et al, 1973).

An important observation was that the maxiumum specific
activity of the exudate sucrose appeared to be a few percent of the
maximum specific activity of the sucrose which could theoretically
be expected to exist in the pool of the treated leaf. The bleeding

shoot had about 25 leaves and each of these will have supplied
non-labelled sucrose to the moving exudate sucrose in the sieve
tube lumina. If these 25 leaves all contributed in the same manner
to exudate production, then the non-labelled sucrose will have
diluted the contribution of the ^{14}C-labelled leaf to the flow of
exudate by a factor of 25. In other words, the movement of sucrose
from the treated leaf to the site of bleeding, over a distance of
30-40 cm., occurs without a change in specific activity.

In the experiment shown in Fig.2 the average exudation rate
was 5.7 ml/24 hrs. Assuming all the 25 leaves of the shoot
contributed equally to exudate production, then (with 17% sucrose
in the exudate) 39 mg of sucrose would be secreted per day by each
leaf. This amount represents 21.6% (i.e. the turnover rate) of
the source pool in an average leaf, which consequently contained
about 180 mg of disaccharide equivalents. A fresh weight of 6.5 g
(2.3 g of dry weight) appeared to be a fair estimate for the size
of an average Yucca leaf. The exudate-sucrose source pool of such
a leaf consequently comprised 7-8% of the dry weight. This value,
being of the same order of magnitude as the non-structural
carbohydrate fraction of a leaf, is not unrealistic and supports
the assumption that all the leaves have exported their
photosynthates in about the same manner.

In another monocotyldedonous plant species, Fritillaria
imperialis L. (Liliaceae), it could be shown that the upper leaves
predominantly supply the inflorescence with photosynthates, while
the bottom leaves mainly supply the growing bulb with assimilates
(van Die et al, 1970; Tietema et al, 1972). In this plant species,
however, the growing bulb is wholly dependent on the photosynthates
of its sole shoot in contrast to Yucca flaccida where a number of
vegetative and some flowering shoots have a common rhizome.

Under normal conditions of bleeding the bulk of the exudate
sucrose is from the current photosynthate of the Yucca leaves. It
could be clearly shown that after-removal of the leaves and the
inflorescence the rhizome and stem stump can produce small amounts
of exudate, which contains 15% sucrose as the only detectable
carbohydrate (Tammes and van die, 1964).

Another theoretically important observation is the
participation of contiguous cells of the sieve tubes of the
inflorescence in the process of bleeding sap production. Isolated
inflorescence segments produce small amounts of exudate. Except
for a lower sucrose content, this sap is very similar in
composition to that found in an inflorescence normally attached to
the plant (van Die and Tammes. 1966). From the extent of bleeding
of the isolated segments, we may estimate the contribution of the
inflorescence axis cells to the total of the exudate sucrose

production of a normally bleeding plant to be about 10%. These experiments show that all the contiguous cells of the sieve tube system of a flowering <u>Yucca</u> shoot are able to act as sources of exudate solutes. It appears that those located in the rhizome are only source cells in case the leaves are unable to supply the sieve tube system with sufficient amounts of sucrose. Probably the sieve tubes surrounding the cells in the inflorescence stem and branches normally act as sinks during all but the last stages of inflorescence (and fruit) development, but if a dominant, artificial sink has been created in the form of a bleeding site, they will become source cells and start to secrete sucrose and other solutes into the sieve tube lumina. The signal for this reversal may be the change in solute content of the neighbouring sieve tube members (van Die and Tammes, 1966; van Die, 1968) or possibly the changes in their turgor. This mechanism might of course also be involved in regulating the source-sink function of the sieve tube surrounding cells of the basal parts of the shoot stem and of the rhizome.

C. The origin of water in the exudate

Is there any theoretical possibility that the exudate water moves mainly through the xylem vessels while the exudate sucrose moves through the sieve tubes? From the rather speculative translocation model of Canny and Phillips (1963), and Canny (1973), one could arrive at this conclusion: the sugar would move by means of its assumed translocation mechanism (moving together with transcellular strands) into the bleeding sieve tube members and these would subsequently osmotially attract water from the xylem. Crucial points in this hypothesis are the existence of continuous filaments and their streaming. According to Parthasarathy (1974c) it is difficult to attribute any such role to P-protein filaments in the sieve tubes of palms since these filaments are not only sporadic in occurrence, but also confined to the periphery of the sieve elements in the species in which they occur. Moreover, it seems highly probable that the sieve plate pores in the metaphloem elements of palms are normally free of filaments (Parthasarathy, 1974c). It also seems improbable that filamentous P-protein plays a part in streaming in immature palm phloem elements because in those palm species where the filaments occur they were found to be randomly oriented (Parthasarathy, 1974a).

Since a temperature of $0^{\circ}C$ is likely to reduce the speed of the alleged strands to a fraction of that which they might hypothetically have at normal temperatures (about $20^{\circ}C$), we did a number of cooling experiments with bleeding <u>Yucca</u> plants. The results unequivocally demonstrated that the longitudinal component of the mechanism of translocation of the exudate solutes in the stalk is temperature insensitive (Tammes et al, 1969). At $0^{\circ}C$ the flow continues at an almost normal rate, but with a significantly higher sucrose concentration. Numerous cooling experiments have shown that, over at least 15 cm of the exudate pathway, water and solutes are kept moving by a driving force located outside the cooled inflorescence stalk and that consequently the translocation model of Canny (1973) is inadequate in explaining the continuous exudate movement through the inflorescence axis of Yucca flaccida.

It is hardly conceivable that sucrose and the other exudate solutes would be translocated through the cooled stem parts in a form other than a solution. The mechanism described by van den Honert (1932) might ,in theory, be able to offer a physical model that would be relatively temperature insensitive, but sucrose is not surface active (with respect to boundary layers) nor is any other major exudate component. The remaining conclusion is, therefore, that during bleeding the aqueous phase of the sieve tube system is a mobile one mainly driven by processes located outside the sieve tube members of the axis. The leaves are the most plausible candidates for this location as they produce and export most of the exudate sucrose.

D. The driving mechanism of sieve tube bleeding

Although the main driving force of the exudate flow seems to be localized in the leaves, there is clear evidence that cells in the stem and in the rhizome are also actively engaged in the maintenance of the flow. There is probably no fundamental difference in the mechanism by which sieve tube loading occurs at all these sites. Some information on the nature of the process that delivers the exudate solutes to the sieve tube lumen could be obtained by two kinds of experiments: (1) detached inflorescence stalk segments of about 15 cm length, which were producing small amounts of exudate, and which were placed with one end in a little water to which a ^{14}C-labelled compound had been added. If ^{14}C-D-glucose, D-fructose, D-galactose, maltose, lactose fructose-1,6-diphosphate and fructose-6-phosphate were fed, ^{14}C-sucrose was the only labelled compound present in the exudate. L-sorbose and glutamine, however, were not at all converted but exuded as such. Apparently in higher plants naturally- occurring sugars can be converted specifically to sucrose and secreted into the sieve tube lumen after being taken up via the xylem (Tammes et al, 1973). (2) Exudation from detached inflorescence stalk segments can be stopped at 0°C and started again after being brought back at room temperature (Tammes et al, 1969).

Summarizing the Yucca data, we conclude that the translocation system involved in the exudation consists of two components: (a) a chilling-insensitive longitudinal flow of the aqueous phase of the sieve elements in the inflorescence stalk to the site of bleeding; (b) a lateral movement of exudate solutes and water from contiguous cells of the sieve tube system into the sieve tube lumina driven by temperature-dependent and rather selective secretion processes.

E. Phloem exudates and the hypothetical nutrient solution
of palm fruits

Does a relation exist between the flow of exudate from the stump of a severed inflorescence and the flow of photosynthates, minerals and water which normally move through the intact inflorescence axis into the developing fruits? Since Yucca plants do not produce fruit in Western Europe we could not try to answer this question by using this plant. Fortunately, however, many details on fruit development and fruit composition have been reported for several palm species in the literature and good knowledge also exists on their phloem exudates.

From the point of view of mass-flow kinetics, the water present in a fruit will be "Münch water" brought into it, together with the transported assimilates. The water content of the ripe fruits of several palm species, however, is considerably lower than that of sieve tube exudate, but this can easily be explained since fruits generally transpire. Transpiration, however, may also imply an influx of xylem water into the fruit.

The dry matter present in the fruit is imported via the vascular bundles, but fruits often photosynthesize and always respire. There only exist a few reports on this matter (e.g. Clements, 1940), and therefore we have assumed in our study that, just as in Citrus (Bollard, 1970), photosynthesis balances the respiration of the investigated palm fruits.

The data on coconut growth reported by Sampson (1923) have been used in a recent paper (van Die, 1974) for the construction of graphs which show that the composition of the growing fruit with respect to total ash, K, Ca, Mg, P and N remains reasonably constant during almost the entire period of fruit development. It looks as if the fruits have been built up from a nutrient solution with a constant composition as far as minerals and nitrogenous substances are concerned. We have called this fluid hypothetical nutrient soluton of the cocos fruit and have made the simple approximation that this fluid enters the developing fruit via the sieve tubes only (Fig.3). The excess of water is assumed to leave the fruit by transpiration, a process that apparently becomes of quantitative significance at a stage just before the final fruit size has been reached and via the xylem. For the development of a ripe cocos fruit of about 1000 g fresh weight, about 3500 ml of nutrient fluid has to be imported and elaborated (van Die, 1974). Since total water loss from a developing cocos fruit will consequently be of the order of 2500 ml per year, and since only the husk can transpire, it is interesting to point to the development of this fruit part. From the data of Sampson (1923) we may conclude that the husk apparently loses most of its original water during the second phase of fruit development (Table 2). However, it probably loses much more water then is apparent since it almost doubles in dry weight during that period. The husk dies a few months before complete fruit ripeness.

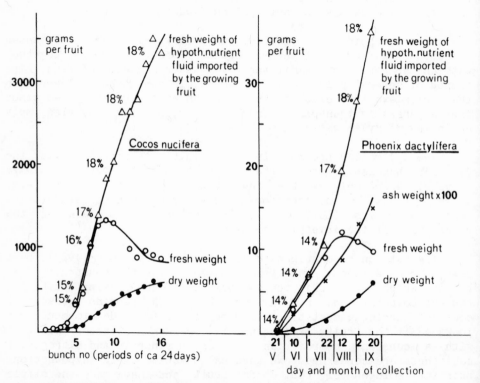

Fig.3. Left: The development of the fruits of successive bunches of regular-bearing cocos tree. The difference in age between two successive bunches is approximately 24 days. Data in fresh and dry weights are derived from Sampson (1923). Right: The development of the fruits of the date palm (Deglet Noor). Data on fresh weight, dry weight, and ash weight of the fruits (without seed) are taken from Haas and Bliss (1935). The lines with triangles show the amounts of the hypothetical nutrient fluids of both fruit species and their assumed dry matter content, from van Die (1974).

Table 2. Water loss from the husk in later stages of cocos
fruit development, according to data from Sampson (1923). The
husks of the youngest fruits (bunches 1-7) could not be separated
from the shell and endosperm. Bunch 16 contained the ripe fruits.
Except for those of bunch 11, the figures are averages of three
fruits.

Bunch Number	Fresh weight of whole fruit	Fresh weight of the husk	Dry weight of the husk	Husk water
1-7				
8	1268 g	841 g	154.9 g	686 g
9	1333	838	187.5	650
10	1295	794	190.8	603
11	1315	789	252.0	537
12	992	541	236.2	305
13	893	472	251.6	220
14	972	486	288.8	197
15	913	333	303.0	30
16	873	329	274.9	54

Fruit size and annual fruit production of a regular-bearing
cocos tree depends, of course, on many factors, expecially on the
annual production of photosynthates. A fair average seems to be an
annual cocos fruit production of 10 bunches of 10 fruits each with
a total dry weight of about 55 kg per tree (van Die, 1974). Since
each fruit will require about 3.5 l of the hypothetical nutrient
fluid, about 350 litres of nutrient fluid with about 17% dry
substances and a constant mineral composition is needed for the
total annual fruit crop. The literature on exudates from cut cocos
inflorescences supports this 350 l value.

In Ceylon, India, Malaya, Indonesia and the Philippines the
exudate ("toddy") from the cocos inflorescence was, and often still
is, collected and used for palm sugar (e.g. "gulah djawa") and
alcohol ("arak") production (for a review of literature on this
subject, see van Die and Tammes, 1974). Cocos palm toddy and
coconut production are alternatives and it depends on several
economic factors whether people prefer to produce fruits or collect
the exudate.   Fortunately the economic significance of palm toddy
has been sufficient to stimulate a number of important studies on
this subject (e.g. Gibbs, 1911; Browning and Symons, 1916; Marsden
and Skilton, 1931; Tammes, 1933; Patel, 1938; Nathanael, 1955).
From these studies we know the amount of exudate which can be

Table 3. The composition of some phloem exudates compared with the postulated composition of the hypothetical nutrient fluids of the fruits of Cocos nucifera and Phoenix dactylifera (Deglet Noor). (1) Tammes and van Die, 1964; van Die and Tammes, 1974; van Die, 1974. (2) Tammes, 1933, 1958; van Die and Tammes, 1974. (3) Nathanael, 1955; Browning and Symons,1916; Gibbs, 1911. (4) Sampson, 1923, with the exception of the dry matter content which is hypothetical. (5) and (7) van Die, 1974. (6) Haas and Bliss, 1935, with the exception of the dry matter content which is hypothetical; the data are based on the author's assumption (van Die, 1974) of an ash content of 20 mg/g dry weight.

	Dry matter content (percent)	mg per g dry weight					
		N	P	K	Mg	Ca	Ash
Phloem exudates from:							
(a) Yucca flaccida (1)	17-19	6.8	1.0	8.4	0.54	0.068	24
(b) Arenga saccharifera (2)	15-18	2.5	0.59	6.8	0.54	0.057	21.6*
(c) Cocos nucifera (3)	15-20	5.5 2-3	0.36-0.61	6.6-9.4	0.20-0.28	0.069-0.086	22.8
Hypothetical nutrient solution imported by fruit while growing:							
(a) Cocos nucifera (4)	15-18	6.0	0.62	6.9	0.6-1.1	1.4	
(5)		4.1	0.52	6.3	0.50	1.06	20.0
(b) Phoenix dactylifera (6)	14-18	3.9	0.64	8.1	0.50	0.68	
(7)		3.5	0.36	8.5	0.66	0.80	20.1

*Ash content derived from Gibbs (1911).

collected per inflorescence or annually per tree, and also how this
fluid is composed.   Table 3 demonstrates that the composition of
the cocos exudate closely resembles that of Yucca flaccida and has
a remarkable resemblance to the composition of the above postulated
hypothetical nutrient solution of the cocos fruit.   There is,
however, one notable exception, i.e. the calcium content.

Analytical data exist on the composition of the developing
fruits of Cocos nucifera and for the date palm Phoenix dactylifera.
For several reasons, however, the results on the fruit species
reported by Haas and Bliss (1935) are more difficult to interpret
than those on the cocos fruit. An important reason for this is
that the date palm produces its fruit only during 6-7 months of the
year, while the regular-bearing cocos palm forms its fruits and
leaves throughout the year. In the latter case the leaf-to-fruit
ratio remains more or less constant, while in the date palm this
ratio increases during fruit ripening (Aldrich and Crawford, 1939).
The supply of photosynthates to the fruits will probably reflect
this ratio and may therefore gradually increase during the ripening
of the date fruit (Aldrich and Crawford, 1939).

From the results of Haas and Bliss (1935) an important
conclusion can be drawn: the mineral composition and the N-content
of the developing date fruits remain strikingly constant during
their whole growth period (van Die, 1974). A second important
aspect is the large degree of similarity between the composition of
the date fruit as reported by Sampson (1923). Fig.3 and Table 3
demonstrate these similarities.

As was found for cocos fruits, water loss from the developing
date fruits would become evident mainly during the later stages of
growth (Fig.3).   Fortunately, Haas and Bliss (1935) have
investigated, for other reasons, the rate of transpiration of
mature date fruits. Water loss per fruit, predicted on the basis
of the present hypothesis, (0.21 ml/24 hr. to 0.52 ml/ 24 hr. for
the period of 22.VII to 2.IX; see Fig.3) falls within the range
reported by Haas and Bliss (1935) for the transpiration of date
fruits under normal climatic conditions (van Die, 1974).

F.   Discussion and some conclusions.

Is it a possibility that hormones present in young ovules
"attract" the assimilates required for fruit growth (e.g. Nitsch,
1970, 1971)? From a severed inflorescence, a fluid very similar to
that normally used for palm fruit development can be obtained
although ovules are absent.   In Arenga saccharifera and Elaeis
guineensis it is the male inflorescence that is tapped for exudate
(Sprecher von Bernegg, 1929; Tammes, 1933). In a recent paper Vonk
(1974) reported the presence of considerable amounts of zeatin
nucleotide in the phloem exudate from Yucca flaccida.   It

therefore, seems a not too daring prediction that cytokinins will be present in the phloem exudates of palm trees too. This may throw some light on the question of the origin of the cytokinins isolated from the liquid endosperm of the coconut (Shantz and Steward, 1955; Letham, 1968; for both references see Nitsch, 1970). The role of these growth regulators in fruit development may be the induction and regulation of the processes which lead to sink activity and to the histological and chemical differentiation within the palm fruit; the young ovule receives the assimilates from the leaves or trunk together with at least some of its growth regulators.

The calcium content of the phloem exudate from palms and the bleeding Agavaceae is very low; probably it is near the theoretical value of an aqueous phosphate-containing solution with a pH of about 8.0-8.2. The calcium content of fruits in general apparently depends on the plant species (Ansiaux, 1959; Bollard, 1970), and also on the Ca-level of the soil (van Schoor, 1957). At first sight it is attractive to consider the Ca-content of the fruit to be a measure of the amount of xylem fluid that has entered the fruit during its development, and which, in its turn, might be determined by e.g. the surface-to-volume ratio of the fruit species, or by environmental conditions. Wiersum (1966), however, could not find an appreciable influx of ^{45}Ca moving in the xylem into developing apple and tomato fruits, while Faust and Shear (1973) have reported that calcium enters the apple fruit via the phloem.

From data reported by Sampson (1923) and also from analyses carried out at our own laboratory (van Die, 1974), it is evident that 90-95% of all the cocos fruit calcium is located in the husk, especially in the numerous fibrous vascular bundles of this fruit part. Since the husk is the transpiring part of the fruit, this observation favours the assumption of an influx of small amounts of xylem fluid into it.

In the literature we can find several reports on a return flow of water out of the fruit to the transpiring leaves (Bartholomew, 1926; Ziegler, 1963). Here, too, we may expect the xylem vessels to be the pathway of the flow. Obviously, depending on e.g. climatic conditions, xylem vessel fluid may move into or out of the developing fruit, taking with it those solutes which normally move through the xylem vessels. Calcium, however, has an ion-exchange type of movement (Shear and Faust, 1970), and therefore will probably stay behind in the vessel walls if the xylem fluid is moving out of the fruits.

The velocity of flow through the sieve tubes of the bleeding Arenga palm, which has been reported to be about 7 m h^{-1} (Tammes, 1952), has given rise to some doubt with regard to the nature of the exudate flow (Canny, 1973). From the growth curves presented in Fig.3 we may infer that about 350 l of hypothetical nutrient fluid may be necessary for annual fruit production and consequently may be assumed to enter the 10 bunches of fruits of either palm species during their development. This means roughly 35 l per inflorescence per year in the case of the cocos palm, or about 100 ml of the hypothetical fluid per inflorescence per day.

The palm exudate literature mentions an average exudate production ranging from about 0.5-0.6 l per inflorescence per day (Browning and Symons, 1916; Patel, 1940; Nathanael, 1955) to values of 1.5 l (Marsden and Skilton, 1931) and about 2.2 l per inflorescence per day (Nathanael, 1955; Browning and Symons, 1916; see also Child, 1964). The rate of bleeding from a cut cocos inflorescence is consequently at least five times, and apparently sometimes up to twenty times, higher than the rate of flow of the hypothetical nutrient solution into the developing palm fruits.

This conclusion has several implications. (1) The loading mechanism of the sieve tube system which normally supplies assimilates to the ten or more bunches with developing fruits, supplies, in the case of a bleeding inflorescence stalk, all these solutes to one artificial sink. (2) Sugars and other solutes can be translocated through the sieve tubes with a mass transfer rate of up to 20 times that of normal dry matter translocation. The Yucca data (Table 1) have shown that in the case of bleeding such a high solution flow velocity does not encounter theoretical objections, while in palms open pores very probably are the normal in vivo situation in differentiated sieve tubes (Parthasarathy, 1974). For a discussion of the question of whether bleeding, or possibly the repeated cutting, has made the sieve pores more accessible for flow, or whether they are always unobstructed, see van Die and Tammes, 1974.

In considering a pressure driven mass flow in sieve tubes, one should be aware that the Yucca experiments have demonstrated that, apart from the main loading sites located in the leaves, auxiliary loading sites exist along the whole translocation pathway, probably in the form of companion cells. Secretion by these cells contributes to exudate flow, albeit for only about 5-10% per axis segment of about 15 cm. Auxiliary loading sites may, in general, be active in all those instances where the main source and the main sink are at a relatively large distance from each other, or where experiments have shown a dependency of the translocation rate on the metabolism of the tissues of the translocation pathway (e.g. Willenbrink, 1968; Qureshi and Spanner, 1973; cf. Harel and Reinhold, 1966).

Evidence obtained from bleeding Yucca opposes the views that (auxiliary) "pumping stations" may be located in the zone of the sieve plates as suggested by Kursanov (1963) and Spanner (1958, 1971), for which the studies of Kollmann (1964, 1973) might supply the structural basis. Our main objection to the sieve plates being "sites of most intense metabolism" (Kursanov, 1963), and to any other force generating system in the translocation pathway, is the outcome of the cooling experiments with bleeding plants. Similar experiments with other chilling-resistent or cold-acclimatized plant species suggest that also the longitudinal component of the assimilate translocation process in intact plants does not depend for its mechanism of flow on chemical processes within the sieve tubes (Swanson and Geiger, 1967; Geiger and Sovonick, 1970; Giaquinta and Geiger, 1973; Webb, 1971; Wardlaw, 1972). It looks more plausible, therefore, to consider the sieve pores and the scarce filaments running through them to be sites of flow control rather than structures involved in pumping (cf. Fensom, 1972). The flow controlling mechanism of the sieve pores, however, comes only slowly into action after the cutting of the inflorescence axis of _Yucca flaccida_ and several palm species. It is this relative slowness, compared with the closing mechanism of the sieve pores in most of the other plant species (cf. Milburn, 1970, 1972), that has made the investigated monocotyledons so useful, although somewhat difficult objects to handle, in the study of translocation of photosynthates.

References

ALDRICH, W.W. and C.L. CRAWFORD. 1939. Proc. Amer. Soc. Hort. Sci. 37: 187-190

ANSIAUX, J.R. 1959. Ann. Physiol. veget. Univ. Bruxelles 4: 53-88

BARTHOLOMEW, E.T. 1926. Amer. J. Bot. 13: 102-117

BOLLARD, E.G. 1970. The Biochemistry of Fruits and their Products, Vol.I (Ed. A.C. Hulme), Acad. Press, London-New York, pp. 387-425

BROWNING, K.C. and C.T. SYMONS. 1916. J. Soc. Chem. Ind. 35: 1128-1142

CANNY. M.J. and O.M. PHILLIPS. 1963. Ann. Bot. 27: 379-402

CANNY, M.J. 1973. Phloem Translocation. Cambridge University Press

CHILD, R. 1964. Coconuts. Longmans, Green Co., London

CLEMENTS, H.F. 1940. Plant Physiol. 15: 689-700

CRAFTS, A.S. and C.E. CRISP. 1971. Phloem Transport in Plants. Freeman, San Francisco.

ESCHRICH, W. 1956. Protoplasma 47: 487-530

ESCHRICH, W., A. HUTERMANN, W. HEYSER, P.M.L. TAMMES and J. VAN DIE. 1972. Zeits, f. Pflanz. Physiol. 67: 468-470

FAUST, M. and C.B. SHEAR. 1973. Trans. 3rd Symp. Accumulation and Translocation of Nutients and Regulators in Plants. pp. 423-436. Warszawa-Krakow-Skierniewice.

FENSON, D.S. 1972. Can. J. Bot. 50: 479-497

GEIGER, D.R. and S.A. SOVONICK. 1970. Plant Physiol. 46: 847-849

GIAQUINTA, R.T. and D.R. GEIGER. 1973. Plant Physiol. 51: 372-377
GIBBS, H.D. 1911. Philip. J. Sci.:A, Chem. Geol. Sci. Ind. 6: 99-206
HAAS, A.R.C. and D.E. BLISS. 1935. Hilgardia 9: 295-344
HAREL, S. and L. REINHOLD. 1966. Physiol. Plant. 19: 634-643
HARTIG, T. 1858. Bot. Zeits. 16: 369-370
HUBER, B. 1953. Ber. Deutsche Bot. Ges. 66: 341-346
IE, T.S., P.M.L. TAMMES and J. VAN DIE. 1966. Proc. Kon. Nederl. Akad. Wetensch. C 60: 660-663
KENNEDY, J.S. and T.E. MITTLER. 1953. Nature 171:528
KOLLMANN, R. 1964. Phytomorphology 14: 247-264
KOLLMANN, R. 1973. Grundlagen der Cytologie (ed. G.C. Hirsch, H. Ruska, P. Sitte) pp. 479-504
KOLLMANN, R. 1973. Trans. 3rd Symp. Accumulation and Translocation of Nutrients and Regulators in Plants. pp. 61-71, Warszawa
KURSANOV, A.L. 1963. Adv. Bot. Res. (ed. R.D. Preston) I: 209-278
MARSDEN, H. and P.L. SKILTON. 1931. Malayan Agric. J. 19: 287-290
MILBURN, J.A. 1970. Planta 95: 272-276
MILBURN, J.A. 1972. Pest. Sci. 3: 653-65
MITTLER, T.E. 1957. J. Exp. Biol. 34: 334-341
MITTLER, T.E. 1958. J. Exp. Biol. 35: 74-84
MUNCH, E. 1930. Die Stoffbewegungen in der Pflanze. Fisher, Jena
NATHANAEL, W.R.N. 1955. Ceylon Coconut Quart. 4: 8-16
NITSCH, J.P. 1970. The Biochemistry of Fruits and their Products, Vol. I (ed. A.C. Hulme), pp. 487-472 Academic Press, London-New York
NNITSCH, J.P. 1971. Plant Physiology (ed. F.C. Steward) VI A, pp. 413-501. Academic Press, New York-London
PARTHASARATHY, M.V. 1974a. Protoplasma 79: 59-91
PARTHASARATHY, M.V. 1974b. Protoplasma 79: 93-125
PARTHASARATHY, M.V. 1974c. Protoplasma 79: 265-315
QURESHI, F.A. and D.C. SPANNER. 1973. Planta 110: 131-144
QURESHI, F.A. and D.C. SPANNER. 1973. Planta 111: 1-12
REYNE, A. 1948. De Landbouw in den Indeschen Archipel (ed. C.C.J. van Hall and C.van de Koppel) II A, pp. 427-525. Van Hoeve, Gravenhage-Bandung
SAMPSON, H.C. 1923. The Coconut Palm. John Bale, Sons and Danielsson, London
SPANNER, D.C. 1958. J. Exp. Bot. 9: 332-342
SPANNER, D.C. 1971. Nature 232: 157-160
SPRECHER VON BERNEGG, A. 1929. Tropische und subtropische Weltwirtschaftspflanzen. Ferdinand Enke, Stuttgart
SWANSON, C.E. and D.R. GEIGER. 1967. Plant Physiol. 42: 751-756
TAMMES, P.M.L. 1933. Rec. Trav. Bot. Neerl. 30: 514-536
TAMMES, P.M.L. 1952. Proc. Kon. Nederl. Akad. Wetensch. C 55: 141-143

TAMMES, P.M.L.   1958.   Acta Bot.   Neerl.   7:   233-234
TAMMES, P.M.L.   and J.   VAN DIE.   1964.   Acta   Bot.   Neerl.   13:
    76-83
TAMMES, P.M.L.   and   J.   VAN   DIE.   1966.   Proc.   Kon.   Akad.
    Wetensch.   C 69:   655-659
TAMMES, P.M.L., C.R.   VONK and   J.   VAN   DIE.   1969.   Acta   Bot.
    Neerl.   18:   224-229
TAMMES, P.M.L., C.R.   VONK and   J.   VAN   DIE.   1973.   Acta   Bot.
    Neerl.   22:   233-237
TIETEMA, T., S.M.R.   HOEKSTRA and J.   VAN DIE.   1972.   Acta   Bot.
    Neerl.   21:   395-399
VAN DEN HONERT, T.H.   1932.   Proc.   Kon Nederl.   Akad.   Wetensch.
    c 35:   1104-1111
VAN DIE, J.   and P.M.L.   TAMMES.   1964.   Acta   Bot.   Neerl.   13:
    84-90
VAN DIE, J.   and P.M.L.   TAMMES.   1966.   Proc.   Kon Nederl.   Akad.
    Wetensch.   C 69:   648-654
VAN DIE, J.   1968.   Vortrage Gesamtgeb.   Botanik n.F.2:   27-30
VAN DIE, J., P.   LEEUWANGH and S.M.R HOEKSTRA.   1970.   Acta   Bot.
    Neerl.   19:   16-23
VAN DIE, J., C.R.   VONK   and P.M.L.   TAMMES.   1973.   Acta   Bot.
    Neerl.   22:   446-451
VAN DIE, J.   and P.M.L.   TAMMES.   1974.   Encyclopedia   of Plant
    Physiology; New Series (ed.   M.H.   Zimmermann and J.A.   Milburn)
    in preparation: Springer, Berlin
VAN DIE, J.   1974.   Acta Bot.   Neerl.   23:
VAN SCHOOR, G.H.J.   1957.   Ann.   Physiol.   veget.   Univ.   Bruxelles
    2:   7-424
VONK, C.R.   1974.   Acta Bot.   Neerl.   23:
WARDLAW, S.F.   1972.   Planta 104:   18-34
WEBB, J.A.   1971.   Can.   J.   Bot.   49:   717-733
WIERSUM, L.K.   1966.   Acta Bot.   Neerl.   15:   406-418
WILLENBRINK, J.   1968.   Vortrage Gesamtgeb.   Botanik n.F.2:   42-49
ZIEGLER, H.   1956.   Planta 47:   447-500
ZIEGLER, H.   1963.   Planta 60:   41-45
ZIMMERMANN, M.H.   1960.   Annu.   Rev.   Plant Physiol.   11:   167-190
ZIMMERMANN, M.H.   1969.   Planta 84:   272-278
ZIMMERMANN, M.H.   1973.   Quart.   Rev.   Biol.   48:   314-321

Discussant: T.E. Mittler
            Division of Entomology
            University of California
            Berkeley, California 94720

Mittler: Dr. Mittler was invited to give his paper on "Exudation from cuts and aphid stylets". The paper follows.

Exudation from Cuts and Aphid Stylets

T.E. Mittler

The preceding paper by Dr. van Die highlights the fact that in some plants exudations from cuts can proceed unabatedly for many hours. The speed with which the sieve tubes of plants get plugged to curtail loss of phloem sap from such injury is possibly one of the most important issues from the theoretical as well as from the applied points of view.

For plants, phloem plugging clearly is a necessity to prevent bleeding to death; for students of translocation, plugging may be a nuisance or an insurmountable obstacle, but sometimes possibly a blessing in disguise; while for people who want to exploit exudations commercially (whether of phloem sap, resin, or rubber latex), plugging imposes severe limitations on yield. If we could determine the causes for plugging, perhaps we can find ways to prevent it.

Most plants, on being cut, appear not to exude any phloem sap, and in those that do exude the exudation is generally insufficient or difficult to collect, of changing composition, and extremely short lived - particularly so in comparison with the lengthy and copious exudations of stable composition obtainable from some Agavaceae and palms. Since phloem sap in all plants normally is under some turgor pressure, it follws that anatomical and physiological differences, whether only quantitative or truly quantitative, must exist to account for rapid blockage in one case and not in the other. Thus, while in most plants exudation from incisions ceases almost immediately because slime, P-bodies, or other matter impinge on the sieve plates and occlude the pores as a results of the surge of sap through them, this clearly is not so in the bleeding Agavaceae and palms.

448   T. E. MITTLER

In these plants, surging and the enhanced sap flow - up to 20 times that in the intact plant - does not appear to cause blockage in itself. For if the rapid sap flow were responsible for blockage also in these plants, then it should not only occur in the cells next to the cut but also in the phloem cells some distance from the cut where the flow rate is probably equally high.

Dr. van Die showed that blockage is much slower when the cut end of a Yucca inflorescence stalk is cooled to 0 than at 20-25°, a finding that strongly supports his conclusion that chemical processes are at work. The formation of callose is such a process, and was shown by van Die and Tammes (1966) to occur in the phloem elements adjacent to the cut.

Do such processes depend on enzymes produced and/or released at the site of injury by the sieve tubes themselves, by the companion or other associated cells? Is oxygen involved? If so, can these physiological and biochemical processes be delayed or blocked by reducing agents, by lowering the oxygen tension, by metabolic inhibitors, or by other treatments?

One treatment that has recently been shown by King and Zeevaart (1974) to prolong exudation of photosynthate from the severed petioles of Perilla and Chenopodium leaves was the addition of 20 mM EDTA or other chelating agents to solutions bathing the cut petioles. Since these materials sequester calcium ions and the exudation was antagonized by supplying the petioles with unchelated calcium ions, the results were interpreted in terms of callose inhibition. The low levels of calcium reported by Dr. van Die in the exudates of Yucca may therefore be important in delaying callose formtion in these plants.

It should not be difficult to determine if plugging could be prevented, or at least delayed, if the cut end of an exuding plant is placed in a nitrogen atmosphere. It has previously been mentioned that it would be nice if we could cut plants inside a pressure chamber. While this presents technical difficulties it could and should be done.

The moment a plant is cut a very steep pressure gradient will arise in the phloem cells closest to the cut, that is in precisely those sieve tube cells that become blocked. It would be interesting and feasible to study the effect on exudation and blockage of reducing this pressure gradient by applying gaseous or liquid pressure to the cut. By providing a kind of feed-back mechanism in this way we may be able to regulate and prolong exudation.

In an intact plant one can envisage an equilibrium between supply and demand. When the demand is low there will be a natural feed-back that will reduce the amount of material mobilized and translocated. When a cut is made an insatiable sink is created and the input-output relationships are drastically changed. The new equilibrium that is attained will now be governed by the ability of the plant to draw on solutes and water, and to overcome the resistance to their flow through the sieve tubes. In most plants this resistance increases immediately near a cut and flow ceases. In others, resistance builds up less rapidly, only after some hours apparently in the case of the bleeding Agavaceae.

Why is it that aphid stylets in some plants have been found to exude phloem sap unabatedly for days without any apparent blockage of the sieve tubes they tap? Has anyone looked for and observed callose formation in the pores of sieve tubes actually tapped by aphid stylets? If blockage did occur in a sieve tube as a results of an aphid feeding on its sap, the intact feeding aphid could merely insert its stylets into another sieve tube and tap its sap until the sap flow was diminished as a result of its blockage. From sections such as the one shown by Dr. Kollmann at this meeting and from others that give the impression of a haustorial-like system of branched stylet tracks, one may infer that aphids do in fact sequentially tap a succession of sieve tubes, whether for this reason alone or not. A cut and exuding stylet bundle cannot do this, however. Lack of blockage of the sieve tubes in this case could be due to such factors as: (1) the lack of severe injury to the sieve tube and its associated cells; (2) free oxygen may not reach the phloem cells; (3) the pressure drop in the stylet-tapped sieve tube is not as severe as that resulting from a cut; (4) the aphid, before it was removed (possibly while it inserted its stylets and probably while it was feeding), introduced some salivary material into the plant that inhibits blockage.

With regard to the pressure and flow relationships I envisage the following to occur when the phloem is cut or tapped by aphids: In the intact plant, I suspect that the pressure gradient profile from source to sink is mostly extremely flat, only transiently being slightly higher at the former and slightly lower at the latter, these differences only occurring when translocation actually proceeds; because it is evident that one cannot have pressure differences in the aqueous continuum of the phloem symplast without causing a flow, and vice versa, of course.

When a plant is cut, a very steep pressure drop occurs within a fraction of a second in the cells cut and adjacent to the cut. If these cells immediately become blocked, this pressure drop will not extend any further from the cut. However, if the plant continues to exude beyond the initial surge, as in the exuding Agavaceae and palms, a more gradual and possibly linear pressure

Fig.1.   Autoradiography of a fully expanded trifoliate leaf of
Phaseolus vulgaris (the older of two such leaves on the plant).
Protocol: Day 1. Intact plant taken from greenhouse; terminal bud
removed; a planchet (25 mm in diam. with sponge rubber sleeve)
attached to underside of each lateral leaflet of oldest expanded
leaf; 30 adult black bean aphids (Aphis favae) placed in one
planchet. Plant left overnight in darkened laboratory.   Day 2.
Apical portion of central leaflet labelled for 15 min in bright
daylight with approx. 8 μC $^{14}CO_2$.   Plant returned to darkened
laboratory.   Day 3.   Planchets and adult aphids removed for
counting; leaf detached for freeze drying and autoradiography. The
aphids had acquired and excreted high levels of radioactivity. The
larvae they had produced remained on the leaves and show up as
spots.   Note the radioactvity in the veins leading to the region
where the aphids had fed and created a sink sufficiently strong to
cause the otherwise exporting leaflet to import assimilate from the
central leaflet.   (From a collaborative study with Dr.   O.A.
Leonard, Botany Department, University of California, Davis.)

gradient will extend back from the cut at zero (ambient) pressure. As the cells near the cut progressively seal, the pressure profile will return to one similar to that of a plant that did not exude (but it will probably have a lower turgor plateau).

When an aphid taps a sieve tube, the turgor drop will be much less than that resulting from a cut. It is conceivable that a aphid prevents initial surging by only gradually letting sap pass up its stylets and into its body when it first pierces a sieve tube. In any event, the stylet food canal (from 0.5 to 2.0 microns in diameter and some 100 to 200 microns in length, depending on the species) through which the sap must pass, represents a considerable resistance to this flow. Therefore the pressure in the sieve cell tapped will drop only partially and pressure gradients will extend from both sides of the pierced cell. These may be symmetrical or skewed depending on the tapped cell´s relative position to the natural sources(s) and sink(s) in exporting or importing leaves, in intact stems or pieces of stem. Suction by an aphid would locally depress this pressure profile further.

On the other hand, one could oppose the pressure that forces sap through a severed stylet bundle by applying pressure to its cut end. This would provide the means not only for measuring the normal turgor pressure of the sieve tube but for injecting materials into it - a technique that may be of great value if it can be developed. One may not, however, have to go to these lengths. I have referred to the injection by aphids of salivary material. Whatever the function of this saliva is in normal feeding, it is clear that in a number of cases sap uptake by aphids is accompanied with the injection into the host plant of some salivary compound different from the gell-like salivary sheath material that is introduced into the plant when the stylets are being inserted.

One of the more recent and dramatic demonstrations of this comes from a study by Forrest and Noordink (1971). These workers found that ^{14}C introduced during feeding by previously labelled aphids into one part of an apple seedling was picked up by previously unlabelled aphids feeding on another part of the seedling.

Making use of aphids to inject materials into sieve tubes rather than for abstracting them from these cells may be of considerable value, and would overcome a highly justified objection to the use of aphids for monitoring normal translocation processes. For as has been pointed out previously here and in the literature (Peel and Weatherley, 1962), and illustrated in Fig.1, aphids create considerable sinks of their own and may therefore grossly distort the normal distribution of assimilates. The extent to which they do so is also evident from the following rather simplistic assumptions and calculations based on uptake data on the pea aphid (Mittler and Sylvester, 1961) and on sieve cell dimensions estimated from scale drawings (Eschrich, 1967) of _Vicia faba_ on which the aphid feeds:

The volume ingested per aphid per hr. is approx. 0.04 l = 4 x $10^7\mu^3$. The volume of _V. faba_ sieve element is approx. 125µ x 42_π = 7 x $10^3\mu^3$. Hence the aphid ingests the volume of approx. 6,000 sieve elements per hr. If it did so from a single continuous sieve tube, it would draw on 75 cm per hr or 37.5 cm per hr from each direction.

This velocity value falls within the 30 - 40 cm per hr range that Eschrich (1967) considered would be needed to bring two labels (^{14}C and fluorescein) simultaneously and in opposite directions to a pea aphid feeding on an internode of _V. faba_.

While the use of aphids appears to impose a degree of artificiality to phloem translocation studies, aphids and other plant sucking insects have evolved a close association with vascular plants, one from which a lot can yet be learned from the botanical as well as entomological points of view.

Financial support from the National Science Foundation of the U.S.A. to attend this meeting is gratefully acknowledged.

References
ESCHRICH. W. 1967. Planta (Berl.) 73: 37-49
FORREST, J.M.S. and NOORDINK, J.P.W. 1971. Ent. exp. and appl. 14: 133-134
KING, R.W. and ZEEVAART, J.A.D. 1974. Plant Physiol. 53: 96-103
MITTLER, T.E. and SYLVESTER, E.S. 1961. J. Econ. Ent. 54: 616-622
PEEL, A.J. and WEATHERLEY, P.E. 1962. Ann. Bot. 26: 633-646
VAN DIE, J. and TAMMES, P.M.L. 1966. Proc. Kon. Nederl. Akad. Wetensch. C, 69: 648-654

Lamoureux: My first question to Dr. van Die concerns the sugar palm, _Arenga_, which is different from palms with terminal inflorescences or lateral inflorescences, because the plant, after growing awhile, flowers from the top down. I wonder if anyone has determined whether the exudate comes from stored material in the stem or from current photosynthate. The second question has to do with tapping. In Southeast Asia, when palm inflorescences are tapped, the tappers, instead of just slicing off the tips, usually crush them. We get much more exudation by crushing the stump of the inflorescence than merely by cutting it. Do you have any comments on this practice?

van Die: The <u>Arenga</u> palm is more or less a transitional stage between a terminal and a lateral-flowering palm. It produces its inflorescence about 3 years before the end of its life span. It has at this stage a small amount of starch in its trunk, and if it is tapped, this amount of starch disappears but the total amount of sucrose you can get from the successive inflorescences is several times higher than the amount of starch that disappears, and probably the difference between these values is supplied by normal photosynthesis. If you make the necessary calculations, you get a good balance sheet.

As far as bruising of the inflorescence is concerned, in most bleeding palms and Agaves you need a kind of pre-treatment. You seldom get exudation directly after cutting. In <u>Cocos</u> palm you have to cut several times and wait a few days in between these several cuts, and then at a certain moment you get bleeding. In our first <u>Yucca</u> experiments, using plants we got directly from a grower in 1960, we often had plants which did not directly bleed and which needed the same treatment as I described for the palms. But in later years we propagated those <u>Yucca</u> plants which were the best bleeders and that means that nowadays we are able to get bleeding sap directly without any pre-treatment. Probably we have made our selections on the bleeding mechanism. As far as the significance of the bruising or kneading is concerned, I believe that it has to do with plugging the xylem vessels. Obviously there will be no exudation if the exudate is pulled into the xylem. As early as 1933, Dr. Tammes observed that, in <u>Arenga</u>, blocking or plugging of the xylem is necessary to get bleeding. If you have an inflorescence stalk that does not bleed, you will find that you can suck or blow air through the xylem vessels, but as soon as bleeding starts, then you cannot blow air through the stalk, and you can also observe that the xylem vessels have been plugged with some kind of material. It is possible that rapid blockage of xylem vessels in inflorescence stalks has something to do with the probability that xylem vessels may not be very functional in the production of fruits.

Mittler: I am a little surprised to hear that there isn't an hormonal involvement in bleeding or in the inhibition of blockage.

van Die: I'm not sure there is no hormonal influence. Indeed, you can get exudate only from the stalk of a plant which is in the flowering state. In my opinion it is not excluded that the ability to produce flowers may have something to do with the potential to produce exudate.

Milburn: Some recent work has been conducted by Zimmermann and myself and, earlier by Zimmermann and Parthasarathy, on some of the issues here and I wish to report these in brief. First, in connection with the role of massage, bending, pounding or whatever

one wants to call it, attempts were made by Parthasarathy and
Zimmermann to see if there was a difference between Cocos
inflorescences which were pounded and those which were not, all
being sliced at regular intervals. They found no difference
between them. Recently, in some work done in Miami, I was
interested in re-examining this problem, since the raison d´etre of
the massage technique as applied to Ricinus originated from the
ideas of Bose who described the tapper experience in India with
palms. The inflorescence of Cocos is surrounded by a very hard
sheath, and although the tapper normally uses a mallet weighing a
couple of kilos, it doesn´t seem to me conceivable that one can
pound the outer sheath so much that the flowers inside would not be
kneaded to a considerable extent.

One of the points that has not been cited is that the
inflorescence is normally bent over like an elbow since the
inflorescence in its immature state stands almost vertical and it
is difficult to collect exudate from it. In bending it over one
kinks the vasculature to a considerable extent. I now believe this
is the most likely reason for the massage-type effect. When one
hits the inflorescence, he actually bends it more and emphasizes
the kink. Here I differ slightly with Dr. van Die, whose
experiments I endorse by-and-large 100%, but I think in this
particular case one can demonstrate that the flow of exudate can be
stopped by bending the inflorescence in this way. When you let it
recover, it continues in the normal way - in other words, the
bending action on the inflorescence stalk is closely parallel to
the experiments we´ve done on Ricinus where, if one squeezes the
phloem, the immediate effect presumably is to cause surging and
hence sealing or blocking.

We have also measured pressures in the inflorescence of 6 to 8
bars. Loss down the xylem is a major problem but one can overcome
it to a certain degree. The trouble with pressure measurements is
that one never knows if he has reached the maximum or merely
approached somewhere towards it.

Crisp: I do not know of any reports on the occurrence of
short- or long-chain fatty acids and steroids in the exudate. Is
this because they have not been analyzed for or because they are
absent?

van Die: We have not encountered fatty acids. I think if
they were present we would have detected them by now in some way or
other. Saponins are present in the sap, though not in large
amounts.

Eschrich: You published a paper with Tammes showing that
various labelled compounds supplied to the stalk are transported
mainly as sucrose, with one exception, as I recall, namely sorbose.

Do you have any explanation why sorbose was not converted? Sorbose no doubt requires several enzymes to change it to one of the hexoses that can be converted to sucrose. When you supply labelled sorbose to the bottom of the stalk and get the same substance out of the phloem exudate, where do you think it is moving? Is it entering the sieve tubes and moving unaltered up the sieve tubes, or is it being carried by the xylem? Can xylem transport be excluded?

van Die: As you know, sorbose does not occur in higher plants (there is one publication on its possible occurrence in higher plants but it is probably of microbiological origin in this case). It is possible, therefore, there are no enzymes available to convert sorbose to another substance. As far as the appearance of sorbose in the exudate is concerned, the danger is always present, of course, that a substance applied at the base may reach the bleeding surface through the xylem. We have always reduced our experiments to 1 or 2 hours, and we have carried out controls with acid fuchsin to see if acid fuchsin applied to the basal part could be detected in the exudate. This has never occurred within a few hours but it does appear after, let us say, 5 or more hours. We cannot exclude the possibility that if a substance is present in the exudate, that it has not moved via the apoplast but, due to the short duration of our experiments, we are almost sure that it has not.

Swanson: Could you elaborate a bit as to how plugging of the xylem occurs? Do I infer correctly from your comments, that in selecting Yucca plants which bleed easily or which are "instant haemophiliacs", to use Dr. Milburn's terminology, you select for plants where the xylem plugs rapidly? How do you explain the mechanism of this plugging action?

van Die: We have no idea as to how it occurs, but it does occur. Perhaps the alkaline pH of the sieve tube exudate irritates the xylem vessels in some way - but that is only an hypothesis.

Kollman: Is there a difference in the composition of the exudate which comes out just after the first cut and after several hours - for instance, in regards to the protein content?

van Die: A difficulty with this problem is that the first cut seldom gives exudate directly, at least in large amounts, and you always get contaminating substances when you make a cut because of the large amount of non-vascular cells which are cut. So we always neglect the first drop. So the difficulty is that by our methods we cannot exactly determine what the real contents are of the aqueous phase of the sieve tubes. We can only determine the replacement of these exudates.

Mittler: This point, the composition of early exudates, has not been established for the first half hour or so in aphids either. The sucrose concentration remains remarkably constant over hours and sometimes days. We do have a technique now for telling when the aphid stylet reaches a sieve tube by cementing a wire to the aphid and passing a current through it - when the aphid inserts its stylets into a plant, the salivations cause a change in electrical conductivity which is recorded on a potentiometric recorder. From these recordings it may be possible to interpret the onset of continuous feeding. Then we would be able to tell almost the instant at which an aphid has reached a sieve tube, and if we cut it at that time, perhaps we could obtain the earliest droplets and find whether it has a different composition.

Moorby: Dr. van die, did I interpret your comments correctly that you expected a lot of water to be transpired through the fruit? Have you measured this rate? In most fruits that I know the transpiration rate is virtually nil, in fact so much so that they are actually a water source for the rest of the plant during transpiration and the fruit shrink. If you do experiments with labelled calcium or strontium at that time, you will find no entry of label and in fact you can pick this up at night.

van Die: Apart from the water loss by transpiration in palm fruits, you always have the probability that part of the water flows back through the xylem.

Moorby: Have you done the actual calculations as to how much is moving and whether this would fit in with the calcium balance?

van Die: We have done some work on the transpiration rate of developing tomato fruits, and we find the rate relatively high in the very young stages.It is just in these young stages that the calcium content is relatively high as a percent of total ash, so it can be that a considerable amount of the calcium present in the mature tomato has come into the fruit in the early stages. If you plot calcium content against dry weight, there is a considerable deviation from linearity.

Moorby: It has been shown that calcium moves into apples at an early stage and actually moves out in the xylem in the later stages.

van Die: That is very strange because the xylem vessel walls have a strong affinity for bivalent ions.

Eschrich: The paper by Ziegler that you mentioned was based on studies with ^{14}C. He showed that ^{14}C moved out of the fruit of the cucurbits, so it must exit through the xylem.

van Die: This is difficult to understand. There are a number of papers which show that calcium moves very slowly through the xylem because of ion exchange. It has been shown that calcium is strongly adsorbed at sites in the lignin of the xylem vessels.

Aikman: If, during growth, calcium had been moving up the xylem and had already reached a high level in the fruits, then these exchange sites would already be saturated. If there is a return flux of water from the fruits in the xylem, then the calcium would be carried down the water stream and partly exchange with the calcium that was already saturating the sites in the xylem. Thus a net flux could still be carried but the tracer might be delayed by exchange.

van Die: Yes, I agree. If a solution of potassium and magnesium salts flows into the fruit, then there will be sites in the xylem vessel walls of the fruit which strongly adsorb the calcium. If after some time, the flow reverses, there is no calcium or magnesium to remove the adsorbed calcium from the sites in the fruit, but potassium, for example, will move straight on back.

Aronoff: Does the diurnal rhythm of the exudation rate in Yucca parallel the photosynthetic rate or root pressure? If the latter, do you postulate any correlation between the two phenomena?

van Die: I can't give an explanation for the diurnal rhythm. We always find the sap to have a considerably higher specific activity during the night than during the day, but the rate of exudation is in most cases higher during the day than at night. If you multiply both, you nevertheless get more ^{14}C-sucrose during the night than during the day. I have no explanation for that.

Elmore: Can you evaluate for us King and Zeevaart's technique for getting phloem exudate?

van Die: The EDTA technique? Perhaps it has something to do with callose formation under the influence of calcium but I cannot give any explanation for it.

Reinhold: I wish to refer back again to the apparent rhythm of ^{14}C production at night. It seems to me that this would be a predictable consequence of nascent ^{14}C assimilate export in preference to stored sugar. You've given us evidence that when you supply ^{14}C for just a short time you get a storage pool building up in the leaf that remains in the leaf for days. After the cessation of ^{14}CO supply, you supply $^{12}CO_2$, so that during the day the current photosynthate will be exported preferentially and will not be labelled. At nights you will have storage sugar exported and that will be the ^{14}C sugar that will be preferentially exported.

van Die: That could be possible but the rate of decline in activity during the day and during the night is the same. Then I have to postulate two pools, one for photosynthates made during the day and another for stored photosynthates, with both pools having the same turn-over rate. I think such an hypothesis is unnecessarily complicated.

Reinhold: The $^{14}CO_2$ was supplied for only a brief period, and at the time it was supplied I would expect some labelled photosynthates to be transported but the moment the labelled CO_2 supply is cut off, then $^{12}CO_2$ assimilation will take place. So in daylight it will be the ^{12}C assimilate that will be transported, and at night, when there is no further ^{12}C-assimilation, then the stored ^{14}C assimilate will be transported.

van Die: Yes, that could be. Starch is present in Yucca leaves but the most abundant stored carbohydrates are fructosans which are completely soluble and present in the vacuole. We have not yet made a complete analysis of the mobilizable carbohydrates of the Yucca leaf.

Hoddinott: I would lke to comment briefly on our experience with EDTA. Several years ago we were trying to get translocation of ^{14}C assimilates throught the phloem loops of Heracleum lanatum and heard about the results of King and Zeevaart with EDTA. We injected the hollow petiole of Heracleum with EDTA solution and while extracting the phloem loops we bathed the loop with EDTA solution. We did not succeed, however, in getting ^{14}C translocation through the loop. When the paper of King and Zeevaart came out and it was clear that their studies referred to detached leaves, we attempted similar studies with H. lanatum. We found the ^{14}C-translocation profiles of the EDTA leaves were indicative of far less ^{14}C being transported down the phloem of detached leaves than through control plants. I don't think that the results with EDTA and callosing are at all clear as yet.

I have one question for Dr. van Die relating to cold temperature on the exudation rate in his winter hardy Yucca. When you cooled the plant to 0°, was there any recovery in the exudation rate at 0°, or did you have to wait until you increased the plant to 20° again?

van Die: There was almost no change in the flow rate. The flow slightly decreased but then continued at that rate. The only effect we observed was a 1% higher sucrose content of the exudate than we had normally at 20°. I cannot explain that 1% increase.

Hall: I have often wondered about the high ATP concentration in the sap, especially with regard to ADP and AMP levels. Is anything known about the turn-over rate of ATP because we have to know that in order to determine whether ATP is being utilized as it is being transported or whether it is just swept along passively with the assimilate stream.

With respect to the EDTA technique, I have also tried this technique on _Ricinus_ leaves. The method I used was to label the leaf with $^{14}CO_2$ and to immerse the cut petiole end in EDTA solution. I found no ^{14}C to come out of the petiole at all in the EDTA solution whereas it does in water.

van Die: As far as the ATP content is concerned, I could never find such high ATP contents as have been reported by other investigators - about 650 µg ml^{-1}. The value we found was not more than half of that. Perhaps our method was not as good as theirs.

Hall: The levels in Ricinus are about 0.6 mM and the level stays high during several hours of exudation. This is till very high in relation to ADP and AMP and suggests that ATP is not being utilized.

van Die: My value was about 0.5 mM.

Willenbrink: Dr. Mittler, I am interested in the point you made about the injection of salivary material into the sieve tubes by aphids. Does this have any effect on callose or proteinase?

Mittler: I can only speculate. There may be some effects on galling. Aphids seem to do better on gall tissue, perhaps because of a greater nutrient supply.

Willenbrink: Your remarks, Dr. Hall, were very interesting to me. I have felt for a long time that it is redundant now to look only for ATP content in the sieve tubes. We must progress now to measuring the energy charge of the ATP system in the sieve tube. We have made calculations on energy charge based on whole tissue - of course that is very, very dangerous - but the energy charge in the whole phloem was nearly equal to that for a photosynthesizing leaf.

Milburn: I just wish to make one point in connection with the EDTA experiments. It seems to me somewhat unreasonable to expect EDTA to have a kind of blanket reacton. Sealing reactions are many and various. They can be oxidative, or proteinaceous, or involve callose. This is only for the few that we know anything about. So I would suggest that if, for example, blockage in _Ricinus_ is proteinaceous, for which we have a good deal of evidence, it would be unreasonable to expect EDTA, which may have a callose-removing effect, to have any major influence.

I´d like to return to an earlier point, that is, the issue of xylem blockage and phloem blockage, as a result of mechanical treatment, because it seems a very fascinating difference between the interpretations of the Tammes-van Die group and ourselves. The mechanism that seems to apply in the case of exudation from palms that I cited earlier was that it is a difficult business cutting a new slice off the end of the inflorescence stalk and you cannot avoid bending the stalk and cutting off the flow. The implication is that when recovery occurs, a massive rate of flow takes place into the deprived sink. This is exactly the same mechanism which has been proposed for the inducement or enhancement of flow in Ricinus. The immediate effect of massage is to interrupt the flow, but the plant in readjusting is able to reverse the sealing effect and to give a massive surge. It is accepted in the folklore of palm tappers that the knife you use must be extremely sharp and I support this from my own experiences with palms. Perhaps Dr. van Die and myself would be able to reach some sort of agreement on the basis of this simple experiment. If one takes an inflorescence stalk and cuts it with a blunt knife, the chance of getting any exudate out is pretty small, but you can cut the same stalk with a very sharp knife and get exudate. It seems to me that xylem blockage cannot explain this type of result.

Swanson: Just to illustrate how species-specific this EDTA effect apparently is, I recently heard a paper by Dr. Dickson (1974 annual meeting of the Midwest Section of A.S.P.P.) in which he reported that excised primary leaves of bush bean, when EDTA treated, translocated 75% as much label out of the leaf as an attached control leaf; but with the Eastern Cottonwood, the EDTA effect on excised leaves was trivially small, about 1% or so.

Johnson: I have not worked seriously with aphids, but out of idle curiosity I have cut a number of stylets and on one occasion rather than exuding and forming a permanent droplet, one of the stylets exuded and then the droplet was withdrawn into the stylet, partly re-exudated, and then withdrawn again. I wonder if this is a common observation and if there is an explanation for it.

Mittler: I have not observed this "panting" phenomenon except in intact aphids which do this occasionally with their honey-dew droplets. In a number of cases I have found the stylet to dry up, so to speak, and I could force a little more sap out by pressing on the willow bark with the blunt end of a scalpel.

Dunford: A recent paper by Milburn has indicated that, at least in Ricinus, cutting the stem caused an increase in the loading rate, presumably due either to decreased turgor or decreased concentration. I was wondering, Dr. van Die, if you have observed any feed-back of cutting the inflorescence stalk on loading rates.

van Die: When the inflorescence of <u>Arenga</u> is cut, then it starts bleeding with a low sugar concentration. The tappers neglect this exudate because it has too little sucrose. The sugar concentration of the exudate gradually increases and after about 35 days it reaches its highest level and at that point the tappers start collecting the exudate. As far as your question on feed-back, I really can´t give a specific answer.

Aikman: I would like to comment on the 1% increase you noted in sugar concentration when the cut stem was cooled to $0^{\circ}C$. Perhaps this increase resulted from the increased viscosity of water. The increase in driving force required to overcome this would results in an increase in concentration.

I would like to ask a question about pressure-driven volume flow. If you simultaneously double-cut to remove a section of the stalk, can external pressure applied to one end be used to drive a flow through the phloem?

van Die: We have not tried such an experiment.

Ho: I would like to direct this question or comment to Dr. Mittler. I seems to me that your calculations indicate that aphid feeding does not put stress on the proteins in the solvent inside the sieve tube, and this indicates to me that the aphid feeding is based on the protein of the solution inside the sieve tube. What I wish to say is that in normal situations the volume moving along a sieve tube is sufficient for aphid feeding, so that aphid feeding does not alter the rate of volume flow.

Mittler: No, feeding does deprive the plant of this volume and associated nutrients. We did experiments using single aphids on young alfalfa seedlings and in this case both the volume of water and the nutrients constituted such a drain on the plants that we could determine appreciable dry weight losses as a result of aphid feeding. Now single aphids as in this case would have an insignificant effect on a larger plant, but multiplied many-fold, a heavy aphid infestation can be a serious drain.

Swanson: I would like to suggest that the "panting" of the aphid stylets may be due to changing water stress in the plants. Dr. Sij, in my laboratory a few years ago, obtained some very tentative data wich suggest that the rate of import of translocate into young sink leaves of the Pinto bean can be increased or decreased over short periods of time by respectively increasing or decreasing the transpiration rate of the sink leaf (by alternatively illuminating and darkening the sink leaf).

Milburn: I was just going to make the same kind of point. Wilting has been used by anatomists prior to fixing, to try to reduce the turgor inside the system. Now unless you know what you are doing you may apply wilting in the wrong manner. The difficulty is that we are dealing with the phloem system which has very marked homeostatic properties. When, for example, we were collecting Cocos exudate and we got a rain shower, there was a rapid surge in the exudation which exactly paralleled what we found in Ricinus, and similar to that which Dr. Swanson has just reported. Our first reaction was that rain water had gotten into the exudate, there was such an increase, but when we calculated the actual amount of solids in the exudate, it was quite evident that as the showery period came on, we got a massive increased loss of sucrose from the system. The difficulty with these wilting attempts or with attempts to affect the exudation rate by controlling the water balance, is that the system does adapt. If you take a severely wilted Ricinus plant, for example, it will exude and this is despite the fact that the xylem has long been emptied through cavitation. Apparently the phloem has the property to readjust its turgor, tending to remain on the positive side. I've seen situations in India with palms which have been tapped, apparently becoming negative in their pressure only transiently during the day due to transpiration, but then the pressure seems to be restored later in the day. So I would just suggest that because anatomists sometimes try to fix sieve tubes in a flaccid condition, we should not infer that it will always help the situation by giving a longer wilt. Back-up experiments are needed to find the most propitious time to catch this flaccid condition.

Heyser: We did some similar experiments with Cucurbita. We actually dried the plants to the point where they were lying on the pots in which they were growing. When we cut them we got the largest size droplets of sieve tube sap. Examining this sap for osmotic concentration by the cryoscopic method, we found the osmotic pressure to be nearly double that of a normal growing plant.

Mittler: Dr. van Die, have your analyses shown the occurrence of any vitamins in sieve tube exudate?

van Die: I have not myself made such analyses, but there is in the literature, a report on a considerable amount of vitamin C in Cocos extract.

Basic Biophysics of Transport

J. Dainty

Department of Botany
University of Toronto
Toronto, Ontario

Introduction

In the translocation system of a plant all kinds of elementary transport phenomena are probably involved. These include diffusion, cyclosis, passive and active membrane permeation and bulk flow. Any long-distance transport system in living organisms must of necessity comprise three steps: a loading step where diffusion and membrane permeation take place; a bulk flow step carrying the transported substances over large distances, and an unloading step where diffusion and membrane permeation occur again. The necessity for the bulk flow step arises from the slowness of diffusion over long distances. Elementary considerations of the random walk, which is what molecular diffusion is, show that the distance, x, diffused, the diffusion coefficient, D, and the time, t, are related by $x^2 \sim Dt$. Most molecules have a diffusion coefficient in water of about 10^{-5} cm^2 s^{-1}; thus a molecule will diffuse 10 μm in about 10^{-1} s, whereas it will take 10^5 s, about a day, to diffuse 1 cm. One can write the formula as $(x/t) = v = (D/x)$, i.e. think of the velocity of diffusion; for a distance of 10 μm, $v = 10^{-2}$ cm s^{-1} = 36 cm h^{-1} whereas over a distance of 1 cm, $v = 10^{-5}$ cm s^{-1} = 0.36 cm h^{-1}. We can thus take for granted the necessity of a rapid bulk-flow step for transport over distances greater than, probably, a millimeter or so.

However, the diffusion steps or, what amount to the same thing, passive or active membrane permeations, are absolutely essential at the two ends of the transporting system. The only way to load and unload the system is by diffusion or its equivalent membrane permeation. This must be made as efficient and as controlled as possible. I will be mostly discussing, in general terms, diffusion and permeation problems, for the other speakers will be talking about the bulk flow process.

463

One or two general remarks can be made about efficiency assuming, without loss of validity, that the end processes are diffusive in nature. The rate of diffusion across a barrier Δx cm thick is given by Fick's Law:

$$J = DA \frac{\Delta c}{\Delta x} \qquad (1)$$

where J is the flow across the barrier in mole s^{-1}, D is the diffusion coefficient in the barrier in cm s^{-1}, A is the area of the barrier in cm^2 and Δc is the difference in concentration of the diffusing substance in mole cm^{-3}. Clearly the larger the area of the barrier and the thinner it is, the greater is the rate of delivery to or removal from the bulk flow.

At the loading end, at least, of the phloem system there seem to be clear-cut situations where diffusion is important. At some stage between its production in the leaf cells and its appearance in the sieve elements, sucrose seems to pass through the apoplast. It must be unloaded, across a membrane - perhaps by "facilitated diffusion" - into the apoplast, cross the apoplast by diffusion and then be actively transported across another membrane into the sieve element-companion cell complex. These transport processes must be so matched that there is no danger of loss of sugar by lateral diffusion in the apoplast. The other place where diffusion is undoubtedly involved is in the plasmodesmata which richly connect most of the cells involved. Tyree (1970) has discussed in a quantitative, theoretical, way diffusion through plasmodesmata; it is a straightforward application of Fick's Law.

A. Passive permeation of membranes

Passive permeation is essentially a straightforward diffusive process either through the membrane material or, for smaller water-soluble molecules, perhaps through narrow aqueous channels crossing the membranes. Whatever the details of the process, for an uncharged solute molecule the flux, ϕ, in mole cm ^{2} s^{-1} across a membrane is given, according to Fick's Law, by:

$$\phi = P\Delta c \qquad (2)$$

where Δc is the concentration difference in mole cm^{-3} and P is a permeability coefficient in cm s^{-1}.

Two warnings should be given about this formula. In the first place, although it is written as if ϕ were proportional to Δc, i.e. P is a constant, this is not necessarily so. P can depend on the concentrations on either side of the membrane even in a purely passive situation. Secondly, the P obtained, say, in an experiment measuring ϕ as a function of the concentration difference in the bulk solutons on either side of the membrane may not be, or only partly be, a property of the membrane itself. For, on each side of the membrane there will be unstirred layers of solution up to 100 μm or so thick (Dainty, 1963). A solute molecule can only move through these layers by diffusion; thus the layers themselves have their own permeability coefficients or resistances, in series with the membrane. P, therefore, is a measure of the combined permeabilities of the membrane PLUS two unstirred layers.

Uncharged solute molecules passively permeate membranes according to Fick's Law and we can say that the driving force for the diffusion is the concentraton gradient (or difference Δc). Somewhat more correctly, according to irreversible thermodynamics, we might say that the driving force is the gradient (or difference) of the chemical potential of the solute. This would lead to the same formula (2) as obtained using Fick's Law. Charged solute molecules, ions, can experience a force additional to that arising from the concentration gradient; this additional force is the electric field which inevitably exists in membranes separating electrolyte solutions. For ions we must formulate the flux as being dependent on the driving force expressed as the electrochemical potential gradient $d\bar{\mu}/dx$. We write:

Flux = mobility x concentration x driving force $(-d\bar{\mu}/dx)$

The electrochemical potential, $\bar{\mu}$, is given for an ideal solution by

$$\bar{\mu} = \bar{\mu}^* + RT\ell nc + zF\psi \tag{3}$$

where $\bar{\mu}^*$ is the electrochemical potential in the standard state, R is the gas constant, T the absolute temperature, C the concentration, z the algebraic valency, F the Faraday and ψ the electric potential. Substituting (3) into the previous equation one gets an equation for the flux containing C, dc/dx and $d\psi/dx$. It cannot be integrated across the membrane without making assumptions about the concentration gradient or the electric potential gradient (field) or the relation between the two. This problem has been extensively treated in the literature (see Dainty, 1962; Higinbotham, 1973a,b) but most of us are content to follow the so-called constant field (Goldman) assumption, i.e. $(d\psi/dx)$ = constant, and recognize its approximate nature. With this assumption one can derive formulae for the passive flux of an ion, for the ratio of influx to efflux of an ion (the so-called Ussing-Teorell equation), and for the electric potential across the membrane. These are well-known relationships which can be found in most treatments of ion transport. They may be more important for animal cell membranes than for plant cell membranes, because of the apparently relatively greater importance of active transport across plant cell membranes.

B. Water transport across plant cell membranes

The movement of water across membranes needs to be treated somewhat separately because it is the bulk constituent of the solutions bathing membranes. Again we take the driving force moving water across a membrane to be the difference in the chemical potential of water between the two sides. It has proved convenient to define a parameter, related to chemical potential, called water potential, and say that the flux of water is proportional to the water potential difference across the membrane. Water potential, Ψ, is defined by the following equation;

$$\Psi = \frac{\mu_w - \mu_w^0}{\overline{V}_w} \tag{4}$$

where μ_w is the chemical potential of the water in the system, μ_w^0 is the chemical potential of pure water at the same temperature and at atmospheric pressure and $\overline{V}_w$ is the partial molar volume of water, approximately 18 cm^3 $mole^{-1}$. The units of Ψ are those of pressure, i.e. bars. The flux, J_v cm^3 cm^{-2} s^{-1}, of water across a membrane, permeable to water only, can now be written as proportional to the driving force, $\Delta\Psi$, as follows:

$$J_v = L_p \Delta\Psi \tag{5}$$

where L_p is the so-called hydraulic conductivity in cm. s^{-1}. bar^{-1}. Water potential, Ψ, seems to depend on three other parameters of the system: the hydrostatic pressure above (or below) atmospheric; P, the osmotic pressure, π; and the so-called matric potential, τ. The dependence is linear:

$$\Psi = P - \pi - \tau \tag{6}$$

The dependence of Ψ on P and τ does not need any discussion here. The inclusion of the matric potential, τ, is more debatable. The term arises in systems like soil and terrestrial plant cell walls in which there is a very large area of interface between water, solids and gases. There is a real sense in which, if one knew the exact pressures and local solute concentrations in such systems, the matric potential could be contained in the P and π terms. However, usually such information is missing and it is best to include the term, τ, in an operational sense. In the plant translocation system, τ can be ignored in practice. We can thus write $\Psi = P-\pi$ and equation (5) thus becomes:

$$J_v = L_p (\Delta P - \Delta\pi) \tag{7}$$

Two implicit assumptions occur in this equation. One is that water moves passively, in response to physical forces only; i.e. there is

no active transport of water. We believe that this is so in the
vast majority of situations. The other assumption is that the
membrane is semi-permeable; i.e. it is permeable to water only and
not to the solute molecules. This is likely to be a good enough
approximation for most naturally-occurring solutes, but may not be
in an experimental situation. Actually the equation can easily be
modified, according to the theory of irreversible thermodynamics,
by the introduction of the so-called reflection coefficient, σ ,
multiplying the $\Delta\pi$ term:

$$J_v = L_p (\Delta P - \sigma\Delta\pi) \qquad\qquad (8)$$

For non-permeating solutes, $\sigma = 1$; for permeating solutes $0 \leqslant \sigma <$
1, although σ can be negative in certain special cases.

The theory of water transport is thus relatively
straightforward and well-understood. However we do seriously lack
quantitative data on values of L_p, particularly for higher plant
cells. Only for a few giant algal cells do we know L_p
unambiguously. In these cells L_p ranges from about 10^{-5} cm. s^{-1}.
bar^{-1} for Characean cells to about 10^{-5} cm. s^{-1}. bar^{-1} for
Valonia. There is reason to guess that 10^{-7} would be an upper
limit for L_p for higher plant cells. Another parameter is hidden
in equations (7) and (8); this is the volume elastic modulus, $\varepsilon =$
$V(dP/dV)\cdot\varepsilon$ is involved because the ΔP term depends upon the cell
volume, V. Again we have little quantitative information about ε;
it approaches 1000 bars for fully turgid Characean cells, but it
may be very low, less than 10 bars, for higher plant cells near
incipient plasmolysis. (Accounts of the water relations of plant
cells and tissues can be found in Dainty (1963), Slatyer (1967),
House (1974).)

C.   Irreversible thermodynamics

Mention has been made above of the theory of irreversible
thermodynamics. This is the correct theory to apply to transport
processes, although it has not had a major impact on transport
studies except in certain special cases such as the introduction of
the reflection coefficient in the equation for water (volume) flow
across membranes. The chief reason for this is that the theory
introduces many additional parameters which, ideally, must be
experimentally determined. Such determinations are too difficult
to be carried out in most systems at present and so the
experimenter tacitly ignores what he guesses are the less important
parameters. Still, it is better to explicitly neglect parameters
and thus know what one is doing; and it is for this reason that a
brief outline of irreversible thermodymics is given here.

Basically the theory, arguing from the fact that entropy always increases in an irreversible process, shows that the flux of any solute species, say, depends on the forces acting on all the components of the system. Thus, for a system consisting of a single salt solution, the cation flux would depend on 3 forces: on the cation; on the anion, and on the water (and maybe a fourth, if active transport is occurring in the system). Thus, even in a simple, passive, single salt system there are 3 x 3 = 9 coefficients to determine. Actually there are only 6 because in 3 pairs of the coefficients, the two members of a pair are equal to each other. Still 6 is a lot of coefficients; 3 of them are the so-called straight coefficients relating the flux of a component to the force acting on that component and the other 3 are cross coefficients, relating fluxes of a component to forces on other components. These latter are the ones which are usually neglected. Active transport can be included in the system by specifying an additional flow (of chemical reaction) and, of course, an additional force (the affinity of the reaction). This would mean for our single salt system 4 x 4 = 16 coefficients, 10 of which are independent of each other (4 straight coefficients and 6 cross coefficients). In this case those cross coefficients which link ion flux to chemical affinity are particularly important.

The theory of irreversible thermodynamics is an important background theory for everybody involved in transport problems to know and understand. Any theory of transport must be formulated so as to be consistent with the theory of irreversible thermodynamics. Some theories, such as electro-osmosis, are inevitably formulated in this way because the cross coefficients are the important ones in the theory. For good accounts and pertinent examples see Kedem and Katchalsky (1958), Dainty (1963), Katchalsky and Curran (1965).

D. Active Transport

Most solutes of physiological interest cross cell membranes "under control", i.e. they are either actively transported or their diffusion is "facilitated". There may be passive movement as well in parallel with active transport or facilitated diffusion. We imagine, although absolute proof is lacking, that in both active transport and facilitated diffusion, the solute molecule moves across the membrane in combination with a carrier molecule, the latter probably being confined to the membrane. The carrier may well be a membrane-bound protein. The carrier-substrate complex is thought to move across, or rotate in, the membrane by diffusion; if the complex is charged it will be subject, of course, to the electrical field in the membrane, which may be particularly important in plant cells.

There are far more well substantiated cases of active transport than there are of facilitated diffusion. In part, the reason for this is that it is easier to set up well-defined and

testable   criteria   for   active   transport   than   for   facilitated
diffusion.   One clear criterion for active transport is that it   is
movement   of   the solute against the free energy (i.e.   chemical or
electrochemical potential) gradient over and above   what   might   be
expected   from random thermal movement.   (There will always be some
of the latter even against a free gradient; it can, in   effect,   be
calculated   from   simple considerations based upon Boltzmann's Law;
actually the formula for calculating this thermal   flux   is   called
the Ussing-Teorell equation by membranologists.)

Movement against a free energy gradient requires energy, which
can only come from metabolism.   Thus another criterion   for   active
transport   is   that it is definitely linked to metabolic energy; if
the appropriate energy supply is   cut   off,   the   transport   stops.
This   energy criterion is not perfectly clearcut; it is possible to
envisage both facilitated diffusion   and   passive   transport   being
affected by metabolism, e.g.   by cessation of carrier synthesis, by
changes   in   the   membrane   structure,   by   changes in the electric
potential difference across the membrane and so on.   Nevertheless,
such   a   link   between   metabolism   and transport strongly suggests
active transport; in the   jargon   of   the   theory   of   irreversible
thermodynamics, we say that the cross-coefficient connecting solute
flux and affinity of the appropriate chemical reaction is non-zero.

Other criteria exist which are common to both active transport
and facilitated diffusion.   They are less clear than the two above;
it is not difficult to envisage purely   passive   transport   obeying
the   same   criteria although, to be fair, it is less likely.   These
other criteria are competitive inhibition, stereospecificity,   high
Q_{10} and   non-linear   (particularly   a   Michaelis-Menten   type)
relationship between flux and external   solute   concentration.   If
all   these   are   present,   then   it   is   at   least   likely   that
carrier-mediated fluxes are occurring.

In what   follows   I   shall   be   mainly   talking   about   active
transport   of   solutes, both charged and uncharged.   I shall try to
tell what the present day   picture   of   solute   transport   (usually
uptake   or influx) in plant cells looks like.   I shall lean heavily
on the   recent   books   and   reviews   listed   in   the   Bibliography:
Epstein   (1972), Higinbotham (1973a,b), Anderson (1973), Zimmermann
and Dainty (1974), MacRobbie (1970, 1971).

Perhaps best known to most plant physiologists is the work   of
Epstein   and others on the uptake of solutes (usually ions) by root
tissue.   Perhaps this work has received most attention because   the
results   and   their interpretation seem simple and straightforward.
The results are   really   rather   remarkable.   Epstein   essentially
measures   short   term   uptake   (over   10-20 min.),   corrected
appropriately for free space uptake, as   a   function   of   external
solute concentration for low salt root tissue.   Over a range of low

concentrations, up to a few mM, the uptake as a function of concentration fits a hyperbolic function of the form of the Michaelis-Menten equation for enzyme activity:

$$V = \frac{c \ V_{max}}{K_m + c} \tag{9}$$

and the good fit is verified by carrying out the equivalent of a Lineweaver-Burk plot (1/V vs 1/C). Such a so-called absorption isotherm has been produced for many ions (cations and anions) and other solutes.

These results have been interpreted by Epstein and others by a simple carrier hypothesis, analogous to an enzyme reaction:

substrate + carrier $\rightleftharpoons$ substrate-carrier $\longrightarrow$ products (inside)
　　　　　　　　outer　　　　　　complex
　　　　　　　membrane
　　　　　　　surface

$$S + C \quad \rightleftharpoons \quad SC \quad \xrightarrow{\ V\ } \ products$$

This hypothesis leads, of course, to the equation given above and from the experimental data both K_m, describing the binding between substrate and carrier, and V_{max}, the maximum influx rate, can be determined. Competition and inhibition by other substrates can be experimentally studied within the framework of this hypothesis and everything seems to fit very nicely. Yet passive transport is ignored, as are electric potentials or any driving force other than external concentration. The tissue used is inhomogeneous, yet practically every substrate fits this simple hypothesis. Most membrane physiologists are pretty sure the situation is much more complex.

In the past few years the absorption isotherms have been extended to higher concentrations by Epstein (1966), Laties (1969) and, particularly, Nissen (1973, 1974). Beyond the low concentration isotherm, with K_m's of the order of 0.1 mM or less, extra absorption takes place and the above authors, particularly Nissen, have tried to fit this extra absorption to carrier mediated transport site(s). There has been much argument as to how many different kinds of sites there are, whether they are all on the plasmalemma or some on the tonoplast, whether they might be a single site which changes its affinity, etc. with concentration, and so on. The reader must judge from the literature how valid it all is, although Cram (1974) has bitterly criticized the whole business.

From the biophysical point of view most of the definitive work on active transport of ions has been done on giant algal cells by people like MacRobbie, Raven, Smith, Dainty, Spanswick, Williams, Hope, Walker and others and on <u>Neurospora</u> by Slayman. The principles established by this work are applicable to all plant cells, even if the detailed facts are not; the work of Higinbotham, Poole, Jeschke, Cram and others on higher plant cells is proof of this.

For giant algal cells the following picture was established by the early 1960's from a careful analysis of fluxes in their dependence on driving forces, and on the effects of light and dark, various light wave lengths and inhibitors. K^+ is either pumped in or is in flux equilibrium, Na^+ is usually pumped out, Cl^- is pumped in. All the main active transport sites (the pumps) are on the plasmalemma, the fluxes are of the order of 10^{-12} mole cm^{-2} s^{-1}, passive fluxes in the appropriate directions exist, and there is a membrane potential (in the Characeae) of about -140 mV, inside negative. At this early stage (Dainty, 1962), the pumps were thought to be neutral, i.e. the substrate-carrier complex was uncharged on both its journeys across the membrane, and the membrane potential was thought to be a so-called diffusion potental, i.e. it arose from the passive fluxes only.

Although much of this basic picture is probably correct, a number of diffuculties began to appear: the membrane potential did not behave in the right way, as a diffusion potential, when the external ion concentrations were varied; the membrane potential was sometimes more negative than any possible diffusion potential could be; the membrane potential (for higher plant cells and <u>Neurospora</u>) rapidly depolarized when the metabolism, and therefore active transport, was very much greater than could be accounted for by the passive fluxes of K^+, Na^+ and Cl^-. These difficulties have been largely resolved by the recognition, long pressed for by Higinbotham (1973a), that at least some of the pumps are electrogenic, i.e. the carrier-substrate complex is charged on one or both of its journeys and therefore contributes directly to the membrane potential, and by the discovery, initially by Kitasato (1968), that large fluxes of H^+ are occurring across the membrane; protons being pumped outwards accompanied by more or less equal passive fluxes of either H^+ inwards or OH^- outwards; these H fluxes are one to two orders of magnitude greater than the K^+, Na^- or Cl^- fluxes. In fact it is now thought that the dominating feature of the ionic relations of plant cells is a powerful, electrogenic, proton extrusion pump operating across the plasmalemma.

Raven and Smith (1974) have cogently argued that such a proton pump must arise naturally in plant cells as an inevitable consequence of the necessity to regulate the internal, cytoplasmic, pH. They have shown that plant cell metabolism produces an excess of protons which must constantly be pumped out of the cell to maintain the internal pH constant. There are many secondary consequences of this, e.g. the possible use of H^+ in "softening" the cell wall and thus maintaining growth. It may be that the other ion pumps are all linked in some way to the proton extrusion pump and its associated passive flux of H^+ or OH^-, i.e. the K^+ pump may be a K^+/H^+ exchange pump, the Cl^- pump a Cl^-/OH^- exchange pump or a Cl^-, H^+ co-transporting pump, and so on.

As lucidly explained by Slayman (1974), the recognition of the central importance of a proton extrusion pump unifies a great deal of transport phenomena at all levels if one accepts Mitchell's (1966) chemiosmotic hypothesis. The essentail events in the energy-conserving processes in mitochondria and chloroplasts are movements of electrons across membranes, i.e. charge separation processes, which create large electrochemical potential gradients, $\Delta\mu_H^+$, for protons. $\Delta\mu_H^+$ drives, by means of an asymmetrical ATPase, the synthesis of ATP and ADP and inorganic phosphate. Such a pump, given sufficient ATP concentration, can be driven backwards, thus pumping H^+ across a membrane. Such is envisaged for the proton extrusion pump of plant cell membranes.

It seems that a H^+ extrusion pump, or its associated passive flux, is coupled not only to the movement of other ions but also to the movement of uncharged molecules such as sugars and amino-acids. Slayman's (1974) work on Neurospora and the work of Komor and Tanner (1974) on Chlorella show this rather nicely in respect of the uptake of glucose. It appears that the carrier-glucose complex will only cross the membrane in combination with H^+, the direct driving force for this being the rather large electric potential difference across the membrane. Of course the necessary external proton concentration must be produced by the proton extrusion pump.

Applying these latter ideas to sucrose loading of the phloem, and including the knowledge that the pH of the translocation fluid is very high and the ratio of K^+/Na^+ is also very high, we should expect a powerful H^+ extrusion pump at the membranes of the sieve element-companion cell complex. And we would expect such a pump to be involved in the co-transport of sucrose and H^+ and the pumping of K^+. Such a pump would also have relevance to the electro-osmotic mechanism of Spanner (1975).

References

ANDERSON, P.W. 1973. Ed. Ion Transport in Plants. Academic Press.
CRAM, W.J. 1974. In Membrane Transport in Plants and Plant Organelles. V. Zimmerman and J. Dainty, eds. Springer
DAINTY, J. 1962. Ann. Rev. Plant Physiol. 13: 379-402
DAINTY, J. 1963. Adv. Bot. Research 1: 279-326

EPSTEIN, E.   1966.   Nature 212:   1324-1327
EPSTEIN, E.   1972.   Mineral Nutrition of Plants: Principles and
    Perspectives.   John Wiley and Sons.
HIGINBOTHAM, N.   1973a.   Bot.   Review 39:   15-69
HIGINBOTHAM, N.   1973b.   Ann.   Rev.   Plant Physiol.   24:   25-46
HOUSE, C.R.   1974.   Water Transport in Cells and Tissues.   Arnold
KATCHALSKY,   A.   and   CURRAN,   P.F.   1965.   Nonequilibrium
    Thermodynamics in Biophysics.   Harvard University Press
KEDEM, O. and KATCHALSKY, A.   1958.   Biochim.   Biophys.   Acta 27:
    229-246
KITASATO, H.   1968.   J.   Gen.   Physiol.   52:   60-87
KOMOR, E. and TANNER, W.   1974.   in Membrane Transport in Plants
    and Plant Organelles.   V.   Zimmermann and J.   Dainty, eds.
    Springer
LATIES, G.G.   1969.   Ann.   Rev.   Plant Physiol.   20:   89-116
MACROBBIE, E.A.C.   1970.   Quart.   Rev.   Biophys.   3:   251-294
MACROBBIE, E.A.C.   1971.   Ann.   Rev.   Plant Physiol.   22:   75-96
MITCHELL, P.   1966.   Biol.   Rev.   41:   445-602
NISSEN, P.   1973.   in Ion Transport in Plants.   P.W.   Anderson, ed.
    Academic Press.
NISSEN, P.   1974.   in Membrane Transport in Plants and Plant
    Organelles.   V.   Zimmermann and J.   Dainty, eds.   Springer
RAVEN, J.A. and SMITH, F.A.   1974.   Can.   J.   Bot.   52:   1035-1048
SLATYER, R.O.   1967.   Plant-Water Relationships.   Academic Press
SLAYMAN, C.L.   1974.   in Membrane Transport in Plants and Plant
    Organelles.   V.   Zimmermann and J.   Dainty, eds.   Springer
SPANNER, D.C.   1975.   this volume
TYREE, M.T.   1970.   J.   Theor.   Biol.   26:   181-214
ZIMMERMANN, V. and DAINTY, J.   1974.   eds.   Membrane Transport in
    Plants and Plant Organelles.   Springer

Horwitz-Type Models of Tracer Distribution during

Unidirectional Translocation

M.T. Tyree

Department of Botany, University of Toronto

Toronto, Canada

I. Introduction

In this lecture I wish to review the basic reasoning and
mathematics that lead to the various models of tracer distribution
in plants as first put forward by Horwitz (1958).   I will then
discuss some possible ways of elaborating the Horwitz-type models.
I will conclude by showing that at least two Horwitz-type models
can be made to fit the tracer profile data reviewed and fitted to
error-functions and called diffusion profiles by Canny (1973).
Thus, fitting an error-function equation to experimental data is
necessary proof, but not sufficient proof, of the Canny-Phillips
(1963) bidirectional translocation model.

Many researchers have administered $^{14}CO_2$ to leaves in the
light and followed the spatial and temporal distribution of
photosynthetically   fixed   tracer.   Unfortunately, the use of
radioactive tracers has not solved the problem of translocation in
spite of many optimistic hopes. The lack of immediate success is
probably due to the complexity of the overall transport machinery.
This machinery consists of a photosynthetic system, a short
distance transport system in the vascular tissue, and another short
distance transport system for consumption, storage and retrieval.
The tracer distribution is determined not only by the transport
process within the sieve cells, but also by the processes of
loading into the sieve cells and unloading to the sink.

A. Some basic assumptions in tracer models

In deriving models for the spatial and temporal distribution
of tracer, everyone assumes an idealized slug-flow in the sieve

tube (or transcellular strand in the Canny-Phillips model). Although flow within sieve tubes is probably laminar, with peak velocities near the center and minimum velocities near the edges, the flow pattern in slug-flow is idealized to be radially uniform. In slug-flow it is assumed: (1) that the velocity of movement equals the mean solution velocity (=the volume flow rate, cm^3 sec^{-1}, divided by the tube cross sectional area, cm^2); (2) that the concentration of tracer in any infinitesimal length of slug is radially uniform. (3) that the slug moves at a uniform velocity throughout the translocation path and (4) that concentration changes are brought about only by loading and unloading of the tracer. On the basis of theoretical considerations and experimental facts (Taylor, 1953) the first two assumptions are valid. But, if the mechanism of translocation is Münch pressure-flow, the spatial profiles of ^{14}C-assimilate will be distorted within the translocation path by the inevitable passive water entry or exit along the sieve cells. Thus, assumptions (3 and 4) may be invalid because the effects of water entry (for example) are to increase the translocation velocity down the path and simultaneously to dilute the translocated sap. Assumptions (3 and 4) may not be too far off the mark if we are looking only at a short part of a Münch pressure-flow path or if the Münch pressure-flow path is working well below maximum capacity.

B.   Unidirectional transport with irreversible lateral loss

1.   Passive loss

Irreversible loss of translocate can be assumed for the purpose of modelling either when the translocate is immediately converted to some other substance when unloaded or when the half-time for isotope saturation of unloaded translocate is very long compared to the time of the experiment. The latter case arises when the unloaded tracer translocate mixes with a large

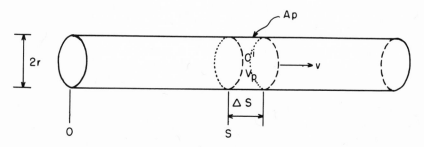

Figure 1.   A diagram of an idealized sieve tube showing a slug of solution Δs thick at distance s. A_p is the cylindrical surface area of the slug, V_p is the volume and the concentration is C^i. The slug moves in the positive direction at velocity v and has radius, r.

quantity of unlabelled translocate so that the back flux of tracer is small and/or when the rate of unloading is small.

The mathematics of the spatial and temporal distribution of tracer in a slug of sap moving down a sieve tube is relatively simple; this is because the fact that the slug moves at a velocity v is only incidental to the problem. Indeed, the velocity need not be constant with distance. Let us imagine that a slug of sap enters a tube at s=0 (Fig.1). The slug proceeds down the tube at a uniform velocity, v (cm sec^{-1}). Let us assume that osmotic dilution or concentration due to water entry or exit is negligible. If the efflux of sucrose is purely passive and irreversible then the flux J^* (mole sec^{-1} cm^{-2}) across the cylndrical slug surface is proportional to the solute concentration, C^i, in the slug;

$$J^* = -P^*C^i \tag{1}$$

where P^* (cm sec^{-1}) is the passive solute permeability. The efflux, J^*, from the slug can be related to the rate of change of concentration, dC^i/dt, because the efflux rate is equal to the rate of change of quantity, Q^i, of solute in the slug per unit area of slug surface area, A_p, i.e.,

$$J^* = \frac{1}{A_p} \frac{dQ^i}{dt}$$

But since Q^i equals the concentration times the slug volume, V_p, we have

$$J^* = \frac{\overline{V_p}}{A_p} \frac{dC^i}{dt} = \frac{r}{2} \frac{dC^i}{dt} \quad \text{because} \quad \frac{\overline{V_p}}{A_p} = \frac{\pi r^2 \Delta s}{2\pi r^2 s}$$

where r is the radius of the slug. Putting this into Eq.(1) we obtain a first order rate equation (2) with the familiar exponential solution (3).

$$\frac{dC^i}{dt} = -\frac{2P^*}{r} C^i \tag{2}$$

$$\frac{C^i}{C^*} = \exp\left(-\frac{2P^*}{r} t^*\right) \tag{3}$$

where C^* is the concentration of the solute in the slug when it began at s=0 and t^* is the time the slug has been in the tube. The reason that the movement of the slug in the tube is incidental to the problem arises from the fact that the distance travelled, s, and the velocity of travel, v, merely determine the time, t^*, the slug has been in the tube because $t^*=s/v$, thus

$$C^i = C^* \exp\left(-\frac{2P^*}{r} \frac{s}{v}\right) \tag{4}$$

From Eq.(4) it is apparent that the concentration profile within the sieve tube is log-linear with distance, s, provided C^* is time dependent and v is constant with time and s.

In Eq.(4) C^i is to be interpreted as the concentration of radioactivity in the tube for tracer flux experiments. What is commonly measured, however, is $Q = Q^i + Q^o$ = the sum of the quantity of solute in the slug (Q^i) plus the amount of solute outside the tube at the position of the slug (Q^o).

We can evaluate Q^o by starting with a conservation statement: $dQ^o/dt = dQ^i/dt$, i.e., the rate of tracer loss from the slug equals the rate of tracer accumulation outside the slug. Therefore, from Eqs.(1 and 4)

$$\frac{dQ^o}{dt} = A_p P^* C^i = A_p P^* C^* \exp\left(-\frac{2P^*}{r}\frac{s}{v}\right) \tag{5}$$

A time $t^* = s/v$ elapses before the front of radioactive tracer reaches a distance s, but from that time until the experiment is complete at time t the rate of tracer accumulation outside, dQ^o/dt, is constant. Therefore, the total accumulation Q^o, by the end of the experiment is $(dQ^o/dt)(t-(s/v))$. Thus,

$$Q^i + Q^o = \overline{V}_p C^* \exp\left(-\frac{2P^*}{r}\frac{s}{v}\right) + A_p P^* C^* \exp\left(-\frac{2P^*}{r}\frac{s}{v}\right)\left(t-\frac{s}{v}\right) \tag{6}$$

The interesting consequence that follows from Eq.(6) has been pointed out by Horwitz (1958); the factor $\exp(-(2P^*/r)(r/v))$ which multiplies the right hand side of Eq.(6) so dominates the situation (except near the advancing front where $t-(s/v)$ is small) that it imposes the well-known exponential fall-off on the longitudinal pattern of tracer distribution.

A spatial dependence in v (as found in Münch pressure-flow models in the steady state) can be incorporated into Eq.(6); this can be done by replacing s/v with

$$\int_0^s \frac{ds}{v}$$

Equation (6) can also be adapted to allow a time dependence in C^*, the concentration of the slug when at s=0. This will allow for the kinetics of phloem loading. For any slug the time of entry in the pipe at s=0 is $t-(s/v)$. If C^* has a known time dependence such that $C^* = C^*(t-(s/v))$, it can be incorporated into Eq.(6) to give,

$$Q^i + Q^o = V_p C^*\left(t-\frac{s}{v}\right)\exp\left(-\frac{2P^*}{r}\frac{s}{v}\right) + A_p P^* \exp\left(-\frac{2P^*}{r}\frac{s}{v}\right)\int_{s/v}^t C^*\left(t-\frac{s}{v}\right)dt \tag{7}$$

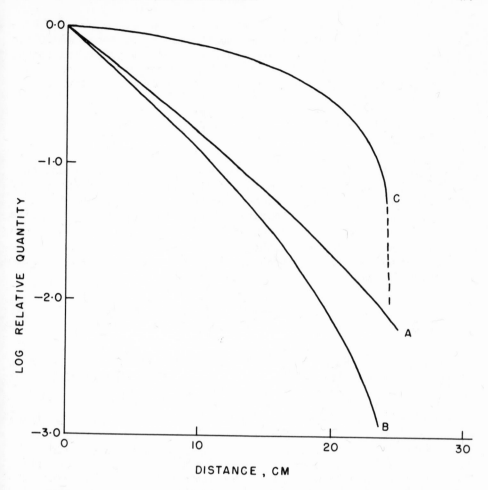

Figure 2. Plots of theoretical tracer profiles arising under three different conditions. Curve A: When tracer loss is passive and irreversible, velocity v is constant, and the concentration in the pipe at s=0, C*, is constant (Eq.7). Curve B: Same as in curve A except now the loading of tracer is time dependent exactly as found in sugar beet leaves; C^* follows Eq.(8). Curve C: No lateral loss, i.e. P^*=0 but the loading of tracer is time dependent as in Eq.(8).

In order to illustrate the behaviour of Eq.(7) I have some sample calculations in Fig.2. In all cases I have plotted the log relative quantities versus distance, where

$$\text{relative quantity} = \frac{(Q^i + Q^o)_{\text{at } s}}{(Q^i + Q^o)_{\text{at } s=0}}$$

In plot A, $2P^*/r = 2 \times 10^{-3}$ sec^{-1} ($P^* = 5 \times 10^{-7}$ cm sec^{-1} and r = 5×10^{-4} cm), v = 1.39×10^{-2} cm sec^{-1} (= 50 cm h^{-1}), and C^* is constant i.e., loading of tracer reaches a maximum concentration at s=0 at the instant of application of $^{14}CO_2$; transport has proceeded for 30 minutes. In this case it can be shown from Eq.(6) that

$$\text{relative quantity} = \exp\left(-\frac{2P^*}{r}\frac{s}{v}\right)\left(1 - \frac{s/v}{\frac{r}{2P^*}+t}\right)$$

In plot B the parameters are the same as in plot A except that now C^* has a time dependence made to match the time dependence of the sucrose source pool reported by Geiger and Swanson (1965a,b) for sugar beet. In this calculation I have a sucrose pool of size B = 3900 μgm dm^{-2}, growing at the rate A = 7.8 μgm min^{-1}; and sucrose is being exported at the rate I = 137 μgm min^{-1} dm^{-2}; the sucrose concentration at s=0 approaches a maximum C^0 as isotopic saturation is reached in the pool. Thus,

$$C^* = C^0 - C^0 \left(\frac{B}{B+A\left[t-\frac{s}{v}\right]}\right)^{\frac{T}{A}+1} \tag{8}$$

In plot C I have calculated the profile of log relative concentration resulting from the kinetics of loading in plot B but now $P^*=0$, i.e., there is no lateral loss of tracer.

It can be seen from Fig.2 that the kinetics of the input concentration, C^*, dramatically influences the kinetics of spatial distribution of tracer. But curve fitting is frequently done by Canny and others on the assumption the C^* is independent of time. Clearly this could be a dangerous assumption especially over the short term - say up to one hour. There is no doubt that it would be possible to fit all of Canny's error-function profiles without assuming lateral loss and without a bidirectional transport model by a suitable choice of input kinetic functions. In fact, this statement has been proved (but not acknowledged) by Canny (1973:13.20) when Canny fit error-function profiles to data on Soya by Fisher (1970). Fisher (1970) has shown that his profiles are due entirely to the kinetics of loading. However, Canny and Phillips in 1963 did correctly point out that some curve fitting attempts may not have any basis in reality or any physical meaning that can be visualized; thus little would be gained from curve fitting with assumed input kinetics without more experimental knowledge. I will not dwell upon the influence of loading kinetics of tracer profiles because this important topic is the subject of the lecture by Dr. D.B. Fisher.

2. Loss by enzyme kinetics
    In his original paper Horwitz (1958) worked out the spatial and temporal kinetics of tracer distribution only for passive

lateral loss, i.e. only when the tracer loss is proportional to C^i (for irreversible loss) or C^i-C^o (for reversible loss). Frequently, however, the flux of organic substances across membranes are seen to follow an enzyme kinetics, e.g. a Michaelis-Menton kinetics in which the solute flux, J^*, is determined by

$$J^* = \frac{r}{2} \frac{dC^i}{dt} = - \frac{C^i M}{K_m + C^i} \tag{9}$$

where M is the maximum rate of solute flux when all catalytic sites are filled and K_m is the Michaelis-Menton constant.

It would be of interest, therefore, to see what kind of tracer distributions would result from translocation mechanism involving irreversible loss governed by Eq.(9). If we apply the boundary condition that the tracer concentration at s=0, C^* is constant from time zero, then Eq.(9) can be integrated to yield an equation that implicitly defines the tracer concentration in the sieve cell, C^i, as a function of distance;

$$s = v \frac{r}{2} \frac{K_m}{M} \ln \frac{C^i}{C^*} + \frac{r}{2} \frac{v}{M} (C^*-C^i) \tag{10}$$

It can be seem from Eq.(10) that the solution for C^i from Eq.(9) has both a logarithmic and linear behaviour. When $K_m \ll C^i$ the efflux rate of tracer is independent of C^i, and C^i decreases linearly with distance. When $K_m \gg C^i$ the efflux rate of tracer is proportional to C^i and C^i decreases exponentially with distance, i.e. the logarithm of C^i is proportional to s.

As before we can determine the quantity of tracer outside the sieve tube, Q^o, by noting that

$$\frac{dQ^o}{dt} = - \frac{dQ^i}{dt} = -A_p J^* \tag{11}$$

Therefore, substituting the right hand side of Eq.(9) into Eq.(11) and integrating from the time of first arrival of tracer, s/v, to the time, t, of the end of the experiment, we obtain,

$$Q^o = A_p \frac{C^i M}{K_m + C^i} (t - \frac{s}{v}) \tag{12}$$

From Eqs.(10 and 12) it is possible to obtain solutions for $Q^i + Q^o (= V_p C^i + C^o)$ as a function of s for given values of r, K_m, M,

c^*, v, and t. Using Eqs.(10 and 12) I have been able to fit the
^{14}C-assimilate profile in the petiole of <u>Tropaeolum</u>; the data in
Fig.3 are reproduced from Fig.7 of Canny and Phillips (1973) and
the data were originally fit by these authors to an error-function
profile. Thus, irreversible loss of tracer from sieve cells
following a Michaelis-Menton kinetics will fit some data at least
as well as an error-function profile. However, I do not believe
the Michaelis-Menton-Horwitz model is a viable alternative
explanation of the tracer profiles fit so successfully by the
Canny-Phillips model; this is because in order for the present
model to fit experimental data the total sugar concentration
(tracer plus non-tracer) must decline rapidly with distance. In
the curve fitted to the data in Fig.3, the total sugar
concentration must drop from 0.5 molar at s=0 to 0.007 molar at
s=20.8 cm. (The parameters to fit the data are $r=5x10^{-4}$ cm,
$K=3x10^{-4}$ mole cm^{-3}, $M=1.3x10^{-10}$ mole sec^{-1} cm^{-2}, $v=5.83x10^{-3}$ cm
$sec^{-1}=22$ cm h^{-1}, and t=3600 sec.) It is very unlikely that such
large drops in total sugar concentration can occur in sieve tubes
without the inevitable hazards of plasmolysis. The tracer
concentration can drop in sieve tubes without a parallel drop in
total sugars by reversible exchange with a large reservoir outside
the sieve tubes; if the reservoir of unlabelled sugar is
sufficiently large the early stages of loss would appear
irreversible. However, in this case the theoretical tracer
profiles would not be the same as in Fig.3, because in enzyme
mediated unloading the tracer efflux is equal to the total flux
times the tracer specific activity and the total flux is governed
by the total sugar concentration. This is in contrast to passive
efflux, where the tracer flux is always proportional to the tracer
concentration. If tracer sugar experienced rapid reversible loss
from sieve tubes during translocation with no decline in total
sugar concentration in the sieve tube, then the kinetics of tracer
loss would revert to the simple passive behaviour discussed by
Horwitz; the theoretical profile would no longer fit the data in
Fig.3.

C. Unidirectional transport with reversible lateral loss

The second Horwitz model for the spatial and temporal
distribution invokes reversible loss of tracer with a volume of
solution outside the sieve cells. Horwitz did not dwell upon his
reversible loss model because at the time of his analysis (Horwitz,
1958) the theoretical tracer profiles resulting from reversible
loss models did not match known experimental profiles. Canny and
Philips (1963) later invoked reversible loss of tracer in their
bidirectional transport model and found that it led to theoretical
profiles of tracer which fit many experiments (Canny, 1973). Since

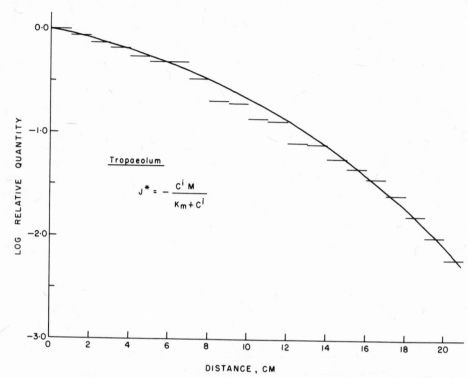

Figure 3. The bars give the ^{14}C-assimilate profiles in the petiole of <u>Tropaeolum</u> replotted from Fig.7 of Canny and Phillips (1963). The solid line is the theoretcal profile resulting from irreversible tracer loss from the sieve tube where the efflux rate depends on C^i according to Michaelis-Menton kinetics Eq.(9). See text for the parameters.

then Choi and Aronoff (1966) and Cataldo, Christy, Coulson and Ferrier (1972) have used reversible loss models with unidirectional transport to analyse THO transport. To date no one has made a concerted effort to compare known ^{14}C-assimilate profiles to a model involving unidirectional transport with reversible lateral loss; I will make this long-overdue analysis the subject of the remainder of my lecture.

1. Some necessary simplifications. For simplicity I will assume (as does Canny and Phillips, 1963) that the transported tracer undergoes reversible passive loss with an outside reservoir of finite volume and in which the tracer is perfectly mixed. In this case the net flux, J^*, between the pipe (e.g. sieve tube) and

reservoir is proportional to the difference between the tracer concentrations inside, C^i, and the reservoir, C^o;

$$J^* = -P^*(C^i-C^o) \qquad (13)$$

The situation is illustrated in Fig.4. Radioactive translocate at concentration C^* begins to enter the pipe at s=0. An imaginary slug of solution at distance s exchanges tracer with the reservoir according to Eq.(13); the slug moves at velocity v. The problem is to determine how the concentrations inside, C^i, and outside the slug, C^o, depend on time and distance. At first the concentrations C^i and C^o are zero everywhere; an infinite time after tracer application the concentrations are C^* everywhere.

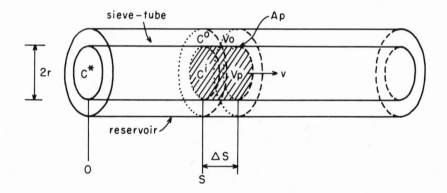

Figure 4. A diagram of an idealized sieve tube showing a slug of solution (shaded) Δs thick at distance s. A_p is the cylindrical surface area of the slug; V_p is the volume of the slug; V_o is the volume outside the slug of corresponding thickness. The concentration in the slug is C^i and outside is C^o. The slug moves in the positive directon at velocity v and has a radius, r. Tracer enters the sieve tube at s=0 at concentration, C^*.

At intermediate times, t, the concentrations at s=0 and at the advancing front at s=vt are easy to determine. At s=0, C^i always equals C^* and we have a single compartment exchange of tracer in the reservoir, thus

$$C^o(s=0) = C^*(1-\exp\left[-\frac{A_p}{V_o}P^*t\right]) \qquad (14)$$

where A_p/V_o is best thought of as the inner-suface to volume rato of the reservoir. At the advancing front at distance s=vt, C^o is

always zero therefore the front experiences an exponential decline
in tracer concentration, i.e.,

$$c^i(s=vt) = C^* \exp\left(-\frac{2P^*}{r}t\right) \tag{15}$$

At intermediate distances the rate of change of concentration
in the reservoir equals minus the net flux rate from the pipe times
the inner-surface to volume ratio of the reservoir, i.e.,

$$\frac{dc^o}{dt} = \frac{A_p}{V_o} P^* (c^i - c^o) \tag{16}$$

In the pipe the concentration c^i at any fixed point changes both
because of lateral loss and because translocaton constantly sweeps
more concentrated solution forward; thus

$$\frac{\partial c^i}{\partial t} = -v \frac{\partial c^i}{\partial \chi} - \frac{2}{r} P^* (c^i - c^o) \tag{17}$$

Because of the complexity of Eqs.(16 and 17) they can be
solved only for the simplest case where the translocation velocity,
v, is constant down the path and where the tracer loaded onto the
tube at s=0 reaches the maximum constant concentration C^* at the
beginning of $^{14}CO_2$ feeding.

2. Curve fitting with the equations governing unidirectional
transport with reversible lateral loss. Equations (16 and 17) were
solved by Schumann (1929) for an analogous heat transfer problem.
The solution involves two dimensionless variables y and z defined
by:

$$y = \frac{2}{r} P^* \frac{s}{v} \tag{18}$$

and

$$z = \frac{A_p}{V_o} P^* \left(t - \frac{s}{v}\right) = \frac{2}{r} \frac{P^*}{\alpha} \left(t - \frac{s}{v}\right) \tag{19}$$

where

$$\alpha = V_o / V_p$$

The solution is a kind of Bessel function which can be computed
through a double infinite summation. The concentration c^i and c^o
are both given as a fraction of C^*;

$$\frac{c^o}{C^*} = e^{-y-z} \sum_{n=1}^{\infty} \sum_{m=n}^{\infty} \frac{y^{m-n} z^m}{m!(m-n)!} \tag{20}$$

$$\frac{C^i}{C^*} = e^{-y-z} \sum_{n=0}^{\infty} \sum_{m=n}^{\infty} \frac{y^{m-n} z^m}{m!\,(m-n)!} \qquad (21)$$

All of the factors needed to obtain a numerical solution of Eqs.(20 and 21) are contained in Eqs.(18 and 19). Exactly the same factors are required in the Canny-Phillips model. These are, the translocation velocity, v, the total time the experiment has run, t, the ratio of the lateral membrane permeability to the sieve tube radius, P^*/r, and the ratio of the reservoir to sieve tube volume, $V_0/V_p = \alpha$. The relative quantities of sugar in the stem at various distances can be deduced from C^0/C^* and C^i/C^* because,

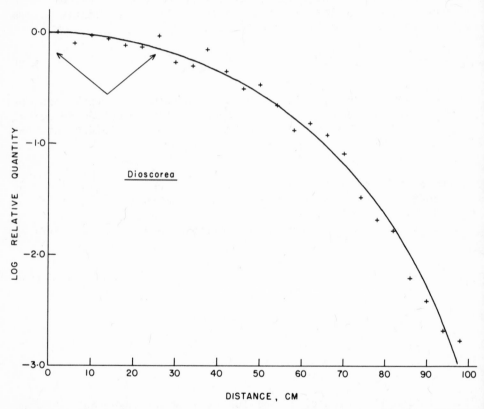

Figure 5. The pluses give the experimental ^{14}C-assimulate profile in the stem of <u>Dioscorea</u> (Figs. 13-18 from Canny, 1973). The solid line is the theoretical profile from Eqs.(18 through 21) for unidirectional transport with reversible loss. The parameters are $P^*/r = 4\times10^{-4}$ sec^{-1}, $\alpha = 1.6$, $v = 1.26\times10^{-2}$ cm sec^{-1}, and $t = 7,800$ sec. Points between the arrows were not fit by the Canny-Phillips model.

$$\text{relative quantity} = \frac{C^i + C^o}{(C^i + C^o)_{s=0}} = \frac{\dfrac{C^i}{C^*} + \alpha \dfrac{C^o}{C^*}}{\left(\dfrac{C^i}{C^*} + \alpha \dfrac{C^o}{C^*}\right)_{s=0}} \qquad (22)$$

Using Eqs.(18 through 22) I have been able to fit every ^{14}C-assimilate profile fit by Canny and Phillips (1963) and Canny (1973) to error-function profiles. The closeness of the fit is remarkable and in some cases even better than those obtained from the Canny-Phillips model. Some representative data and fits are shown in Figures 5 through 12. In the case of _Dioscorea_ in my Fig.5 Canny (1973) was not able to fit the points between s=0 and

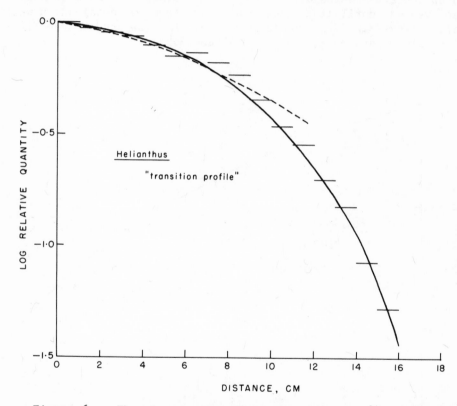

Figure 6.   The bars give the experimental ^{14}C-assimilate profile in the stem of _Helianthus_ (Fig.8 from Canny and Phillips, 1963). The solid line is the theoretical profile from Eqs.(18 through 12) for unidirectional transport with reversible loss. The parameters are $P^*/r = 5.7 \times 10^{-4}$ sec^{-1} , $\alpha = 0.95$, $v = 6.7 \times 10^{-3}$ cm sec^{-1}, and $t = 2{,}400$ sec. The dashed line gives the fit from the Canny-Phillips model.

s=25 cm; similarly Canny and Phillips (1963) could not fit the data
for <u>Helianthus</u> (my Fig.6) between s=1: and s=16 cm. In the latter
case Canny and Phllips (1963) ascribed the non-fit to be a
consequence of the experimental curve being a "transition profile".
But in every case the Horwitz reversible loss   model   for
unidirectional   transport   can   be made to fit all the data points.
The exact values of the parameters needed are to be found in the
captions.   In each case the velocity, v, was taken to equal the
maximum advance of the tracer profile divided by the time, t, of
the experiment;   this probably led to an under-estimate of v. The
ratio P^*/r required to fit all the curves varied from a low of
4×10^{-4} sec to a high of 3.4×10^{-3} sec^{-1}; if all the radii were 5 μm
this would correspond to P^*'s between 2×10^{-7} to 1.7×10^{-6} cm sec^{-1}.
The   curves   were   fitted by first choosing a P^*/r ratio that would
match the attenuation of the wave front with α=1.   Subsequently α
was varied until the appropriate curvature was obtained; usually
minor changes in P^*/r were then needed.   All the curves were fitted
with volume ratios, α, between 0.95 and 3.5.

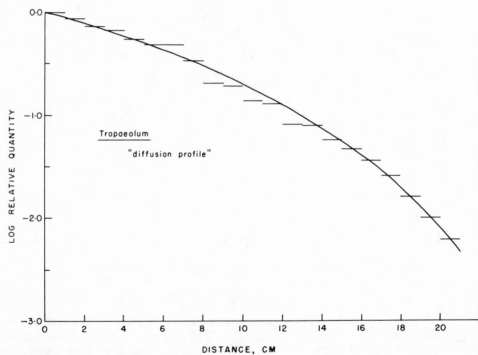

Figure 7. Same as Fig.4 but now the solid line is the
theoretical profile from Eqs.(18 through 21). The parameters are
$P^*/r = 5.7 \times 10^{-4}$ sec^{-1}, α= 3.5, v = 5.83×10^{-3} cm sec^{-1}, and t =
3600 sec.

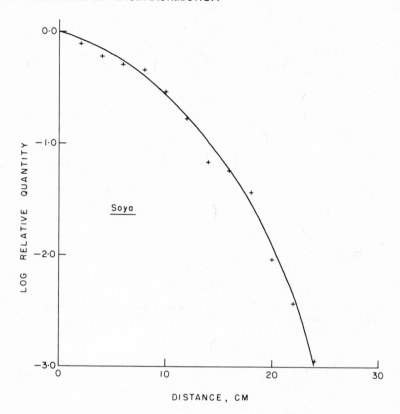

Figure 8. The pluses give the experimental ^{14}C-assimilate profile in the stem of <u>Soya</u> (Fig.13.9 from Canny,1973). The solid line is the theoretical profile from Eqs.(18 through 21) The parameters are $P^*/r = 2.4 \times 10^{-3}$, $\alpha = 2.5$, $v = 2 \times 10^{-2}$ cm sec^{-1}, and t = 1200.

D. The relative merits of unidirectional versus bidirectional reversible loss models

        When Canny and Phillips (1963) published their mathematical treatment of the bidirectional tranport model, they remarked that "the ease with which the theoretical curves fit the experimental ones is very striking". They were aware, however, of the pitfalls associated with drawing too many conclusions from curve fitting because they went on to say, "The fitting of experimental curves by theoretical equations is a process that can be performed in several ways. Empirical equations containing several arbitrary constants can be made to fit almost any curve, and the constants required do not have any basis in reality or any physical meaning that can be visualized. But rational equations, based on the behaviour of a logical model, contain parameters which are important physical

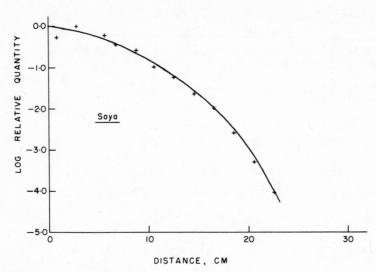

Figure 9. Similar to Fig.8 (Fig. 13.12 from Canny, 1973). $P^*/r = 3.4 \times 10^{-3}$ sec^{-1}, $\alpha = 3.0$, v = 1.88×10^{-2}, and t = 1200 sec.

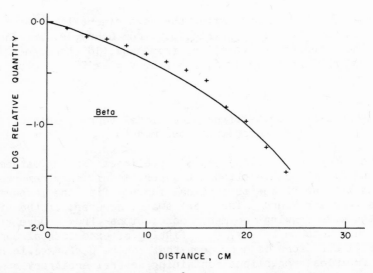

Figure 10. Similar to Fig.8 except for Beta (Fig. 13.14 from Canny, 1973). $P^*/r = 6.4 \times 10^{-4}$ sec^{-1}, $\alpha = 2.5$, v = 1.33×10^{-2} cm sec^{-1}, t = 1800 sec.

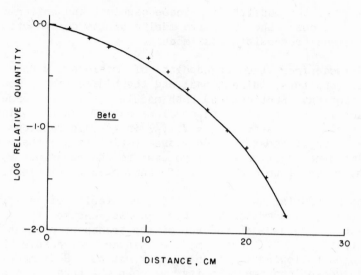

Figure 11. Similar to Fig.8 except for <u>Beta</u> (Fig. 13.15 from Canny, 1973). $P^*/r = 9 \times 10^{-4}$ sec^{-1}, $\alpha = 2.0$, $v = 1.33 \times 10^{-2}$ cm sec^{-1}, t = 1800 sec.

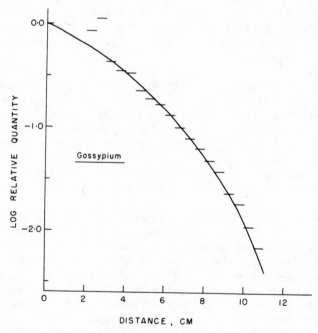

Figure 12. Similar to Fig.8 except for <u>Gossypium</u> (Fig. 13.22 from Canny, 1973). $P^*/r = 6 \times 110^{-4}$ sec^{-1}, $\alpha = 3.0$, v = 3.06×10^{-3} cm sec^{-1}, t = 3600 sec.

constants of the model." With these cautions and criteria in mind,
I will discuss the relative merits of the unidirectional versus
bidirectional reversible loss models.

Viewed from the standpoint of rational versus empirical
models, both the unidirectional and the bidirectional models are on
equal footing. Furthermore, both models invoke the same physical
constants: v, r, P^*, and a volume ratio. In the unidirectional
model the volume ratio is $\alpha = V_o/V_p$; in the bidirectional model the
volume ratio (= cross sectional area ratio) is ρ = the ratio of the
positive plus negative strand volumes to the reservoir volume (and
is assumed equal to one in all of Canny's curve fitting).

Both models suffer equally the fault of a number of
questionable assumptions.   Both models assume that the $^{14}CO_2$
immediately appears at the top of the transport path at a constant
concentration, C^*.   Both models assume reversible loss with
reservoirs following linear passive kinetics. Both models assume a
constant translocation velocity, v, down the tube. And both models
require negligible lateral storage from the reservoir.

The fact that both models give a good fit to experimental data
is not too surprising since both models have a number of physical
and structural properties in common. Both assume reversible tracer
exchange between pipes carrying tracer in the positive direction
and a reservoir of limited volume. The only difference is that the
bidirectional model assumes a second pipe carrying tracer in the
opposite direction.

Both models require a certain amount of special pleading
regarding the anatomy of phloem. The special pleading about the
structure appropriate to the bidirectional model is well known.
Since it has not been recognized that the unidirectional model of
reversible tracer loss gives equally good curve fitting results,
perhaps some speculation would be appropriate here concerning
possible structures and mechanisms of translocation that might be
consistent with the unidirectional reversible loss model.

The structure requirements imposed by the unidirectional model
of reversible tracer loss are pipes translocating tracer in the
positive direction together with a reservoir of finite volume which
exchanges tracer passively and reversibly and remains perfectly
mixed within the reservoir. Also, the rate of loss from the
reservoir to the surrounding tissue must be small compared to the
exchange rate between the reservoir and pipe.   Because of these
requirements the total (chemical) sugar concentration in the
reservoir must be the same as the pipe adjacent to it.
Furthermore, the requirement of perfect mixing suggests that the
radial extent of the reservoir must not be large compared to the

radius of the pipe. These requirements virtually eliminate the possibility that the pipes are the sieve tubes and the reservoir is the apoplast because the apoplast is extensive and its sugar concentration is surely less than that in the sieve cells.

Another possibility is that the reservoir is the surrounding cells and that the cells accumulate sugar from trace quantities in the apoplast which is in turn fed by the sieve cells. This would give the kinetic appearance of passive tracer exchange if both the sieve cells and the surrounding cells were in flux equilibrium with a small quantity of sugar in the apoplast. In order to match the assumptions in the model the sugar in the surrounding cells would have to remain there unmetabolized during the experiment; I feel this is unlikely. Furthermore, the volume ratio of surrounding cells to sieve cells is far in excess of the volume ratios required (0.95 to 3.5); the shape of the theoretical curves are strongly dependent on the volume ratio.

Transcellular strands like those in the Canny-Phillips model would be ideally suited to fit the requirements provided all the strands were surrounded by a membrane reducing the rate of sugar permeation and provided all the strands transported in the same direction. The reservoir would be the sieve tube lumen outside the transcellular strands. If the strands were 1 μm in diameter the P^*'s would range from 2×10^{-8} to 1.7×10^{-7} cm sec^{-1} for curve fitting purposes. This kind of transcellular strand structure could fit equally with two theories of how translocation is motivated; one is the streaming mechanism as discussed by MacRobbie (1971, provided streaming occurs within membrane bounded transcellular strands), and the other is the peristaltic pumping model (Aikman and Anderson, 1971).

There is one other possible structure that would fit the unidirectional reversible loss model; the pipe could be the sieve tube and the reservoir could be the companion cells and phloem parenchyma connected to the sieve tube by plasmodesmata. This interpretation would assign the role of the permeability barrier to the plasmodesmata; in this case the permeability would be identified as $P^* = a\widetilde{D}/\ell$ where a = the relative wall area occupied by plasmodesmata, D = the diffusion coefficient of sugar in the plasmodesmatal fluid and ℓ = the length of the plasmodesmata. If D = 2×10^{-6} cm^{-2}sec^{-1} and ℓ = 10^{-4} cm then a would need to be 10^{-4} to give a $P^* = 2 \times 10^{-6}$ cm sec^{-1}. This final possible structure fits nicely with the recent work on solute distribution in sugar beet leaves (Geiger, Giaquita, Sovonick, and Fellows, 1973). In their study they found a sharp increase in solute concentration at the membrane of the sieve element-companion cell complex. Geiger et al (1973) also observed plasmodesmatal connections between sieve tubes and companion cells and between companion cells and phloem parenchyma and mesophyll. This suggests that there are at least

two symplastic systems in plants - one connected with the
translocation system and the other independent of the translocation
system.   The   companion   and phloem parenchyma cells would then be
both the reservoir and the site of phloem   loading   and   unloading.
The   appealing   thing about this last interpretation is that it now
makes known tracer profiles consistent with the kind   of   structure
needed   for   the Münch pressure-flow mechanism of translocation; it
is for this reason that I favour this final interpretation.

      I wish to close with one main   point   to   be   borne   in   mind
concerning   mathematical   modeling.   While   modeling   of   tracer
kinetics can be of great heuristic   value,   the   theoretician   must
examine all possible models before drawing far-reaching conclusions
regarding   the merits of any one model.   Also, models are never any
better than   the   information   upon   which   they   are   based;   more
communication   is   needed between theorists and experimentalists to
assure   that   modeling   assumptions   have   some   foundation   in
experimental   fact   and   to assure that experiments are designed to
test models critically.

References
AIKMAN, D.P. and ANDERSON, W.P. 1971. Ann. Bot. 35: 761-777
CANNY, M.J. and PHILLIPS, O.M. 1963. Ann. Bot. 27: 379-402
CANNY, M.J. 1973.   Phloem Translocation.   Cambridge   University
   Press, London
CATALDO, D.A., CHRISTY, A.L.,   COULSON,   C.L.,   and   FERRIER,   J.M.
   1972. Plant Physiol. (Lancaster) 49: 685-689
CHOI, I.C. and ARONOFF, S. 1966. Plant Physiol. (Lancaster) 41:
   1119-1129
FISHER, D.B. 1970. Plant Physiol. (Lancaster) 45:   107-113
GEIGER, D.R, GIAQUINTA, R.T., SOVONICK,   S.A.   and   FELLOWS,   R.J.
   1973. Plant Physiol. (Lancaster) 52: 585-589
GEIGER,   D.R.   and   SWANSON,   C.A.   1965a.   Plant   Physiol.
   (Lancaster) 40: 685-690
GEIGER,   D.R.   and   SWANSON,   C.A.   1965b.   Plant   Physiol.
   (Lancaster) 40: 942-947
HORWITZ, L. 1958. Plant Physiol. (Lancaster) 33: 81-93
SCHUMANN, T.E.W. 1929. J. Franklin Inst. 208: 405-416
TAYLOR, G. 1953. Proc. Royal Soc. London (A) 219: 186-203

Kinetics of Tracer Efflux from Leaves[1,2]

Donald B. Fisher

Department of Botany
University of Georgia
Athens, Georgia 3062

I must admit to ambivalent feelings in presuming to address this group on some of the more theoretical aspects of translocation. My approach to translocation is primarily experimental and my contributions to theory have been only modest, to put it generously. To some degree, this is a matter of choice, because I think the most pressing problems for translocation are experimental ones. A theoretical approach, and I am speaking here essentially of a modelling approach, quickly leads to a need for information which is either shaky or or simply non-existent. But one of the most important reasons for this state of affairs is that experimentalists frequently lack a sufficient knowledge of theoretical aspects to make the most useful kind of observations or to appreciate that some required observations have not been made at all. Consequently, I feel that there should be an intimate relationship between experimentation and modelling. For that reason, I've had some difficulty in dividing the material into separate presentations. To the degree to which that difficulty becomes apparent, I will have made a point, although I hope that it's not one that gets overdone.

The problem with which I will be mainly concerned is trying to describe, by modelling, the effect of various factors in the leaf which influence the rate of tracer input into the translocation pathway. The presentation will be in terms of a leaf which is exposed to radioactive CO_2, although ways in which the model might be modified for some other cases could be anticipated from its development. Quite frankly, the models are fairly straightforward

1. Some of the unpublished work cited herein was conducted jointly by the author and William H. Outlaw, as indicated.
2. The author's work has been supported by National Science Foundation Grants GB14719 and GB33903.

and I would not normally regard them as worth 60 minutes of
anyone´s time. But in the context of the conference goals, I think
it´s appropriate to include some commentary about modelling
generally, using the material I will be covering as a particular
example. In addition, I will also comment on a few basic points of
tracer theory and methodology which are fundamental but not fully
appreciated.

There are at least two factors which must be taken into
consideration in trying to account for the rate of tracer efflux
from a leaf (10). Probably the most obvious is the rate at which
the tracer enters the translocation stream. In the case of
photosynthetically-labelled sucrose, this rate would be determined
by whatever might happen between its site of production in the
mesophyll cytoplasm and its arrival in the veins. This is
obviously an important factor in contributing to tracer efflux from
the leaves, but I want to put it aside for the moment and consider
another factor first.

The fact that a leaf is not a point source, but is extended in
space causes a problem when we start trying to relate the loading
kinetics in the translocation stream to the rate of tracer efflux
from the leaf. If a leaf is labelled with $^{14}CO_2$ under uniform
environmental conditions, it seems reasonable to expect that
whatever the factors might be that affect the movement of labelled
sucrose to the veins, they would be the same in all areas of the
leaf. The problem is, although it´s reasonable to suggest that the
kinetics of ^{14}C-sucrose appearance in the translocation stream
should be similar in each area of the leaf, there would be no such
synchrony in kinetics as their contributions exited from the leaf
at the petiole. Contributions from nearest the petiole would get
there first and these would be followed by tracer which had been
added to the translocation stream in more distant parts of the
leaf. Obviously some way is needed of calculating the total
contribution of tracer by each leaf area to the translocation
stream as it exits from the leaf at the petiole. When I thought of
this, it was apparent that leaf size, leaf shape and translocation
velocity would all enter into the problem, but it was difficult to
envision in quantitative terms what their effects might be.

Because of its geometry, a linear leaf is the easiest example
to start with. The configuration that was assumed for a linear
leaf model is shown in Fig.1. (Except for some added detail, the
development which follows is the same as previously published (9).
It consists essentially of a translocation stream bordered by a
source pool which secretes sugar into the translocation stream.
The flux of sugar into the translocation stream is J_o. (The J´s in
the present equations are identical to the n´s in the original
equations (9); the change reflects the generally used notation for
a biological flux.) This is the number of grams entering per square
centimeter of leaf area per minute. For mathematical convenience,

the translocation stream is assumed to be continuous rather than as discrete veins. The source pool is meant to represent only that pool of sugar which is the immediate precursor to translocated sugars. Other sugar pools are not represented. In effect, the diagram represents only the sieve tubes with their bordering transfer cells. The orientation of the coordinate system is such that the x-direction is along the length of the leaf, and the z-directon is perpendicular to the leaf surface. With this orientation, note that the velocity will have a negative value.

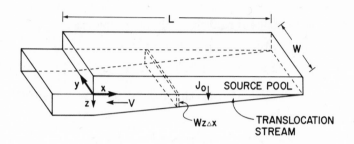

Fig.1.   Geometry for the linear leaf model.

The derivation of the equations for this model will be presented in some detail, not because the result is a particularly memorable one, but because the general procedure is one that is universally applied in approaching transport problems. At the same time, the relative importance of the steps in arriving at the final equation can be perhaps more readily appreciated.

In dealing with a mass transport problem, the differential equation for the problem simply arises from a statement of the law of conservation of matter as applied to some incremental volume. In this case, the incremental volume is $Wz\Delta x$, and the mass to be conserved is that of the tracer, R. The mass balance for R may be stated as:

(Rate of accumulation of R) = (Flow of R in) - (Flow of R out)
    + (Transport of R in from the source pool) - (Loss of R)

In this model there will be transport only in the x-direction, diffusion will be ignored, and loss will be ignored. The first two assumptions probably aren´t too difficult to accept, but the third is more of a problem. However, the case against an important kinetic role for loss was presented earlier (10), and the safest approach here is to assume that the same argument holds for movement in the leaf. This is a particularly important assumption because it says that what happens in the translocation stream doesn´t influence the kinetics outside of it.   This allows the kinetic size effect to be considered entirely separately from the effect of source pool kinetics.

With these assumptions, this verbal statement may be written as

$$\frac{\Delta \rho_R}{\Delta t} (Wz\Delta x) = (WzV\rho_R\Big|_x - WzV\rho_R\Big|_{x+\Delta x}) + J_{RO}W\Delta x$$

(Read "$WzV\rho_R\Big|_x$" as "$WzV\rho_R$ evaluated at x"   (1)

where ρR = concentration of R, [=] dpm cm^{-3}

V = velocity, [=] cm min^{-3}

J_{RO} = flux of R from the source pool, [=] dmp $cm^{-2}min^{-1}$

The next step is to divide through by the incremental volume, $Wz\Delta x$, to get

$$\frac{\Delta \rho_R}{\Delta t} = \frac{(WzV\rho_R\Big|_x - WzV\rho_R\Big|_{x+\Delta x}) + J_{RO} W\Delta x}{wz\Delta x}$$ (2)

$$= \frac{zV\rho_R\Big|_x - zV\rho_R\Big|_{x+\Delta x}}{z\Delta x} + \frac{J_{RO}}{z}$$

Note that z cannot be eliminated from the first expression on the right, since

$$\text{since } z\Big|_x \neq z\Big|_{x+\Delta x}$$

If V is assumed to be constant, it can be factored out of the parentheses. Taking the limit as Δx and Δt approach zero,

$$\frac{\partial \rho_R}{\partial t} = \frac{J_{RO}}{z} - \frac{V}{z}\frac{\partial(z \rho_R)}{\partial x}$$ (3)

This procedure for obtaining the differential equation for a mass transport problem is a general one. It amounts simply to a statement of the law of conservation of matter for some molecular substance in an incremental volume of the system. The change in the amount of substance in the incremental volume is equal to the net result of those processes which contribute the substance to or remove it from the volume. The whole statement is divided by the incremental volume and all the Δ's are allowed to approach zero to obtain the differential equation. The resulting equation is called "The Equation of Continuity" for that particular substance, and a completely general version can be derived by applying the same approach to an arbitrary incremental volume instead of one that

depends on the geometry of a particular system. This general
approach to the derivaton is worth emphasizing because all of the
basic equations describing transport phenomena are derived in much
the same way. The Equation of Motion, which describes movement
within a fluid, is derived by applying the principle of momentum
conservation and the Equation of Energy applies the law of energy
conservation. Because the equations are completely general it is
possible in theory, at least, to simplify them for each particular
system. One glance at the general equations, however, would be
enough to convince anyone that transport phenomena is a very
difficult field. But I think that it's important to realize that
such a field exists, that it's reasonably well-defined, that it is
taught as such in most chemical engineering departments, and that
many of the answers we need for translocation must come from that
general subject area (for a standard text on the subject, see (1);
also see (17) for biological applications).

If this equation is to be solved, we have to know z as a
function of x. This can be taken care of by the Equation of
Continuity for total sugar:

$$\frac{\Delta\rho}{\Delta t} Wz\Delta x = W (L-x) J_o - \rho VzW (=0) \tag{4}$$

Since we will assume the system is in a chemical steady state,
the above expression equals zero so, solving for z,

$$z = \frac{J_o}{\rho V} (x-L) \tag{5}$$

This is an interesting equation in itself since, if both sides
are multiplied by W, it can be seen that it predicts that the
cross-sectional area of the translocation stream should be
proportional to the leaf area upstream from it. This
proportionality was verified for a soybean leaf and for a maple
leaf (9). Although the phloem area per unit leaf area was the same
for both, the agreement was probably fortuitous. Geiger, Saunders
and Cataldo (12) found a 2 1/2-fold variation in the total
cross-sectional area of sieve tubes in sugar beet leaves which had
similar leaf areas and Canny (3) cites data from several sources
which also show variation in the phloem area per cm^2 of leaf
surface.

If we make the substitution for z, the differential equation
becomes

$$\frac{\partial\rho_R}{\partial t} = \frac{\rho V J_{RO}}{(x-L)J_o} - V \frac{\partial\rho_R}{\Delta x} - \frac{V\rho_R}{(x-L)} \tag{6}$$

500   D. B. FISHER

which illustrates why modelling so easily becomes frustrating.
Although it may be possible to come up with the proper differential
equation, it may be quite difficult to solve even for a relatively
simple situation. Too much time can be wasted simply in trying to
find a method for solving the equation. One gets lost in
mathematics in a way which is of little use in understanding the
problem at hand.

Equation 6 can be solved by the Laplace transform method,
which offers a fairly useful approach to a variety of systems which
are describable by linear differential equations with constant
coefficients (probably the type most frequently encountered). In
the present context, essentially what this transformation
accomplishes is to convert the partial differential equation into
an ordinary differential equation. If the transformation is
applied instead to an ordinary differential equation it is reduced
to an algebraic equation. The simplified equation is solved and
its inverse transform gives the desired solution. The process is
analogous to using logarithms to solve multiplication or division
problems and then taking the antilog to get the answer. Just as
with logarithms, the use of Laplace transforms can be "cookbooked,"
but not as easily. (For an explanation of the Laplace transform
method, see (16); this is an excellent reference for applied
mathematics in general.)

The solution to Equation 6, evaluated at x=0, is

$$\rho_{RP}(t) = \frac{-\rho V}{J_o L} \left[\int_0^t J_{RO}(u)du - U(t + \frac{L}{V}) \int_0^{t+L/V} J_{RO}(u)du \right] \qquad (7)$$

where u is a dummy varible of integration and U(t-a), the unit
step function, is defined as

$$U(t-a) = \begin{cases} 0, & t<a \\ 1, & t \geq a \end{cases} \qquad (here, a = L/V)$$

By looking at the kinetic terms in the equation, one can see
how the solution accounts for size effect. The constant outside
the parentheses won't have any kinetic effect but will determine
the absolute magnitude of the changes. The first integral simply
adds up all of the tracer contributions from leaf areas at
successive distances from the petiole, but these distances are in
terms of time. As an illustration, suppose that J_{RO} is a linear
function of time, V = 1 cm min^{-1} and that ten minutes have passed
(Fig.2a). At this time, the amount of ^{14}C leaving the leaf will be
proportional to the amount of tracer added to the translocation
stream by the closest area during the 10th minute, plus the amount

added by the area one minute away during the 9th minute, plus that added by the area 2 minutes away during the 8th minute, etc. Obviously this is the integral of J . But suppose the leaf is only 10 minutes long, and 15 minutes have passed (Fig.2b). Now the first integral is including tracer from parts of the leaf that don't exist; for example, tracer added during the 4th minute by an area 11 minutes away, etc. This non-existent tracer has to be subtracted from the first integral; this is accomplished by the second integral. It is similar to the first, but is displaced along the time axis by L/V. The sum of the two integrals is the same as the definite integral between (t + L/V) and t. It could be expressed as such for a linear leaf, but not for the other models.

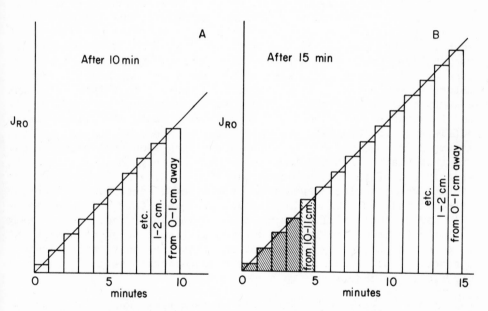

Fig.2.   Graphic explanation for the significance of the integrals in Eq.7, when J_{RO} is proportioal to time, V=1 cm min^{-1}, and L=10 cm.   The bars represent contributions from leaf areas which are 1 min., 2 min., 3 min., etc. distant from the petiole. The first integral represents the total area under the curve at some time, t. A: Contributions from various areas when t=10 min. B: When t=15 min., the second integral subtracts the shaded portion, since it represents contributions from non-existent leaf areas.

At this point we have, or seem to have, what we started out for: an equation which describes the effects of source pool kinetics, leaf size, and translocation velocity on the rate of tracer efflux from a linear leaf. But I think that much more is accomplished in the formulation of a mathematical model than can be appreciated from the final equations. Simply being forced into a systematic step-by-step analysis of a problem considerably sharpens one's insight into it, suggests new relationships and inevitably demonstrates specific needs for more experimental data. Because it is based on mathematics, the process forces the issue on alternatve explanations; a choice has to be made if the equations are to be simplified to a solvable form. For the same reason, it forces internal consistency. It is easy, in a process as complicated as translocation, to fall into the trap of using one explanation for one observation and another for another observation, but to have used what are in fact contradictory explanations. For all this, it remains quite obvious that modelling does not diminish controversy. But at least the arguments are more explicit.

How are the equations which arise from a mathematical model to be evaluated, particularly by a non-mathematician? Testing their predictions against experimental data is an absolute necessity, but what about the process of the derivation itself? It is important to realize that the equations which describe a model are nothing more than shorthand statements of how the model supposedly works. Literally speaking, the assumptions are everything. If the assumptions for a model are given, its mathematical formulation is an inevitable consequence. From that point of view, it is sufficient to evaluate a model solely on the basis of the assumptions made for it, without worrying about the mathematics. For that reason, I think that it is particularly important to state clearly all of the assumptions made for a model, and that one should not be discouraged from evaluating a model by a thicket of mathematical equations. In practice, however, this approach can be only partly successful, because the assumptions made are rarely completely true and it is sometimes difficult to judge the amount of error introduced by an assumption that is only approximately correct. Not infrequently, these judgements have to be based on experience in modelling.

Table I lists the assumptions for this model. Most are probably not very controversial, and I have already discussed those which might be (the last three). Note that all of these assumptions had been made by the time we arrived at the differential equation. For this reason, simply for the sake of clearing up one's ideas about what's going on, the process of deriving the differential equation may be worthwhile even if it's incomplete or there's not much hope of solving it.

Table I.   Assumptions for leaf models describing the effect of kinetic size on tracer efflux.

1. ρ = constant
2. Diffusion ignored
3. No isotope effect
4. Chemical steady state
5. J_o = constant
6. V = constant
7. No removal of translocate from the translocation stream

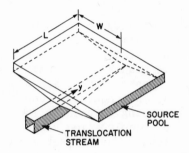

Fig.3.   Geometry for the rectangular leaf model.

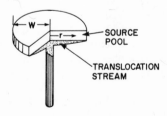

Fig.4.   Geometry for the peltate leaf model.

On the basis of the same assumptions, it is possible to derive equations which describe the rate of tracer efflux from leaves with peltate and rectangular shapes. The assumed geometry for these models is shown in Figures 3 and 4. The process for a peltate shape is very similar to that for a linear leaf except the equations are derived in a cylindrical coordinate system. The model for the rectangular shape consisted essentially of two linear leaves at right angles to a central channel, or midrib. Theoretically, the latter approach might be used for a variety of leaf shapes by using a series of linear leaves of various lengths along the same midrib, but it's not possible to invert the resulting Laplace transforms.

The solutions for these leaf shapes may be found in a previous publication (9). They are somewhat complicated, but the importance of "kinetic size" (their dimensions divided by the velocity of translocation) clearly appears in their terms.

Figure 5 shows some specific solutions to these models for the case where the specific activity of tracer in the source pool increases linearly with time. This might approximate the source pool kinetics during the beginning of a steady state labelling experiment (or, in some plants, the first 15-20 min. after pulse labelling). The values chosen for leaf size and translocation velocity are meant to be roughly representative of several experimental species including soybean. The leaf area for the models is 24 cm and the translocation velocity is 1 cm min^{-1}. The linear leaf is 10 cm long. There's not much effect of leaf size, shape and translocation velocity on the rate of tracer export from the leaves under these conditions. After a short interval, the kinetics of tracer efflux reflect the source pool kinetics.

But if the kinetic size is increased, its effects extend over a longer time. Figure 6 compares the same leaves, but with a velocity of only 0.1 cm min^{-1}. Here the effects of leaf size and shape are much more pronounced. It seems reasonable to believe that the effect of kinetic size has contributed appreciably to the shape of some translocation profiles in the literature. This is true of Canny's data for willow (2) and Qureshi and Spanner's data for Saxifrage (19), where the translocation velocities were slow, and in Mortimer's experiments with sugar beet (18), where the leaves were fairly large.

As I mentioned earlier in connection with our morning glory experiments (10), there's another way to account for the effect of kinetic size. It is more accurate, because it can include effects due to a variety of leaf shapes and to patterns of venation. Basically, the previous solutions divide the leaf area into small squares and sum their individual contributions to the radioactivity in the translocation stream as it leaves the leaf. Essentially,

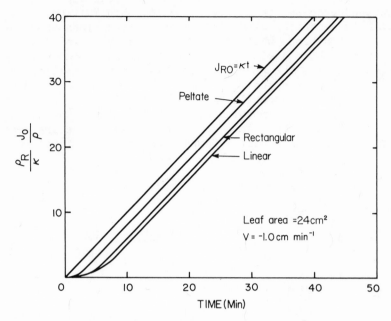

Fig. 5. Predicted efflux for the three leaf models when J_{RO} is proportional to time and $V=1$ cm min^{-1}.

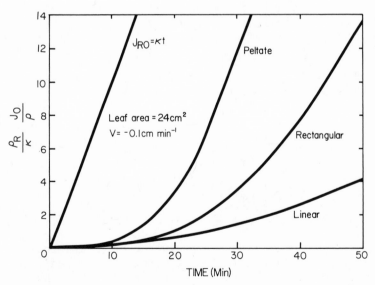

Fig.6. Same as for Fig.5, but $V=0.1$ cm min^{-1}.

the same results can be obtained from a numerical approach which is based on that process, in a literal sense. Fig.7 illustrates that approach with a morning glory leaf. The dotted lines show the leaf outine and venation while the quadrangles represent 1 cm² areas. The numbers of these areas at various distances from the petiole are tabulated at the right. This model is the one I referred to earlier (10), in trying to account for the shape of the morning glory transcloation profile on the basis of source pool kinetics and kinetic size. As I mentioned, we simply summed the amounts of tracer contributed by each of these areas to the translocation stream as it left the leaf. Comparison between this numerical

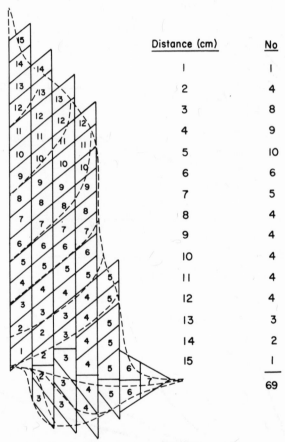

Distance (cm)	No
1	1
2	4
3	8
4	9
5	10
6	6
7	5
8	4
9	4
10	4
11	4
12	4
13	3
14	2
15	1
	69

Fig.7. Determination of the number of 1 cm² areas at various distances from the petiole of a morning glory leaf, for use in a numerical calculation of the kinetic size effect. The actual leaf outline and major venation are shown by the dotted lines. (See 10, Fig. 14, for application of the model.)

approach and the earlier analytical approach illustrates two common observations. First, a numerical approach can be much simpler, particularly when an analytical solution becomes complicated. Second, a numerical approach usually can be adjusted much more easily than a strictly analytical approach to fit the actual parameters of the system being modelled. For these reasons, numerical methods have become of considerable importance in applied mathematics (see 16).

There are two practical problems in actually using the kinetic size equations to describe tracer efflux from a leaf. The first is that it may be difficult to estimate the velocity of translocation in a leaf. In soybean, for example, the velocity is almost certainly different in the leaf and in the stem (7). The cross-sectional area of the translocation stream in the stem was apparently about 1 2/3 times that in the petiole. This would imply that the velocity in the petiole should be about 1 2/3 that in the stem, where it was actually measured

The second problem brings us to the subject of source pool kinetics. Quantitative autoradiography of ^{14}C in the transfer cells seems to offer a reasonable approach to determining those kinetics (10) but it´s not a simple one. Furthermore, it´s not much help, by itself, in explaining the kinetics.

The simplest explanation for the source pool kinetics would be that all of the leaf sucrose is well-mixed, with the sucrose specific activity in the source pool simply reflecting that of sucrose in the leaf as a whole. It is a matter of experimental observation that in several species leaf sucrose specific activity does not reach a maximum for 15-20 minutes after pulse labelling with $^{14}CO_2$ (10,13). In fact, Geiger and Swanson were notably successful in accounting for the translocation kinetics in sugar beet on just this basis (14). However, our experimental data demonstrate very noticeable sucrose compartmentation in soybean and Vicia faba leaves. I want to apply some basic compartmental analysis to some of that data but there is another question that should be considered first. The possibility that the shape of the translocation profile, particularly the first part, might be caused by some diffusion-limited transport step during movement to the veins has been suggested by Vernon and Aronoff (23), by Evans Ebert and Moorby (6) and by others. In fact, it is fairly easy to imagine several places where transport might be diffusion-limited and to assign to that step a diffusion coefficient which would be sufficiently small to generate translocation profile-like kinetics. However, it must be kept in mind that the diffusion coefficient cannot be made so small that it would not allow the observed rates of movement of total sucrose. With this restriction on the diffusion coefficient it becomes impossible to envisage any diffusion-limited transport step which would affect appreciably the rate of tracer movement to the source pool.

If transport delays can be eliminated as a possible cause of difference in specific activity, then it is appropriate to apply compartmental analysis to the system. I want to do this for both soybean and <u>Vicia faba</u>, since they seem to be fairly different. My first experience was with soybean, where the data on sucrose specific activity and translocation kinetics forced the conclusion that the kinetics of whole leaf sucrose specific activity were different from those in the source pool (10). It seemed reasonable to propose a two-compartment model for leaf sucrose, since photosynthetically labelled sucrose obviously had to pass through essentially non-photosynthetic cells before reaching the veins. Fig.8 illustrates the model and its response to pulse labelling. The first compartment represented the spongy and palisade parenchyma, and the second compartment represented sucrose mostly in the paraveinal mesophyll, but also in the bundle sheath and phloem parenchyma. Movement was assumed to occur only from the photosynthetic to the non-photosynthetic compartment. The kinetics predicted by the model were in reasonably good agreement with observed behaviour because pulse labelling resulted in a linear increase in the second compartment for about 25-30 min., followed by a slower decline. With steady state labelling there would be an exponential increase during the first 25-30 min., followed by a more linear phase.

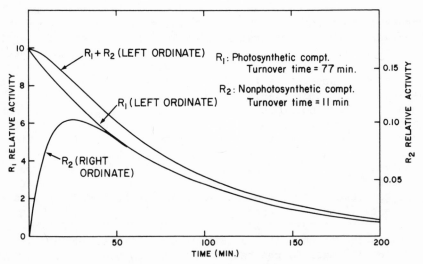

Fig. 8. Kinetic response to pulse labelling of the 2-compartment model for sucrose compartmentation in soybean. The relative turnover times for the two compartments reflects their estimated relative cell volumes.

In 1969, Geiger and Cataldo (11) demonstrated the accumulation of high concentrations of soluble ^{14}C, almost certainly sucrose, in the minor veins of sugar beet leaves. This seemed to provide additional support for the basic validity of a two-compartment model, since its most important feature was the presence of a substantial sucrose pool in a non-photosynthetic tissue between the mesophyll and the sieve tubes. This would help to explain the common appearance of an exponential translocation profile during the first 30 minutes or so of translocation during steady state labelling. But it also implied that the export kinetics in sugar beet should be different from the leaf sucrose kinetics, which is untrue.

There may still be some basic validity in a two-compartment model in some leaves, including soybean, but our picture of sucrose compartmentation has been considerably complicated by our work with Vicia faba leaves (10). We have attempted to simulate this data by a compartmental system, but before describing it I should point out that our reason for this approach was not that we thought there was sufficient data for an unequivocal description of sucrose compartmentation. We have a fair amount of data but it is nowhere near that complete, and many assumptions have to be included in the model. On the other hand, we felt that there was reasonable evidence for nine compartments. But how would a nine-compartment

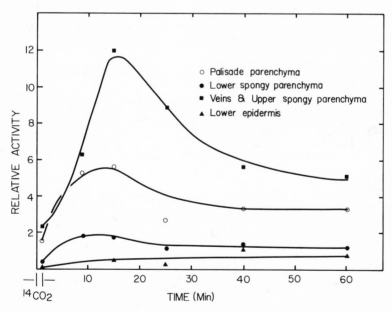

Fig. 9. Kinetics of total ^{14}C-sucrose in the measurable compartment combinations in a Vicia faba leaf. Calculated from their specific activities (see 10, Fig. 18) and the amounts of sucrose in each.

model respond in a labelling experiment? With that much complexity
it´s difficult to guess at its behaviour with any reasonable degree
of accuracy. Consequently, there is no effective way of evaluating
the appropriateness of possible alternative forms of the model.

The data we were more concerned with accounting for are
illustrated in Fig.9. Except for the lower epidermis, these curves
were calculated from the specific activity (10) and the total
sucrose in each of these areas. Total sucrose was calculated from
the sucrose concentration and tissue volume. For the epidermis,
the pool size came from the peeling data, which was corrected for
broken cells.

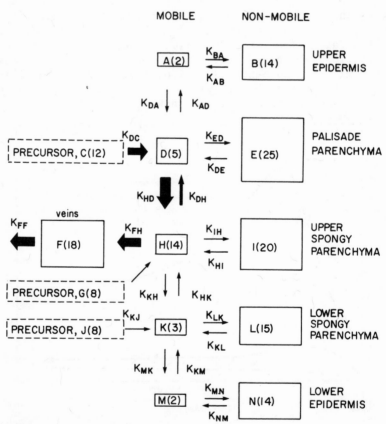

Fig. 10. Diagram of the compartmental model assumed for a
Vicia faba leaf. Pool sizes are given by the number in parentheses
for each compartment. Width of the arrows gives a rough indication
of the transport rates. Based on data discussed in (10).

The form of the compartment model which in our estimation was
the most satisfactory in accounting for the sucrose kinetics, is
shown in Fig.10. Each histological compartment was separated into
a mobile and non-mobile compartment, except for the veins. The
relative sizes of mobile and non-mobile pools were assumed to be
similar in all tissues. The compartments are identified by
letters, followed by the pool size, which also is indicated by the
block area. The size of the precursor pools was made large enough
to cause the measured delay in reaching maximum sucrose specific
activity. The pattern of sucrose movement between compartments is
indicated by the arrows, whose widths give a rough idea of their
magnitude. The turnover time for all non-mobile pools was assumed
to be the same. The rate of photosynthesis in the palisade
parenchyma was assumed to be 5 times that in the spongy, based on
the relative amount of ^{14}C found there after short fixation times.

As constructed in Fig.10, the main flow of
photosynthetically-fixed carbon is into the palisade cytoplasm,
through the cytoplasm in the upper layer of spongy parenchyma, and
then to the veins. This would account for the generally high
specific activity of sucrose in the veins. However, if the spongy
parenchyma were not separated into two layers, its overall specific
activity rapidly becomes much higher than the observed values.
This separation into two layers is a realistic one, since there
were always at least two layers of cells present, with our "spongy
parenchyma" sample always coming from the lower one, where there
were no veins.

Our approach to testing various alternatives for the model was
simply to solve the set of simultaneous differential equations for
the compartments, using whatever pool sizes or transport
coefficients the alternative called for. The actual solutions were
carried out by a digital computer. The behaviour of the model,
using the coefficients we judged to be the most satisfactory, is
shown in Fig.11. These curves describe the kinetic behaviour of
compartment combinations which are equivalent to our experimental
data. We were able to mimic the data reasonably well, except that
we couldn't get a hump in the curve for the spongy parenchyma
between 10 and 15 min.

We tried a variety of changes in the model, of which I will
mention only a few. With some exceptions, the response was not
very sensitive to moderate changes in pool sizes and transport
rates. However, smaller mobile pool sizes resulted in a more rapid
turnover of sucrose in the leaf. Increasing the exchange between
the non-mobile and mobile compartments caused a rapid depletion of
tracer from the parenchyma. If movement was more rapid between the
mobile compartments, the peak in the veins was broadened and too
much activity entered the lower spongy parenchyma. The possibility
that a sucrose precursor might move was investigated by simulating

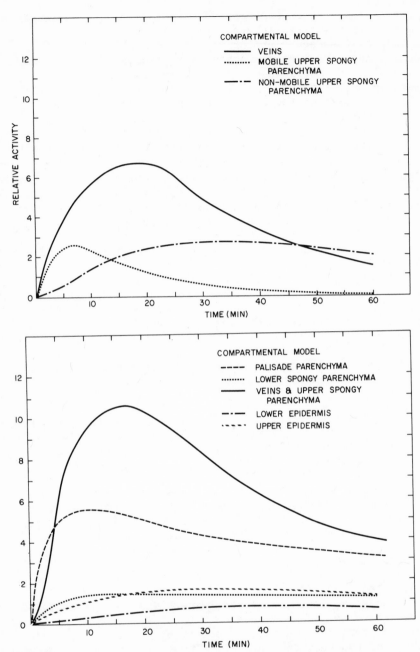

Fig.11. Kinetic response to pulse-labelling the precursor pools of the compartmental model in Fig.10, using transport rates judged to be "most satisfactory" in reproducing the observed kinetics. A: Response of the measurable compartments (Fig.9). B: Response of the compartments comprising the veins and upper layer of spongy parenchyma.

exchange between precursor pools. Even low rates substantially diminished the difference in specific activities between the palisade and spongy parenchyma. We recognize that the incompleteness of our data lends ambiguity to the model. Still, we regard it as a useful and informative approximation of sugar compartmentation in a Vicia leaf.

As I pointed out earlier (10), there is probably no universally applicable pattern of sugar compartmentation in leaves. The most useful aspect of this work probably lies simply in making the notion of compartmentation an acceptable one. But our knowledge is fragmentary, at best, and some more work will be necessary before its relationship to translocation kinetics is clear.

At this point I would like to digress somewhat from source kinetics in particular to discuss a few basic points about tracer experiments in general. Somewhat to my own surprise, it turns out to be less of a digression than I had anticipated.

To attempt a reasonably rigorous treatment of tracer kinetics in a biological system it is necessary either to be able to assume it is in a chemical steady state, or to measure changing pool sizes with time. The latter data are usually not easy to obtain, so, in general, the safest approach is to run an experiment under reasonably constant environmental conditions and hope that it will be possible to finish before endogenous variations become important. This is not meant to imply that a chemical unsteady state is uninteresting. Very much the opposite is true, because it gets at the problem of controls on translocation. But that is another and even more difficult problem than the one I am trying to address now. The point is that we are not prepared to interpret an isotopic unsteady state which is imposed on top of a chemical unsteady state.

For purposes of illustration, I'd like to formulate my discussion around one compartment in a compartmental system. The differential equation for the compartment is easily written simply by setting the set change in R per unit time equal to the amount of tracer coming in minus that going out in the same time interval:

$$\frac{dR_1}{dt} = (k_{12}\frac{R_2}{S_2} + k_{13}\frac{R_3}{S_3} + \cdots + k_{1n}\frac{R_n}{S_n}) - (k_{21} + k_{31} + \cdots + k_{n1})\frac{R_1}{S_1} \qquad (8)$$

where R_n = cpm in the nth compartment

S_n = amount of substance in the nth compartment, [=] gms

k_{1n} = transport from the nth compartment into the 1st compartment, [=] gms min^{-1}

k_{n1} = transport from the 1st compartment into the nth compartment, [=] gms min^{-1}

This equation is valid whether or not the system is in a chemical steady state. However, since we are assuming a chemical steady state, both k and S are constant, so the equation is first order, with constant coefficients. To a mathematician, this says a considerable amount about how the system should behave, but there is one property in particular that I would like to illustrate. Imagine that some other compartment in the system were pulse labelled and the kinetics of tracer in our compartment were some function F(t). But what if it weren't possible to "pulse label" in the strict sense? Strictly speaking, a "pulse label" is one in which the label is initially in only one compartment. In more general terms, what would the response be to an arbitrary input of tracer?

Suppose that the rate of tracer input into the labelled compartment were some function of time, I(t). The kinetics in the compartment under observation can be envisaged as resulting from a series of pulse labelling experiments, with the total response being the additive effects of the individual pulses. This may not be apparent immediately, but imagine for the moment that the first three pulses involved different isotopes: ^{14}C, ^{13}C and ^{11}C. Clearly the kinetics of total non-^{12}C isotope would be the sum of the effects from the individual pulses, as it also would be if the same behaviour of R arises from the fact that the differential equation describing it has constant coefficients.

If there are n pulses, numbered consecutively 1, 2, 3,...i, i+1,...n, and each pulse lasted for a period t, then the amount of tracer in the ith pulse would be $I_i(i\Delta t)\Delta t$. (Instead of an "input" function, I(t) is sometimes regarded as a "weighting" function, since it indicates the "weight" of each "impulse".) At this point we run into a notational problem because time is being referred to in two separate ways. One is the time at which a pulse was added, and the other is the time at which an observation is made. The time at which tracer was added will be indicated as "τ", and the time of observation as "t". At any given t, the amount of tracer in a compartment will be the sum of tracer aded by pulses entering at all preceding τ's. The amount of tracer added in a given pulse is now indicated by $I_i[i\Delta\tau]\Delta\tau$, and each pulse would produce a response similar to F(t), but the size of the response would be proportional to the amount of tracer added $[I_i(i\Delta\tau)\ \Delta\tau]$ and it would not start until $t=i\Delta\tau$, the time at which the ith pulse was added. That is, the response to the ith pulse would be

$$R_i(t) = I_i\ (i\Delta\tau)\ F\ (t-i\Delta\tau)\Delta\tau \qquad\qquad (9)$$

The total kinetics of R would be the sum of the responses due to all the impulses $I_i(i\Delta\tau)\Delta\tau$:

$$R(t) = \sum_{i=1}^{n} R_i(t) = \sum_{i=1}^{n} I_i(i\Delta\tau)\ F\ (t-i\Delta\tau)\Delta\tau \qquad (10)$$

If Δt is allowed to approach zero, this is the integral

$$R(t) = \int_0^t I(\tau) \, F(t-\tau) \, d\tau \tag{11a}$$

or, equivalently,

$$R(t) = \int_0^t I(t-\tau) \, F(\tau) \, d\tau \tag{11b}$$

This integral is the "convolution integral" of the input function, $I(t)$ with the "impulse response function" $F(t)$. The convolution integral can sometimes be evaluated directly, but it is usually evaluated by means of Laplace transforms. Numerical evaluation is also useful and straighforward.

The convolution integral points out several basic aspects of tracer experiments when they are run under chemical steady state conditions. Most importantly, it emphasizes the central role of the impulse response function, and consequently of pulse labelling experiments, in describing the kinetic properties of the system. Once this response is known, all other responses resulting from other inputs can be predicted. However, if a labelling protocol is used which does not give the impulse response function it is difficult to predict from the results how the system would respond to another input function. It becomes impossible to make such predictions if a chemical unsteady state has been introduced as well. A case in point is the common practice of enclosing a leaf for several hours in a small chamber with an unspecified amount of radioactive CO_2. Not only are the input kinetics vague, but there will be considerable variation in the rate of photosynthesis, with perhaps much of the experiment proceeding near the CO_2 compensation point.

The convolution integral also provides a quantitative basis for judging what labelling period might be expected to produce a reasonable resemblance to an impulse response function. If the kinetic changes in a compartment are relatively small, say about 5% or so, during the labelling period, the observed response will be a good representation of the impulse response function. Pools of phosphorylated intermediates would not usually show impulse response kinetics to a 2-minute labelling period, for instance, whereas the total sucrose pool would, but only during its decay phase.

Obviously one would hope not to have to resort to a convolution integral in order to interpret his experimental results. However, there is one particularly important relationship which should be clearly recognized. In a steady state labelling experiment, the input function is a constant. It is immediately apparent from the convolution integral (particularly the form in 11b) that the kinetics obtained from a steady state labelling experiment will be proportional to the integral of those obtained from a pulse labelling experiment. If the amount of activity introduced per unit time were the same in both experiments, the relationship would be an equality. That this relationship is real has been indicated previously in pointing out the relationship between linear and exponential translocation profiles in soybean. Some examples for particular compounds in leaves are shown in (9).

My original purpose in presenting the convolution integral lay in the points that I've already made. However, I realized on reviewing this material that our numerical solutions for the effect of the kinetic size of a morning glory leaf and for profile spreading (10) simply amounted to numerical solutions of convolution integrals. For the kinetic size effect of morning glory leaves,

$$R(t) = \int_{x=0}^{x=L} J_{RO} \left(t - \frac{x}{V}\right) W(x) \, dx \tag{12}$$

As earlier, J_{RO} represents the flux of tracer from the source pool, and $R(t)$ is the amount of tracer leaving the leaf per minute. $W(x)$ dx is the incremental area at a distance of x cm from the petiole. Note that Equation 12 can be converted to a form similar to Equation 11a by the relationship $\tau = x/V$.

For the profile spreading model,

$$\rho_R (x, t) = \int_{V_{min}}^{V_{max}} \rho_{RO} \left(t - \frac{x}{v}\right) A(v) \, dv \tag{13}$$

where $\rho_R(x,t)$ is the density of R at some distance x from the source, $\rho_{RO}(t)$ is the density of tracer at the source and $A(v)$ is the fraction of the cross-sectional area moving with velocity v. Again, Equation 13 can be expressed in terms of time by the relationship $\tau = x/v$.

From these examples, it seems quite apparent that the convolution integral should be a useful mathematical tool for interpreting several aspect of translocation kinetics. For additional discussion of the role of the convolution integral in problems involving tracer transport, see the texts by Sheppard (20) and by Simon (21).

Since I have claimed that the kinetics of tracer efflux from the leaf are the major factor in generating the shape of the translocation profile, it seems logical to wonder why it hasn't received the emphasis that it apparently deserves. In part, the answer is that it has by some, but I am referring particularly to theoretical approaches to translocation kinetics. The need for specifying input kinetics arises automatically in deriving the equations for tracer kinetics along the path. As a matter of mathematical necessity, this would seem to be sufficient by itself to generate considerable emphasis on the input kinetics. But the experimental emphasis has never been there except in a few instances such as Geiger and Swanson's and my own. I think that in gauging the prevailng attitude toward the role of the source in translocation kinetics, it is fair to point to the number of translocation profiles which have been published in comparison to the number of studies in which the kinetics of the translocated sugar have been followed in the source leaf over the same period. I have always been somewhat puszzled by this state of affairs, but I think that much of the reason lies in the way in which source pool kinetics have been dealt with in models of translocation kinetics.

All modeling attempts go back to Horwitz' paper in 1958 (15) and justifiably so. However, in spite of this success in aproaching some of the basic aspects of movement along the path, Horwitz had trouble with the source. This seemingly arose from a determination to use physical analogs as much as possible, because his "source" was a definite tissue volume containing tracer which surrounded a permeable pipe of uniform diameter. "Loading" amounted to diffusion of tracer from the tissue into the pipe, and he calculated the tracer distribution along the length of pipe in the source which resulted from this loading process. The only two cases he considered were one in which the tracer concentration in the tissue was constant, in which case the output from the pipe was also constant, and one in which the tissue was pulse labelled, with the output kinetics from the pipe showing an exponential decline. The mathematics required to calculate the output kinetics were somewhat complicated, and these two examples were the only ones he actually used as input into the translocation path. This was unfortunate, because formally speaking the equations for the path kinetics only require an arbitrary function for input kinetics. His analyses of the path kinetics were therefore unnecessarily restricted to only two input functions, without experimental justification for either. Since neither resembled the profile shape, he was led to emphasize possible patterns of movement along the path as explanations for the profile shape. Although the

emphasis turns out to be misplaced, I don't think that Horwitz can
be strongly faulted for it. Not much thought had been given to the
subject at that time and there were very little data to go on. But
his interpretations of what might be happening along the path have
implied restrictions, because of the assumed source kinetics, that
have not always been clear to subsequent workers.

In 1962, Spanner and Prebble (22) described the movement of
^{137}Cs in <u>Nymphoides</u> petioles and presented a mathematical analysis
of the kinetics. They ignored the input kinetics in their analysis
because their observations concerned periods which were mostly a
long time after introducing the tracer. This is justifiable, but
only because at long times the amount of tracer outside the
translocation stream so much predominates over that inside it, and
changes in the stream are not very noticeable. This point is
confused, however, by their mathematically based discussion of
kinetics near the front in which they use unrealistically high
rates (for sugars) of loss and assumed source pool kinetics. The
net result of their treatment is to leave the impression that you
can somehow get along without knowing the source kinetics. This is
reasonably true for their experimental system, but it is not at all
true for their illustrative examples, which concern the early parts
of the profile.

Evans, Ebert and Moorby (6), in their analysis of
translocation kinetics in soybean, described the source pool as a
single compartment. However, their equations were derived for a
compartment which could grow or shrink with time. Consequently,
their equation describing its kinetics seems unfamiliar, even for a
chemical steady state. There is no term which contains "e" to some
power, as one would expect for a compartmental treatment. When one
substitutes values which are appropriate for a chemical steady
state, which they believe is appropriate for most of their
experiments, one of the terms becomes 1^{∞}. Obviously something is
wrong, but the term can be evaluated by using L'Hospital's rule and
turns out to be the familiar $e^{-t/T}$, where T is the turnover time.
Their source pool kinetics for pulse-labelling are therefore
identical to Horwitz' time variable model. However, they still did
not resemble the early part of the translocation profile, while
they therefore attributed to a "delay distribution" arising either
from a long saturation time for monosaccharide pools or from
varying diffusion times of sucrose to the sieve tubes. That
explanation is essentially correct, in my view, but its
mathematical significance was lost because the effect was not
accounted for in their equations. One gets the impression in
reading their two papers that these authors had a very
sophisticated knowledge of tracer kinetics. Their attitude seems
to be that it just wasn't worth cluttering up the mathematics to
account for an obvious but minor effect. But that "minor effect"
is in fact the cause of the exponential profile shape, and probably

few translocation physiologists will be satisfied until its cause is more positively identified and accounted for in a model.

Canny and Phillips' equations (4) were mentioned previously (10). The problem there, of course, is that Canny (3) uses a solution which is applicable only when the level of activity at the path beginning doesn't change much. That assumption differs from reality by several orders of magnitude. When the source kinetics have to be considered, the appropriate solution is the wave equation, but he has essentially eliminated it from consideration.

After examining the role of the source kinetics in these various models of translocation (with the exception of Evans et al), I feel left with the implication that since translocation kinetics are obtained from measurements made along the pathway then something going on in the pathway should necessarily account for those kinetics. Given this attitude, the importance of source kinetics automatically receives cursory consideration. This is accentuated by the fact that the seemingly all-important exponential profile can be generated by diffusion kinetics or by loss from the translocation stream. But the kinetics required of the source are totally unrealistic in both cases. In fact, the relative emphasis on the role of the source versus that of the path in explaining translocation kinetics, has to be reversed. To a first approximation it is the source that is of primary importance; the path can be treated lightly.

References
1. BIRD, R.B., W. STEWART, and E.N. LIGHTFOOT. 1960. Transport Phenomena. J. Wiley and Sons, Inc., New York.
2. CANNY, M.J. 1961. Ann Bot (NS) 25: 152
3. CANNY, M.J. 1973. Phloem Translocation, Cambridge University Press, London
4. CANNY, M.J. and O.M. PHILLIPS. 1963. Ann. Bot. (NS) 27: 379
5. CHRISTY, A.L. and J.M. FERRIER. 1973. Plant Physiol. 52: 531
6. EVANS, N.T.S, M. EBERT and J. MOORBY. 1963. J. Exp. Bot. 14: 221
7. FISHER, D.B. 1970. Plant Physiol. 45: 107
8. FISHER, D.B. 1970. Plant Physiol. 45: 114
9. FISHER, D.B. 1970. Plant Physiol. 45: 119
10. FISHER, D.B. (This volume, p. 327)
11. GEIGER, D.R. and D. CATALDO. 1969. Plant Physiol. 44:45
12. GEIGER, D.R., M.A. SAUNDERS and D.A. CATALDO. 1969. Plant Physiol. 44: 1657
13. GEIGER, D.R. and C.A. SWANSON. 1965. Plant Physiol. 40: 685
14. GEIGER, D.R. and C.A. SWANSON. 1965. Plant Physiol. 40: 942

15. HORWITZ, L.   1958. Plant Physiol. 33: 81
16. KREYSIG, E.   1973.   Advanced Engineering Mathematics (3rd
     Edition), John Wiley and Sons, Inc., New York
17. LIGHTFOOT, E.N.   Transport Phenomena and Living Systems.
     1974. John Wiley and Sons, Inc. New York
18. MORTIMER, D.C. 1965. Can J. Bot. 43: 269
19. QURESHI, F.A. and D.H. SPANNER. 1973. Planta 110: 145
20. SHEPPARD, C.W.   1962. Basic Principles of the Tracer Method.
     John Wiley and Sons, Inc. New York
21. SIMON, W. 1972. Mathematical Techniques for Physiology and
     Medicine
22. SPANNER, D.C. and J.N. PREBBLE. 1962. J. Exp. Bot.   13:
     294
23. VERNON, L.P. and S. ARONOFF. 1952. Arch Biochem. Biophys.
     36: 383

DISCUSSION

Discussant: J. Dainty
              Department of Botany
              University of Toronto
              Toronto, Ontario

This discussion occurred after papers by M.T. Tyree on "Horwitz-type Models for Tracer Profiles" and by D.B. Fisher on "Kinetics of Tracer Input".

MacRobbie: spoke of some work by Passioura in manuscript. He took all the experimentally observed input kinetics he could find in the literature and calculated the profiles for different translocation times using all the models of translocation proposed, e.g. Horwitz (reversible and irreversible), Canny and Phillips. He found he could get a good fit with any model by choosing the right parameters.

Fisher: stressed that a good fit with a profile is not kinetics; the key is what happens with time. He pointed out the fact that the specific activity of tracer in the petiole is approximately equal to that in the leaf after a short period of time. To reconcile this with the Canny and Phillips model, the specific activity of exporting sucrose would have to be several times higher than it is in the leaf.

Spanner: referred to some work by Harmon Davidson, in manuscript. He worked with Nymphoides, which has a long thin petiole. He monitored the movement of tracer down the exposed central bundle at a number of stations; he also followed the accumulation in the root stalk. From these observations he was able to derive very complete kinetic data for computer analysis.

Aronoff: pointed out that Fisher's four component model is identical with the biochemical model for the aconitase reaction which he, Aronoff, solved a long time ago. He said that the analogies could be extended further to, say, electrical engineering type networks, and the problem could also be handled as an open system rather than a closed one.

Fisher: in reply said that his four component model should not be taken too specifically. It did demonstrate why you get differential equations with constant coefficients, which one gets with any steady state system.

Aikman: referred to Tyree´s problem of summation near the input stage in his calculations and said that this might be avoided by taking the process there as a combination of bulk flow with apparent diffusion.

Miller: referred to work reported by Turgeon in an earlier paper which implied discrete but not continuous sources in the leaf. He pointed out that diffusion and mass flow from these sources are quite different.

Tyree: agreed with Miller and said he couldn´t see how a diffusion analogue, such as the Canny Phillips model, could lead to transmission of a peak.

Moorby: whose work with ^{14}C had been referred to complimentarily by Fisher, said that at the time they had not considered compartmentation.

Fisher: put on record that one would get a different picture if the two pools are put in series.

Steady State Models of Münch Pressure Flow

M.T. Tyree

Department of Botany, University of Toronto

Toronto, Canada

I. Introduction

Since 1972 several papers have appeared that deal with the mathematical modelling of Münch pressure flow systems. The first paper (Eschrich, Evert and Young, 1972) dealt with Münch pressure flow relaxation phenomena in man-made tubular semi-permeable membranes. The next two papers treated models applicable to Münch pressure flow in sieve tubes. Young, Evert and Eschrich (1973) derived the mathematical formalism applicable to steady state translocation in sieve tubes for an arbitrary distribution of solute sources and sinks; while these authors discussed the qualitative behaviour of their equations they published no sample calculations. Christy and Ferrier (1973) published another paper at about the same time in which they independently derived a model that applied to Münch pressure flow for an arbitrary distribution of solute sources and sinks both in the steady state and in the time-dependent state; a number of sample calculations were published for sugar beet. Shortly thereafter, Tyree and Dainty (1975) and Tyree, Christy and Ferrier (1974) derived yet another steady state model that produced numerical results in substantial agreement with Christy and Ferrier (1973). This model had the advantage of being computationally simpler, and it was applied to obtain steady state solutions of translocation over long distances, e.g. 50 meters. Anderson (1974) appears to have independently arrived at a steady state model using the formalism of standing gradient osmotic flow which was applied first to a kind of Münch pressure flow system in animals (Diamond and Bossert, 1967).

Although the Christy-Ferrier model (1973) is computationally the most cumbersome, I believe it will be ultimately of greatest value to an understanding of the theoretical behaviour of the Münch pressure flow system. Two papers are now in an advanced state of

preparation   concerning   time-dependent   solutions   of   the
Christy-Ferrier model. One by Ferrier and Christy deals with the
time-dependent reduction and then resumption of translocation in
sugar beet after the application of a localized cold block. In the
other   paper   by Ferrier, Tyree and Christy demonstrate the
time-dependent behaviour of a 15-meter long transport path in
response to diurnal variation of the (xylem) water potential and
sugar-loading rate in the source.

So much has, in fact, been written about steady state Munch
pressure flow models that I do not see the point of presenting yet
another long theoretical treatment. I will   confine myself to a
brief review of the basic concepts needed to derive steady state
models of Munch pressure flow together with a review of the
conclusions that can be drawn from sample calculations.

Since 1972 there has also been an on-going debate in the
literature over what ought to be a relatively minor point of
terminology, i.e., whether we should refer to the phenomenon that
interests us all as Münch "pressure flow" or "volume flow"
(Eschrich et al, 1972; Weatherley, 1973; Young et al, 1973; Christy
and Ferrier, 1973; Tyree et al, 1974). Or, indeed, whether by
analogy we should call it "standing gradient osmotic flow"
(Anderson, 1974).   I do not believe the debate over this point of
nomenclature is serious enough to warrant much more written debate.

II. Some basic concepts behind Münch pressure flow models

In order to talk about approaches to mathematical modeling we
must keep in mind the generalized outlines of a sieve tube as shown
in Fig.1.   The sieve tube begins at distance s=0 and transports
from left to right in the direction of increasing s.   For
simplicity we can assume that only water and sucrose are being
transported, although it is a relatively simple matter to permit
the inclusion of other solutes in the analysis. The sieve tube of
radius r is surrounded by a semi-permeable membrane across which
water (but not sucrose) can move passively in response to the water
potential difference across the membrane. The water potential is
ψ^o, and on the inside the water potential, ψ^i, is governed by the
sucrose concentration, C, and the hydrostatic pressure, P.   The
water flux rate across the membrane is J_w^*, and the sucrose is
either loaded or unloaded at the membrane flux rate J^*. The water
and sucrose entering or leaving the tube combine to produce a
certain specific mass transfer rate, J, and a mean fluid velocity,
v, at any arbitrary distance s. ψ^o is a function of distance
because it reflects the decline in water potential associated with
transpiration. P must decline in the direction of volume flow, and
C is a functon of distance because C is determined in part by J^*
and J_w^* which are both variable down the sieve tube. By a similar
line of reasoning it follows that J and v could also be dependent
on s.

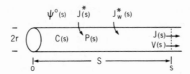

Fig.1.   A   diagram   of   a   sieve   tube   showing   the   computed
variables.   The   tube is of radius r, and the sieve plates are not
shown.   See the text for the meaning of the symbols.

In order to arrive at a steady state model of   Münch   pressure
flow it is necessary only to clearly define the interrelationships
of   the   various   parameters   illustrated   in   Fig.1.   One
interrelationship   is   a   statement that the specific mass transfer
rate at any   point   equals   the   sucrose   concentration   times   sap
velocity at that point.

$$J = Cv \qquad\qquad (1)$$

A second   statement   that   can   be   made   is   that   the   sugar
transport   rate ($=\pi r^2 J$, mole sec^{-1}) at point s must equal the total
loading rate from   s=0   to   s,   otherwise   sugar   would   either   be
accumulated   or   depleted   with   time which is not permitted in the
steady state.   The sucrose entry rate through the end wall   at   s=0
is $\pi r^2 J^*(0)$ and the entry rate through the cylindrical surface is
$2\pi r \int_0^s J^* ds$, thus

$$\pi r^2 J^*(o) + 2\pi r \int_0^S J^* ds \;=\; \pi r^2 J \qquad (2)$$

Also a similar conservation statement applies   to   the   volume
flow   rate   ($=\pi r^2 v$)   at s which must equal the total volume loading
rate from s=0 to s, i.e.,

$$\pi r^2 J_W^*(o) + \pi r^2 \overline{V}_s J^*(o) + 2\pi r \int_0^S J_W^* ds + 2\pi r \overline{V}_s \int_0^S J^* ds = \pi r^2 v \quad (3)$$

where $\overline{V}_s$ is the volume occupied by one mole of aqueous sucrose.

Only two more interrelationships are needed and these   are   in
the form of transport equations.   The first states that the average
sap velocity, v, is porportional to the pressure gradient, dP/ds.

$$v = -L\,\frac{dP}{ds} \qquad\qquad (4)$$

where L is the hydraulic conductivity of the sieve tube. The other transport equation recognizes the porportionality between the water flux across the sieve tube membranes, J_w^*, and the water potential difference, $\Psi^1 - \Psi^0$.

$$J_w^* = -L_p(\Psi^i - \Psi^0) = -L_p(P - RTC - \Psi^0) \tag{5}$$

where L_p is the sieve cell membrane hydraulic conductivity.

Equations (1 through 5) completely define the Münch pressure flow system in the steady state. From these equations one can derive the differential equations that can be used for iterative solutions of the model system. Although I have used the symbols and general approach of only one paper (Tyree el al, 1974), basically the same kind of reasoning is used in developing all of the steady state models. The only major variation on this approach is to be found in the time-dependent model of Christy and Ferrier (1973). This approach is outlined below.

Christy and Ferrier (1973) divided their tubular transport path into discrete compartments Δs long where Δs is the length of one or several sieve cells. Let us concentrate on the ith compartment; see Fig.2. (Where necessary I have altered the symbols to make this presentation internally consistent.) The water flux into the ith compartment is

$$J_{w_i}^* = L_p(\Psi^0 - P_i + RTC_i) \tag{6}$$

where L_p is the sieve cell membrane hydraulic conductivity. The mean velocity (=the volume flux) down the ith compartment to the i+1 compartment is

$$v_i = L'(P_i - P_{i+1}) \tag{7}$$

where L' is the hydraulic conductivity of the compartment (e.g. one sieve cell).

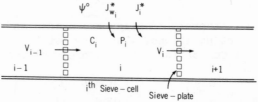

Fig.2. A diagram of a sieve tube element showing the computed variables and the relationship of the ith compartment to the i+1 and i-1 compartments. See the text for the meaning of the symbols.

Since water must be conserved it follows that the water flow rate into the ith compartment must equal the water flow rate out $[=v_iA_i(1-v_sC_i)]$, where A_i is the cross-sectional area of the ith sieve cell);

$$v_{i-1} A_{i-1} (1-\overline{V}_sC_{i-1}) + J^*_{w_i} A_i = v_iA_i (1-\overline{V}_sC_i) \qquad (8)$$

where A_i is the membrane surface area of the ith compartment. From Equations (6, 7 and 8) the pressure in the ith compartment can be calculated in terms of all the other quantities. The time dependence of the concentration change in the ith compartment can also be deduced by allowing conservation of matter, i.e., the concentration at time $t + \Delta t$ equals the concentration at time t plus the amount by which the concentration changes in time Δt due to solution entry and exit from neighbouring compartments and due to sugar loading or unloading.

$$C_i(t + \Delta t) = C_i(t) + \frac{J^*A_i + v_{i-1}C_{i-1}A_{i-1} - v_iC_iA_i) \Delta t}{V_i} \qquad (9)$$

where V_i is the volume of the ith compartment and J^* is the sugar flux rate across the membrane; $v_iC_iA_i$ etc. is the mass transfer rate (mole sec^{-1}) from the ith compartment. Equations (6 through 9) are all that is needed to carry out iterative calculations for the temporal and spatial evolution of the Münch pressure flow system.

III. The parameters needed to carry out calculations

Quite a few parameters about which we have relatively little data must be fed into the mathematical models before sample calculations can be executed. These parameters are: 1) the distribution and intensity of solute sources and sinks, $J^*(s)$; 2) the sieve tube radius, r; 3) the water potential profile outside the sieve tube, $\Psi^0(s)$; 4) the sieve cell membrane hydraulic conductivity, L_p; 5) the sieve tube hydraulic conductivity, L, which in turn depends on the sap viscosity and quantitative ultrastructure of the sieve cells; and 6) some kind of boundary condition or initial condition defining sap concentration and/or pressure in the sieve tube.

Although little is known about the exact distribution and intensity of $J^*(s)$, limits can be placed on the values of J (s) because the values must be such as to produce the appropriate specific mass transfer rate beyond the source. Also acceptable limits can be placed on $\Psi^0(s)$ from published water-relations studies. Probably the biggest unknowns are L and L_p. There are no direct measurements of L_p in sieve cells and only a few reliable figures are available for the L_p of other plant cell membranes. The greatest unknown is the true value of L since no direct measurements are available and since calculations of L turn on one's interpretation of sieve plate ultrastructure.

Despite this large number of unknowns there is something to be gained by doing calculations that cover a wide but reasonable range for the unknown parameters. In this way it is possible to envisage the general behaviour and the physical limitations of Münch pressure flow.

This kind of calculation has been carried out by Tyree et al (1974) and the following conclusion was reached. A Münch pressure flow system can meet all the transport requirements over long distances (e.g. 50 meters) provided: 1) the sieve tube is surrounded by a semi-permeable membrane; 2) sugars are actively loaded in one region and unloaded at another; 3) the sieve pores are unblocked so that L is high (up to 4 cm^2 sec^{-1} bar^{-1}). The values of L computed from sieve cell dimensions for a number of tall plants are 4.3 for <u>Yucca flaccida,</u> 20 for Sabal palmetto, 8.3 for <u>Tilia americana</u>, 1.5 for <u>Robina pseudoacacia</u> and 5.2 for <u>Vitis</u> sp. 4) The value of L_p must not be much less than 5×10^{-7} cm sec^{-1} bar^{-1}. 5) The sugar concentration is high in the source region (around one molar), and 6) the average sap velocity is low (around 20 to 50 cm h^{-1}).

Most of these conclusions follow quantitatively from the basic requirements for Munch pressure flow, i.e., that the pressure gradient, dP/ds, must be kept small enough to prevent plasmolysis of the sieve tube. From Eq.(4) dP/ds = -v/L; since dP/ds is inversely proportional to L, the hydraulic conductivity must be high, but I feel it unlikely that most trees will have a sieve tube hydraulic conductivity over 4 cm^2 sec^{-1} bar^{-1}. More quantitative anatomical work needs to be done to arrive at better estimates of L. From Eq.(1) it can be seen that v = J/C. In order to keep v low at a reasonable specific mass transfer rate (J = 12-18 gm h^{-1} cm^{-2}) C must be in the range of one molar in the source region.

The reader should refer to Christy and Ferrier (1973) and Tyree et al (1974) for several numerical solutions. (Three examples from the latter paper were discussed by Tyree but are not shown here.)

References

ANDERSON, W.P. 1974. In Transport at the Cellular Level, Society for Experimental Biology Symposium No. XXVIII. M.A. Sleigh, Ed.

CHRISTY, A.L. and FERRIER, J.M. 1973. Plant Physiol. 52: 531-538

DIAMOND, J.M. and BOSSERT, W.H. 1967. J.   Gen.   Physiol.   50:
    2061-2083
ESCHRICH, W., EVERT, R.F. and YOUNG, J.H. 1972.   Planta (Berl.)
    107: 279-300
TYREE, M.T., CHRISTY, A.L.   and FERRIER, J.M.   1974.   Plant
    Physiol. In press.
TYREE, M.T. and DAINTY, J.   1975.   In Encyclopedia of Plant
    Physiology: New Series Section 5. In press.
WEATHERLEY, P.E. 1972. Planta (Berl.) 110: 181-187
YOUNG, J.H., EVERT, R.F. and ESCHRICH, W. 1973.   Planta (Berl.)
    113: 355-366

DISCUSSION

Discussant: J.M. Ferrier
              Department of Physics
              Ohio State University
              Columbus, Ohio 43210

Dr. Ferrier was invited to give his paper on "Time Dependent Behaviour of Münch Translocation in Sugar Beet and Trees: Theory vs Experiment." The paper follows.

Time-Dependent Behaviour of Münch Translocation

in Sugar Beet and Trees: Theory vs. Experiment

Jack M. Ferrier

The most straightforward way to solve a complicated system of equations is to use a numerical computer method, particularly if some equations are non-linear, and if time dependent behaviour of the system is desired.   In the case of the difference equations describing the Münch pressure flow model for translocation, there are four equations and four variables per sieve tube section, except at the ends (1).   A careful analysis shows that there is one more variable than independent equation, so that the concentration at one end of the tube is not unique, but can be set subject to physical limitations.

A solution to a system of equations is a set of numbers which are self-consistent when substituted for the variables.   If, following an increment of sucrose loading and unloading via the appropriate equations, the other equations are iterated ten to twenty times, a solution can be obtained for each increment of time during the time evolution of the system. This time evolution of the mathematical model can then be compared to the time-dependent behaviour of the experimental translocation system.

Experiments involving recovery from inhibition of translocation by cold block in the translocation path in sugar beet show that, for a temperature above a "critical temperature" of about $0°C$, translocation recovery is complete within 90 minutes, as is translocation velocity recovery (2). Model results show that a resistance factor of fifty to two hundred is required to produce translocation inhibition comparable to experimental results, indicating that inhibition is due to sieve plate pore blockage rather than viscosity. The model results give a complete recovery of translocation in less than 90 minutes, but in some cases the velocity recovery is not complete (3).

The recovery time and velocity recovery are affected by the unloading rate: the higher the unloading rate, the faster the recovery and the higher the velocity recovery. The unloading rate has a lower limit equal to the rate of sucrose entering the sink, and an upper limit equal to the pre-inhibition rate.

Sucrose concentration also has an important effect: the higher the overall concentration, the lower the velocity required for a given translocation rate, and thus the lower the pressure gradient required, resulting in faster recovery. Recovery is also faster the closer the cold block is to the source.

The model mechanism of recovery involves a build-up of sucrose on the source side of the cold block, and a possible draw down of sucrose concentration on the sink side of the cold block, depending on the unloading rate. The resulting increased osmotic pressure gradient across the cold block produces the increased hydrostatic pressure gradient required for recovery.

Model results for the time-dependent sucrose loading and time-dependent xylem water potential gradient have also been obtained (4). The effect on sucrose concentration as a function of position and time is substantial for a time variation in loading rate, but there is little effect due to a time variation of xylem water potential gradient.

A sinusoidal time variation in loading results in a sinusoidal time variation of concentration at every point. There is a concentration wave superimposed on a concentration gradient, such that the concentration gradient is negative at every point. The phase velocity, which is the velocity with which the concentration peak moves, is quite high. For a 15 meter tree, the phase velocity of the concentration wave produced by a diurnal variation of loading rate is about ten times higher than the velocity of the sucrose solution. The phase velocity has an average value of 2.2 m hr^{-1}, ranging from 1.1 to 4.0 m hr^{-1} at different points, while the solution velocity varies with time between 0.15 and 0.50 at the top of the tree, between 0.07 and 0.31 at the middle, and goes to zero at the bottom, which would be at the root tip. (This is with a combined loading variation and xylem water potential gradient variation, with maximum loading and maximum xylem water potential gradient at the same time. With loading variation only, the phase velocity averages about 3.2 m hr^{-1}.)

These results are in accord with experimental results. The model results agree with the 1937 data of Huber et al (5,6) on variation of concentration with time and position, and with their measurement of phase velocity, which gave 1.5 to 4.5 m hr^{-1}. The model results for solution velocity are in approximate agreement with the 1968 measurements of Zimmermann (6), which gave 0.30 to 0.70 m hr^{-1}. Another point of agreement is that, if it is assumed that maximum loading occurs at sunset and minimum loading at sunrise, the model result gives maximum concentration at night, as in the data of Huber et al.

Since the parameters in these calculations are based on measured sieve tube dimensions, permeabilities, and translocation rates, as well as on the simple assumptions of the Münch hypothesis, the agreement between experimental and model results gives some support to the Münch hypothesis.

(NOTE: The parameters used in the 15 meter tree calculation are: sieve tube radius = 10μ; tube conductivity = 0.30 cm sec^{-1} atm^{-1} for 10 cm. membrane conductivity = 5×10^{-7} cm sec^{-1} atm^{-1}; specific mass transfer rate = 11 g cm^{-2} hr^{-1}; the loading rate varied from 1.62×10^{-2} to 0.32×10^{-2} µg sec^{-1}; the xylem water potential gradient varied from 0.3 to 0.1 atm m^{-1}; loading was in the first 10 cm with uniform constant unloading in the rest of the system.)

References
1. CHRISTY, A.L. and J.M. FERRIER. 1973. Plant Physiol. 52: 531-538.
2. GIAQUINTA, R.T. and D.R. GEIGER. 1973. Plant Physiol. 51: 372-377.
3. FERRIER, J.M. and A.L. CHRISTY. 1974. Submitted for publication.
4. FERRIER, J.M., M.T. TYREE and A.L. CHRISTY. 1974. Submitted for publication.
5. HUBER, B., E. SCHMIDT and H. JAHNEL. 1937. I. Tharandt. forstl. Jb. 88: 1017-1050.
6. ZIMMERMANN, M.H. 1969. Planta 84: 272-278.

Weatherley asked: how does the total cross-sectional area of the phloem pathway vary as one moves from the leaf down to and through the trunk of a tree? He would guess that it is much smaller in the trunk than in, say, the petioles and therefore the translocation velocity would be much greater in the trunk than in the petioles.

Walsh: said there are very few relevant studies, but generally the sieve elements in and near the source are much narrower than in the stem.

Spanner: pointed out that not all the   leaves   are   exporting down the trunk, some are exporting towards the crown.

Lamoureaux: said that   so   far   as   he   knew   no   appropriate anatomical   study   had   been   done   on trees to answer Weatherley's question.   Such a study   had   been   done   about   90   years   ago   on Cucurbita.

Geiger: said that in sugar beet the cross-sectional area   was about   the same from the source downwards, the sieve elements being smaller in the source than in the stem.

Milburn: referred to the relative ease of obtaining exudation at the bottom of a tree   compared   to   the   twigs,   supporting   the general notion that the sieve elements get much narrower the closer to the source.

Aikman: said that for trees the driving force on water should contain a term for the gravitational potential energy.

Tyree: replied that he had not included such a   term   in   his models.   He   noted that $d\Psi/dx$ in the xylem did change according to the time of day from a static $1/10$ bar metre^{-1} to a maximum of $3/10$ bar meter^{-1}.

Milburn: strongly disagreed with Aikman about   the   necessity of   including   a   gravitational   term.   He   explained   that   the phloem/xylem   system   behaved   as   a   kind   of   siphon   in   which gravitation played no explicit part.

Aikman: was aware of the siphon picture but said that his own view   was   the   correct   one   when   referring   to   experimental measurements of pressure in phloem and xylem.

Weatherley: supported Milburn's point of   view,   saying   that gravity plays no part in the movement of water.

Some Aspects of the Münch Hypothesis

Paul E. Weatherley

Botany Department, University of Aberdeen

Aberdeen, Scotland AB9 2UD

In the first part of this contribution I will consider the evidence for and against the Münch pressure flow hypothesis and in the second part I will discuss the structure of the sieve tube in relation to the Münch hypothesis.

I. Evidence for and against the pressure flow hypothesis

A. A priori evidence for mass flow
The velocity of translocation is well established since quite distinct ways of estimating it lead to similar results. The passage of dry matter into developing tubers or fruits (Dixon and Ball, 1922), the movement of isotope tracer fronts (Kursanov, 1963) and the progress of oligosaccharide ratios in phloem exudates down the trunks of photosynthesizing trees (Zimmerman, 1969) all indicate that velocities up to at least 200 cm h^{-1} can be attained. As Dainty reaffirmed in his introductory lecture, such velocities of long distance transport preclude diffusion and since the sieve tube does not seem to be a suitable system for massive surface movement as put forward by Van den Honert (1932), there remains only some form of mass flow.

B. Experimental evidence in favour of mass flow
1. Heat pulse method. This was successfully applied to the sieve tube by Ziegler and Vieweg (1961) using Heracleum mantegazzianum. In this plant the vascular strands lie near the surface of the inner cavity of the petioles so that the phloem can be separated readily from the xylem and ground tissue whilst remaining intact above and below. A heat pulse was applied to a given spot and its movement detected by thermocouples placed above and below the spot. Convincing evidence was obtained of the pulse being carried along by mass flow in the sieve tubes. Recently, however, Hoddinott and Gorham (1974) have shown that even slight

disturbance of the vascular strands of H. mantegazzianum leads to a loss of translocatory function. It is not clear how these different findings can be reconciled.

2. Non-specificity of transport. Whilst it is clear that sucrose is the main solute which is transported, it has been established that a great variety of substances such as endogenous and exogenous hormones, ions, viruses and even mycoplasms - which are particulate bodies - are also rapidly transported within the sieve tubes. It appears that any substance or particle which can enter the sieve tube, is then carried along passively. This is consistent with a mass flow of solution along the sieve tube.

3. Hydraulic properties of the sieve tube. If the sieve tube is behaving like a pipe through which a solution is flowing, then restricting the flow (turning a tap) at one point should affect the flow almost instantaneously all along the pipe. The properties of the sieve tube in this respect were examined in willow by Walding and Weatherley (1972). They use rooted cuttings of willow with a leafy branch near the top. Radioactive phosphate was applied to the abraded bark of the stem just below the point of insertion of the branch and two G.M. tubes were set up lower down the stem, one 14 cm below the point of application of the tracer and the second 10 cm further down still. When the leaves were illuminated, translocation down the stem occurred and the phosphate which had entered the phloem was carried down the stem, the radioactive front arriving first at the upper Geiger tube and some two hours later at the lower tube (apparent velocity of translocation about 5 cm h^{-1}). When the leafy shoot was plunged into iced water there appeared to be a simultaneous change in the rate of arrival of tracer at both of the G.M. tubes and the response at both of the G.M. tubes occurred almost at once following the cooling treatment. However, statistical analysis of the results of nine experiments gave the confidence limits (P = 0.05) to be ± 550s i.e. the response followed the cooling treatment by not more than 9 min. This is a rapid response in relation to the rate of translocation.

More recently Watson (unpublished) has attempted to reduce the errors in this type of experiment. Sunflower plants were fed with ^{32}P by the leaf flap method and two G.M. tubes were sited down the stem below the fed leaf and 20-30 cm apart. When both the G.M. tubes showed a steady rise in radioactivity the stem just above the G.M. tube was ringed with hot wax and the response in arrival of radioactive tracer at the two G.M. tubes compared. It was found that the response at the second tube was in fact later than the first tube, the mean interval (20 experiments) being 64 sec. This implies a velocity of propagation of the effect of about 25 cm min^{-1}, almost two orders of magnitude greater than the normal rate of translocation. This is consistent with the concept of mass flow along an elastic tube, the sudden closure of the tube by ringing causing a rapid wave of contraction to move away from the ring.

4. Exudation. It was on the basis of exudation   from
incision in the back that Münch put forward his pressure flow
hypothesis. Subsequently many workers (Huber, Tammes, Zimmermann,
Milburn, Baker and Hall) have made incisions in the bark and have
obtained exudation in copious amounts from the cut phloem and this
certainly indicates that the sieve tube system is under pressure,
and is capable of maintaining a very considerable outflow of sap
for many hours. This is consistent with the mass flow hypothesis.
However, it can be argued that this outflow is brought on so
suddenly, and the rate of flow so great, that it may be that the
structure of the sieve tube is altered - in fact the exudate can
contain filamentous structures and organelles which would not be
expected to be carried along in the intact system. This being so
it may be that the exudation from aphid stylets is more relevant to
normal translocation.

Sections cut through aphids feeding on woody stems show that
the stylet bundle penetrates a single sieve tube (Mittler, 1953;
Zimmermann, 1963). Using techniques originally devised by Kennedy
and Mittler (1953) the sap can be collected and the rate of flow
can be measured. The rate of flow (around 1 mm^3 h^{-1}) is equivalent
to a velocity of between 1 and 2 mh^{-1} down the sieve tube. This is
high but within the limits given above. The composition of the
exudate is consistent with its being the true translocatory medium
containing about 10%-15% sucrose and a fraction of 1% amino acids,
about the same concentration of potassium and much lower
concentration of other phloem mobile electrolytes. However, it may
be asked whether this really represents a longitudinal flow down
the stem. The exudate might represent an outflow from a single
sieve element in which the water and solutes are continously
replenished by an influx from surrounding cells. But for this to
happen the permeability of the cell membranes would have to be
several orders of magnitude greater than is normally found in plant
cells. The implication of this is that flow into the punctured
sieve element comes mainly via the sieve plates where it does not
encounter a membrane. This was supported experimentally by
Weatherley et al (1959) who found that stylet exudation was little
affected by razor cuts made longitudinally and only 2 cm away from
the stylet on either side, whereas a similar cut made transversely
and 10 cm vertically above the stylets resulted in an immediate
fall in rate of exudation.

Further evidence for the movement to cut stylets being
longitudinal was obtained by Walding (1968). A leafless and
rootless length of willow stem was supplied with water at either
end. Radioactive phosphate was applied to the abraded bark at a
site near the upper end. If this was left for even as long as
several days, little movement of phosphate occurred, as evidenced
by autoradiography of the bark vertically below the site of the
applied radioactivity. If now a feeding aphid was established 20
cm below the isotope source, it was found that after some hours the

honey dew became radioactive. At this juncture a strip of bark between the source and the aphids was excised and frozen for autoradiography. A radioautograph obtained using two closely spaced aphids is shown in Plate 1. A narrow band of radioactivity now stretches between source and aphid. This, it is suggested, represents the pathway of movement which occurred in response to the aphid's feeding. Thus the aphid can act as a sink in its own right, presumably by the turgor pressure in the sieve tube falling as a result of the outflow through the stylet canal, and this in turn inducing a mass flow along the sieve tube.

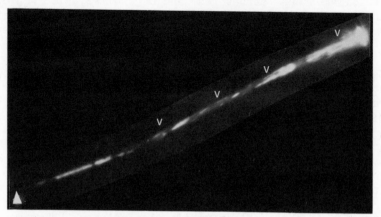

Plate 1. Autoradiograph of willow bark showing track of ^{32}p from point of application at the top left had corner to two feeding aphids sited at the bottom right hand corner. Arrowheads indicate places where there is a suggestion of two separate tracks.

C. Essential features of the pressure flow hypothesis

Münch's turgor pressure flow is the simplest mass flow hypothesis. If sugar is secreted into the sieve tube at one end water will enter osmotically and the turgor pressure there will rise. Conversely if sugar moves out of the sieve tube, at the other end, water will follow osmotically and the turgor pressure there will fall. Thus a difference in turgor pressure (hydrostatic pressures) between the terminal regions will arise and a flow of solution will take place. Energy will be expended in the transfer of sugar into and out of the sieve tube, but flow along the tube is passive. It should be noted in passing that the pools of sugar from and to which sugar is transferred need not be at a higher concentration in the source region than in the sink region. The sugar secreting pumps at source and sink acting in tandem could readily transfer sugar from source to sink against a source:sink concentration gradient. A corollary to this is that the direction in which flow in the sieve tube takes place depends on the direction (whether into or out of the sieve tube) that the pumps are working at either end. Thus, provided ingress and egress pumps can function within the same region, we can envisage movement between two pools by two parallel sieve tubes to be in opposite directions.

Normally in the source region the secretion of sugar will be into, and in the sink region out of, the sieve tube. It would appear, however, that reversal is possible, i.e. into storage cells during one period (e.g. summer) and out of these cells in another period (e.g. spring). Also a change from a quiescent to an active state in willow stems has been demonstrated by Peel and Weatherley (1962). They found that exudation from aphid stylets can remain constant in rate and concentration when the stem is ringed below a leafy branch which is acting as a source. There appears to be a switch-over to an alternative source - the storage cells surrounding the sieve tubes in the region below the ring. What controls the direction or bringing into action of the secreting pumps is unknown.

D. Evidence for the flow being impelled by turgor pressure
    1. Plasmolysis. It is a pre-requisite of the Münch hypothesis that the sieve elements should be surrounded by a differentially permeable membrane through which osmosis can occur and so a turgor pressure generated. If this is so the sieve elements should be plasmolysable. Plasmolysis was first demonstrated by Currier, Esau and Cheadle (1955) and has been confirmed a number of times since.

    2. Turgor pressure. This would be expected to have a high enough value to allow a sufficient gradient to arise between source and sink regions. The turgor pressure of plant cells is usually measured as the difference between the water and osmotic potentials. Application of this method involves abscission of the phloem following which considerable fall in turgor pressure might occur. A more direct way has been used by Mittler (1957). If the rate of flow of sap through an aphid proboscis is measured together with the viscosity of the exudate and the dimensions of the stylet canal ascertained, the Poiseuille equation can be applied and thus the hydrostatic pressure inside the cell calculated. The difficulty here is that the radius of the stylet canal is not uniform. Mittler obtained a figure of 0.6 µm at one end and 1.8 µm at the other. Since these radii are raised to the fourth power in the equation the resulting calculated turgor pressure is very sensitive to the radius chosen. A mean value gives a figure of around 20b. This is consistent with the high plasmolytic value (approx.c̄ 30b) for sieve elements in the leaf veins found by Geiger et al (1973).

3. Demonstration of pressure gradients along the sieve tube.
This has been attempted by Hammel (1968) who inserted needle
micromanometers into the phloem of oak trees. He compared two
sites 5 m apart and whilst there was considerable variation in the
values obtained within each site, there are certainly indications
of higher values at the upper site. The results suggest a gradient
of something less than 0.5 bm^{-1}. This is the kind of gradient to
be expected on the Münch hypothesis.

4. The effect of lowering the water potential in the xylem.
Weatherley et al (1959) investigated this using isolated willow
stem segments near the centre of which an exuding aphid proboscis
was sited. After a steady exudation was achieved the water in the
xylem was displaced by a mannitol or sugar solution of known
osmotic pressure. It was found that the higher the concentration
of osmoticum in the xylem the lower was the rate of exudation and
the higher the concentration of the exudate. This is consistent
with the pressure flow hypothesis, for a lower water potential in
the xylem would reduce the turgor pressure in the sieve tubes, and
hence lower the efflux through the stylet canal. This does not
necessarily mean that normal translocation flow would be affected

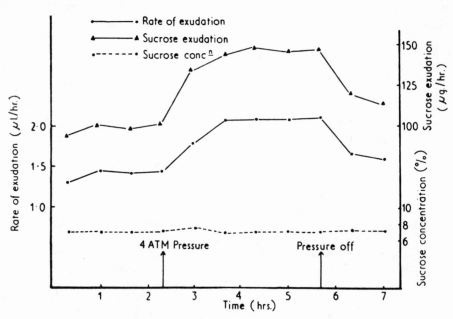

Fig.1. The effect of a pressure gradient in the xylem on
sieve tube exudation. Water under pressure was applied to the wood
at one end of a segment of stem (about 15 cm long) with an aphid
proboscis situated near the opposite end. (From Peel and
Weatherley, 1963.)

in the same way, but it is what might be expected of a pressure flow system with a simple outflow sink such as provided by a stylet canal.

5. The effect of a hydrostatic pressure gradient in the xylem. Whilst it is not possible to alter the turgor pressure gradient in a sieve tube directly this might be achieved indirectly in a length of woody stem by creating a gradient of hydrostatic pressure in the xylem. This possibility was investigated by Peel and Weatherley (1963) using isolated lengths of willow stem. An aphid proboscis was located near one end of the segment, and water under pressure applied to the xylem at the other end. Water then flowed through the xylem and dripped out at the aphid end. Thus water would tend to enter the sieve tubes at the high pressure end with a resulting increase in turgor pressure at that end. On the pressure flow theory there should be a translocation along the sieve tubes from the high pressure to the low pressure end and an increase in exudation from the proboscis should result.

The results of such an experiment is shown in Fig.1. The application of 4b caused an increase of exudation of about 10%. It should be noted that the concentration of sucrose in the exudate remained constant, thus an increase in sucrose secretion into the sieve tube had been induced. The simplest explanation of this is that the entry of water into the sieve tube at the high pressure end would lower the concentration of sugar there and the secretion of sugar would increase in response to this fall in concentration. It was found that if the pressure was raised at the proboscis end the rate of exudation was reduced. Superficially this seems to be simply the reverse situation, but it is clearly not as simple as that, for if the water potential surrounding the stylet-pierced

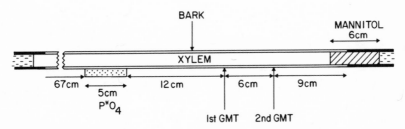

Fig.2. Experimental set-up using a stem segment of willow for investigating the effect of introducing mannitol solution in the tracheae of the xylem at one end, on the translocation in the bark of $P*O_4$ applied near the other end. GMT = Geiger-Muller tube.

sieve element is raised, one would expect an increased influx of water into the sieve element and hence an _increase_ in rate of exudation. Thus whilst the effects of pressure gradients in the xylem are not simple, they do at least indicate that the system is sensitive to changes in water potential and this suggests that the flux of water into and out of the sieve tube is an osmotic process.

     6. The effect of an artificial osmotic sink. Experiments have been carried out similar to those described above except that instead of raising the pressure in the xylem a solution of mannitol was introduced. Preliminary experiments were carried out by Walding (1968) and extended by Mackilligan (unpublished). Isolated segments of willow stem were used and tracer introduced on the side of the stem through an abrasion in the bark as already described. The arrangement is shown in Fig.2. 0.5 m mannitol solution containing a dye (basic fuchsin) was forced into the xylem until it reached a notch cut into the xylem on the upper side and 6 cm from the end. Translocation of the radioactive phosphate was monitored by the two G.M. tubes spaced at 12 and 18 cm from the source. It will be seem in Fig.3 that in the absence of mannitol (control) there was no evidence for movement of the tracer, but with the

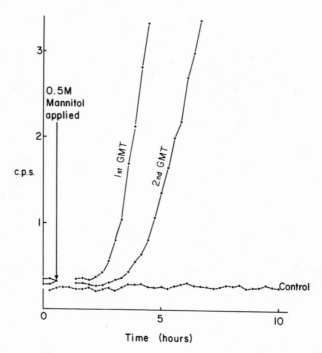

Fig.3. Readings (counts per second) on the two Geiger-Müller tubes shown in Fig.2 plotted against time. In the control no mannitol solution was introduced.

mannitol   translocation   occurred at a velocity of about 10 cm h^{-1}.
Bark ringing between the   source and   the   G.M.   tubes entirely
prevented   the   movement.   The   suggested   explanation for these
results   is   that   the   mannitol reduces the water potential in the
xylem so that water moved out of the sieve tubes   by   osmosis and
this led to a fall in turgor pressure in response to which pressure
flow occurred along the sieve tubes.

E. Physiological evidence against pressure flow
Certain features of translocation are implicit with the Münch
pressure flow mechanism.   If translocation is a flow of solution
then water and all the   substances   dissolved   in   it   should   move
together and so have identical velocities of transport. Also such
a picture of translocation rules out simultaneous movement in both
directions   within   a   single sieve tube. Lastly the necessary
expenditure of energy is concerned with the loading and unloading
of solutes in the source and sink regions, the longitudinal flow
being passive. For the most part these topics are   dealt   with   by
other   contributors   to this volume and only a brief resumé will be
given here. (For references see   Crafts   and   Crisp,   1971;   Peel,
1974.)

1. Simultaneous transport of various solutes and water.   On
the face of it, it should be possible to see whether water and
various solutes all move together by applying appropriate tracers
to   an   exporting leaf simultaneously and following their progress
through the stem. Unfortunately the results are complicated by
differential   leakage   or   adsorption during their passage down the
sieve tube. Thus THO appears to move more slowly than sugars and
other solutes, a finding which would seem to preclude mass flow.
However, THO is demonstrably a rapidly penetrating and mobile
molecule   (Cataldo,   Christy   and Coulson, 1972).   This same
difficulty of interpretation applied to the recent experiment using
aphids (Peel et al, 1969) in which a rectangular strip of willow
bark   was   fitted   to a double chambered reservoir so that one half
(a) could be irrigated with a solution containing ^{32}P, ^{14}C sugar
and THO, whilst the other half (b) was irrigated with pure water.
Exudate from a proboscis on (a) contained all three tracers whilst
that   from   a   proboscis on (b) contained ^{32}P and ^{14}C only. It was
concluded that ^{32}P and ^{14}C had moved from (b) to (a) (several cm)
without being accompanied by water (non-arrival of THO at proboscis
on (b).   An alternative hypothesis however is that THO leaked out
of the translocating sieve tube into the water irrigating (b) so
rapidly that there was no detectable level of this tracer when it
got to proboscis (b).   In further experiments (Peel,   1970)
gradients   of   THO, ^{35}S (sulphate) and ^{32}P (phosphate) were
established in an isolated willow stem segment. A single aphid was
then sited near the high concentration end and its honeydew was
found as expected to contain all three tracers. These would have

entered the tapped sieve tube in part from the local region of high concentration and in part from the more distant regions of lower concentration of tracers. Cutting off this more dilute supply by ringing the stem a few cm from the aphid caused an expected rise in activity of ^{35}S and ^{32}P whose supply was not restricted to the local region of high concentration. In contrast THO activity remained the same suggesting that entry was "local" even before ringing, i.e. no long distance movement of water was occurring before ringing. Here again, however, rapid equilibrium of the THO in the single tapped sieve tube with surrounding cells in its passage to the feeding aphid, would mean that the THO activity in the honeydew would reflect the "local" level both before and after ringing, hence there would be no change in THO activity of the exudate before and after ringing.

2. Simultaneous bidirectional transport within a single sieve tube. This is discussed elsewhere in this volume. Considerable experimental ingenuity has been devoted to demonstrating simultaneous two-way transport in a single sieve tube. But anatomical complexities and the permeability of the sieve tubes to the tracers used has so far precluded an unequivocal demonstration of this phenomenon.

3. Effect of metabolic inhibition. In the Münch hypothesis the flow along the sieve tube is passive and should not be sensitive to metabolic inhibition. However, practically, it is not easy to obtain results which give an unequivocal answer, for if the application of an inhibitor has no effect it may be this is because it has failed to penetrate the sieve tubes. If on the other hand there is inhibition this may be because the pathway is altered structurally and not because the pumps have been slowed or stopped. Callosing of the sieve pores or redistributon of P-protein are such structural changes which may lead to blockage, following application of an inhibitor. The first difficulty is overcome by the application of low temperature rather than a chemical inhibitor. Here again if there is inhibition it might be due to structural changes, but little inhibition would argue strongly against the existence of pumps. Experiments using low temperatures are discussed later in this article.

F. Conclusion

The evidence for long distance transport being a mass flow of solution is strong, let alone the fact that its only rival, diffusion, is totally inadequate to explain the measured velocity of translocation. However, the evidence that the flow is impelled by a gradient of turgor pressure is not particularly strong, but the evidence against it is weak and pressure flow does offer a model which is consistent with most of the physiological characteristics of translocation.

II. The Münch hypothesis and the structure of the sieve tube

A.   The maximum density of filamentous contents compatible
with pressure flow

Pressure flow is acceptable from the   point of view of the
structure of the sieve tube provided that it is free from
obstructions, for an application of the Poiseuille equation to the
empty sieve tube system with a velocity of translocation of 1 mh^{-1}
shows that a pressure gradient of little more than 0.25 bm^- is
required (Weatherley, 1972). some cytological studies however
reveal rather considerable obstructions which take the form of
filaments sometimes fairly densely packed in the sieve pores,
running out into the lumen on either side. In spite of this it
does seem that the sieve pores allow free passage of solid
particles. For example Tammes and Ie (1971) and Barclay and Fensom
(1973) have demonstrated that carbon black particles (200-700 $\overset{\circ}{A}$
diam) introduced into a sieve tube can move from one sieve element
to the next across a sieve plate. More remarkable is the movement
of mycoplasms which are relatively large objects (2,000 $\overset{\circ}{A}$ diam) but
are readily transported in the sieve tubes and demonstrably pass
through the sieve pores (Doe, Y. et al, 1967; Giannotti et al,
1970; Worley, 1973; Jacoli, 1974). so it is quite evident that the
sieve pores are not filled with a gel through which such things
could not pass but rather with a system of parallel filaments
between which solutions and suspensions can flow.   The question
arises then as to what density of filaments in the sieve pores and
lumen is consistent with the pressure flow hypothesis.   The
relevant calculation was carried out using a formula derived by
Fensom and Spanner (1969) with which the pressure drop in a system
of longitudinally running rods can be calculated knowing the
dimensions of the rods and their distance apart.   In these
calculations I have taken first the quite modest value of 10 cm h^{-1}
for the velocity of translocation, i.e. 20 cm h^{-1} in the sieve
pores, if they occupy half of the cross-sectional area.   The
available turgor pressure gradient is taken to be one bm^{-1} and
other, structural dimensions are given in Figure 4.

The minimum distance between the filaments in the sieve pores
which is consistent with the Münch hypothesis works out as 1000
$\overset{\circ}{A}$.* This means that there would be about 2,000 filaments passing
through each sieve pore (diam. 5 μm) and a longitudinal section
1000 $\overset{\circ}{A}$ thick would display 50 filaments.   A   diagramatic
representation of such a longitudinal section is given in Figure 4.
If the section were 50 nm thick, only 25 filaments would be seen

---

* Tammes et al (1971) have applied a similar analysis to Yucca
flaccida and have found that with filaments 250   diameter and 1000
nm apart in the sieve pores, the bleeding process is compatible
with the Munch hypothesis.

in the sieve pore and they would appear twice as far apart. It is interesting to note that at this spacing the filaments would occupy no more than 1% of the cross-sectional area of the pores.

Turning to the lumen it could well be that the filaments are uniformly distributed through the cross section as shown in Figure 4. In this case they would be 1,300 Å apart and the pressure gradient necessary to impell the solution at 10 cm h^{-1} would be 100 bm^{-1}. Clearly this is quite incompatible with the turgor pressure flow theory.

The above considerations refer to a velocity of translocation of 10 cm h^{-1}. As stated earlier values an order of magnitude greater than this are commonly encountered. The filaments in the pores must be around 200 nm apart to allow such a velocity of flow with a pressure gradient of 1 bm^{-1}. It follows that there would be about 500 filaments running through each sieve pore and of these on

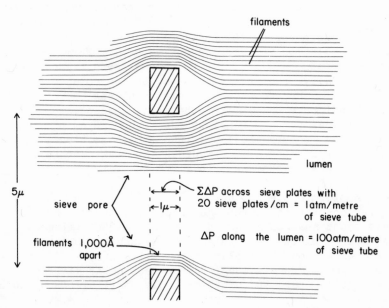

Fig.4. Diagrammatic representation of a longitudinal section through a sieve plate showing a uniform distribution of filaments (diam 10 nm) in the sieve pore and lumen. the pressure gradients shown are those necessary to drive a 10% sucrose solution through the system at a velocity of 10 cm h^{-1}. For further information see text.

average around 12 would be seen in a longitudinal section 100 nm thick (see Figure 5). If as before the filaments are uniformly distributed in the lumen, they would be 280 nm apart therein and the required pressure gradient needed to impel the solution at 100 cm h^{-1} would be 120 bm^{-1}, again an impossible value for the pressure flow theory. The point may be made in passing that if the same amount of filamentous material as shown in the lumen in Figure 5 were distributed to form an irregular three dimensional network, the resistance to flow would probably be no less, but its appearance in a section would be a series of short lengths of filament randomly distributed and would appear not unlike some electron micrographs of the lumen which are described as "open" or "empty".

At this point it is interesting to consider what density of uniform parallel filaments in the lumen <u>is</u> compatible with the

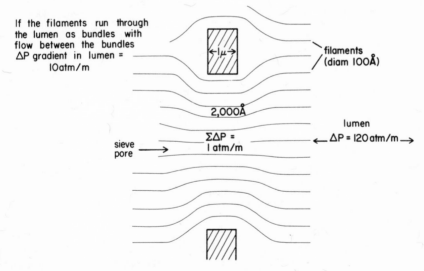

Fig.5. Diagrammatic representation of a longitudinal section through a sieve plate showing a uniform distribution of filaments (diam 10 nm) arranged 200 nm apart in the sieve pore. With a velocity of translocation of 100 cm/h and with 20 sieve plates/cm, this gives $\Sigma\Delta P$ across sieve pores = 1 atm/m. Corresponding ΔP gradient along lumen = 120 atm/m.

pressure flow theory. Taking once again a pressure gradient of 1 bm^{-1}, an application of Fensom and Spanner's formula shows that with a velocity of 10 cm h^{-1} the filaments would need to be 1 μm apart and at 100 cm h^{-1} 2 μm apart. These densities are shown in transverse section in Figure 6. It can be shown that median longitudinal sections 100 nm thick would on average each contain 2.5 filaments in the first case and 1 filament in every two sections in the second case.

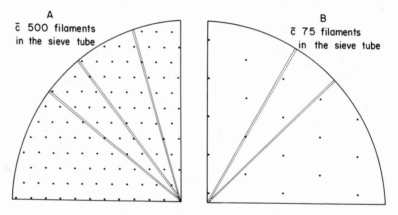

A
c̄ 500 filaments
in the sieve tube

B
c̄ 75 filaments
in the sieve tube

NB. The points representing the filaments are not scale they would be approx. ¹/₁₀ the diameter shown.

Fig.6. Diagrammatic transverse sections showing the maximum density of filaments (diam 10 nm) evenly distributed through the lumen, which is consistent with the pressure flow hypothesis (Pressure gradient 1 atm m^{-1}). A. Velocity of translocation 10 cm h^{-1}. Distance between filaments 1μ. On average, 2.5 filaments would be included in a median longitudinal section 1000Å in thickness. B. Velocity of translocation 100 cm h^{-1}. Distance between filaments 2.5μ. On average, 1 filament would be included in every two sections.

It is concluded that the Münch hypothesis is compatible with a uniform system of parallel filaments running through the sieve pores provided that they are at least 100 nm apart for low velocities of transport or 200 nm apart for high velocities. Viscous drag on similar filaments running the whole length of the lumen is so considerable that they can only be very sparsely distributed and electron micrographs of the lumen would show it to be virtually empty. Many electron micrographs however show density of obstructing structures much greater than this and if these are believed to give a true picture, the Münch hypothesis must be rejected as an adequate mechanism in such cases. Here recourse must be had to some form of pumping not just at source and sink, but "all along the line".

B.   The effect of low temperature - evidence against activated
mass flow.

Calculations based on the measured rates of respiration of phloem tissues suggest that metabolic pumps could create a pressure gradient of about 3 b cm^{-1} (Weatherley and Johnson, 1968) and such a gradient would be consistent with a much greater density of occluding material which might itself be part of the pumping mechanism. Now if the energy from such pumps comes from respiration, translocation whould surely be sensitive to temperature. As mentioned earlier the demonstraton of a reduction in rate of translocation of lowering the temperature to around 0^0C might result from structural changes or effects unrelated to the action of a pumping mechanism. On the other hand if there is demonstrably little effect of lowering the temperature, this argues strongly against the existence of respiration dependent pumps. In fact a number of workers have shown that the effects of lowering the temperature of the petiole or stem has a slight or transient effect (see reviews by Crafts and Crisp, 1971 and Wardlaw, 1974). Weatherley and Watson (1969) found that translocation in willow was lttle affected by lowering the temperature of the bark over a length of 10 cm down to -4^0C below which temperature the phloem froze and translocation stopped. At -4^0C there was a 95% inhibition of the rate of respiration of the bark tissues.

In most of these experiments the cooled zone of petiole or stem was little more than 2 or 3 cm and it could be argued that even if the pumps were stopped by the cooling, those remaining active above and below could push the solution passively through such a short inactive region. Two pieces of work involving cooling over much longer distances are therefore of particular interest.

Lang (1974) worked with a water plant Nymphoides peltata the leaves of which have petioles as long as 60-70 cm. The experimental plants were reduced to a rootstock bearing a single leaf. A solution containing ^{137}Cs was applied to a 7 cm length of the petiole where it penetrated the phloem and was translocated to the rootstock and its accumulation therin was continuously recorded. The temperature of a 30 cm length of the petiole between the ^{137}Cs reservoir and the rootstock could be controlled. When a steady temperature was maintained the activity in the rootstock showed a linear rise over a period of many hours indicating a constant mass transfer down the petiole. Lowering the temperature from 23^0C to 4^0C caused an abrupt pause in translocation followed by a recovery over about 5 h to reach a steady value only a little lower than the original rate. When the temperature was raised there was an immediate change to a higher steady rate with no intermediate period of adjustment. Comparison of the steady rates before and after the temperature change gave a mean Q_{10} of about 1.2. This is slightly less than the Q_{10} expected from the change in viscosity (1.35) but if the fact that only half of the total length of the petiole was cooled or heated, the expected Q_{10} would be 1.17. The author concludes that the Q_{10} for translocation in these experiments is explicable in terms of viscosity alone - much greater effects of temperature being expected with metabolic pumps all along the line. The cause of the transient fall in rate on lowering the temperature is explicable in terms of structural changes in the sieve tubes (see also Geiger, 1969 and Giaquinta and Geiger, 1973) but consideration of this phenomenon is incidental to our present discussion.

Watson (in press) worked with rooted cuttings of willow each bearing a single leafy branch. The temperature of a 65 cm length of the stem between branch and roots could be lowered by means of a cooling coil and monitored by thermocouples inserted beneath the bark. The branch was enclosed in a glass vessel through which air was circulated and via which a 20 minute pulse of $^{14}CO_2$ could be administered. The aim was to measure the translocation of tracer through the cooled stem and to compare this with controls at normal temperatures. Fourteen plants were used, 7 in which the stems were cooled to 0^0C and 7 with no cooling. In the former group the stems were cooled for 18 h before the administration of the $^{14}CO_2$ pulse and the low temperature maintained for a further 8 h. Then the bark between branch and roots were stripped off, cut into 5 cm lengths and eluted with 80% ethyl alcohol. The radioactivity of the extracts was measured by scintillation counting. The fed branch was also eluted and the residual activity measured. Estimates were also made of activity in the wood and also that remaining in an insoluble fraction. In both cases the amounts were negligible compared with those recovered by aqueous alcohol extractions of the bark. The control plants were similarly treated but were not cooled. It was found that the total uptake of tracer by the cooled and uncooled plants were similar (within 5% of each other). The mean results of the seven experiments are represented in Figure 7 which shows the distribution of tracer down the stem. It is evident that cooling had lttle effect either on the distance travelled by the tracer during the 8 h duration of transport or on

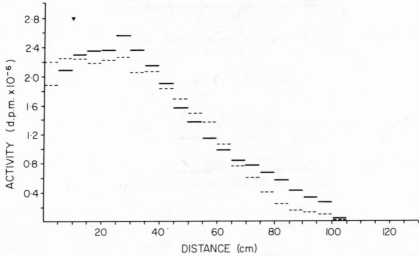

Fig.7. Profiles of activity in cooled (---) and uncooled (-)
stems of willow 8 h after pulse feeding a leafy branch higher up
the stem with $^{14}CO_2$. Cooling to 0^0C was applied over the 65 cm
length indicated by the arrows. d.p.m. = disintegrations per
minute. (From Watson, in press, Ann. Bot.)

the total quantity transported. With respect to the latter, a
comparison of the quantity of tracer remaining in that part of the
plant above the cooling coil expressed as a proportion of the total
incorporated, indicates a 10% hold up of photosynthate by the
cooling treatment. This is less than the expected increase in
viscous drag, for a temperature drop from 20^0 to 0^0C results in a
doubling of the viscosity. Adjustment for the proportion of the
total pathway actually cooled (65 cm in about 1.8 meters), gives an
expected reduction in rate of translocaton of about 30%. However
on the Münch hypothesis a smaller effect is explicable in terms of
a build up of hydrostatic pressure on the up-stream side of the
cooled region giving a compensatory increase in rate.

    I submit that the small effect of low temperature applied over
such great lengths makes it unlikely that there are pumps along the
sieve tube. We are thus thrust back on to the Münch hypothesis as
the only alternative not requiring such pumps. As we have seen the
only stumbling block to turgor pressure flow is the presence of
intracellular obstructions. A fair density of filaments in the
sieve pores is acceptable (Figure 4) if the velocity of flow is not
high (10 cm h^{-1}). An equivalent uniform distribution of filaments
in the lumen on the other hand imposes a considerable resistance -
quite unreconcilable with turgor pressure flow. This difficulty
can be overcome however by supposing that the filaments are

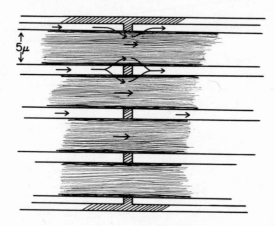

Fig.8. Diagrammatic representation of a longitudinal section through a sieve plate showing parallel bundles of filaments running through sieve pores and lumen. Flow would be <u>within</u> the bundles through the sieve plate and <u>between</u> the bundles through the lumen. $\Sigma\Delta P$ for sieve pores = 1 atm/meter. ΔP gradient for lumen = 1 atm/meter. The bundles occupy 50% of cross sectional area of the sieve tube.

aggregated into bundles running through the lumen. If these bundles were of the same diameter as the sieve pores, they would occupy about 50% of the cross-sectional area of the lumen (Figure 8) and the pressure necessary to drive the sap along the lumen <u>between</u> the bundles is only 1 bm^{-1} for a velocity of 10 cm h^{-1} in contrast to 100 bm^{-1} for a uniform distribution of filaments (Figure 4). Thus flow would be <u>within</u> the bundles through the sieve pores and largely <u>between</u> the bundles in the lumen. With a velocity of translocation of 100 cm h^{-1} however the necessary pressure gradient for flow down the lumen between the bundles would be 10 bm^{-1}, too high a figure to be consistent with the pressure flow theory. We must therefore envisage that the bundles group themselves or even aggregate to form a peripheral layer leaving the central region of the lumen empty. Alternatively they could become condensed in the lumen to form narrower strands of less diameter than the sieve pores. If they were 1 μm in diameter rather than 5 μm (as assumed above) the necessary pressure gradient is reduced to 0.5 bm^{-1} and pressure flow is feasible.

C.   Construction of a model sieve tube consistent with
the Münch hypothesis
What is attempted now is to 1st the basic structural
requirements of of the pressure flow hypothesis and to see how
these could be achieved by evolution from an unspecialized
parenchyma cell.   (a) Cell membrane. This has two roles to play.
First as a semi-permeable boundary for osmosis leading to a state
of turgor. Second, to mediate in the active transfer of sugars and
perhaps other solutes into and out of the sieve tube.   (b) Mass
flow continuity along the sieve element and between sieve elements
with minimum drag from static structures.

It might be thought that the simplest solution would be to
have vacuolar continuity through the sieve pores giving free flow
all along the sieve tube. But unlike the plasmalemma there is no
continuity of the tonoplast through the plasmodesmata and to
achieve this would mean a fundamental change of structure
phylogenetically unlikely.   The alternative is to develop the
symplast as a mass flow pathway. Now the vacuole becomes an
obstruction and is eliminated.   However, to have the lumen with
cytoplasm would hardly be conducive to mass flow.   Why not
eliminate it also, leaving the cell wall lined wtih a naked
plasmalemma the lumen and plasmodesmata being filled with solution?
This would represent an ideal from the pressure flow standpoint,
but no doubt cannot be achieved since the plasmalemma probably
needs a minimum layer of cytoplasm for its continued maintenance.
The cytoplasm in the lumen and sieve pores is therefore present as
a minimal lining leaving an unobstructed central volume along which

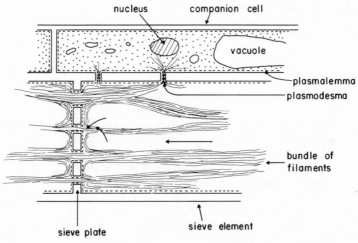

Fig.9. Hypothetical structure of the sieve tube and companion
cell consistent with turgor pressure flow.

solution can flow. This envisages a membraneless interface between peripheral cytoplasm and rapidly moving solution. This might require the cytoplasm to be a gel or of a filamentous nature such as P-protein appears to be, which would be anchored to the side walls and their plasmodesmata. To reduce the bulk of the peripheral cytoplasm the nucleus and many of the organelles are lost and it is tempting to see their function as taken over by those in the companion cells which are intimately connected to the sieve elements via the lateral plasmodesmata. We might almost see the residual cytoplasm of the sieve elements as a symplastic extension of the companion cells specialized to service the plasmalemma of the sieve elements. As commonly accepted the sieve pores are regarded as enlarged and simplified plasmodesmata which like the lumen must be lined with cytoplasm and here its filamentous nature is important in reducing drag especially if the necessary lining in the pores more or less fills them. However, we have seen that a certain density of filaments is tolerable on the pressure flow hypothesis. The filaments in the sieve pores would extend out into the lumen as bundles running longitudinally with the main flow in the space between them. This hypothetical picture of the sieve tube system is represented in Figure 9. It has been constructed to be consistent with the Münch hypothesis. The more it corresponds with actual cytological findings the less is it necessary to go beyond the Münch hypothesis to explain translocation in the sieve tube.

Acknowledgement

My warmest thanks are due to Dr. Barrie Watson for his advice in the preparaton of the script of this paper.

References

BARCLAY, G.F. and FENSOM, D.S. 1973. Acta Bot. Neerl. 22: 228-232

CALALDO, D.A., CHRISTY, A.L. and COULSON, C.L. 1972. Plant Physiol. 49: 690-695

CRAFTS, A.S. and CRISP, C.E. 1971. Phloem transport in plants. Freeman, San Francisco.

CURRIER, H.B., ESAU, K. and CHEADLE, V.I. 1955. Amer. J. Bot. 43: 68-81

DIXON, H.H. and BALL, N.G. 1922. Nature (Lond.) 109: 236-237

DOI, Y., TERANAKA, M, YORA, K. and ASUYAMA, H. 1967. Ann. Phytopath. Soc. Japan 33: 259-266

FENSOM, D.S. and SPANNER, D.C. 1969. Planta 88: 321-331

GEIGER, D.R. 1969. Ohio J. Sci. 69: 356-366

GEIGER, D.R, GIAQUINTA, R.T., SOVONICK, S.A. and FELLOWS, R.J. 1973. Plant Physiol. 52: 585-589

GIANNOTTI, J., DEVAUCHELLE, G. and VAGO, C. 1970. Congr. Int. Microsc. Electron., Grenoble III: 353-354

GIAQUINTA, R.T. and GEIGER, D.R. 1973. Plant Physiol. 51: 372-377

HAMMEL, H.T. 1968. Plant Physiol. 43: 1042-1048
HODDINOTT, J. and GORHAM. P. 1974. Can. J. Bot. 52: 349-353
HONERT, T.H. VAN DEN. 1932. Proc. K. Akad. Wet. (Amsterdam)
    35: 1104-1111
JACOLI, G.G. 1974. Can. J. bot. 52: 2085-2088
KENNEDY, J.S. and MITTLER, T.E. 1953. Nature 171: 528
KURSANOV, A.L. 1963. Adv. Bot. Res. 1: 209-274
LANG, A. 1974. J. Exp. bot. 25: 71-80
MITTLER, T.E. 1953. Nature 172: 207
MITTLER, T.E. 1957. J. Exp. Biol. 34: 334-341
PEEL, A.J. and WEATHERLEY, P.E. 1962. Ann. Bot. 26: 633-646
PEEL, A.J. and WEATHERLEY, P.E. 1963. Ann. Bot. 27: 197-211
PEEL, A.J., FIELD, R.J., COULSON, C.L. and GARDNER, D.C.J. 1969.
    Physiologia Plant. 22: 768-775
PEEL, A.J., FIELD, R.J., COULSON, C.L. and GARDNER, D.C.J. 1970.
    Physiologia Plant. 23: 667-672
PEEL, A.J., FIELD, R.J., COULSON, C.L. and GARDNER, D.C.J. 1974.
    Transport of nutrients in plants. Butterworths, London.
TAMMES, P.M.L. and IE, T.S. 1971. Acta Bot. Neerl. 20:
    309-317
TAMMES, P.M.L., DIE, J. VAN and IE, T.S. 1971. Acta Bot. Neerl.
    20- 245-252
WALDING, H.F. 1968. Studies in sieve tube translocation. Ph.D.
    thesis, University of Aberdeen
WALDING, H.F. and WEATHERLEY, P.E. 1972. J. Exp. Bot. 23:
    338-345
WARDLAW, I.F. 1974. Ann. Rev. Plant Physiol. 25: 515-539
WATSON, B.T. 1975. Ann. Bot. (in press)
WEATHERLEY, P.E., PEEL. A.J. and HILL, G.P. 1959. J. Exp. Bot.
    10: 1-16
WEATHERLEY, P.E. and JOHNSON, R.P.C. 1968. Int. Rev. Cytol.
    24: 149-192
WEATHERLEY, P.E. and WATSON. B.T. 1969. Ann. Bot. 33: 845-853
WEATHERLEY, P.E. 1972. Physiol. Vég. 10: 731-742
WORLEY, J.F. 1973. Photoplasma 76: 129-132
ZIEGLER, H. and VIEWEG, G.H. 1961. Planta 56: 402-408
ZIMMERMANN, M.H. 1963. Scientific American 208: 132-142
ZIMMERMANN. M.H. 1969. Planta (Berl.) 84: 272-278

DISCUSSION

Discussant:　　J. Dainty
　　　　　　　　Department of Botany
　　　　　　　　University of Toronto
　　　　　　　　Toronto, Ontario

Cronshaw: opened the discussion by stressing the need to be aware of biological variability and the necessity of proper statistical treatment. He found Weatherley's diagram of a sieve element designed, by Weatherley, to be good for Münch flow. So far as the distribution of P-protein in the lumen is concerned, no electron microscopist can be absolutely definite about it.

Spanner: an unrepentant rejector of the Münch hypothesis, said that Weatherley had calculated a pressure drop of 2 bar meter for a velocity of 10 cm h^{-1} with a sieve pore diameter of 5 μm. But the velocity may be 100 cm h^{-1} and the sieve pores nearer 1 μm. Then before we know where we are, we are right back at an unacceptably high pressure drop for the Münch hypothesis. With respect to inhibitor experiments, he was sorry that the studies with chemical inhibitors had not also been stressed. These indicated that energy is required in the phloem conduits for more than just "care and maintenance". He thought that in the low temperature experiments the cold block may have resulted in a local build-up of the pressure gradient; this would compensate for the lost power of a normal active mechanism over the chilled region. Low temperature studies on mitochondrial respiration had shown a parallel behaviour to that of translocating axes. "The system is far more complex than our simple explanations suggest."

Weatherley: replied that he did not deny the complexity. Although he favoured Münch, he was simply trying to pinpoint the problems; we now needed some critical measurements.

Tyree: brought up the problem of gymnosperms, whose sieve areas containing asastomosing plasmodesmata going into a central median cavity. Here one has not filaments but membrane sheaths. How does Münch work here?

Walsh: strongly supporting Tyree's point, said that he had great difficulty in reconciling what he sees with mass flow. He sees lots of endoplasmic reticulum and a tonoplast. How can one get high velocities of translocation given such structures?

Weatherley: asked, in reply, whether endoplasmic reticulum has to be anchored on a tonoplast? He then raised the question as to whether an even distribution of filaments in a pore, in the static state, could be bundled together by hydrodynamic forces, and perhaps swept into a peripheral position, once flow starts?

Johnson: then described a working model he has made of a sieve tube. He filled a series of short lengths of 3 mm diameter glass tubing with an aqueous suspension of a filamentous gum and placed them end to end, each separated by electron microscope grids to simulate the sieve plates. He then induced flow (electro-osmotic) by an electric current and found after a time that the flow had produced transcellular, clumped filaments of the gum.

van Die: commenting on Spanner's criticism, said that in Yucca where the sieve pores were only 0.4 μm in diameter, the sieve plate was only 0.4 μm thick. He had calculated that with 10 strands in each pore, each strand separated by 100 nm, a flow rate of 40 cm h^{-1} was produced by a ΔP of 5.6 bar m^{-1}.

MacRobbie: in reply to Walsh, said that a tonoplast is not necessary to hold cytoplasmic structures peripherally; a gel layer would do it. She agrees with Weatherley that, whatever the sieve tube structure, we must get rid of the tonoplast if any mechanism is going to work.

Walsh: insisted that the tonoplast is there. It may not be doing a holding job, but he is sure there is a loose membrane system running from cell to cell. It may be loose assemblages of endoplasmic reticulum surrounded by tonoplast, as if the cell had several vacuoles. They may be transcellular strands in Thaine's sense.

Willenbrink: agreed that the cooling experiments are most critical with respect to finding out the significance of metabolism, but he agrees with Spanner about the relevance of the effect of inhibitors. He does not trust experiments on excised systems or on exudation.

Young: gave a presentation of his views at this point.

He first commented on terminology saying that during a problem's formative stages it is rather important to be careful about terminology because it influences the way one thinks. When he was introduced to the translocation problem two years ago by Evert he recognized that the essence of the problem is: what comes in must go out. He felt that this idea of continuity could be better expressed by the term "volume flow" than by "pressure flow". Now the problem is at a different stage; terminology is less important, providing there is only one.

With respect to the present models as presented by Tyree, we are only really using the equation of continuity. The general hydrodynamical equations of motion, with the exception of Poiseuille's Law as a reasonable approximation, have not been used. At some stage it will be necessary to consider the implications of this neglect. He does think that it is not other models versus Münch flow, but other models on top of Münch flow.

He briefly discussed limits to transport in Münch type flow, commenting that the physics is utterly simple and readily comprehensible by all. He pointed out that there must be a characteristic distance, ℓ, involved in the transport given by $\ell = \overline{OP}/\overline{dP/dx}$, where OP is the mean osmotic pressure and $\overline{dP/dx}$ is the mean hydrostatic pressure gradient. Because pressure (and Ψ) are falling along the path of translocation, there is a constant influx of water which will cause the velocity to increase. This will cause a more rapid drop in pressure and eventually the velocity will start to run away. We must obviously stay out of this region, the condition for this being that the length of the tube, $b \ll \ell$. This of course, means that ΔP must be a small fraction of P. The equation for ℓ given above can also be written as $\ell = \overline{c}RT/(\overline{V}/L_p)$, where $\overline{c}$ is the mean solute concentration and $\overline{V}$ a mean velocity. Looked at in this way, the condition $b \ll \ell$ agrees with Tyree's point that translocation over long distances can occur if (1) the concentration is sufficiently high, (2) the velocity is sufficiently low and (3) L_p is sufficiently high.

In the first Planta paper (with Evert and Eschrich) a relaxation system was studied in which an initial non-equilibrium system relaxed towards equilibrium. The relaxation time, τ, depended on the hydraulic conductivity of the membrane, the radius of the tube and the mean osmotic pressure. In a steady state system the relaxation time is only of peripheral importance, but in the time dependent systems described by Ferrier, τ will be relevant in determining the behaviour of the system.

Finally, with respect to models of the type Ferrier described, Young suggested that one cannot arbitrarily and independently specify the sources and sinks. The fluxes are presumably enzymic processes and will therefore be concentration-dependent, the simplest being Michaelis-Menton type. If this is done, certain absurdities, such as no sucrose beyond the cold block, can arise in these models.

Lee: then reported on some work he had recently done in Aberdeen, attempting to measure the water potential, Ψ, and the osmotic pressure, π, for phloem tssue of Heracleum. He took phloem tissue from the top and bottom of one-metre-long petioles and measured both Ψ and π by standard psychrometric techniques. His results say that $\Psi_{top} < \Psi_{bottom}$ and $\pi_{top} < \pi_{bottom}$. The turgor pressures were all positive, but $P_{top} < P_{bottom}$! This is, of

course, the wrong way to Münch flow. He discussed the possible
objections to his measurements: (1) as for cutting releasing
turgor pressure, he argued that the cutting would only release the
turgor pressure over a 2 cm length, whereas he used 20 cm lengths
of phloem tissue; (2) against the fact that π was measured 3-4 hrs.
after cutting, he pointed out that other authors found this
allright; (3) as for the argument that the plant was not
translocating at the time, he checked on this with ^{32}P. He also
pointed out that Hammel's data on phloem turgor pressures in trees
scarcely supports the Münch flow hypothesis.

Aikman: now pointed out that there are three features in the
Munch model which are accessible to experiment and people tyically
only look at two. These features are (1) a concentration gradient
of sucrose in the phloem, (2) a hydrostatic pressure gradient in
the phloem and (3), commonly ignored, the return flow of water in
the xylem requires that Ψ at the bottom of the phloem > Ψ at the
bottom of the xylem > Ψ at the top of the xylem > Ψ at the top of
the phloem.   This latter requirement has not been investigated.
Hammel's data show that Ψ at the higher point in the trunk > Ψ at a
lower point, which is the reverse of what one might expect.   Also
$Ψ_{leaf}$, which should be ≃ Ψ at the top of the xylem, > $Ψ_{phloem}$ which
is also "wrong". Hammel's data support Münch on the first two
requirements but not on the third; more good experimental work
needs to be done.

Spanner: suggested that phloem translocation in mangroves,
where the xylem tension can be regularly 30-60 bars, needs looking
into.

Swanson: supported the mass flow hypothesis by citing some
experiments done in his laboratory several years ago on sugar beet.
They measured translocation and respiratory rates at low
temperatures and calculated that about $4x10^6$ sucrose molecules
moved through a sieve element at a velocity of 54 cm h^{-1} per ATP
molecule turned over in that element. Even accepting the
approximate nature of the calculation, Swanson finds it very
difficult to see how there is enough ATP to provide pumping at the
sieve plates, say, for all these sucrose molecules.

Milburn: criticized Lee's experiments on the grounds of the
unreliability of the techniques used, particularly the
psychrometric. On the other hand he supported Hammel's data, the
only ones of their kind.

Lee: replied that there are no values of Ψ and π together in
the literature; it was his idea to get some.

Milburn: agreed with the value of such data, but the
difficulty is that present techniques are very uncertain.

Tyree: thought that psychrometric methods used, by Lee for instance, greatly underestimate π. The freezing and thawing method of killing used means that the cell contents are much diluted with apoplastic fluid and further diluted, in the case of phloem tissue, by the contents of parenchyma cells which have a lower osmotic pressure. π could be wrong by a factor of two!

Currier: said that it seemed that Weatherley's only doubts about Münch flow stemmed from whether the sieve pores are too severely closed with P-protein. He quoted observations on open pores by Cronshaw, Evert and Eschrich, Fisher and himself. He submitted that the P-protein problem was like the callose problem. He declared that the sieve pores are quite open.

Johnson: found Currier's remarks pleasantly provocative. He had found that if a translocating bundle is frozen to -50^0C in $1/10$ s, filaments, close together, are found in the pores. What would Currier recommend as a method for preparing sieve plates?!

Currier: replied: sneak up on the plant, keep it in equilibrium with its environment, don't dissect out vascular bundles, use Cronshaw's technique of having a degree of wilting, make little longitudinal slits so that the fixative moves into the phloem (i.e. no cross-sections), use the the techniques of Evert, etc.

Eschrich: thought that Johnson had dissected out the phloem bundle which was probably somewhat bruised before being killed with freon. If the petiole were just cut and fixed in glutaraldehyde, it would give fewer filaments than Johnson got. Any handling of a plant will stimulate it and give both P-protein and callose. He said that the surging effect and wound callose occur over only about 20 sieve elements. There must be something else going on which we as yet don't know.

Currier: brought up another problem which is difficult to assess, namely why does the P-protein "fray" out of a sieve pore in a symmetrical way on both sides? It certainly does not look like a surging effect. He thought this, and the whole P-protein, a major challenge and urged hard experimental work to overcome the distressing uncertainty about P-protein.

Mittler: asked Weatherley why, in designing an ideal sieve element, he should go to the length of creating filaments? Is it not simpler to have a cell evolve without them? If the filaments fulfil the function of blocking sieve pores, after injury, why is such an elaborate structure as lots of filaments needed to achieve this result?

Weatherley: replied that if a certain amount of material is needed inside the plasmalemma for various reasons, filamentous material would be best because the drag would be minimal and it is just possible that the flow might push aside these longitudinal filaments.

Reinhold: addressed a question to Young, who had said that it was necessary to make unloading a function of concentration while vein loading can be treated as a constant. She said that there is lots of evidence that the unloading exerts a feedback control on vein loading. What factor would he pick out to allow for this feedback?

Young: could not give a definite answer to this.

Geiger: said that there are two present who have seen more open pores than closed ones, viz. Giaquinta and Fisher. He pleaded for the possibility that the technique they use, rapid freezing, freeze (?) substitution, might have some very good answers. There is very good preservation of microtubules, dictyosomes, polyribosomes, messenger RNA, mitochondria, etc., though not of P-protein. Most of the pores are open, but occasional ones are closed with material in the form of strands or bundles which extend to the side walls. If the preparations are mistreated then the sieve pores are full.

Johnson: said that one could just as easily argue that open pores are a result of damage. He thought it made little sense to wilt a plant to see what the pores are like if one expects a pressure-driven mechanism. He also thought the Currier's "fray" pattern could arise if normally the filaments were closely packed together and they relaxed with time during the fixation.

Kollmann: finally reminded the audience of the gymnosperms. Their sieve plates have plasmodesmata-like pores and yet translocation velocities are about 60 cm h^{-1} as in angiosperms. Are plasmodesmata open sieve pores?

The Electro-Osmotic Theory

D.C. Spanner

E.M. Unit
Bedford College
London, England NW1 4NS

I. Introduction

The other day I was talking to Dr. Charles Hebant and he told me he was hoping, after the conference, to go to Vancouver in search of a "beautiful little moss". I was incautious enough to ask him if it really was a very beautiful moss. "All mosses are beautiful" he replied, with great feeling.

I think sometimes we forget what as scientists we are doing. We are scrutinizing, analyzing and theorizing about the syntax and grammar of Shakespeare. We are chipping away fragments from a Rembrandt and subjecting them to X-ray analysis or mass spectrometry. Or again, we have our oscillographs and computers and are busy with Fourier analysis to find the numerical relationships that constitute a harmony of Beethoven. And while we are doing this we are necessarily, for the moment, out of touch with the far more important aspect of significance and meaning; we are uncomprehending, and blind, and deaf to what Shakespeare and Rembrandt and Beethoven are saying to us.

Of course, because drama and painting and music give us enjoyment we find it easy enough to leave behind the scientific approach and to become more fully human, especially in the company of friends; but my experience of scientific meetings, including this one, gives me a real fear that we do not let this process go far enough. You see, the plant also is a work of art, a marvellous one - "fearfully and wonderfully made". We take this frail thing and subject it to all sorts of indignities - we burn it up, dissolve it in acids, cut it into the thinnest of thin shreds for the electron beam, fill it with radioactive tracers and so on. It never complains, but continues to do its best; nevertheless, all the while it challenges us to rise above our science and to

comprehend it steadily and as a whole; to understand the message it
carries.

I say all this because it is all too easy for science to
become arrogant and exclusive. The spirit that claims that man has
"conquered Everest" because he has crawled up it is present with us
too; and we need to remember that when our science has said its
last word, and conferences on phloem have been overtaken by the
march of progress, the green plant like our own lesser works of
art, will still remain with its message to all who will pause and
consider.

I have to speak tonight on the subject of mechanism.
Physiologists often regard this as the ultimate question.
Cytologists are all right in their way, physiologists sometimes
think (below their breath, of course), but they are the ones to
whom the real issue belongs. I want to dissociate myself from this
attitude. We are all in this thing together and we need and we
serve each other. For my own part I try to approach our mutual
problem in the spirit of that great physicist Lord Rayleigh, who
prefaced one of his papers, I believe, with the quotation from the
Psalms, "The works of the Lord are great, studied by all those who
have pleasure in them." It should after all serve to humble us that
the problem of sugar transport, which taxes our powers of
investigation to the limit, has been mastered by the smallest blade
of grass from infancy. With these preliminary remarks therefore
let me embark on my subject, asking your forgiveness if in my
enthusiasm I forget to exercise the humility and reverence that I
have been speaking about.

A. Presuppositions of the theory

Every theory has its presuppositions. What are those of the
potassium or electro-osmotic theory? I would list them as follows:
(1) The first is presented in Fig.1, which shows occluded sieve
plate pores from Salix alba (by courtesy of Dr. Parthasarathy).
The theory has grown out of such evidence as this rather than the
other way round. Of course such a micrograph can be regarded as a
misleading artefact, the plugging of the pores with P-protein being
due to turgor release. To me it seems, however, that turgor
release would be more likely to blow pre-existing plugs out then to
introduce them, thus into pores previously quite open. Note the
almost crystalline transverse alignment of the "beads" on the
fibrils in the pore; the fine preservation of the delicate
plasmalemma and ER membranes; and the thick callose cylinder,
probably formed during fixation and compressing the pore contents.
Such micrographs as this are found far too commonly, and after
much-too-careful attention to preparative techniques for them to be
dismissed lightly as irrelevant to the problem of normal transport.
However, I wish to be quite explicit about the nature of this
presupposition. It is not that every sieve plate is so occluded,
nor that even a majority are. It is simply that Fig.1 represents

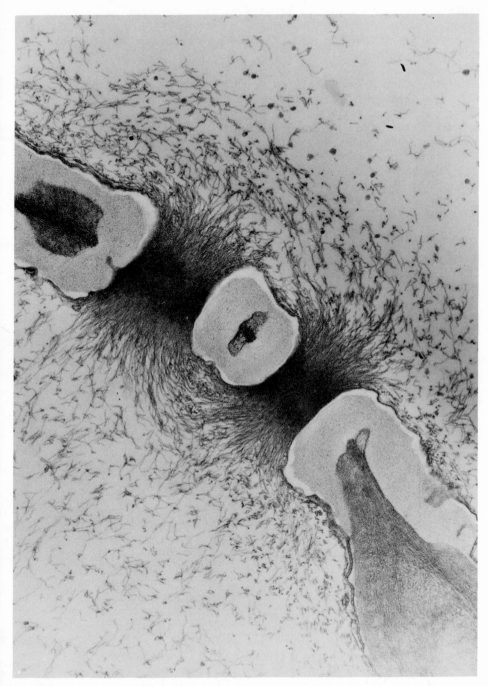

Figure 1. Sieve plate pores from phloem of <u>Salix</u> <u>alba</u>.   The
pores have probably undergone some constriction during fixation due
to the deposition of extra callose material. By courtesy of Dr.
M.V. Parthasarathy. x36,000.

(substantially) a state of affairs in the functioning sieve tube which theory must accept and accord a significant place. It will be satisfied if perhaps only 20% of the plates are so occluded in the operating state.

The other presuppositions can be passed over more quickly: (2) Abundance of K^+ in the sieve tubes; (3) A negative charge on the P-protein at the pH of the moving sap; (4) Arrangements for rapid active ion uptake probably by the sieve-tube membrane, and perhaps by pinocytosis; (5) An adequate supply of (e.g.) ATP.

How these factors enter will be evident as we proceed. Let me now go back to the beginning.

B. History

So far as I know, the first suggestion that electro-osmosis might be involved in sieve tube function was made by D.S. Fensom in 1957. His scheme, which was a bare outline only, invoked the hydrogen ion with electro-osmosis operating apparently through the open and unobstructed pores of the sieve plate; Fensom has long since abandoned it. Independently in the following year, the present author (Spanner, 1958) put forward a scheme in rather greater detail. This invoked the potassium ion acting electro-osmotically in the much finer channels of the sieve plate pores plugged with P-protein, evidence for such plugging having then recently been provided by the pioneering electron microscopical work of Hepton, Preston and Ripley, 1955. This scheme remains today unchanged in essentials, though modifications and elaborations have been made to meet criticisms and to incorporate further evidence.

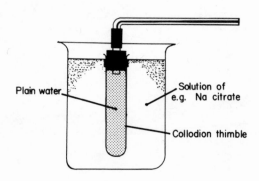

SIMPLE ELECTROOSMOSIS

Figure 2. A simple system for demonstrating "anomalous osmosis". The collodion is pretreated by soaking in dilute gelatine.

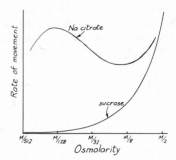

Figure 3. Typical results obtained with the apparatus shown in Fig.2.

C. What is electro-osmosis?

It is first necessary to understand the phenomenon of electro-osmosis in simple systems. It is a common phenomenon, and in the form of "anomalous osmosis" was rediscovered about 125 years ago by Dutrochet, (in Hober, 1946). Anomalous osmosis occurs when a membrane of pig's bladder or a plug of clay or fine sintered glass separates two compartments containing electrolyte solutions of   unequal   strengths.   It is important to recognize that electrolytes must be present. A simple system for demonstrating anomalous osmosis (Fig.2) can be made by casting a collodion thimble in a test tube, soaking it in dilute gelatine solution, and then attaching it to a glass capillary as shown. If it be filled with   distilled   water   and   then   immersed in sucrose solution exosmosis of water takes place and its rate can be followed by observing the meniscus in the capillary. If sodium citrate or another suitable salt be substituted for the sucrose exosmosis again takes place; but the variation of rate with concentration of the external solution is dramatically different (Fig.2). Quite low strengths of the salt produce effects equal to that of strong sucrose, and a mechanism other than simple osmosis is plainly at work. While the movement is taking place a difference of electrical potential develops across the membrane. At the same time anions and cations pass across in equivalent amounts. It is the resultant dissipation of the concentration difference between the two sides which supplies the free energy for the process. A more manageable system for studying the basic phenomenon can be set up by using the cation- or anion-exchange resin membranes to form a division between two compartments containing identical solutions (Fig.4). In this case no mechanism exists for spontaneously producing a difference of electrical potential across the membrane (since the solutions on the two sides have the same concentration). Accordingly movement only occurs if a potential difference be applied, and in the apparatus this is done by inserting electrodes into each compartment and connecting them to a battery. The electrodes are ideally reversible, that is, they can go on functioning without changes such as polarization occurring. When a current is passed it is found that a movement of water takes place. With a membrane of anion exchange resin flow is towards the anode; with one of cation exchange material it is towards the cathode. How is it to be explained?

Mechanism of electro-osmosis

The mechanism of electro-osmosis as currently understood can be illustrated with reference to Fig.5. This depicts a membrane of a material (such as cation-exchange resin, or gelatine-treated collodion) which carries fixed negative charges on its surface. In the case of the resin these may be provided by $-SO_3H$ groups; in the case of the treated collodion by $-COOH$ groups. Through the membrane extend pores, and the walls of these pores also bear the

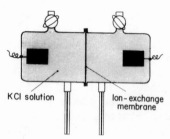

KCl solution   Ion-exchange membrane

Figure 4. Apparatus for investigating electro-osmosis. The electrodes are reversible Ag-AgCl. Movement is registered in the capillaries.

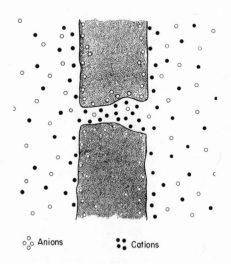

$^{\circ}_{\circ}{}^{\circ}$ Anions $\bullet\!\!\bullet\!\!\bullet$ Cations

Figure 5. Pore in an electro-osmotic membrane (the drawing is not to scale).

same fixed negative charges. The whole is flooded with a solution of an electrolyte like KCl. Provided the pores are not too large (and this is an important proviso) the space within them will be dominated electrically by the charges on their walls; as a consequence the electrolyte ions of opposite sign (in this case K^+) will be attracted in, and those of similar sign (Cl^-) repelled out. Thus the small fluid columns within the pores carry a net charge; and the greater their preponderance of mobile ions of one sign over those of the other the more powerful and efficient will the subsequent electro-osmosis be.

If to such a system as this a potential difference is applied, it can readily be appreciated what will happen. The cations in the pores will move one way, and the anions the other; but since (in the case illustrated) the cations outnumber the anions the liquid in the pore will experience a net frictional drag from the movement of ions towards the negative compartment. It is this movement that constitutes electro-osmosis. In the case of the collodion thimble system discussed earlier the potential difference arises from salt diffusion and the pattern of movement is more complex and hard to understand; nevertheless, the principle is believed to be the same.

Requirements
    The requirements for an effective system thus appear to be four: (1) There must be suitable electrolytes in the solution; (2) There must be a divisional membrane of material bearing a high density of fixed surface changes; (3) This membrane must be traversed by pores neither too large nor too small; and (4) Arrangement must exist for maintaining a continuous and adequate electrical potential across it. These four requirements will be discussed briefly in turn.

The electrolyte
    Since electro-osmosis depends on frictional interaction between the ion and solvent it will obviously be more in evidence with ions in which this interaction is high. Another way of putting this is to say that per unit current, movement of water will be greater with ions of lower conductivity. The potassium ion (which the theory invokes) is not particularly favourable in this respect; the sodium ion is about 50% better, and the lithium ion almost twice as good. Nevertheless, potassium is nearly five times as effective as the hydrogen ion. A second point concerns the concentration. Looking at the pores in Fig.4 it can readily be imagined that when the concentration of the electrolyte is low the dominance of the cations may well be almost absolute, and there will be virtually no anions present at all. But when the concentration of the electrolyte is high the influence of the fixed charges will tend to be swamped and the cations and anions inside will practically balance each other. This means that per unit current traversing the membrane water will flow more rapidly when

the electrolyte concentration is low than when it is high. Some figures obtained with excised phloem illustrate this. The transport of water in moles per Faraday or water molecules per K$^+$ ion (no anions involved) was 120 in 10^{-4} M KCl, 83 in 10^{-3} M and only 29 in 10^{-1} M (Fensom and Spanner, 1969). However, what is meant by "high" and "low" concentration in this connection depends on the density of the fixed charges. If this be raised, other things being equal, the transport figures will be correspondingly improved.

Membrane material

So far as membrane material is concerned the requirement is that it should be as highly charged as possible. That is why ion-exchange materials are so effective. Among biological structures those composed of strongly acid - or basic proteins, phospholipids or of polyuronides are possibilities; neutral substances like cellulose will be ineffective unless they adsorb ions. The pH will be important; acid proteins for instance will be more strongly charged at high pH's and so will provide a more effective structure.

Size of Pores

The best size for the pores depends on the thickness of what is called that "electrical double-layer" at the interface between the solution and the solid material of the membrane. This in turn depends on the concentration of the electrolyte. Some typical figures for an electrolyte like KCl are about 10 nm for for 10^{-3} M, 3 nm for 10^{-2} M and 1 nm for 10^{-1} M. Very roughly therefore we could say that, as some sort of guide, the pores should ideally be perhaps five times these thicknesses in diameter. This would mean that in 10^{-3} M KCl they should be very roughly 50 nm; in 10^{-1} M about 5 nm. This is, of course, quite small compared with the unobstructed pores in the sieve plate which are about 20 to 100 times larger.

D. Energy relationships

Of course continuous electro-osmosis implies a continuous input of energy to drive the ionic current and any acceptable theory must suggest how this is accomplished. This is one of the most difficult points for the present theory, and will be discussed later. However, a related question of immediate interest is this: is electro-osmosis a sufficiently powerful mechanism to do what is required of it? Energy apart, can it develop the necessary forces? Reference to Fig.2 would suggest that it can, and calculation confirms this. Consider further the pore shown in Fig.6. If we assume that the membrane charge is sufficient to virtually exclude all anions from the pore but to attract in cations to a normality "x" (or if, alternatively, x is its excess normality of cations over anions) then clearly the small column of liquid in the pore will carry a charge equal to $\pi r^2 \ell$ x.F, where F is the Faraday.

Situated in a potential gradient of E/ℓ it will experience a force of $\pi r^2 \ell x F . E/\ell$, and this averaged over the cross-section becomes xFE. This represents what may be regarded as a sort of equivalent pressure (P) tending to move the liquid column through, and it is interesting to calculate its magnitude in a simple case. If we put $x = (1/25)N$, $E = 5mV$ and $F = 96,500$ coulombs equiv.$^{-1}$, we find P to be 0.2 atmos. This is really very considerable. If it is legitimate to regard the sieve plate as such a membrane, and if one in four plates is operative, then with say 2000 plates per meter the electro-osmotic forces would be capable in these circumstances of contributing an effort of 100 atmospheres per meter. There seems no doubt therefore, that as a mechanism, electro-osmosis is capable of developing the very considerable forces required.

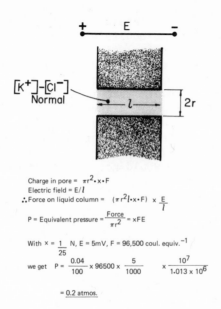

Figure 6. Calculation of the "pressure" due to electro-osmotic forces.

Magnitude of current

Before we describe a workable physical model of an electro-osmotic pump there is a further point which must be discussed. How large is the electric current that must flow to promote a given flow of water? This is an important matter because it constitutes the main problem for the theory. Experiment shows (Tyree, 1968; Barry and Hope, 1969) that in preparations such as Characeae cell walls the passage of a Faraday of electricity is accompanied by the transfer of about 50 moles of water in a solution of KCl of strength 0.01 N. Roughly the same is true for excised pieces of phloem (Fensom and Spanner, 1969). If all the

Velocity of water towards membrane = 100cmh^{-1}

Volume flow per cm^2 of membrane = $100 \text{ cm}^3 \text{h}^{-1}$

$$= 100/18.0 \text{ moles h}^{-1}$$

$\therefore$ Potassium current $= \dfrac{1}{200}$ (say) $\times \dfrac{100}{18.0}$ moles $\text{h}^{-1}\text{cm}^{-2}$

Electric current $= \dfrac{1}{200} \times \dfrac{100}{18.0} \times \dfrac{96500}{3600}$

$$= 0.75 \text{ amp. cm}^{-2}$$

Figure   7.   Calculation   of   the   current   accompanying
electro-osmotic movement.

current is carried by the potassium ions this means that each ion,
in effect, drags 50 water molecules with it; but if mobile anions
are also involved this is not a fair estimate for our purposes, for
the anions add to the current while detracting from the flow. In a
system in which anions are immobilized the figure might be
considerably higher. However, even if we allow 200 water molecules
per K^+ the current is still very large. Suppose the water is
moving through at the linear velocity (measured before it enters
the pores) of 100 cm per hour, then the corresponding current
density is no less than 0.75 amp per cm^2, which is very high.   It
constitutes a difficulty I would not wish to deny.

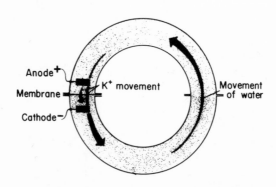

CLOSED CIRCUIT ELECTROOSMOTIC PUMP

Figure 8.   Design for an electro-osmotic pump.   The tube is
filled with fine granular cation-exchange resin. The membrane is
of the same material.

E.   A physical model

A working model of a system in which electro-osmosis can cause the continous circulation of water might be suggested as follows. Referring to the diagram (Fig.8) two tubes formed into semicircles are connected together by flanges to form a closed circuit. Across one junction is placed a cation-exchange resin membrane, and on either side of this are inserted reversible electrodes. The tubes are packed with fine granular cation-exchange material in the K^+-form, and filled with a very dilute solution of KCl. Although it would be difficult to construct them, it is legitimate to suppose that the electrodes are such that at the cathode there is no action except that K^+ ions are removed from the system as they arrive; while at the anode K^+ ions are simultaneously added, (this is in fact what happens to the Ag^+ ions with reversible Ag-AgCl electrodes).

On applying a potential to the electrodes potassium ions migrate through the membrane moving water with them. On reaching the positive cathode they are removed, others being simultaneously released at the negative anode to repeat the process. (The anions partnering the potassium are of course non-motile, being represented by the granular resin.) However, not all the ions which cross the membrane are discharged at the negative cathode. Some are carried on by the stream past the electrode and continue around the ring. Thus the effect of the applied voltage is not only to promote a current via the electrodes; in addition it sets up a circulation of water and potassium around the system. The electrode current is in fact only a proportion of the total potassium current crossing the membrane. How small a proportion it might be is rather an important question for the theory. It cannot yet be answered, but the phenomenon is sufficiently complex to leave the hope that perhaps it may be found to be in some circumstances a small one. This remains, however, a hope.

The few small mobile anions in the solution would be swept along with the stream (the electrical gradient assisting) away from the cathode and towards the anode, where in the event of their being prevented from discharging they would accumulate. This is an important point.

F.   The electro-osmotic or potassium theory

We are now in a position to state the electro-osmotic - or as it might better be called - the potassium theory of sieve tube mechanism. It derives the sieve tube column from the model just discussed in the following way. The circular tube is imagined separated at its open junction and straightened out. It is then in thought placed end-to-end with other similar units all oriented in the same direction to give a repeating structure (Fig.9). This assemblage represented the sieve tube.

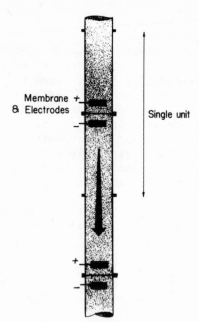

Membrane +
& Electrodes

Single unit

Figure 9. Linear pumps connected in series to simulate a sieve tube.

Corresponding to the cation-exchange membranes we have the sieve plates, with their pores occluded with negatively-charged P-protein fibrils, the latter forming a system of fine parallel channels of the right dimensions.   The granular filling of the model conduit is represented by the P-protein network occupying part at least of the sieve tube lumens. The anode is replaced by a region of the sieve tube plasmalemma above the sieve plate across which an inwardly-directed flux of K^+ is maintained by active uptake processes; the cathode becomes a similar region below the plate through which K^+ leaks out into the apoplast.   This postulated difference of behaviour of the cell membrane above and below the plate (active uptake on one side and leakage on the other) is explained by the fact noticed previously that mobile anions will accumulate in the region of the anode and disappear from that of the cathode. This phenomenon will affect ATP, known to be fairly abundant in the sieve tube sap, and of course a strong anion; and it is not difficult to show mathematically that the effect of the hydrodynamic forces of flow on such an ion wll be to pile it up exponentially above the plate to a concentration enormously greater than that below the plate. If active uptake of K^+, and indeed plasma membrane integrity requires a supply of ATP then it is very plausible to suggest that this phenomenon results in a marked polarization of the sieve plate in accordance with the theory.   The scheme is illustrated in Fig. 10.

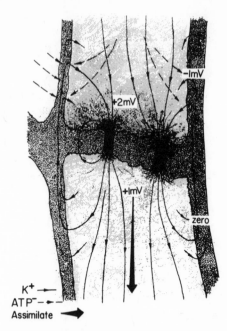

Figure 10.   The electro-osmotic theory of sieve plate action.
(The flow lines are partly convectional and partly diffusional.)

Magnitude of the sieve plate potential
     It should be clear from the description given so far that the
theory implies no overall gradients of either electrical potential
or even of pressure, though the latter is   probably required for
initiation of transport, a point which will be returned to.   The
electrical potential invoked follows a saw-tooth pattern, and the
same is broadly true of the pressure.   In the diagram given,
electrical potentials are indicated which are purely hypothetical
are they likely to be in any sense realistic; We must digress for
a moment, for this point requires an estimate of the sort of
resistance to flow which would be offered by pores plugged with
P-protein.   Now a formula for the flow of a viscous liquid along an
array of parallel filaments has been published (Fensom and Spanner,
1969) and application of this to a typical situation (as judged
from electron micrographs) is given in Fig.11.   It can be seen that
calculation indicates that electro-osmotic forces will need to
contribute an "effort" equivalent to about 0.1 atmos. per plate in
the situation described. Combining this with the calculation in
Fig.6 we conclude that a mere few millivolts might be adequate.
Bearing in mind the influence of ion concentration (Fig.6) and
fibril spacing (Fig.11) on "effort" and "resistance" respectively,
it can be appreciated that even this is as likely to be an
overestimate as an underestimate.   It constitutes a major revision
of the suggestion previously published (Spanner and Jones, 1970).

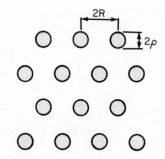

HEXAGONAL ARRAY OF MICROFILAMENTS

"Poiseville"—type equation for axial flow:

Mean velocity $V = \dfrac{R^2}{8\eta} \cdot \dfrac{\Delta P}{\Delta x} \cdot$ (factor depending on ratio R/ρ)

Example: with 2ρ - 100Å $2R = 300$Å
            sieve plate thickness $\Delta x = 1\mu m$
            mean pore velocity $V = 200$ cm h^{-1}
            viscosity $\eta = 0.015$ poise

we get,

$$\Delta P = \left(\frac{200}{3600}\right) \times \frac{8 \times 0.015 \times (1 \times 10^{-4})}{(150 \times 10^{-8})^2} \times \frac{1}{2.44} \times \frac{1}{(1.013 \times 10^{-6})}$$

$$= \underline{0.12 \text{ atmos.}}$$

Figure 11. Calculation of the resistance to flow down an array of parallel fibrils such as that in the sieve plate pores. (For formula see Fensom and Spanner, 1969.)

"Measurement of viscosity of liquids in Quartz capillaries "
    by Churayev, Sobolev & Zorin (special discussion Faraday Soc. 1970)

Down to a radius of 0.04 μm (diam. = 800 Å)
Poiseville's equation is accurate for non-polar liquids   (e.g. CCl_4).
For water it underestimates the viscosity (by 40% for d = 800 Å)
unless an adsorbed layer of thickness 80 Å is assumed.
Thus there seems no escape from the implications of the
Poiseville equation with minute capillaries.

THE STOKES' EQUATION APPROACH

Application of the Stokes equation for the steady velocity
of a sphere moving through a viscous fluid to the substituted
quarternary ammonium ions — too large to hold a water shell
and hence of calculable dimensions — gives the ionic mobility
accurately as far down as the tetra-amyl ion:

$$(C_5H_{11})_4\,N^+ \quad \sim \quad \text{radius} = 5.29 \text{ Å}$$

Below this the Stokes equation overestimates the mobility.
Again, no hope is offered of a resistance to flow lower in the
microcapillary region than that indicated by the ordinary
macroscopic laws.

Figure 12. Evidence for the validity of macroscopic equations of flow for macromolecular-sized channels. (For the Stokes equation see Robinson, R.A. and Stokes, R.H. "Electrolyte Solutions", Butterworths, London, 1959.)

An objection that needs to be noted in passing, concerns the validity of applying macroscopic viscosity equations to channels of such minute size; might such application not lead to serious error? The brief answer to this would seem to be "No". Evidence supporting this conclusion is given in Fig.12; it seems to demonstrate that the Poiseuille and similar equations will under-estimate rather than over-estimate the resistance to flow in microscopic channels, but that the error will not be crucial so far as translocation theory is concerned.

The potentials suggested will not of course stand in isolation. They will be superimposed on Donnan potentials arising from the non-uniform distribution of colloidal matter in the sieve tubes, and this may well make them, small as they probably are, almost impossible to verify experimentally. Certainly they will be well within the physiological range.

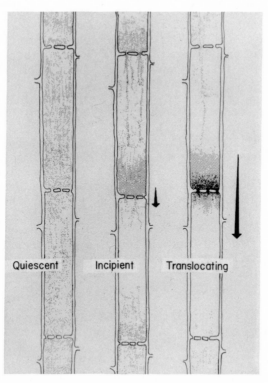

Figure 13.   Initiation of the conducting process according to the   electro-osmotic   theory.   The   shaded regions represent P-protein.

Initiation of transport

The principles of the electro-osmotic theory can be better understood by considering how it envisages the initiation of sieve tube movement. This is set out diagrammatically in Fig.13. We start with a sieve tube which, either because it is only just mature, or because there has been for some time an absence of mobile carbohydrate, is quiescent (left). The P-protein in such a tube may be considered as more or less uniformly dispersed in the lumens, the result of Brownian agitation. Not every sieve element, however, will possess an equal amount of this material. Translocation begins when mobile carbohydrate is loaded into the sieve tubes at a sink region. Such loading operates at once to initiate flow in the fashion made familiar by description of the Münch theory. The slow movement thus started carries the P-protein down the elements towards the sieve plates (Fig.13, centre). What happens next depends on local circumstances. Where P-protein becomes early well-threaded through the pores the streaming fluid may drag it through in the form of long streamers (top centre). In this case it may go to augment the material in the next element. On the other hand, where there was abundance of P-protein initially present in an element it may build up on the plate too quickly to allow it to be dragged through (middle centre). Where the P-protein was initially sparse it may again be displaced through the plate below, or even remain little-moved (bottom centre). Among these possibilities the interesting situation is the one where the plate becomes occluded. As soon as this is appreciable, lowered anion permeability sets in, anions (e.g. ATP) begin to pile-up above it and drain away below it. This initiates the radial movement of potassium and by establishing an electrical gradient through the plate consolidates and regularizes the density and alignment of the P-protein fibrils in the pores (right centre). In effect they "crystallize" along the electric lines of force. Once the plate has "caught on" in this way strong electro-osmotic action sets in the commitment of that particular plate to electro-osmotic action becomes more pronounced. Meanwhile adjacent plates in tubes with short elements have lost their chance and remain inoperative. The reason why the increased flow which results when a plate begins to function electro-osmotically does not sweep out the P-protein from its own pores lies in the fact that the electrical forces responsible for it operate in the reverse direction on the negatively-charged fibrils. They hold them therefore against the rising hydrodynamic drag.

What this account implies is that electro-osmosis operates as a sort of servo-mechanism assisting weak pressure-flow. As such, it is consistent with the evidence that the sieve tubes have no inherent polarity but can conduct in either direction as determined by the relative positions of source and sink. It is important to note that it does not imply that every sieve plate must be plugged with P-protein; but it locates in those that are the major force for long-distance transport. It allows for the initial impulse to conduct, derived osmotically at the loading sites in the source sieve elements, to be transmitted progressively without attenuation down a column of indefinite length toward the sink elements. And it provides for the process of conduction to be sustained by the injection of metabolic energy at suitably-spaced sieve plates en route.

Path of the potassium

The circuit for the potassium is completed by a return path which has been the subject of several suggestions. The one I now favour makes use of the thick sieve tube cell wall, i.e. it is in the apoplast. Besides the circulating potassium however, experiment indicates that we must allow for a considerable longitudinal transport of this element. This from one point of view helps the theory; for all the potassium crossing the plate acts frictionally on the sap, whereas only the circulating needs to be dealt with by the membrane transport apparatus, and the magnitude of this membrane transport is the principal difficulty. However, from another point of view a new problem emerges. This may be briefly stated by saying that the long-distance transport of K^+ short-circuits the membrane battery and prevents it building up a useful potential difference across the plate. Interaction between the K^+ ion and sucrose is one suggestion for overcoming this; but it certainly presents a problem.

Energy input

The electro-osmotic theory is one of the group which requires an input of metabolic energy for transport in the actual sieve tubes. It is clear how it envisages this. The point of application is at the plasmalemma were K^+ is being actively transported inwards. This, it is suggested, utilizes the energy of ATP derived presumably from the companion cells. Not only will the ATP be swept towards the proper region by the hydrodynamic forces; it will also be attracted electrically by the very fact of cation uptake. In view of the low sieve plate potentials that appear to be adequate it can be assumed that the level of respiratory energy required constitutes no serious problem.

The fate of anions

It has been argued that small mobile anions would not be transported in a system following the lines described. However, this is probably not so. Such anions would be swept up to an exponential profile of high concentration against the sieve plates. Inevitably, therefore, they would force their way through, and at a rate ultimately determined by their speed of entry into the sieve tube system. But would they not, by forcing an entry into the negatively charged channels of the sieve plate, cancel the preponderance of the cations there and render electro-osmosis void? This conclusion would not seem to follow. High local anion

concentration would imply high local cation concentration. One can
appreciate that the relationships in the pores would become rather
less straighforward. Complex formation under conditions of high
concentration might occur to convert anions into uncharged species;
and in any case anomalous osmosis shows that equal transport of
ions of both signs is a real possibility. For the moment judgement
must be suspended on this point.

G.   An evaluation of the theory

It is unquestionably difficult to be objective in evaluating
one's own theory; nevertheless, the attempt must be made to sum up
its strengths and weaknesses.

On the structural side is the consideration that it does
obvious justice to the fact of the very widespread occurrence of
P-protein in sieve tubes. This is a major point. It builds too on
the evidence that P-protein plugs many of the sieve plates. No
doubt this is a controversial subject, but I remain of the opinion
that the electron microscope evidence, judged on its own merits and
without physiological prejudice, is firmly in favour of the view
that fairly dense, semi-crystalline plugging of many at least of
the plates is a natural state of affairs and must be taken
seriously in any theorizing. Such a picture as that in Fig.1 is
far too common in careful preparations, it turns up far too often
in different subjects and with different techniques, and I may add
is far too elegant to be dismissed easily as due to the trauma of
sudden turgor release. Then there is the fact of the conspicuous
presence in many specimens of endoplasmic reticulum aggregates.
The theory is able to offer at least the suggestion that they are
connected somehow with the necessity for heavy ion traffic into and
out of the sieve tubes. Similarly with the noticeably thick walls
of the tubes. These are hardly necessary to resist the high turgor
pressure, since the sieve tubes are commonly quite narrow; the
theory offers an explanation in terms of an adequate return path
for the K^+ ions (Fig.10).

On the physiological side the strong points of the theory are
that it provides a driving force of unquestionable adequacy for
even the longest systems, and it does so without overlooking the
P-protein. Further, it meets the very strong evidence that this
driving force is dependent on metabolic energy released not merely
at the source sink terminals, but in the sieve tubes themselves.
This evidence, already strong, has I believe been rendered fairly
conclusive by the recent experiments on the stolon of Saxifraga
sarmentosa by a fine worker now in Islamabad, Dr. F.A. Qureshi
(Qureshi and Spanner, 1973). The evidence that such agencies as
anoxia, KCN and dinitrophenol reversibly depress transport without
rendering the sieve tubes leaky or causing obstruction of the sieve
plates tells strongly in favour of some theory of active mass flow.

There is a considerable amount of work (see Hartt, 1969) indicating that potassium has an important effect on the translocatory process, and this is circumstantial evidence of significance. Of course, many of the most important arguments adduced in support of the Münch hypothesis discriminate equally in favour of the electro-osmotic: the evidence for mass flow as the mode of movement, for the source-sink arrow as the indicator of direction, and for sugar as the prime mover. In fact it can be claimed that the theory is consistent with all the major well-establshed characteristics of the translocation process. The only exception which might be adduced is that of simultaneous bidirectional movement in individual sieve tubes. My personal opinion, however, is that the evidence for this is very problematical; that against it, very strong. Except as a very minor possibility I have never believed in it.

However, strong as the claims of the electro-osmotic theory are, it faces formidable difficulties. First amongst these I would place the sheer magnitude of the K^+ current it seems to imply. It almost seems out of the question, but I remain hopeful. Secondly, there is the question of the transport of anions. Undoubtedly this takes place; one of the clearest instances is that of the transport of the non-metabolized Br^- ion found by Qureshi (Qureshi and Spanner, 1973b) in my own laboratory. But again I am hopeful. Energy considerations I do not feel are critical, especially in view of the recognition that only a very small potential drop over the sieve plate is required, and that not all sieve plates need be operational.

II. Conclusion

In conclusion I would point out one or two interesting possibilities. If the sieve plate does constitute an electro-osmotic membrane then it is an extraordinary one indeed, quite unlike any artificial membranes readily available. For instance, it is extremely thin (about one tenth of the thickness of the flimsiest polythene film). This maintains the electrical forces at a high level while reducing the hydraulic resistance. Then, its channels are straight and very uniform - compare the scanning electron micrograph of a cation exchange resin membrane (Fig.14) with those of a sieve plate pore (Fig.15). Again, they are of labile character capable possibly of adjusting themselves to an optimum width. Finally, they are constituted not as spaces within walls, but as spaces around fibrils. Arguably, this is a better arrangement for favouring the electrical forces as against the hydraulic resistances.

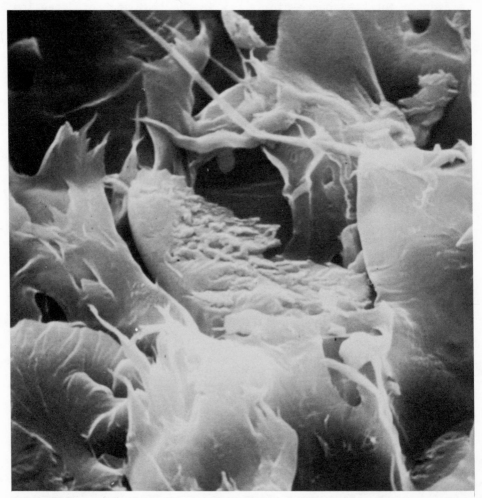

Figure 14.   Scanning electron micrograph of transverse section
of cation exchange resin membrane material.   By courtesy of Miss V.
Cowper.   x 1000.

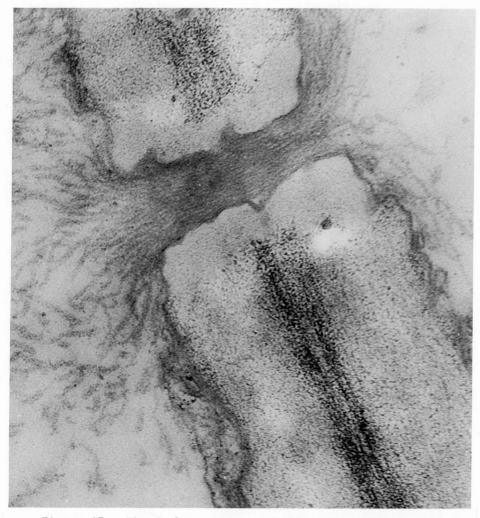

Figure 15.   Sieve plate pores occluded with P-protein from the petiole of <u>Nymphoides</u>.   By courtesy of R.L.   Jones.   x 120,000

Then one needs to remember that the sieve tube sap contains not only sugars and potassium ions, but a variety of other things. Among these, amino acids are important, with their striking physicochemical properties. It is not impossible that these or other solutes might influence the process of electro-osmosis in unexpected ways through the phenomena of interaction in microscopic pores (Bishop and Richards, 1968).

The P-protein has not yet been shown conclusively to possess the necessary charged property. This remains therefore a rather speculative element; but everything considered, the degree of speculation involved in the theory cannot be said to be "unacceptably high".

References

BARRY, P.H. and HOPE, A.B. 1969. Biophys. J. 9: 729-757

BISHOP, W.H. and RICHARD, F.M. 1968. J. Mol. Biol. 38: 315-328

FENSOM, D.S. 1957. Can. J. Bot. 35: 573-582

FENSOM, D.S. and SPANNER, D.C. 1969. Planta 88: 321-331

HARTT, C.E. 1969. Plant Physiol. 44: 1461-1469

HEPTON, C.E.L., PRESTON, R.D. and RIPLEY, G.W. 1955. Nature 176: 868-870

HOBER, R. 1946. Physical chemistry of cells and tissues. J. & A. Churchill, London.

QURESHI, F.A. and SPANNER, D.C. 1973a. J. Expt. Bot. 24: 751-762

QURESHI, F.A. and SPANNER, D.C. 1973b. Planta 112: 121-128

SPANNER, D.C. 1958. J. Expt. Bot. 9: 332-342

SPANNER, D.C. and JONES, R.L. 1970. Planta 92: 64-72

TYREE, M.T. 1968. Can. J. Bot. 46: 317-327

Activated Mass Flow: Surface Flow

E.A.C. MacRobbie

Botany School
University of Cambridge
Cambridge, England

I.   Introduction

A number of other theories of translocation have been put forward, starting from the view that certain features of phloem raise difficulties or unresolved questions for the Münch hypothesis, and that therefore something "better" (or more elaborate) is required to explain the process of translocation. At the moment it is not clear that these features can either be accommodated in the Münch hypothesis, or failing this, be explained away as preparative artefacts. But equally, it is not clear that these difficulties are serious enough, and well enough established as real, to rule out the Münch hypothesis and demand a feasible alternative. The verdict would still seem to be "not-proven".

In this paper I would like to examine translocation theories other than the two already discussed (Münch and electro-osmosis), and in the next paper I would like to consider the relation between some of these alternatives and the Münch theory.

Any translocation hypothesis must accommodate the two characteristic features of sieve elements, the existence of numerous sieve plates, and the elaboration (in many species) of their contents during differentiation, with the proliferation of P-protein in whatever configuration it takes in vivo. The evolutionary retention of sieve plates, and indeed their increase in number in "advanced" phloem (Esau, 1969), suggests that they have a function in translocation. The same may be argued for P-protein, which has generally been thought to be a specific constituent of sieve elements synthesized during their differentiation into functional translocating units; this view has, however, been questioned recently (Weber et al, 1974), in the light of evidence that the great bulk of P-protein in Cucurbita exudates

585

is very similar to ribosomal protein, and might arise as a product
of the degeneration of ribosomes during differentiation. The
question remains however, whether a satisfactory role for both
sieve plates and P-protein can be found within the framework of the
Münch hypothesis. The role suggested is in blocking the conducting
system after damage (Anderson and Cronshaw, 1970; Milburn, 1971)
and the gelling properties of a small component of P-protein
(Walker and Thane, 1971) or of the major fractions of P-protein
(Weber et al, 1974), are consistent with this suggestion. The
other view has been that rapid callose deposition (Eschrich, 1970)
provides an adequate blocking mechanism, and that some other role
for P-protein in its fibrillar forms must be found.

One suggestion of a passive role of P-protein in translocation
flow can be made, that of lowering the viscosity of the sieve tube
contents (see Hand and Williams, 1970). While this is feasible,
the arrival of large amounts of P-protein at the receiving end of
the files of sieve elements poses very great problems of removal,
and only two solutions seem to be possible - either P-protein is
normally anchored in some way and does not flow, or the system is a
two-way flow, although the evidence is against this and suggests
one-way flow. But neither of these conditions is consistent with
this suggestion of the function of P-protein.

In a search for possible roles for P-protein various
suggestions have been made that P-protein may be involved in the
generation of the motive force for whatever type of flow is
envisaged, giving it a direct role in the translocation process.
These include suggestions of a surface flow along P-protein
filaments, tubules or sheets, and of a variety of forms of
activated mass flow. I would like in this paper to review the
variety of proposals, in greater or lesser detail of their
arrangements, which have been put forward, to consider the
experimental evidence which may, or may not, be offered in their
support, and to try to identify definitive characteristics of such
models which might be used to distinguish them from Münch flow.

A.   Surface Flow

It was suggested by van den Honert (1932) that sucrose might
spread by surface diffusion along suitable interfaces in the sieve
elements - by analogy with his demonstration of the rapid spread of
alkaline potassium oleate along an ether/water phase boundary.
More recently Lee (1972) has pointed out that the P-protein
filaments might provide an enormous surface area for this type of
surface diffusion; he calculates an area of $3.4 \times 10^5 \text{cm}^2$ in 1 cm^3
of sieve element volume, assuming sieve elements 250 µm by 20 µm
with 10% of the lumen occupied by filaments. But the arguments
against this type of mechanism are very strong. In the first
place, the gain by restricting diffusion to one dimension instead

of three is not nearly enough to account for the net transfers found; the benfits of such restriction are assessed by Adam and Delbrück (1968), but it is quite clear that random thermal displacements of sugar molecules, even if confined to one dimension by affinity for sites on a linear structure, cannot be responsible for the steady state transfers observed over path lengths of many cm. In the second place, the evidence is that the phloem is relatively non-specific in its longitudinal transfer, and will transport anything that can get in - sucrose, amino acids, inorganic anions and cations, organic anions, a range of growth substances, a range of herbicides, viruses, mycoplasmas. The only way of achieving this non-selective transfer would seem to be by convective flow, a mass movement of sieve element contents, impelled by a non-specific force. Surface diffusion cannot do what we require of the translocation mechanism.

B. Activated mass flow

The essential feature of these theories is that P-protein is held to be responsible for the generation of motive force for bulk flow of the sieve tube contents, by reason of its affinity with one or other of the proteins concerned in generating movement in other biological systems. Thus it is worth discussing the general properties of other motile systems before considering possible morphological arrangements of similar molecular systems in phloem. Two structural components are recognized as having an association with movement in cells, namely microtubules and microfilaments. There is a very large literature on these structures; for the purpose of this brief discussion it is enough to note that their properties and their role in various types of movement in plant cells are reviewed in detail by Hepler and Palevitz (1974). Although the two systems are structurally and chemically quite distinct, there are similarities in the chemical cycles thought to be involved in the generation of motive force by each system. It seems likely therefore that any other process in which chemical events in the cell generated a force for movement, might be associated with one or other of these systems, or with structures having some general similarity with microfilaments or microtubules.

(1) Microtubules

A detailed description of the chemistry and properties of microtubules is given in the review cited (Hepler and Palevitz, 1974). Microtubules are tubules 24 nm in diameter, with a dense cortex and lighter core; the cortex is made up of subunits of the protein tubulin, each 4-5 nm in diameter, arranged as 13 laterally associated protofilaments, each staggered with respect to its neighbours. Cross-bridges, 2-5 nm thick and 10-40 nm long, are associated with the outer wall, and provide links between adjacent microtubules, or between a microtubule and the membrane of another cell structure (vesicles, mitochondria, nuclear envelope, endoplasmic reticulum, plasmalemma). The ATP-ase activity associated with the tubule may arise from a component similar to the dynein ATP-ase from flagella, and which may be the cross-bridges. It is argued that a chemical cycle involving cross-bridge ATP-ase activity could be responsible for generating motion. The composition and properties of tubulin are now well characterized, and the role of microtubules in a range of cell processes in both plant and animal cells is well established. If we are looking for similarities in function in an unknown system three properties of microtubules should be borne in mind - the importance of Ca levels in their formation (the assembly requiring $Ca < 10^{-6}$ M), their ATP-ase activity, and its activation by Ca, and their sensitivity to colchicine and to vinblastine sulphate.

(2) Microfilaments

Again the review by Hepler and Palevitz (1974) provides a detailed description, and the properties of microfilaments are also covered in a review on the role of actin and myosin in cell movement by Pollard and Weihing (1974). It is now established that a wide variety of cell movements, in both animal and plant cells, are mediated by 5-8 nm microfilaments, and that these may be identified with the actin part of a cytoplasmic system, with properties similar to those of actomysin in muscle.

The essential features of all such systems are the presence of actin and myosin, and the operation of a repeated chemical cycle involving their interaction and the splitting of ATP, by which motion is generated. The function form of actin, F-actin, consists of two helically wound chains of globular G-actin sub-units (molecular weight 46,000); the polymerization requires high ionic stength and the presence of ATP, and each monomer in the filament binds 1 mole of ADP. The resulting filament is polar, and has binding sites for myosin with a spacing corresponding to the half pitch of the double helix, namely 35-37 nm. The best diagnostic for the involvement of actin in such structures is the ability of the filaments to bind the protein sub-unit, heavy meromyosin (HMM), the head unit of the monomer of muscle myosin. Because of the polarity of the actin filament, and the fact that the actin-myosin bridges form at an acute angle, the attachment of HMM to actin filaments forms a characteristic arrowhead pattern, a "decorated" filament, which can be visualized in the electron microscope. By this means the role of actin filaments (and therefore presumably of myosin) in a number of motive forces has been established. Apart from muscle these include movements in a number of vertebrate cytoplasmic systems (in platelets, granulocytes, fibroblasts, brain), amoeboid movement, and cytoplasmic streaming. For the purpose of this discussion the two most relevant examples are streaming in the slime mould Physarum polycephalum and in the giant characean internodes such as Nitella or Chara. Observations which show the association of fibrils with streaming in the slime mould are summarized by Wohlfarth–Botterman (1964), Kamiya (1968), and Komnick, Stocken and Wohlfarth–Botterman (1937). The suggestion is that a network of microfilaments in the cortical cytoplasm might contract, thereby generating a pressure gradient within the cell. The filaments can be decorated with HMM (Nachmias, Huxley and Kessler, 1970), and both actin and myosin have been extracted from Physarum for biochemical characterization (Hatano and Oosawa, 1966a,b; Adelman and Taylor, 1969a,b; Hatano and Ohnuma, 1970). In Nitella a parallel array of bundles of anchored microfilaments, lying in the interface between the cortical gel layer and the streaming sol, seems to be responsible for the generation of a shearing force, by which the contents within the gel layer are maintained in continuous circulation (Kamiya and Kuroda, 1956; Donaldson, 1972; Nagai and Rebhun, 1966). The filaments have recently been identified as actin by decoration with HMM (Palevitz, Ash and Hepler, 1974; Williamson, 1974), and the presumption is that myosin is also present, although neither component has yet been extracted and purified for biochemical work.

The essential common feature of such actomyosin systems seems to be the cyclic interaction between myosin cross-bridges and an actin filament; the cycle involves the dissociation of actin from myosin in the presence of ATP, the splitting of ATP to give a myosin-products complex, the release of ADP and phosphate from myosin as the myosin rebinds to the actin filament, the further dissociation of myosin from the actin filament at the initiation of the next cycle. Thus, ATP serves as the initiator of the cycle by dissociating actin and myosin and also as the energy source; the cycle of myosin detachment from the actin filament, and transfer of the myosin from the myosin-products complex to form a new acute-angled actomysin link, involves a rotation of the myosin head unit with respect to the actin filament, a translation of myosin with respect to the actin. The reactions of the cycle are often sensitive to Ca^{++} levels, but in different systems the sensitivity may be conferred, either by a light chain of the myosin molecule, or by a troponin-tropomyosin complex on the actin filament. The relative movement of actin and myosin is responsible for whatever type of motion is generated in the system, depending on the organization of the two interacting components. Bipolar myosin fibrils interacting with two attached, oppositely polarized actin filaments in striated muscle, produce relative sliding of the two types of fibril, and hence contraction. Other arrangements could produce other results (Pringle, 1968). Myosin dimers interacting with polarized actin filaments attached to the cell surface could produce a shape change; myosin dimers (or bipolar aggregates) interacting with a loose meshwork of actin filaments could produce a contraction of the network and generate an internal pressure (as in the slime mould streaming); myosin monomers reacting with actin filaments could simply give an impulse to the surrounding solution, and generate shearing force along an array of similarly polarized actin filaments as in Nitella.

Thus it is possible to imagine a variety of arrangements of the essential components of the actomyosin system which could drive flow of one kind or another, and the question is how such a system could be recognized. Relevant attributes would include the presence of actin filaments, recognizable by HMM binding, and containing also bound nucleotide, the presence of a myosin component though not necessarily in a highly organized form, ATP-ase activity, sensitivity to ionic concentrations and particularly to Ca^{++}, precipitation by vinblastine sulphate, and sensitivity to cytochalasins. However, the question of specificity of cytochalasin inhibition is controversial - see reviews by Wessells et al (1971), Carter (1972), Hepler and Palevitz (1974), Pollard and Weihing (1974). While cytochalasin undoubtedly does inhibit actomyosin-mediated processes in cells (locomotion, phagocytosis, pinicytosis, cytokinesis, cytoplasmic streaming, various morphogenetic movements involving thin filaments) there are also reports of effects on membranes, and on specific transport processes. A clear example of this kind is the inhibition by cytochalasin B of the photopolarization, germination and apical hair initiation in developing eggs of _Pelvetia_ _fastigiata_ (Nelson and Jaffe, 1973); these are thought to be surface-mediated processes and the cells do not show cytoplasmic streaming. Among specific transport processes inhibited by cytochalasin B are the uptake of a variety of small molecules including glucose, deoxyglucose, methylglucose, glucosamine, uridine, thymidine (references in the reviews of Hepler and Palevitz (1974) and Pollard and Weihing (1974)). Thus it is likely that an actomyosin-mediated system will show sensitivity to cytochalasin _in vivo_, but a sensitivity to cytochalasin may or may not indicate a role for actomysin in the process.

(3) Possible arrangements of motile systems in sieve tubes.

A number of authors have suggested generally that P-protein filaments might be involved in the generation of motive flow, and have discussed how such motile machinery might be organized. The suggestions differ in the type of flow pattern which could be produced, but since there is no universal agreement over what is the real pattern of flow in the phloem there is no restriction on such speculation.

Weatherley and Johnson (1968) suggest that the fluid in the sieve tubes might be propelled by contraction or movement of the filaments in some way, and Wooding (1971) also speculated on the possibility that P-protein filaments might produce directed movement. In an earlier review (MacRobbie, 1971) I discussed possible arrangements in sieve elements, and suggested that the two most likely possibilities were (a) a strongly longitudinal orientation of P-protein fibrils, capable of producing a shearing stress in the fluid surrounding them, providing a system working basically like that in <u>Nitella</u>, or (b) an arrangement of P-protein filaments more like that in the slime mould, in which a contraction of a fibrillar network generates local internal pressure gradients. In either case some means of imposing control of the induced flow is required, since we know that the rate and direction of translocation are controlled by the source/sink relations of the phloem. Thus it is essential in any model in which the force is generated internally to include some provision for controlling the orientation and polarity of the force-generating system. I suggested that one way in which this might be achieved is by using a Münch-type mechanism to orient the filaments. If sucrose is pumped into the sieve elements in the source region, and is removed in the sink region, then inevitably there will be a mass flow within the conduit of the Münch-type. If polarized motile filaments then orient themselves in the flow (Huxley, 1963) then the initial Münch flow can be held responsible for polarizing the force generating system for much more effective movement. In the absence of any net gradient a more random arrangement of filaments might be found, with the possibility of regions of local order, and two-way movement by the presence of oppositely directed fibrils in the same sieve element, or in different sieve elements. But with a marked source/sink gradient imposed such a system would be a one-way system, with a strongly polarized force-generating system. In either of these arrangements the force-generating filaments are anchored to some lateral structure and do not move with the stream, thereby avoiding delivery problems at the sink end.

If the filaments are organized as a loose network and generate internal pressure gradients it is less easy to see how the flow pattern can be controlled by the source/sink relations. In the slime mould's rhythmic streaming some sort of cycle of polymerization/depolymerization of the actin filaments seems to be envisaged, the transition being sensitive to some factor of the local internal environment (Ca^{++}?). In phloem we might imagine waves of the cycle of polymerization/depolymerization travelling down the sieve elements, the result of travelling waves of some factor affecting the transition this could be action potentials and consequent changes in Ca^{++}, for example. It might then be possible to argue that these were initiated at the source end of the sieve tube by a pressure-induced instability of the membrane potential. This provides a type of peristaltic network rather than peristaltic tubule. Alternatively we might argue that the internal pressure-generation simply provides a means of maintaining and extending the pressure generated at the source end by the Münch-type entry of water to the sieve element, and avoiding its rapid dissipation in driving the flow through a high longitudinal resistance.

    Another proposal for force generation within sieve elements is
based on a very different organization of the sieve tube contents -
that put forward by Thaine. His proposals are that translocation
takes place through transcellular strands 1-7 µm in diameter, which
pass through the sieve plates and are continuous over many sieve
elements. The existence of such strands was proposed by Thaine
(1961, 1962), disputed by Esau et al (1963), and defended by
Thaine, Probine and Dyer (1967) and by Jarvis, Thaine and Leonard
(1973). A mechanism for translocation involving such strands was
put forward (Thaine, 1964, 1969; Thaine and Preston, 1964), the
suggestion being that contractile protein filaments in the outer
wall of a complex strand (5 µm in diameter with internal
endoplasmic tubules 1 µm in diameter) are responsible for
peristaltic pumping of the tubule contents. Thaine believes that
P-protein as normally visualized in the electron microscope, should
be identified with the damaged residue of the strand system.

    Aikman and Anderson (1971) have made a theoretical analysis of
such a peristaltic system which shows that if reasonable values are
taken for the variables of the peristaltic wave, reasonable values
for the flow velocity and rate of energy dissipation are obtained.
Hence they argue that the system could work, but recognize that the
hypothesis stands or falls on fine structural evidence for or
against the existence of the structures required. The kinetics of
distribution of material in a system of this kind have been
considered by Canny and Phillips (1963), and by Canny (1973);
however, examination of such tracer profiles, in the light of the
observed kinetics of loading in the phloem (Evans et al, 1963;
Fisher, 1970), suggests that the boundary conditions chosen by
Canny and Phillips are invalid (i.e. the concentration in the
phloem at the imput end cannot be taken as constant over the
experimental times) (MacRobbie, 1971). The mathematical form of
such profiles, and their close fit to complementary error
functions, cannot be taken as evidence for the strand model of
translocation.

    A peristaltic model is also proposed by Fensom (1972) with
further elaboration. He believes that P-protein, or
"microfibrillar material", exists in intact sieve elements in the
form of lipoprotein helices, forming a system of tubules, capable
of peristalsis (Robidoux et al, 1973). Fensom envisages two types
of flow - the movement of pulses of sucrose and amino acids within
these tubules, in an inner compartment of the sieve element, driven
by peristaltic contractions of the tubule wall, but also a mass
flow of the rest of the sieve tube contents impelled by the
peristaltic waves passing down the tubule. In support of this
hypothesis he cites the results of micro-injection experiments in
phloem strands of Heracleum, intact in the sense of being attached
to the plant at both ends, and of ^{14}C-sucrose feeding to a region
of such phloem strands (either attached at both ends, or detached
from the petiole); in both instances he claims the existence of a

pulsed flow of ^{14}C label, moving ahead of a mass flow component. There are problems with the injection experiments, in that the injection of between 0.002-0.5 µl of 0.3 M sucrose in 10-15 s into a sieve element of diameter 20 µm is not a trivial event; this volume represents 20-5000 sieve elements, and implies a linear velocity of some 150-3600 cm h^{-1}. High rates of mass flow might therefore be expected, irrespective of translocation, provided the pores were open. The demonstration that carbon black particles (20-70 nm in diameter) injected in this system can travel through sieve plates indicates that the pores are indeed open (Barclay and Fensom, 1973). The evidence for pulses is not strong in the published figures; in the injection experiments pulses seem to be doubtfully significant, given background counting levels, and the possibilty of unequal blocking of different sieve tubes in the feeding experiments might lead to unevenness in the tracer profiles. Further and more sensitive experiments for the detection of pulses seem to be required before we can accept them as firmly established. In any case there is controversy over the ability of such manipulated phloem strands to translocate; Hoddinott and Gorham (1974) report that although isolated, but attached, phloem loops will take up and transport tritiated sucrose applied to the loop, they are not involved in the transolocation through the petiole of ^{14}C-assimilates from the leaf to which they are attached.

Fensom and Williams (1974) have further elaborated on this type of model, using the system of undulating filaments proposed for <u>Nitella</u> by Allen (1974), who has also suggested that similar systems might be involved in phloem. Fensom and Williams propose a system of axial hollow tubules, to which flagella-like branches are attached; the waving filaments are held to generate mass flow in sieve element, while microperistalsis in the hollow tubules produces pulses of sucrose moving ahead of the mass flow. They argue that the energy dissipation in this system is much less than that for peristalsis in 60 nm tubules alone, and that there is also better evidence for fibrillar material in sieve elements than for large membrane-bound tubules.

A rather different view of the role of P-protein in generating flow was put forward by Hejnowitz (1970). He suggested an electrical wave travelling along P-protein fibrils or tubules, generating an electro-osmotic flow. This avoids the difficulties of ion recirculation outside the phloem inherent in Spanner's model, but if we imagine ion circulation across the tubule walls then there are still problems. An alternative is to postulate a wave of surface charge travelling along a solid fibril, akin to the proposals of Ambrose (1972) for the actomyosin system. Ambrose suggests that changes in surface charge inherent in the cyclical actin/myosin/Ca^{++}/ATP interactions can lead to a travelling electrical disturbance on the actin filament, generating electrokinetic forces in the fluid around; he considers that this is a more likely mechanism for force-generation in the actomyosin system than the mechanical forces of rotating myosin heads. If the P-protein were endowed with Ca-dependent ATP-ase activity then the model could be transferred directly.

The problem remains of how an activated mass flow could be achieved in gymnosperms in the absence of P-protein, but in the presence of abundant ER through the pores. It would be feasible to imagine filamentous protein material associated with the ER which acted in a similar way. Organelles can become associated with the microfilaments in Nitella and move along their length, and it appears that the actomyosin system can therefore be membrane-associated, providing at least transient associations between structures. Short microfilament lengths bound to the ER might generate motive force. There is no evidence for this, but since Münch flow problems are more serious in gymnosperms, it remains a possibility worth exploring.

(4) Evidence for or against activated mass flow.

I would like in this section to consider first the evidence that P-protein does, or does not, have the properties expected if it is to play a role in generating motive force for the translocation flow. Most of the speculations over such mechanisms have been derived from the existence of large amounts of this specific protein (or proteins) in phloem, but closer examination of its properties, as these emerge, does not always strengthen the speculation. The review by Hepler and Palevitz (1974) gives a brief discussion of the current state of the comparisons between P-protein and the other "motile" proteins, tubulin and actin.

P-protein is frequently aggregated into fibrils, although it also exists in tubular or crystalline forms (Cronshaw and Esau, 1967; Parthasarathy and Mühlethaler, 1969); it is polymerized by KCl, it is precipitated by Ca^{++} and vinblastine sulphate (Kleinig et al, 1971; Weber and Kleinig, 1971). More important, however, are the use of more specific reagents for actin or tubulin. Thus, Wooding (1969) and Kleinig et al (1971) found no colchicine binding to P-protein, suggesting it is not tubulin. Williamson (1972) was unable to decorate P-protein filaments with HMM, and Hepler and Palevitz (1974) quote unpublished results in which crystalline P-protein was similarly unable to bind HMM. However, the report at this meeting of the successful decoration of P-protein filaments with HMM is very exciting for the hypothesis of activated mass flow (Ilker and Currier, 1974). The suggestion is that P-protein filaments in glycerinated hand sections of Vicia faba and Xylosma congestum can form regular arrowhead complexes with HMM; such P-protein filaments are seen in bundles with an unidirectional polarity, and those near the surface are bound to the plasmalemma. It is suggested that only the filamentous form of P-protein can bind HMM, whereas the tubular form cannot, and that this may explain the previous negative reports.

In other respects the reported properties of P-protein extracted from phloem exudates are also deficient for the purposes of these hypotheses. Kleinig et al (1971) found no bound nulceotide and no ATP-ase activity in P-protein from Cucurbita exudates, and the more extensive study by Weber et al (1974) reaches the same conclusions. On the other hand Gilder and Cronshaw (1973a,b) argue from the results of their cytochemical studies on tobacco phloem that there is ATP-ase activity associated with P-protein in situ. There is therefore, a degree of uncertainty over whether the activity is real, but belongs to a fraction which is left in the sieve elements and does not appear in the exudate, or whose activity is lost in the extraction or under the assay conditions, or whether the cytochemical demonstration is unreliable, reflecting non-specific adsorption of particulate phosphate materials by P-protein, as argued by Weber et al. The first argument, that the essential fraction is left behind, could probably only be made with respect to the very high protein exudate from Cucurbita, and is in any case a weak one.

The fractionation studies on P-protein also provide little or no evidence for actin in phloem exudates (Walker and Thaine, 1971; Walker, 1972; Weber and Kleinig, 1971; Weber et al, 1974). Walker found only a small fraction of gelling protein, PG-protein, in his extracts (up to 4%), and one small component of the structural PF protein has a molecular weight similar to actin. Weber et al (1974) found most or all of the protein in their extract from pumpkin fruit exudates to consist of basic gelling proteins, and none of their eight fractions resembles actin. A striking feature of the last work is the high isoelectric point of the protein fractions, at pH 9.6 to 10.4, implying that they will be positively charged at the normal pH in sieve elements. This contrasts with the existence of an acidic protein in the Cucurbita stem exudate studied by Walker, and the acidic nature of the protein extracted by Yapa and Spanner (1972) from Heracleum. But the most important aspect of the detailed analysis of the fruit exudate by Weber et al is the very close similarity (both biochemical and immunological) between P-protein and the weakly basic components of ribosomal protein. The significance of this for the role of P-protein remains to be established, but it certainly does not suggest a type of "motile" protein for the main bulk of protein in Cucurbita.

On the physiological side we might look for evidence that interference with metabolism along the path does, or does not, interfere with translocation. With this aim a number of workers have looked at the effects of inhibitors of various kinds, or of local temperature changes, applied to the translocation path.

There is one report of cytochalasin sensitivity, but as we have already seen, this is not necessarily a reliable diagnostic of a role for actomyosin in the sensitive process. Thompson and Thompson (1973) found that cytochalasin B inhibited translocation, when applied 10 cm from the region of feeding ^{14}C-sucrose to isolated but attached phloem strands of Heracleum - but only if applied for 10 minutes before tracer feeding. In contrast, Williamson (1972) found no effect of cytochalasin, applied to the translocation path, on translocation of ^{14}CO$_2$-labelled assimilate in Lepidium phloem. It may well be, therefore, that the effect in Heracleum was on the loading process, the entry of externally applied sucrose into the phloem, rather than on translocation as such, and further work seems to be required.

Other inhibitor work has been held to support the hypothesis of some sort of activated mass flow, and to be difficult to reconcile with the Münch hypothesis. The effects of anoxia along the path vary. Thus Willenbrink (1957) and Ullrich (1961) found no inhibition by anoxia of fluorescein transport in the phloem of Pelargonium, although Mason and Phillis (1936) argued from their work on cotton that oxygen was required for translocation. Qureshi and Spanner (1973a) have looked in detail at the effects of nitrogen on translocation in the stolon of Saxifraga, from parent plant as source to daughter plant as sink; they find that a nitrogen sleeve along the path does inhibit translocation, that the effect is on the path and not on source or sink, that it is reversible, and that callose blockage does not seem to be involved (Qureshi and Spanner, 1973b). Similar inhibitions in the same system by cyanide or DNP applied to the path have also been shown (Qureshi and Spanner, 1973b,c,d); the effects were again reversible, and again not attributable to effects on source or sink, or to callose formation. In the work on cyanide it was shown that labelled cyanide was not transported to the source, and only in a very limited amount to the sink where its concentration was only 1/2000 of that in the zone where it was applied. The translocation path beyond such inhibited regions seems to be unimpaired, in that it is capable of taking up ^{14}C-sucrose and delivering it to the sink. There are also earlier reports of inhibition of translocation by cyanide (Willenbrink, 1966, 1968; Ho and Mortimer, 1971), although earlier effects of DNP are more variable. There are also reports of stimulatory effects of light in the path region on translocation (Hartt and Kortschak, 1967; Spanner, 1973a). Thus, the conclusion from a number of studies is clear, that normal metabolism in the translocation tissues along the path is necessary for translocation, and the effect is not mediated by source or sink activity, or by callose deposition. If we do not wish to accept this as evidence for the operation of an activated mass flow mechanism then we must produce a credible explanation in some other terms compatible with the Munch hypothesis.

In contrast to this inhibitor work a range of studies, in which localized temperature changes are made along the path, seem to show that normal metabolic activity in the conduit is not necessary for the flow, that the effects of temperature on translocation are physical rather than chemical. The effect of cooling a length of the path is frequently transient, with recovery to original rates of flow in a relatively short time (Geiger, 1969; Geiger and Sovonick, 1970; Webb, 1971; Lang, 1974). There are two effects of cooling. In chilling-sensitive plants at low temperatures, for example bean below 10°C, there is physical obstruction of sieve plates, seen by comparison of the pores in freeze-substituted sieve elements (Giaquinta and Geiger, 1973); in chilling-resistant plants (such as sugar beet or willow) very low temperatures are required for this response. Giaquinta and Geiger suggest that the critical temperature is a function of the composition of membrane lipids, and that the response is the result of cytoplasmic changes following on a temperature-induced phase change in the membrane lipids, which then induce a flow of cytoplasmic contents (including organelles, P-protein, membraneous material) into the pores. The second effect is of more interest for this discussion, and represents the effect of temperature changes less drastic than those required to effect this blockage. Here the effect is agreed to be relatively small; for example, after the transient response to cooling, or on warming, the temperature coefficient for translocation in Nymphoides petiole was consistent with the physical effects of temperature on viscosity (Lang, 1974). Thus, even above the critical temperature, the effects are most easily explained in terms of an increase in resistance, by a degree of blocking of the pores, and a change in viscosity; they do not lend support to the hypothesis that metabolic energy along the path is required. The results of Coulson et al (1972) suggested this strongly; they measured rates of translocation, respiration, and ATP levels, in the petiole of sugar beet after its cooling, and found that the recovery of translocation to near its former rate was not associated with any recovery of respiratory rate. Thus, transloation was unimpaired even with the respiration reduced to 10% of normal.

There is therefore, an unexpected divergence of the results of metabolic interference by inhibitors and by cooling. If we believe in Münch flow we have to explain the reduction of flow by inhibition along the path, remote from source or sink, but on the other hand if we believe in activated mass flow we have to explain its insensitivity to cooling the path.

C. Final comments

This paper has tried to present a summary of those mechanisms for translocation (other than electro-osmosis) which involve P-protein directly in the force-generation, to discuss how such a process might work, and to consider the experimental evidence for or against such activated mass flow. This evidence is not conclusive. If the sieve pores are open then Münch flow is adequate for translocation, but there remain features of phloem behaviour which require explanation. If the pores are not open, then we do need an alternative to Münch flow, and in my view the hypothesis that P-protein is involved in force generation, of a kind similar to that in other motile systems, offers the best alternative. Although I feel the verdict is still "not proven", it is fair to say that the studies over the past few years designed to test hypotheses of this kind have not on the whole, strengthened them. In the next paper I would like to try to assess the relations between such activated mass flow and the inevitable Münch flow in the conduits, to examine the problems of trying to distinguish them, and to consider the aspects of phloem behaviour that remain unresolved or unexplained on each of these hypotheses.

References

ADAM, G. and DELBRUCK,M. 1968. Structural Chemistry and Molecular Biology, ed. A. Rich and N. Davidson, W.H. Freeman. p.198

ADELMAN, M.R. and TAYLOR, E.W. 1969a. Biochemistry 8: 4964

ADELMAN, M.R. and TAYLOR, E.W. 1969b. Biochemistry 8: 4976

AIKMAN, D.R. and ANDERSON, W.P. 1971. Ann. Bot. 35: 761

ALLEN, N.S. 1974. J. Cell Biol., in press.

AMBROSE, E.J. 1972. Motile Systems of Cells, ed. S. Dryl and J. Zurzycki, Polish Academy of Sciences, Warsaw. p. 9

ANDERSON, R. and CRONSHAW, J. 1970. Planta 91: 173

BARCLAY, G.F. and FENSOM, D.S. 1973. Acta Bot. Neerl. 22: 228

CANNY, M.J. 1973. Phloem Translocation. University Press, Cambridge.

CANNY, M.J. and PHILLIPS, O.M. 1963. Ann. Bot. 27: 379

CARTER, S.B. 1972. Endeavour 31: 77

COULSON, C.L., CHRISTY, A.L., CATALDO, D.A. and SWANSON, C.A. 1972. Pl. Physiol. 49: 919

CRONSHAW, J. and ESAU, K. 1967. J. Cell Biol. 34: 801

DONALDSON, I.G. 1972. Protoplasma 74: 329

ESAU, K. 1969. The Phloem. Encyclopedia of Plant Anatomy V.2, Gebruder Borntraeger, Berlin and Stuttgart.

ESAU, K., ENGLEMAN, E.M. and BISALPUTRA, T. 1963. Planta 59: 617

ESCHRICH, W. 1970. A. Rev. P. Physiol. 21: 193

EVANS, N.T., EBERT, M. and MOORBY, J. 1963 . J. Exp. Bot. 14: 221

FENSOM, D.S. 1972. Can. J. Bot. 50: 479

FENSOM, D.S.   and DAVIDSON, H.R.   1970.   Nature 227:   857
FENSOM, D.S.   and WILLIAMS, E.J.   1974.   Nature 250:   490
GEIGER, D.R.   1969.   Ohio J.   Sci.   69:   356
GEIGER, D.R.   and SOVONICK, S.A.   1970.   Pl.   Physiol.   46:   847
GIAQUINTA, R.T.   and GEIGER, D.R.   1973.   Pl.   Physiol.   51:   372
GILDER, J.   and CRONSHAW, J.   1973a.   Planta 110:   189
GILDER, J.   and CRONSHAW, J.   1973b.   J.   Ultrastruct.   Res.   44:
     388
HAND, J.H.   and WILLIAMS, M.C.   1970.   Nature 227:   369
HARTT, C.E.   and KORTSCHAK, H.P.   1967.   Pl.   Physiol.   42:   89
HATANO, S.   and OHNUMA, J.   1970.   Biochim.   Biophys.   Acta 205:
     110
HATANO, S.   and OOSAWA, F.   1966a.   Biochim.   Biophys.   Acta
     127:488
HATANO, S.   and OOSAWA, F.   1966b.   J.   Cell Physiol.   68:   197
HEJNOWICZ, A.   1970.   Protoplasma 71:   343
HEPLER, P.K.   and PALEVITZ, B.A.   1974.   A.   Rev.   Pl.   Physiol.
     25:   309
HODDINOTT, J.   and GORHAM, P.R.   1974.   Can.   J.   Bot.   52:   349
HO, L.C.   and MORTIMER, D.C.   1971.   Can.   J.   Bot.   49:   1769
HUXLEY, H.E.   1963.   J.   Molec.   Biol.   7:   281
ILKSER, R.   and CURRIER, H.B.   1974.   Planta, in press.
JARVIS, R., THAINE, R.   and LEONARD, J.W.   1973.   J.   Exp.   Bot.
     24:   905
KAMIYA, N.   1968.   Symp.   Soc.   Exp.   Biol.   22:   199
KAMIYA, N.   and KURODA, K.   1956.   Bot.   Mag.   (Tokoyo) 69:   544
KLEINIG, H., DORR, I., WEBER, C.   and KOLLMANN, R.   1971.   Nature
     New Biol.   229:   152
KOMNICK, H., STOCKEM, W.   and WOHLFARTH-BOTTERMANN, K.E.   1973.
     Int.   Rev.   Cytol.   34:   169
LANG, A.   1974.   J.   Exp.   Bot.   25:   71
LEE, D.R.   1972.   Nature 235:   286
MACROBBIE, E.A.C.   1971.   Biol.   Rev.   46:   429
MASON, T.G.   and PHILLIS, E.   1936.   Ann.   Bot.   50:   455
MILBURN, J.   1971.   Planta 95:   272
NACHMIAS, V.T., HUXLEY, H.E.   and KESSLER, D.   1970.   J.   Molec.
     Biol.   50:   83
NAGAI, R.   and REBHUN, L.I.   1966.   J.   Ultrastruct.   Res.   14:
     571
NELSON, D.R.   and JAFFE, L.F.   1973.   Develop.   Biol.   30:   206
PALEVITZ, B.A., ASH, J.F.   and HEPLER, P.K.   1974.   Proc.   Natn.
     Acad.   Sci.   U.S.A.   71:   363
PARTHASARATHY, M.V.   and MUHLETHALER, K.   1969.   Cytobiologie 1:
     17
POLLARD, T.D.   and WEIHING, R.R.   1974.   C.R.C.   Critical Reviews
     in Biochem.   2:   1
PRINGLE, J.W.S.   1968.   Symp.   Soc.   Exp.   Biol.   22:   67
QURESHI, F.A.   and SPANNER, D.C.   1973a.   Planta 110:   131
QURESHI, F.A.   and SPANNER, D.C.   1973b.   Ann.   Bot.   37:   867
QURESHI, F.A.   and SPANNER, D.C.   1973a.   J.   Exp.   Bot.   24:   751

QURESHI, F.A.   and SPANNER, D.C.   1973d.   Planta 111: 1
ROBIDOUX, J., SANDBORN, E.B., FENSOM, D.S. and CAMERON, M.L.
    1973. J. Exp. Bot. 24: 349
THAINE, R.   1961. Nature 192:772
THAINE, R.   1962. J. Exp. Bot. 13: 152
THAINE, R.   1964. J. Exp. Bot. 15: 470
THAINE, R.   1969. Nature 222: 873
THAINE, R. and PRESTON, R.D. 1964.   The Formation of Wood in
    Forest Trees, ed. M. Zimmermann, Academic Press, New York. p.
    259
THAINE, R., PROBINE, M.C. and DYER, P.Y.   1967.   J.   Exp.   Bot.
    18: 110
THOMPSON, R.G. and THOMPSON, A.D.   1973. Can. J. Bot. 51: 933
ULLRICH, W.   1961. Planta 57: 402
van den HONERT. 1932. Proc. Ned. Akad. Wetensch. 35: 1104
WALKER, T.S.   1972. Biochim. Biophys. Acta 257: 433
WALKER, T.S. and THAINE, R.   1971. Ann. Bot. 35: 773
WEATHERLEY, P.E. and JOHNSON, R.P.C.   1968.   Int.   Rev.   Cytol.
    24: 149
WEBB. J.A.   1971. Can. J. Bot. 49: 717
WEBER, C. and KLEINIG, H.   1971. Planta 99: 179
WEBER, C., FRANKE., W.W. and KARTENBECK, J.   1974. Exp.   Cell
    Res. 87: 79
WESSELS, N.K., SPOONER, B.S., ASH, J.F., BRADLEY, M.O., LUDUENA,
    M.A., TAYLOR, E.L., WRENN, J.T. and YAMADA, K.M.   1971.
    Science 171: 135
WILLENBRINK, J.   1957. Planta 48: 269
WILLENBRINK, J.   1966. Z. Pflanzenphysiol. 55: 119
WILLENBRINK, J.   1968. Vortr. Gesamt. Bot. 2: 42
WILLIAMSON, R.E.   1972. Planta 106: 149
WILLIAMSON, R.E.   1974. Nature 248: 801
WOHLFARTH-BOTTERMANN, K.E.   1964. Primitive Motile Systems in Cell
    Biology, ed. R.D.Allen and N.   Kamiya, Academic Press, New
    York. p. 79
WOODING, F.B.P.   1969. Planta 85: 284
WOODING, F.B.P.   1971. Phloem.   Oxford Biology Readers, Oxford
    University Press, Oxford.
YAPA, P.A.J. and SPANNER, D.C.   1972. Planta 106: 369

Mechanisms: Comparative Behaviour

E.A.C. MacRobbie

Botany School, University of Cambridge

Cambridge, England

A.   Introduction

The aim of this paper is to consider the relations between the different hypotheses, the problems of trying to distinguish between them by critical experiments, and the aspects of phloem behaviour, or of the characteristics of the sieve elements, which remain unresolved or unexplained on each of these hypotheses.   In particular I would like to examine the relation between any form of activated mass flow and the inevitable Münch flow in the conduits.

The first critical question is that of the site of generation of motive force for the flow in the sieve tubes - is it at the ends of the path, or is it also augmented along the path?   There is general agreement that solute pumping into the phloem in the source region, and its removal in the sink region, will create conditions for Münch flow; there will inevitably be generated a pressure gradient within the sieve tube moving its fluid contents as a whole.   The question is whether this will be adequate for the rates of flow observed, and the answer is determined by the longitudinal resistance of the conduits, by the nature of contents of the pores in the sieve plates.   If the pores are open then their longitudinal conductance is high enough, given reasonable values for the water permeability of the lateral membranes of the conduit, for the observed solute and volume flows to be achieved by feasible values for concentration and pressure gradients within the sieve elements (Christy and Ferrier, 1973).   A problem arises only if the pores are blocked with P-protein or other material, and have a very high longitudinal resistance, so that the flow generated in the sieve elements in the source is dissipated very rapidly by frictional resistance.   In this case we do need some mechanism for enhancing the Münch flow, for maintaining the driving force over the length of the path, and this must come from within the sieve tubes by some

601

sort of activated mass flow - by electro-osmosis or by the activity of some sort of "motile" protein. It is worth considering the differences between these hypotheses, and looking for ways of distinguishing them. It turns out to be more difficult than one might think - partly because one can rewrite the explanations for a variety of observations on phloem in different forms, depending on one s belief for the mechanism.

B. Münch Hypothesis

We may first consider the Münch hypothesis, since if we can believe that the phloem properties are compatible with the prerequisites for Münch flow, then our problems are solved. For this we require effectively open pores, and we envisage the energy for flow supplied by the activities of the source and sink at the ends of the path, by processes of active solute flux at source and perhaps also in the sink; no direct energy input along the path is required. We are then required to provide explanations for certain specific features of the phloem, in which it differs from other tissues of the plant. In my view these include the very odd ionic relations of the sieve elements, the presence of P-protein, and the question of its role in phloem.

C. Ionic relations

The question of the ionic relations of the sieve elements seems to me to be an important one. Sieve elements are very alkaline (pH 7.4-8.7) and are also characterized by odd ionic ratios. Thus they have K/Na ratios much higher than those in other cells of the plant, and the Mg/Ca ratio is equally extreme. Thus in phloem exudates K values are 20-85 mM, but Na levels are only 0.06-0.3 mM, much lower than in most plant cells; Ca is even more striking, with levels of only 0.25-0. mM, as against Mg of 2.3-23 mM (references in MacRobbie, 1971). It has been recognized for many years that Ca is immobile in phloem, by which we mean that it does not enter the translocation stream. However, in most cells the K/Na and Mg/Ca ratios are maintained by active transport, by the active extrusion of Na (with or without also the active uptake of K), and by the active extrusion of Ca, and the extreme ionic ratios seen in sieve elements would then seem to be an expensive luxury to maintain, unless there were some good reason. We need therefore to suggest some credible role for the very low Ca and Na levels in sieve elements, compatible with whatever mechanism for translocation we accept. In this connection also it is interesting that K deficiency in sugar cane has a strong inhibitory effect on translocation before there are any effects on net photosynthesis in the leaf (Hartt, 1970; Amir and Reinhold, 1971), so that some role for high K levels in maintaining translocation may also be involved.

It is easier to find possible roles for such strong ionic regulation within the sieve elements in terms of the various theories of activated mass flow, but we may consider possibilities in terms of the Münch hypothesis, with the aim of future experimental studies in mind. As far as Ca is concerned, some interaction between Ca and P-protein is likely to be involved, and this question should therefore be left until the role of P-protein has been discussed. It is possible also that the high K level is also related to the properties of P-protein, but the other possibility is that it reflects the activities of the loading system in the source. The role of Na^+ gradients in the uptake of sugars and amino acids by a number of animal cells is well recognized (see for example reviews by Stein (1967), Schultz and Curran (1970)), and there are indications that the gradients generated by active H^+ extrusion from cells may also be coupled to solute uptake processes; for example the β-galactoside transport system in E. coli has been linked to H^+ movement by West and Mitchell (1972), and a hexose-proton symport is postulated as the means of sugar uptake in Chlorella (Komor, 1973; Komor and Tanner, 1974a,b). Thus it is possible that the active transport systems, by which sucrose is loaded into the translocation system in the source, involve also ion movements which are then reflected in the K concentrations and/or pH of the sieve elements.

D.   Role of P-protein

A major consideration in any hypothesis for translocation must be to provide a role for P-protein, and in the Münch hypothesis the role suggested is in blocking the conducting system after damage (Anderson and Cronshaw, 1970; Milburn, 1971). But it seems to me that supporters of the Münch hypothesis must expand this suggestion before we can accept it as a solution to the problem. We need to have this explanation elaborated in terms of the special properties of P-protein in its pleiomorphic forms, and of the internal environment of the sieve element, as these become better known. Even if the pores are normally open in translocating sieve elements, then it is clearly very easy to block them, to a greater or lesser extent, with P-protein material - either by experimental manipulation of one kind or another, or by external influences in the intact system. Thus we need to postulate a very labile blocking agent, with the blocking process closely regulated and responsive to changes in the internal environment of the sieve elements resulting from their damage. The finding of Hoddinott and Gorham (1974) that isolated but attached Heracleum strands are cut off from the assimilate stream from the leaf, although still able to translocate internally, suggests that damage can result in the activation of the blocking response in sieve elements at a considerable distance from the region affected. Thus a stimulus which can be transmitted rapidly over considerable distances seems to be involved.

The possibility that a wide range of transmitted responses to external influences might be mediated by propagation of action potentials in plant tissues has been discussed in a review by Pickard (1973). It is perhaps relevant that action potentials in phloem parenchyma are thought to be involved in transmission of the motor response in Mimosa petioles (Sibaoka, 1962), and that propagated electrical responses in phloem tissue may be a more general feature of plants (Sinyukhin and Gorchakov, 1968; discussion in Pickard, 1973). It may be therefore that the extreme ionic gradients in sieve elements contribute to an ability to respond to external interference by a propagated electrical disturbance. Equally they may be involved also in the consequent response in the receiving tissue, for example in blocking of the sieve pores by P-protein. It is again easy to speculate that Ca might be important, that the state of P-protein might be sensitive to Ca levels; given the very low level of Ca in sieve elements there is no doubt that damage will result in its drastic change. Thus it is possible to elaborate on the blocking role hypothesis, in ways which will accommodate, and provide a role for, some of the oddities of sieve element properties. But what is required in some experimental indication that such speculations have any basis in fact.

One other problem should be discussed here, and that concerns the state and location of P-protein in intact sieve elements, on the basis of the Münch hypothesis and open pores. Large quantities of protein in the flow will pose problems in receiving sieve elements in the sink; the transit times for the longitudinal flow in sugar beet sieve elements come out to be about 5 sec per element; hence the pile-up in the sink end of the file will be very rapid. Thus we must envisage P-protein anchored, somehow, to the lateral structures of the sieve elements (membrane or organelles themselves anchored in gel). Again this could be under control by the internal environment in the element, but we need to consider how we believe this could be achieved.

The other postulate of the Münch hypothesis is that no direct energy input is required along the translocation path. As we say in the previous paper the experimental evidence bearing on this question is equivocal. Lowering the temperature produces a transient inhibition in rate, but the sustained effect is consistent with a physical rather than a chemical effect on the flow. We are however left with the problem of explaining the transient, in terms of a partal blocking of the pores, and have to explain why P-protein should behave thus. We also have to explain the inhibitor effects - that interference with energy production (rather than both production and consumption) does reduce translocation, again presumably by an effect on the physical state or distribution of P-protein, consequent on a change in the internal environment of the inhibited sieve elements.

The conclusion would seem to be therefore that the Munch hypothesis is tenable if the pores are open, but that it still needs a good deal added to it before it provides a complete explanation of the behaviour of translocating systems.

E. Activated mass flow

The essential feature of these hypotheses is that they include mechanisms for the generation of force within the sieve elements, but mechanisms which are sensitive to the gradients created by the distributions of source and sinks; in effect they include mechanisms for increasing the flow as the gradients increase, and for maintaining high rates of flow in the face of partly plugged pores, whose longitudinal resistance to Munch flow would be considerable. This is true whether we are discussing electro-osmosis in the form Spanner has now proposed, or systems of Nitella type fibrils oriented in the Munch flow, or a contracting fibrillar network with propagated waves of contraction sensitive to pressure in the source region. For the purposes of comparison with the Munch hypothesis with various mechanisms may be considered together; their effect is that volume flow throughout the conduit is no longer a linear function of the pressure gradient, but the longitudinal conductance (L_s) increases with the volume flow J_v, or with the gradient ΔP.

Before considering how this works in relation to source/sink activity we may simply note that the mechanisms of activated flow do provide obvious explanations for the oddities of phloem properties. The ionic relations, and particularly the low Ca level, are important for these hypotheses; in fact for any system involving "motile" proteins it would probably be necessary to invoke an even lower free Ca^{++} level in the sieve elements, achieved by a calcium-sequestering membrane system of the kind found in muscle or in the slime mould. Further, the great merit of these theories is that they give P-protein, the specific protein elaborated during the differentiation of sieve elements, a central role in the translocation mechanism.

The physiological experiments are, however, again equivocal; the inhibitor experiments are to be expected, but this time the temperature effects pose the problem. It would be expected that force generation within the sieve tubes would be much more affected by temperature than we find, although we shall see that we may be able to argue a way round this.

What I would like now to do is to consider how such Munch-polarized force generating systems would behave, and in particular to point out that in the end the flow is still controlled by the source/sink relations. For the purposes of this argument we may consider the simplified system in Figure 1, in which a source, with pressure P_1, and solute concentration C_1, accumulates solute actively from surrounding tissue of water potential ψ_{01} ; it is connected by a pipe impermeable to both water and solute, to a sink region, with pressure P_2 and solute concentration C_2, which supplies solute to the surrounding tissue of water potential ψ_{02}. For the purpose of examining the general behaviour, in qualitative terms, it is convenient to use this fiction, that there are neither water nor solute fluxes across the lateral membranes in the path region. In the Münch hypothesis the longitudinal conductance, L_s, is independent of flow, and is a function only of the geometry of the sieve elements. In the activated mass flow system there are mechanisms for generating force along the path, but these are sensitive to flow, or to the overall gradient. This means that the apparent value of L_s is a function of J_v, of the pressure gradient ΔP; J_v is no longer a linear function of ΔP, but increases more steeply.

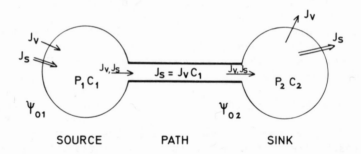

Fig.1. Diagram of steady state flow model where J_s is the solute flux, J_v is the volume flux, P is the pressure, C is the concentration, and ψ is the external water potential.

But consider how the system will work, and how a steady state flow will be achieved, starting from rest. The flow starts with the active influx of solute, J_s, into the sieve elements of the source, thereby increasing their concentration C_1. A volume flow is induced and the hydrostatic pressure in the sieve tubes increases. At the source end we may write:

$$(J_v)_{in} = L_p (\psi_{01} - P_1 + RT\,C_1)$$

where L_p is the hydraulic conductivity of the loading membranes. The volume and solute flows along the path are given by:

$$(J_v)_{path} = L_s(P_1 - P_2) = f(J_v) \cdot (P_1 - P_2)$$

since we are considering L_s, the apparent longitudinal conductivity, to be a function of the flow.

$$(J_s)_{path} = (J_v)_{path} \cdot C_1$$

With the arrival of this flow at the sink end of the conduit, both pressure and concentration will increase there, and outward flows of both volume and solute will follow; the volume flow is determined by the water potential gradient between phloem and sink tissue, the solute flux will be a complex function of solute concentration C_2. Thus we have:

$$(J_v)_{out} = L_p' \, (P_2 - RTC_2 - \psi_{02})$$

$$(J_s)_{out} = g(C_2)$$

The whole system (if given time) will then adjust to values of P_1, P_2, C_1 and C_2 by which steady flows of both J_v and J_s are achieved throughout the system. In effect C_2 will go on rising until the efflux of solute from the sink end balances the input J_s, and the other variables will adjust in turn to equalize the flows.

As a result, the system is entirely source/sink controlled. The gradients of concentration and pressure required to achieve this will depend on the effective longitudinal conductivity, that is on the force-generating system, and may be very different from those required by the equivalent Münch flow, but the qualitative behaviour of the system will be very similar. Thus it will respond osmotically (to ψ_{01} or ψ_{02}, for example), just as we would expect to find in a Münch flow, and it will respond at once to changes in the input from the source, or in the output to the sink tissues, the function $g(C_2)$.

We might also predict transient changes of the kind seen in the cooling experiments. On cooling the path the effective value of L_s in that region of the path will fall, and the flow diminish out of the region; but this will induce changes in P_2 and C_2, and the gradients in the system will readjust until a new steady state of solute flow J_s is achieved through the path. Provided that the pressure generated in the source is capable of driving a passive flow through the cooled region, (so long as the passive value of L_s is not extremely low, and the cooled region not extremely long), then the flow can be restored.

Equally the system will respond to interference with the sink region, in two ways. As flow continues into the phloem in an inhibited sink region, with no output from the phloem, a back pressure will develop to oppose further flow. The force in the path will then diminish, as say the fibrils become more randomly oriented.

It will therefore be difficult to distinguish such mechanisms from Münch flow. Their qualitative behaviour will be very similar, but they will simply be faster than we expect - if we knew normal state of the sieve pores, and a firm idea of what we expect L_s to be for unassisted flow.

It seems, therefore, that if we are to distinguish between Münch flow and a form of activated mass flow, it must come from a better understanding of the state and distribution of P-protein in intact sieve elements, and of the chemical and biochemical properties of P-protein. The properties required of P-protein on the two views of the mechanism are very different. If we cannot find evidence that P-protein is similar to actomyosin (or an analogous "motile" protein) then perhaps we can show that it has properties appropriate for the specialized, controlled, blocking agent required in the Münch behaviour. Alternatively we may show that it has properties appropriate for electro-osmosis - although in my view the quantitative difficulties in potassium-linked electro-osmotic flow remain overwhelming. The suggestion that P-protein is positively charged at the pH in sieve elements (Weber et al, 1974) would also rule out this form of electro-osmotic mechanism, but again there are opposing views of the characteristics of P-protein from different studies.

In summary, until we are clearer about the nature and distribution of P-protein in a range of conditions and translocation states, the problem of mechanism for translocation will remain unsolved. We need chemical characterization of P-protein as well as electron microscopical investigation of sieve elements. We shall then be in a position to discuss mechanisms in a less speculative manner.

References

AMIR, S. and REINHOLD, L. 1971. Physiol. Plant. 24: 226
ANDERSON, R. and CRONSHAW, J. 1970. Planta 91: 173
CHRISTRY, A.L. and FERRIER, J.M. 1973. Pl. Physiol. 52: 531
HARTT, C.E. 1970. Pl. Physiol. 45: 183
HODDINOTT, J. and GORHAM, P.R. 1974. Can. J. Bot. 52: 349
KOMOR, E. 1973. F.E.B.S. Letters 38: 16
KOMOR, E. and TANNER, W. 1974a. Eur. J. Bioch. 44:219
KOMOR, E. and TANNER, W. 1974b. A. Pflanzenphysiol. 71: 115
MacROBBIE, E.A.C. 1971. Biol. Rev. 46: 429
MILBURN, J. 1971. Planta 95: 272
PICKARD, B.G. 1973. Bot. Rev. 39: 172
SCHULTZ, S.G. and CURRAN, P.F. 1970. Physiol. Rev. 50: 637
SIBAOKA, R. 1962. Science 137: 226

SINYUKHIN, A.M.   and GORCHAKOV, V.V.   1968.   Soviet Pl.   Physiol.
    15: 400
STEIN, W.D.   1967.   The Movement   of   Molecules   across   Membranes.
    Academic Press, New York.
WEBER, C., FRANKE, W.W.   and KARTENBECK, J.   1974.   Exp.   Cell Res.
    87: 79
WEST, I.C.   and MITCHELL, P.F.   1972.   J.   Bioenergetics 3:   445

DISCUSSION

Discussant: D. Aikman
School of Biological Sciences
University of East Anglia
Norwich, England NOR 88C

Discussion after papers by D.C. Spanner on "Electro-osmotic Theories" and by E.A.C. MacRobbie on "Other Theories" and "Summing-up of Biophysical Approach". The following invited discussant paper opened the proceedings.

Discussion of Electro-osmotic Theories of Phloem Transport

D. Aikman

There are two types of electro-osmotic model of phloem transport. The one, based on standing gradients, has been proposed independently by Fensom (1957) and by Spanner (1958). The other type of model assumes a travelling electrical wave and has been suggested by Hejnowicz (1970). I wish to discuss that described in Professor Spanner's paper.

Electro-osmosis itself, as Prof. Spanner has very clearly illustrated, is a well established phenomenon in the physical sciences. The postulated model is consistent with an interpretation of the anatomy of phloem, and with the suggestion that the metabolism of the phloem pathway is involved in the transport process. However, the important question at this meeting is to consider whether this phenomenon does indeed provide the motive power for phloem transport. In my search for an answer I will explore the energetics for such a system in general terms, comparing the energy available from metabolism with that required to drive flow. I will then examine the model in relation to some other known quantitative physiological relationships.

In electro-osmosis a flow of volume is driven by a difference in electrical potential rather than by a difference in pressure. Prof. Spanner has shown the mechanism by which this interaction would occur in his proposed potassium electro-osmotic system. However, it is an inescapable feature of such interacting flows, that the cross-coupling process is less efficient than the direct one. In relation to electro-osmosis, Dainty, Croghan and Fensom (1963) note that the gradient of the electrical field is less efficient energetically than that of the pressure in the same system as the motive power for volume flow. Hence, if we calculate

611

what would be required to drive the observed flow rates by
pressure, we know that electro-osmosis would require more power.
Following the approach of Weatherley and Johnson (1968) we can
compare this with that available from metabolism.

Taking the upper limit of sieve tube respiration of
carbohydrate to be 0.05 g (carbohydrate) cm^{-3} day^{-1}, and the free
energy of combustion, ΔH, of carbohydrate to be 3.8 x 10^3 cal g^{-1},
and assuming 100% efficiency of conversion, the maximum total
energy released from the oxidation of carbohydrate in phloem can be
calculated to be 3.4 x 10^8 erg cm^{-3} h^{-1}. (But note that conversion
of carbohydrate to ATP conserves only about 50% of the free energy
and that this metabolism must cater for the other needs of the
tissue as well.)

Taking a pressure drop, to drive flow through the
micro-filament matrices of the pores, of 0.12 x 10^6 dyne cm^{-2} per
sieve plate and a volume flow of 100 cm^3 cm^{-2} h^{-1}, one can
calculate that the rate of energy dissipation for pressure driven
flow is 0.12 x 10^8 erg cm^{-2} h^{-1}. Assuming that one quarter of the
20 or so plates per cm are operative, then the energy requirements
for pressure driven flow will be 0.6 x 10^8 erg cm^{-3} h^{-1}.
Electro-osmotically driven flow would require more energy than
this. Spanner's suggested values for an electrochemical potential
difference equivalent to 1 mV per plate and a potassium current of
0.75 amp cm^{-2} would suggest the addtion to this of the energy
required to drive the return flow of potassium through the plasma
membrane and through the apoplast. From Fig. 9 in Spanner's paper
in this volume, a further factor of 3 is appropriate. Thus we can
see that, on energetic grounds alone, potassium electro-osmosis is
inadequate by at least a factor of two.

Even if it weren't, the model is still in serious disagreement
with physiological parameters. For example, Dr. Enid MacRobbie
(1971) has raised the point that, since the potassium must drag the
water and the sucrose, its velocity must be greater than that of
water and sucrose. That is, the number of molecules of sucrose and
of water dragged per ion of potassium through the pore must be
lower than those in the solution in the lumen. Perhaps, to meet
this objection, Spanner has now modified his model to give a higher
potassium current but lower driving force. This preserves the
physical energetics, but what of the potassium flux he has now
suggested in relation to the ATP turn over rate?

From the metabolic rate per unit volume given earlier,
assuming 50% efficiency of conversion to ATP ($\Delta G = 7000$ cal $mole^{-1}$),
the ATP turn over rate per cm^2 of sieve plate can be calculated to
be 3.2 x 10^{-8} mole cm^{-2} s^{-1}. A K^+ flux of 0.75 amp cm^{-2} is
equivalent 0.8 x 10^{-5} mole cm^{-2} s^{-1}. Therefore the ratio of K^+
ions moving to ATP molecules hydrolysed is 250. For the

well-investigated Na^+/K^+ transporting ATPase, the accepted ratio is
2 K+ per ATP. Hence the coupling ratio of the postulated model is
not acceptable physiologically.

Other features are also unphysiological. The measured
electrical conductivity of plant cell membranes is of the order of
0.1×10^{-3} mho cm^{-2}. The return current flow through the membrane
(0.2 amp cm^{-2} if the length of lateral wall involved is equal to
the diameter of the sieve plate) would require a potential
difference of 2 kV rather than the 1 mV allowed by Spanner. Even
taking a conductance found during excitation of nerve (30×10^{-3}
mho cm^{-2}), and even if an increase in membrane area by a factor of
100 assuming a brush border type of anatomy is allowed, then the
voltage is still a couple of orders of magnitude too small. I will
not comment in detail on such quantitative features as the implied
magnitudes of the active ion flow rates per unit area, on the
surface charge density of the fibrils in the pore and on the
conductivity of the apoplast; nor on such qualitative features as
the transport of anions and of mycoplasma, nor on the results of
attempts to induce the phenomenon experimentally in phloem.
Suffice it to conclude that, despite the fact that the model is
consistent with a possible interpretation of the structure of
phloem and with the suggestion that the metabolism along the phloem
pathway is coupled to transport, and despite the interesting work
that the theory has stimulated, potassium electro-osmosis cannot be
an acceptable model. It is with regret that I must say to Prof.
Spanner that not only will his entrant not win the race; it cannot
even complete the course. It is physical, but it is not
physiological.

References
DAINTY, J., CROGHAN, P.C. and FENSOM, D.S. 1963. Can. J. Bot.
    41: 953-966
FENSOM, D.S. 1957. Can. J. Bot. 35: 573-582
HEJNOWICZ, Z. 1970. Protoplasma 71: 343-364
MACROBBIE, E.A.C. 1971. Biol. Rev. 46: 429-481
SPANNER, D.C. 1958. J. Exp. Bot. 9: 332-342
WEATHERLEY, P.E. and JOHNSON, R.P.C. 1968. Int. Rev. Cytol.
    24: 149-192

Eschrich: opened the discussion by referring to
Parthasarathy's picture of the "double-brush" structure in sieve
pores. This electron micrograph showed callose in the pores.
Eschrich thinks that a constriction of protein in the pores causes
callose formation, which is the opposite of the point of view of
Spanner who considers that the callose constricts the protein to
give the right filament spacing for electro-osmosis to work.
Eschrich also thought that the water-binding capacity of callose
might play a part in the water relations of sieve elements.

Spanner: replied at some length to Aikman's criticisms. He said that the sort of calculations, on energy, etc., done by Aikman were familiar but contain many possible holes. At this stage of our knowledge to be out by a factor of only 2 was hardly alarming. For instance, a tremendous weight is placed on the initial rate of glucose oxidation this may be widely out. The figure of 0.75 amp cm^{-2} may be reducible; it depends on the figure of 200 water molecules per K^+. He also put forward the suggestion that the K^+ ions which are doing the driving may pass through the system less rapidly than the water! At this stage he mentioned a mechanical system consisting of a stationary rigid wall, a wheel and a slider, a system in which the slider moves twice as fast as the wheel; thus from a purely formal point of view "what drives needs not necessarily move faster than what is driven." Further, if pinocytosis is the mechanism of K^+ uptake the ratio of K^+ to ATP may be quite large; so may be the membrane conductivity.

He also reminded us of the possibly self-optimizing character of the system. Charged linear molecules in an ionic atmosphere adjust their relative separations such that the potential energy is a minimum. If the ionic atmosphere alters, the separation will change and there might be space for further molecules to thread their way in; the sieve plate could optimize itself in this way. Because of these self-optimizing possibilities, one must always make sure in calculations one is using exactly the actual parameter values - those at the "crest", so to speak.

MacRobbie: said that if she were designing an electro-osmotic system in a sieve tube she would use H^+ rather than K^+. Since, according to Spanner, there may be a large concentration difference in ATP above and below the sieve plate, one could have different rates of electrogenic H^+ without the problem of K^+ recirculation. There are now strong reasons for thinking that H^+ pumps are strong and ubiquitous, and since phloem is so alkaline they are likely to be highly efficient there.

Price: raised some questions about K^+. He said that the high concentration of K^+ in the sieve elements implies high rates of loading and unloading of K^+ on the Munch model and less on the electro-osmotic model. He also said there should be some provision for recirculation of K^+ via the xylem. He asked whether there is a lower K^+ concentration in those species which don't have P-protein?

Dainty: didn't think that K^+ fluxes were more easily handlable in the electro-osmotic theory than in the Münch theory. He pointed out that the K^+ fluxes at the loading and unloading ends would be about the same as the sucrose fluxes, i.e. about 20 pmole $cm^{-2} s^{-1}$, which is fairly high but not too unlikely a value for a K^+ flux. (Root cortex cells may have similar K^+ fluxes.)

Kleinig: pointed out that for electro-osmosis P-protein must
be negatively charged.   But in Cucurbita maxima it is positively
charged at the sieve element pH of about 8.5; the P-protein
isoelectric point is about 9.5 or greater.

Spanner: suggested that <u>Cucurbita</u> is rather unusual with wide
open pores and a special gelling protein.

Aronoff: said that the protein being basic did not preclude
the presence of carboxyl side chains, even if the net charge was
positive. He later said that in a heterogenous system the overall
pH of 8.5 or so may not have much meaning. If appreciable amounts
of organic acids are present, it is quite possible that the local
pH at the sieve plates could be more than the 9.5 needed to make
the net charge on the P-protein negative.

Aikman: at this point explained the necessity to
destructively analyse any model.   He illustrated this with the
analysis he and Anderson made of a peristaltic model of phloem
transport; they failed to disprove peristalsis but - does it exist?
This is the converse of his conclusion about electro-osmosis: even
if the structure exists, it couldn't work for phloem.

Reinhold: said she was surprised that MacRobbie had said that
inhibitor experiments need explaining in connection with the Münch
theory.   On the contrary, in her opinion, they need explaining in
respect of contractile protein or electro-osmotic theories.   These
latter mechanisms should be particularly sensitive to uncouplers of
oxidative phosphorylation, but Willenbrink failed to find any
effect of DNP or CCCP and in her laboratory DNP had its effect on
translocation via the loading process, not along the path. The
only experiment where DNP seems to act along the path is dubious
because of the prolonged soaking in high concentrations of DNP.

MacRobbie: agreed that the DNP results are not clear-cut.
However, she was referring to results with anoxia and cyanide by
Spanner and Willenbrink, which seemed to show their effects on the
path and not on either source or sink. She also referred to recent
results reported by Willenbrink using valinomycin.

Willenbrink: also agreed that DNP was odd.   However, he
thought they had a positive result on the path with CCCP.

Geiger: then referred to some work by Sih and Swanson in
which anoxia of a long section of petiole of squash caused no
inhibition over a very long period. Calculations showed that ATP
could not support any active mechanism in the path.

Christy: said that anoxia had no effect on translocation in
sugar beet.

Spanner: said that we may have to recognize that different plants respond differently; he is a little suspicious of squash with its large sieve tubes. He said that in the work referred to by MacRobbie, they did get an effect of DNP as well as of anoxia and cyanide. They used the Saxifraga stolon which had a different translocation problem with its very narrow sieve tubes, all plugged with P-protein.

Schmitz: asked whether anyone had an explanation of what the fairly high amount of ATP in sieve tubes is doing. There seem to be no good estimates of turn-over rates, etc.

MacRobbie: replied that there are several different versions of what the ATP is doing depending on the mechanism one believes in. There is no difficulty in envisaging roles for ATP in any activated mass flow theory such as protein contraction or electro-osmosis. Münch supporters might say it is partly there to wait for wounding, etc., when it will go into action synthesizing callose or changing P-protein configuration; and unloading, needing ATP, is probably going on all along the path. She knows of no measurements of the rate of turn-over of ATP.

Dainty: pointed out that ATP, in the Münch model, will be heavily involved in the pumping of sugar and K^+ into the sieve tubes at the loading end. This ATP will have to diffuse to the sites of the pumps on the membranes and it will inevitably be swept down by the mass flow: diffusion fighting mass flow. Thus the ATP seen lower down is in part the remnants of the ATP used in the loading process.

Spanner: suggested that there is something unusual about respiration of phloem tissue, for instance there are no intercellular air spaces and perhaps peroxidases are involved. With respect to actin he suggested that any actin appearing in phloem exudate may arise from the companion cells and not the sieve elements. He also referred to the experiments of the Osaka School (Kamiya et al) on isolated cytoplasm of Nitella. In the cytoplasmic drops actin filaments form little polygons which execute rotatory movements. It would be worth looking for such phenomena in phloem exudate.

Kleinig: said that actin is well known to be invariable throughout the plant and animal kingdom, and the properties of P-protein must rule out the possibility that the bulk of the P-protein is actin-like. Of course there is the possibility, as MacRobbie pointed out, that some of the minor proteins are actin-like.

MacRobbie: pointed out that the report on actin by Currier was not on exudate but on glycerinated hand sections; it was actin in situ. So far as she knows, no one has looked for rotating polygons in exudate.

Weatherley: said that theorizing is much easier than doing experiments, which are all too few! The trouble with activated mass flow as a supplement to Münch flow is that it doesn't make much difference; it is like increasing the conductance or shortening the pathway. However, MacRobbie said that the flow might not be linearly related to the pressure difference; this might be something one might test.

MacRobbie: agreed on the need for critical experiments. She suggested getting the system into a state of stress in which Münch flow would have the greatest of difficulties (use of inhibitors, or cooling, or water stress, or a plant like the mangrove), vary the loading rate and measure the parameters of interest (pressure, concentration, etc.) and compare them with the same parameters in an unstressed state.

Milburn: thought that evidence might slowly be accumulating in favour of the Münch hypothesis, other hypotheses being gradually phased out. However, at the moment we can't measure all the necessary parameters for Münch flow. He predicted a long period of increasingly refined measurements to show eventually that other mechanisms have negligible effect. One thing he would have liked to have seen more emphasis on is the experimental side; the key issue is that of technique. In this connection he criticized the two knife blade technique of getting a specimen by simultaneous cuts. As the blades are going in, because of the high loading per unit area of blade, there will be a pressure surge ahead of them which will invalidate the method.

Aikman: pointed out, in conclusion, that we had not answered Kollmann's question: what about the conifers?

Eschrich: had the last word by suggesting that the next conference should have a session on "phloem science fiction.""

Summary of the Conference

Paul Weatherley

Regius Professor of Botany

University of Aberdeen
Aberdeen, Scotland

"Myself, when young, did eagerly frequent doctor and saint and heard great argument about it and about, but evermore came out by the same door wherein I went."

For those of you for whom the words of Omar Khayyam apply, I suppose this conference has been a waste of time. For those of you who came thinking you were right, and are going away still thinking you are right, be assured you are probably wrong.

When Sam Aronoff first convened the organizing committee of this Institute, it occurred to us that what we should aim at was a dialogue between the anatomists and the physiologists, because it was clear that a great gap, a chasm I might almost say, exists between the two and, that the physiologists should speak not to their physiological colleagues but to the anatomists: and likewise, the anatomists should speak not to their brother anatomists, but to the physiologists. Insofar as this aim has not been achieved, the Institute has failed. However, you may think we have had at least a partial success and this may be even more complete when what has been spoken is set down in print to be more slowly digested.

Now may I examine this gap between physiologists and anatomists a little further, because phloem transport studies exemplify so clearly the classic antipathy between form and function. The organism is, according to the philosophers, a four-dimensional event. It has three dimensions of space and one of time (1). The anatomists study the spatial patterns in abstraction from time. They study a partial event, what may be called a time-slice.

They are not interested in changes in time so much as in focusing attention on the spatial pattern at an instant in time. Furthermore, the degree of abstraction from the whole is increased by their restricting their attention to one part of the pattern, e.g. to a single cell or part of a cell, or restricting the three dimensions of space to two, as in a thin section. Again, in preparation for sectioning, the material has to be fixed - as R.D. Preston has said - "It has to be boiled or fried", and in the end, the microscopist looks at something which is highly abstract and very far removed from the original four-dimensional event from which it was extracted. We all know from learning elementary wood anatomy, how difficult it is to synthesize a three-dimensional picture from a study of transverse and longitudinal sections.

What about the physiologist? He studies the temporal extension of the three-dimensional event, i.e. the process - what is going on in time. But he invariably restricts himself to one narrow aspect of the spatial pattern - to some single process, often at the molecular level, and examined in abstraction from the whole. This partial event is just as remote from the whole as the anatomist's section prepared for the electron microscope.

To exemplify the difference in approach of the anatomist from that of the physiologist, I will take the imaginary study of a clock which is, after all, a kind of simple organism. The anatomist discovers that there is a pattern of cogs, but these are usually stationary, for the clock stops when the back is opened, or when it is dipped in glutaraldehyde. The "fixed" clock is then sectioned, and I ask you to contemplate the problem of reconstructing the clock from one millimeter sections taken in various planes through the mechanism - a fairly difficult synthetic exercise! The physiologist, for his part, only hears the ticks, and sees the hands go round. He administers fairly crude treatments. He plunges the clock into water and it stops. He drops it on the floor, and the alarm goes off, though with some clocks it does not. Do these possess an alarm, or are they just not sensitive to being dropped? Each of these studies has its own fascination, and each may become an end in itself. Stationary clock collecting is a very fascinating thing to occupy yourself with, as is investigating the question of how long does the clock go with one wind, two winds, etc. Perhaps it is not surprising that these different studies attract different kinds of people, each equally erudite, but each developing his own language. Each, therefore, is not understood by the other, and mutual ignorance heightens the barrier between them. Each may become fearful of the other, or simply be happier ignoring the other. Worst of all, are those who become half familiar with the other and think they know it all.

Let us now turn to more specific considerations, but still keeping the clock analogy as our background. First of all, the structure of the sieve tube. The difficulty of not disturbing the structure before it can be seen, is a very considerable problem. Not only are the structures extremely labile, but the functioning sieve tube probably has solution passing through it at a rate of a hundred or more micrometers per second. The complete elimination of some shift seems impossible to avoid. However, a picture has emerged. Phylogenetically and ontogenetically, there seems to be a move towards increasing simplification from a basic parenchyma cell. In the most advanced forms, the structures seem to be adapted to streamlining of flow, giving a minimum of drag. Vacuoles, nucleus, and so forth, are dispensed with. The only remaining structure of indisputable function is the plasmalemma. The companion cell is an inseparable servant, and undoubtedly both it and the sieve tube function as a unit. Most people seem to believe that the P-protein is anchored. But what it does, or whether it does anything at all, is not clear. Nor is its location in the translocating sieve tube certain. The majority opinion seems to favour sieve pores containing rather little P-protein, i.e. with a low resistance to flow. On this view dense filling arises from surging or slow fixation. Those who believe in this, look upon sieve pore blockage as safety mechanism analogues to the clotting of blood on wounding. Those, on the other hand, who see dense packing of P-protein in the sieve pores as the true natural state, favour the view that they are instrumental in impelling the translocation stream. Several hypotheses have been presented in this connection. Electro-osmosis still seems to be a possibility, but not a probability. An actin-myosin mechanism seems possible, but the evidence for it is as yet slight, to non-existent. Peristalsis is energetically possible, and more elaborate models have been proposed.

As belief in unobstructed pores increases, and with it the inevitability of Münch flow, these pumps "along the line" are being relegated to the status of ancillary mechanisms, though they may be important in less advanced systems such as those of the gymnosperms, algae, and other lower plants where the conducting cells remain more complex, the pores smaller and the contents more persistent. If this view is correct, the phylogenetic simplification of the structure of the sieve tube was accompanied by a concomitant simplification of mechanism that resulted in an increased efficiency, i.e. a decreased consumption of energy.

Phloem transport entails three physiological processes: loading, long-distance transport, and unloading. Loading is recognized as active, but whether the pathway of movement from the photosynthesizing mesophyll cell to the sieve tube is via the symplast, the highest concentration of sugar would presumably be in the photosynthesizing mesophyll cell, giving a downhill movement towards the sieve tubes. However, it appears likely that the concentration of sugar in the sieve tube is much higher than in the mesophyll, pointing to the active step as near to the sieve tubes. If active transfer processes occur only across membranes, it follows that entry into the sieve tube must be across the plasmalemma, i.e. from the apoplast into the symplast of the sieve tube (see below).* In other words, sugar must arrive at the sieve tube via the apoplast. If this is so, it must of course pass into the apoplast from the symplast of the mesophyll cells wherein it originates. Thus, sugar must cross the plasmalemma twice - once from symplast to apoplast in the mesophyll (presumably downhill), and again from apoplast to symplast in the phloem (uphill). This does appear rather devious. It suggests that either the loading is better controlled in the phloem itself, rather than in the mesophyll cells, or the apoplast provides a better pathway than the symplast. On the face of it, the latter is surprising for movement here is against the transpiration stream. However, this does not seem to present an insuperable problem. The necessary build-up of concentration at the evaporating surface to cause sufficient diffusion against the flow of water, is not very great. One further point in connection with apoplastic movement. Greater efficiency would be achieved if back diffusion from the phloem to mesophyll along the symplast was minimal. This would be achieved if there were few plasmodesmata between mesophyll cells and sieve tubes, and this seems to be so.

What factors control loading? Apart from its being an active process, little is known about it. On the one hand, there is some evidence that the concentration of sugar in the sieve tube is important, as is no doubt, that in the pool. Does the turgor pressure of the sieve tube play any part? We simply do not know. About unloading even less is known, though it appears to be active also. We know virtually nothing about the relationship between unloading and loading, except that at a steady state they must be equal and somehow or other, they adjust themselves so that this comes about.

Turning now to long-distance transport. There are rather few facts available with which to discriminate between the different hypotheses. That translocation is a mass flow is certain, but as between the two main mass flow theories, that of Münch and that involving travel along some cytoplasmic vehicle. There is no unequivocal decision. As an example, we may take the Canny-Phillips protoplasmic streaming hypothesis. Evidence for

---

* Sieve tube in this context is meant to include companion cells, transfer cells, etc. which may be concerned in the loading process.

this is based solely on its fitting profiles of labelled assimilate moving down the stem. But we have seen that these profiles are equally consistent with the Munch model. Indeed, it appears that isotope profiles are not diagnostic with respect to mechanism and so, presumably, interest in them will languish as will the Canny-Phillips hypothesis. The Münch hypothesis seems, then, to be rising to a position of unassailable glory. However, it should be remembered that it is still based on few facts, and its position still derives more from the weakness of its adversaries than from its own inherent strength. The new formulations of the Munch flow model have not yet proved themselves in providing critical experiments to test their validity. I suspect the discriminating experiments will be very difficult to do. Theorizing is much easier than carrying out experiments.

The future. It is a matter of the highest priority to arrive at a precise and unequivocal description of the structures within the sieve pore and lumen of the translocating sieve element. That there is still controversy over this is a sign of uncertainty. At present there is a tendency to refer to sieve pores as either "open" or "blocked", whereas in reality the functioning sieve pores are likely to have some contents. If this is so, further quantitative data, e.g. the number, dimensions, and distribution of filaments and other structures, are needed to allow calculations to be made on their resistance to flow, and so the feasibility of the pressure flow hypothesis to be tested more rigorously.

I have referred already to the gymnosperms and "lower" plants which have less simplified conducting cells. Here we should surely resist the basic human urge for uniformity - for a single mechanism in preference to two - and recognize that in these plants "pumps along the line" may play a key role. I suggest we need a little more comparative physiology to go with our comparative anatomy.

If the structure of the sieve tube and the effects of low temperature and inhibitions are consistent with the pressure flow hypothesis, then further search for pumps is not worthwhile, and further hypothesizing can cease. We can then bend our efforts to the crucial problems of loading and unloading and exchange along the pathway. Also, a further study of the effects of factors such as temperature and water stress on translocation may be important in their own right and not just to provide evidence for, or against a hypothetical mechanism. Such investigations lead inevitably to the question of the integration of translocation with other physiological processes in the whole organism. To what extent does it regulate, or is it regulated by photosynthesis, growth, etc.? In the long term, the translocation and respiration of a leaf must equal its photosynthetic gain, since whilst diurnal changes in dry weight are manifest, there may be little general trend in dry weight over a period of many days (unpublished data for cotton).

Does this mean that photosynthesis controls translocation, or vice versa?   Would the yield of a crop plant increase   if   its translocatory   capacity   were   increased; in other words, is translocation ever a limiting factor for growth?   If not in herbaceous plants, it might be so in trees where the distance between source and sink is considerable. The answers to some of these questions are already being found and may be of considerable applied interest. For example, if translocatory efficiency could be defined and measured, it might be selected for in a tree breeding programme.

It is to be hoped that those who did not have the good fortune to be present at this Institute will find the printed volume useful.   Unavoidably, it consists of the contributions of only a few of the scientists who are engaged in research on phloem transport.   Coverage is therefore incomplete, but most aspects of the anatomy and physiology of the phloem have been touched upon, and a fair picture of the "state of play" has, I think, been achieved. Those of us who were present were enriched by making new friendships, and renewing old ones.   The mountains also played their part, not only in providing relaxation, but also in helping us to keep things in proportion.

References

(1) WOODGER, J.H.   1948. Biological Principles, Routledge & Kegan Paul, Ltd.

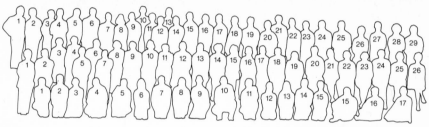

FRONT ROW: 1. Illo Gauditz, 2. Wolfgang Heyser, 3. Rosemarie Heyser, 4. Christopher Lobban, 5. Klaus Schmitz, 6. Maleeya Krautrachue, 7. Paul Weatherley, 8. Herbert Currier, 9. Sam Aronoff, 10. Elmo Beyer, 11. James Cronshaw, 12. Shelagh Hall, 13. Enid Britt, 14. Frank Campbell, 15. Bernadette Fondy, 16. Kent Crookston, 17. Walter Eschrich, 18. M.V. Parthasarathy.

MIDDLE ROW: 1. Lee Oh, 2. John Oh, 3. Charles Hébant, 4. John Troughton, 5. Richard Johnson, 6. Charles Lamoureux, 7. Lalit Srivastava, 8. David Aikman, 9. Carroll Swanson, 10. John Hoddinott, 11. Jeff Moorby, 12. Bruno Quebedeaux, 13. Roger Lee, 14. Poul Hansen, 15. Barbara Bentwood, 16. Robert Thompson, 17. Roland Dute, 18. Lim C. Ho, 19. Nicholas Lepp, 20. D.M. Miller, 21. Susan Sovonick Dunford, 22. Clive Price, 23. Enid MacRobbie, 24. John Milburn, 25. J. van Die, 26. Leonora Reinhold.

BACK ROW: 1. Michael Walsh, 2. Jack Dainty, 3. Jack Ferrier, 4. Robert Turgeon, 5. Thomas Mallory, 6. Heinz Behnke, 7. R. Kollmann, 8. Inge Dörr, 9. Hans Kleinig, 10. Douglas Spanner, 11. Johannes Willenbrink, 12. Paul Gorham, 13. Rudolf Christ, 14. Thomas Housley, 15. Thomas Mittler, 16. William Outlaw, 17. Carl Crisp, 18. Mel Tyree, 19. Lawrence Christy, 20. Robert Giaquinta, 21. Robert Fellows, 22. Donald Geiger, 23. Dom Cataldo, 24. C.D. Elmore, 25. Eli Romanoff, 26. James Fung, 27. Donald Fisher, 28. Barrie Watson, 29. Richard Fischer.

SUBJECT INDEX

Activated mass flow, 587
  actin and, 588
  essential feature of, 605
  evidence for and against,
    594
  low temperature and, 549
  microfilaments and, 588
  microtubules and, 587
  and Münch flow,
    relation between, 601
  and surface flow, 585
Active loading
  evidence for, 262, 285
Active transport, 468 ff
  criteria for, 469
Agavaceae and Palmae
  translocation in, 427
Alaria, 216
Albuminous cells, 11, 164
  description of, 10
Algae, translocation in, 187
Angiosperms, phloem in, 14
  sieve tube, arrangement of
    filamentous P-protein in,
    236
Aphids and bidirectional
  movement, 403
  injections of salivary
    material by, 451
  rate of ingestion, 452
Aphid stylets
  exudation from, 447
Aphid taps and turgor drop, 451
Apoplastic movement, 283
Areoles, development of, 310
ATPase, 588

Autoradiography, 345
  and transport, 403
  variable retention of
    sucrose in, 346

Bidirectional spreading of
  tracer, 405
Bidirectional transport, 401 ff
  anatomical aspect, 405 ff
  mechanism of, 411
  and recycling, 409
  regulation by source and
    sink, 412
Bryophytes, callose in, 200
  leptome, 211 ff
  plasmodesmata in, 200
  sieve elements of, 199
    differentiation, 200
Bundle sheath cells, 308

Callose, 65, 138
  dormancy, 17
  inhibition and calcium
    ions, 448
  in Laminaria, 216
  synthesis, CN treatment
    and, 383
  types of, 5
^{14}C-arrival in sink leaves,
  kinetics of, 335
^{14}C-sucrose, availability of
  exogenous sucrose for
    translocation, 216
Carrier hypothesis, 470

627